INTRODUCTION TO
STATISTICS
A CALCULUS-BASED APPROACH

HOWARD B. CHRISTENSEN
BRIGHAM YOUNG UNIVERSITY

SAUNDERS COLLEGE PUBLISHING
A HARCOURT BRACE JOVANOVICH COLLEGE PUBLISHER

FORT WORTH   PHILADELPHIA   SAN DIEGO   NEW YORK   ORLANDO   AUSTIN   SAN ANTONIO
TORONTO   MONTREAL   LONDON   SYDNEY   TOKYO

Associate Editor: Pamela Whiting
Manuscript Editor: Joan Harlan
Freelance Manuscript Editor: Marji James
Cover Designer: Linda Cable
Production Manager: David Hough

ISBN: 0-15-545965-1

Library of Congress Catalog Card Number: 91-73615

Printed in the United States of America

PREFACE

This book is designed for a one-semester course in introductory statistics for students with a background in single-variable calculus. It could also serve as a textbook for an introductory course in mathematical statistics. It is less mathematically rigorous than some books used for a mathematical statistics course, but most of the same material is included (i.e., this book does not take the "theorem–proof" approach to mathematical statistics characteristic of some texts).

A preliminary version of this book has been used since 1986 at Brigham Young University, and comments and suggestions from the students have been incorporated in this edition. The audience for which this text was prepared consisted of engineers, economics majors, computer science majors, and math and statistics majors along with a few others. Most of these students tend to question the need for another "math-type" couse and want to know why a course in statistics is required of them. I have attempted to address those concerns by making the course, and the book, as relevant to their needs and interests as possible, while preserving the insight as to the "whys" of statistics that a calculus prerequisite would allow.

This book takes a data-oriented approach to the teaching of statistical principles. I have found that most students are interested in how decisions are made using data. However, to properly understand how such decisions are made, an understanding of how data are collected is necessary. Thus, the principles of random sampling and the properties of chance and probability become important concepts to examine. And, in order to understand the sampling process, it is necessary to know how a probability model of a population is used. Once the link between the data and a probability model is made, the process of estimating the parameters of the probability model, using the data, can be understood. This is the approach taken in the text. The text concludes with the data-analysis techniques using point and interval estimation and tests of hypotheses.

The data-orientation of the book begins in Chapter 1, with the introduction of some of the common principles of descriptive statistics, both graphical and numerical, including boxplots and stem-and-leaf diagrams. At the beginning of the chapter the linkage between a "population" and a "sample" from the population is established in order to make a "statistical inference" about the population. This linkage is then restated in each chapter.

Chapter 2 presents the fundamentals of probability necessary to understanding the principles that chance introduces into a data-analysis setting when a ramdom sample is selected.

Chapter 3 introduces the theoretical framework for modeling a population from which data is obtained by using a probability density function. Means and variances of probability density functions are first defined as an "expected value" concept; they are then tied to the mean and variance of a set of data in a sample by viewing them as the theoretical means and variances of the population from which the data were obtained.

Chapter 4 introduces some of the most common continuous and discrete probability models (probability density functions) that might be encountered in practice. One of these might be selected to model a set of sample data that is being studied. The Table of Contents lists the specific probability density functions described in this book.

Chapter 5 introduces the techniques needed to determine the sampling distribution of sample statistics when samples are drawn from a specific probability model and when functions of those sample observations are created (e.g., computing the mean and variance of the sample). The use of the distribution function approach and the moment-generating function approach are presented in this chapter.

Chapter 6 discusses the most commonly encountered sampling distributions and introduces the central limit theorem. Thus, the reader will be introduced to the normal distribution, the chi-square distribution, the T-distribution, and the Fisher's F-distribution as specific distributions of interest to describe the properties of certain statistics computd from a sample of observations from a population. A short presentation of order statistics is presented as optional material.

At this point in the course, the basic theoretical foundation has been laid and the inferential ideas of statistics (e.g., point and interval estimation and tests of hypotheses) can be introduced. These ideas are the subject of Chapters 7 and 8.

Chapters 9 and 10 can be viewed as supplementary material wherein the applications of the results of the preceding eight chapters are presented. These last two chapters briefly introduce the chi-square goodness-of-fit test and the chi-square analysis of two-way contingency tables, and provide a brief introduction to analysis of variance, some nonparametric tests, and a short section on linear multiple regression.

To keep from diluting the data-oriented approach, most theorems are stated without proof. If the proof can be easily developed by the student and is instructive, it will be part of the exercises. Some important proofs have been placed in an appendix to the chapter in which the theorem was introduced. The intent is to emphasize the application of the theorem rather than to dwell on its sometimes distracting proof.

In several chapters, when sufficient background has been presented, examples of applications of statistical principles are introduced. Thus, you will find a section on statistical quality control at the end of Chapter 6 and a section on the principles and formulas for sample survey designs at the end of Chapter 7.

Each chapter begins with a statement of objectives, an introduction to the chapter, a description of where we are going in the chapter and a chapter exercise that proposes a problem that can be solved with the concepts to be learned in the chapter. Each chapter is then divided into sections of logically grouped concepts. Each section has a short introduction, a list of definitions, an extended discussion of the ideas of the section, example problems, and concludes with a list of exercises. At the end of each chapter is the solution to the chapter exercise that was introduced at the beginning of the chapter, a summary, and a set of exercises that cover the broader ideas covered in the chapter.

An Instructor's Manual with Solutions contains the solutions to the exercises found throughout the text and test questions.

The values in the tables in the Appendix were produced by S, a statistical software program available on many computer systems. Many of the graphs and plots were also generated using statistical software such as Data Desk, JMP, Mathematica, and Minitab. Appreciation is given to the creators of these powerful statistical software programs, especially in light of the work involved in trying to find an alternative. I wish to thank colleagues, students, and friends for their suggestions, for their patience with the slow process of developing these materials, and for their insights and improvements. Among those I want to thank are Kenneth Boback, Pennsylvania State University Wilkes-Barre Campus; Frank Gunnip, Oakland Community College; Thomas King, California State Polytechnic University-Pomona; Ray Lindstrom, Northern Michigan University; Benny Lo, Ohlone College; R.D. Pandian, North Central College; Leonard Presby, William Paterson College; Eric Schonblom, University of Tennessee at Chattanooga; Don Shriner, Frostburg State College; Harrison Wadsworth, Georgia Institute and Technology; Joe Walker, Georgia State University. I also express thanks to the Department of Statistics at Brigham Young University, and the department chair, Lee Hendrix, particularly, for encouraging the development of these instructional materials for our use at BYU and for supporting me in seeking publication. Lastly, I express thanks to my wife and family who gave up a summer of activities so that this book could finally come to completion.

Howard B. Christensen

CONTENTS

1.
METHODS FOR DESCRIBING DATA   1

1.1 The Statistical-Inferential Model   4
1.2 The Measurement Problem   6
1.3 Summarizing Data—Graphical Methods   12
1.4 Summarizing Data Numerically—Preliminaries   21
1.5 Numerical Methods—Measures of Location   30
1.6 Numerical Methods—Measures of Variability   36
1.7 Graphical and Numerical Methods for Bivariate Data   47
1.8 More Graphical Methods   59
1.9 Analysis of the Chapter Exercise   65
     Chapter Summary   70
     Chapter Exercises   71

2.
INTRODUCTION TO PROBABILITY   74

2.1 Terminology Applied to Probability and a Review of Set Theory   76
2.2 The Probability Postulates   80
2.3 Addition Rule of Probability—Probability of the Union   85
2.4 Conditional Probability, the Multiplication Rule, and Independence   89
2.5 Bayes' Theorem   97
2.6 Analysis of the Chapter Exercise   103
     Chapter Summary   105
     Chapter Exercises   106
     Appendix A: Some Basic Counting Rules   108

3.
RANDOM VARIABLES, PROBABILITY FUNCTIONS, AND DISTRIBUTION FUNCTIONS   114

3.1 The Probability Density Function and the Cumulative Distribution Function   116
3.2 "Expected Values" and the Mean and Variance of a Probability Density Function   131
3.3 Other Measeures of Location and Variability   140
3.4 Chebyshev's Inequality   147
3.5 The Moment-Generating Function   152
3.6 Joint Probability Density Functions   155
3.7 Marginal and Conditional Distributions and their Properties   160
3.8 Expectations and Moments for Bivariate Distributions   170

3.9   Statistical Independence   175
3.10   Some Useful Relationships of Linear Combinations of Random Variables   179
3.11   Analysis of the Chapter Exercise   184
        Chapter Summary   186
        Chapter Exercises   187
        Appendix A: Proof of Chebyshev's Inequality   189
        Appendix B: Double Integration   191

4.
DISCRETE AND CONTINUOUS PROBABILITY MODELS   192
4.1   The Bernoulli Distribution   194
4.2   The Binomial Distribution   197
4.3   Other Distributions Based on Bernoulli Trials   204
4.4   The Poisson Distribution   212
4.5   The Discrete Uniform Distribution   216
4.6   The Continuous Uniform Distribution   221
4.7   The Gamma Distribution   224
4.8   The (Negative) Exponential Distribution   230
4.9   The Normal Distribution   235
4.10   The Lognormal Distribution   245
4.11   The Bivariate Normal Distribution   248
4.12   Generating Random Values from Various Distributions   252
4.13   Analysis of the Chapter Exercise   262
        Chapter Summary   264
        Chapter Exercises   266

5.
THE DISTRIBUTION OF A FUNCTION OF RANDOM VARIABLES   269
5.1   The Distribution Function Technique   271
5.2   The Change of Variable Technique   278
5.3   The Moment-Generating Function Technique   283
5.4   Analysis of the Chapter Exercise   286
        Chapter 5 Summary   287
        Chapter 5 Exercises   288
        Appendix: Approximating the Mean and Variance of Functions of Random
        Variables—The "Delta Method"   289

6.
SAMPLING DISTRIBUTIONS   294
6.1   The Concept of a Sampling Distribution   296
6.2   The Sampling Distribution of the Mean and the Central Limit Theorem   298

6.3   The Chi-Square (C-2) Distribution   312
6.4   Student's T Distribution   320
6.5   Fisher's F Distribution   328
6.6   The Normal Approximation to Binomial Probabilities   334
6.7   A Brief Look at Order Statistics   338
6.8   Analysis of the Chapter Exercise   342
6.9   Statistical Quality Control
      Chapter 6 Summary   343
      Chapter 6 Exercises   345

7.
ESTIMATING POPULATION PARAMETERS   353

7.1   Parameter Estimation by the Method of Moments   355
7.2   Parameter Estimation by the Method of Maximum Likelihood   360
7.3   What Makes a "Good" Estimator   366
7.4   Introduction to Interval Estimators   376
7.5   An Interval Estimator of the Mean of a Normal Distribution   379
7.6   Interval Estimators of the Difference in Means   387
7.7   Interval Estimators for a Variance or Ratio of Variances   394
7.8   Interval Estimators for the Proportion of a Bernoulli Distribution   399
7.9   Analysis of the Chapter Exercise   405
7.10 Scientific Survey Sampling   409
      Chapter 7 Summary   406
      Chapter 7 Exercises   407

8.
TESTING HYPOTHESES I: NORMALLY DISTRIBUTED POPULATIONS   421

8.1   Tests of Hypotheses and the Scientific Method   423
8.2   Errors and Their Probabilities   433
8.3   The Power Function and the Sample Size   443
8.4   The One-Sample Test of Means in Normally Distributed Populations   452
8.5   Two Sample Tests of Means for Normally Distributed Populations   458
8.6   Test of Means from Paired Observations   466
8.7   Tests of Variances in Normally Distributed Populations   472
8.8   The Neyman-Pearson Lemma and Likelihood Ratio Tests   478
8.9   Analysis of the Chapter Exercise   490
      Chapter 8 Summary   492
      Chapter 8 Exercises   494

9.
TESTS OF HYPOTHESES II: LARGE SAMPLE APPROXIMATE TESTS   497

9.1   A Large Sample, C2 Goodness-of-Fit Test   499
9.2   Large Sample Tests of Proportions: One and Two Populations   510

9.3   Two-Way Contingency Tables   517
       Chapter 9 Summary   529
       Chapter 9 Exercises   531

10.
OTHER APPLICATIONS   534

10.1   Experimental Design, the One-Factor Experiement, and the Randomized Block Design   535
10.2   Some Nonparametric/Distribution-Free Tests: the Sign Test and the Wilcoxon-Signed Rank Test   548
10.3   More Nonparametric Tests: The Mann-Whitney Test and the Kruskall-Wallis Test   559
10.4   Linear Regression Analysis   569
       Chapter 10 Summary   584

APPENDIX: STATISTICAL TABLES   586

INDEX   608

1

METHODS FOR DESCRIBING DATA

OBJECTIVES

When you have completed this chapter you should be able to summarize a set of data graphically, using a histogram or bar graph, or numerically, using appropriate measures of location or variability. The numerical values you should be able to compute are (1) quantile values, (2) empirical moments, (3) the mean, (4) the median, (5) the mode, (6) the range, (7) the variance, and (8) the standard deviation. You should be able to relate the numerical values you compute to properties of central tendency, variation, skewness, and symmetry.

For a data set having pairs of measurements, you should be able to make a scatterplot or bivariate histogram and be able to compute the covariance or the correlation of the paired measurements. You should be able to relate the numerical properties of the data to graphical characteristics possessed by the histogram or bar graph of the data set.

INTRODUCTION

To someone who has never taken a course in statistics, the work of a statistician is often misunderstood. Many think the job of a statistician involves nothing more than clerical operations such as summarizing data, computing averages and percentages, and making charts and graphs. Others see the statistician as a special kind of mathematician dealing with a subject most people can't understand and all find boring.

However, the typical statistician finds fulfillment in the profession, not boredom. The field is one with great variety and wide application. It is a constantly growing profession, with each new problem studied spawning numerous other questions to be answered.

With such a variety of perceptions about the work of the statistician, my challenge is to correct the misconceptions you may have, while creating some of the same excitement for you that the statistician finds. First, however, we must provide a fairly brief definition of what statistics is as a field of study. That is complicated by the fact that there is as much variation in definitions as there are people who attempt the task. Therefore, my definition will simply serve as the definition for this textbook and will provide the framework within which we work.

What Is Statistics?

Statistics deals with the development and use of methods and techniques to understand and explain chance phenomena. Typically, data are collected from a population. The data are then examined for patterns and characteristics which would help explain the population being studied.

Any process which has outcomes that are unpredictable or uncertain can be better understood by using the principles of statistics and probability. Consider the following typical examples:

1.   The weather at a particular location varies from day to day and eludes exact prediction. Meteorologists are interested in the study of weather patterns to try and identify the variables that are most closely linked to the varying weather patterns. If such patterns can be found, more accurate weather forecasts can be made. Data concerning high and low temperatures, wind speed, barometric pressure, cloud cover and type, humidity, etc. are the likely characteristics to be studied.

2.   Quality control engineers recognize that in any production process there is item to item variability. For example, the quality of a particular metal part may depend on the temperature of the metal when it is produced, the metal thickness, malleability, the speed of the cutting tool and its temperature, and so forth. Data concerning these characteristics might be collected in order to produce a more uniform and higher quality product.

3.   A sociologist is studying the causes of divorce and its impact on the family. A sample of divorced families is randomly selected along with a sample of families having stable marriages. Data are collected relative to family size, ages of children, income of husband, working or non-working status of the wife, years of marriage prior to divorce, socioeconomic status of the family, adjustment problems of the children, the children's success in school and society, and so on. It is hoped that this data will reveal factors that might be related to divorce and its associated problems.

4.   At a large time-sharing computer facility the number of users and the length of their computing jobs vary from hour to hour. The system operator wants to know how many terminals to provide the users, how much disk storage users need, and what kind of rate structure to provide its users to encourage efficient use of facilities. Data is collected on the pattern of usage, length of jobs, CPU time used for each job, used and unused disk storage space. The data are analyzed so that decisions can be made that will help to run the facility more economically and efficiently.

5.   A politician, running for elective office, wants to know the likelihood of election as well as what issues are of concern to the electorate. A sample of registered voters is selected from her district. Data about their voting preference, likelihood of voting, age, sex, political persuasion, political issues of concern to the prospective voter, and other related items are collected. Analyzing the data will help the politician better understand the political views of the registered voters in her district.

In these examples, first note the diversity of the settings of the problems—from weather to politics to business to computers operations. In fact, most fields of study, at some phase, collect and interpret data—thus utilizing the concepts of statistics.   Secondly, each of the examples discussed above, dealt with uncertainty—unpredictable weather, variation of product quality, unknown political attitudes, and so on. Thirdly, to help solve the problem, data were collected and analyzed to detect any patterns, trends and relationships the data might show. The formal methods that are developed to collect, analyze and interpret data are the subject matter of a statistics course. This book will introduce the terminology and vocabulary of statistics and lay the foundation for developing data-analytical skills.

Where We Are Going in This Chapter

The general objective of any statistical investigation is the collection and analysis of data so you can better understand the process that produced the data or the population from which the data came. The data are usually represented in numerical terms. In this chapter we introduce common methods of summarizing and representing data for easier interpretation and decision making.   Both graphical and numerical methods will be presented.   In addition, many new terms will be introduced that you will need to make a part of your personal vocabulary and new concepts will be presented that you will need to learn to apply.

Chapter Exercise

[The "chapter exercise" presented in each chapter represents a problem typical of the problems that would be solved using the concepts that you learn in the chapter. It is suggested that as you look at the exercise you think of how you might solve the problem. Then as you study the chapter, look for specific concepts that would help you in the solution. This will help you study with a purpose, and ought to provide more "directed study" for the entire chapter.]

The following data are scores in a recent semester for 104 students taking a course in statistics. For each student, their final exam score (F) is recorded along with their total points earned (T). (Total points are expressed as a percentage of total points possible).   Some questions to be answered are

1. What is the average score on the final exam?
2. What is the average score for "total points earned"?
3. How do the two sets of scores compare on an overall basis?
4. Does one set of scores show more variation or spread than another?
5. How strong is the relationship between a person's final exam score and his overall score in the course?
6. If these scores could be considered typical of all current or future students, what are the implications for you and the grade you might get in the course?

The following data occur in pairs: A total of 96 student scores on the final (F) and their corresponding "total %" score (T) are reported below. For instance student 1, shown in bold type, had a 77 "total %" score with a score on the final exam of 71, student 2 had a 61 and 42, etc.

T	F	T	F	T	F	T	F	T	F	T	F
77	**71**	91	99	79	73	81	70	65	69	82	64
61	42	79	65	85	75	87	94	75	63	66	59
55	77	59	50	83	74	26	20	90	92	92	94
71	61	70	55	65	27	79	73	85	84	80	71
99	100	73	71	96	96	86	84	73	80	93	94
91	72	78	85	68	74	64	72	75	64	70	65
71	64	67	53	91	100	97	99	70	59	58	55
93	90	75	71	94	90	82	84	75	80	75	43
82	89	72	42	71	47	91	83	69	47	67	54
63	67	94	94	94	94	73	66	73	66	68	77
90	84	83	87	98	99	100	99	46	30	83	78
97	94	87	84	75	71	60	54	76	56	71	55
76	74	65	43	54	30	98	99	81	71	80	79
89	84	80	75	96	96	79	76	83	89	65	42
83	83	59	79	70	68	70	67	81	79	89	79
92	97	69	35	78	76	91	96	94	99	87	84

1.1   THE STATISTICAL-INFERENTIAL MODEL

Introduction

In the introduction to this chapter some examples of statistical problems were given. Looking at those examples, we can identify several traits which provide a common structure to all studies of a statistical nature. This common structure can be very helpful in understanding statistical problems. It is also useful in keeping you oriented in the course. This structure is specified in this section as it applies to the entire text and also to this chapter exclusively, and is referred to as the "statistical-inferential model" or SIM In this context, the concept of a (1) population, (2) sample, and (3) statistical inference is explained. A diagram that shows the relationship among these three ideas is also presented.

Defintions

DEFINITION 1.1: A *population* is a collection of units. A *sample* is a subset of the units in the population.

DEFINITION 1.2: A *random sample* is a sample of units in the population such that each unit in the population has an equal chance of being a part of the sample.

DEFINITION 1.3: A *statistical inference* is a conclusion about a *population* based on data drawn from a *sample* of the population.

DISCUSSION

A statistical study or investigation consists of three simple components. First, a *population* of units possessing information of interest is defined. Second, a random sample of units from the population is selected. Information (in the form of data) obtained from each unit in the sample is summarized and analyzed. Thirdly, if good sampling procedures are used when selecting the sample, then the information and characteristics of the sample are inferred to be typical of the entire population. (We define a *good* sampling procedure as one which guarantees each element in the population a known chance of selection. The phrase—a *random sample*—means that the chance of selection is equal for each element in the population. The use of random numbers from a table or computer is essential in selecting such sample.) Such an inference is called a statistical inference. These three components are displayed in Figure 1.1 which provides a perspective of where we are going for the entire course.

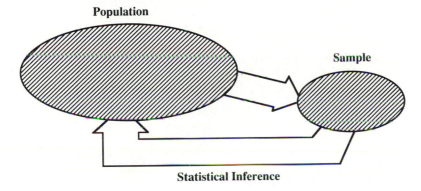

Figure 1.1 The Statistical-Inferential Model (SIM).

To be of value, the statistical inference will report not only what characteristics the population is thought to have, but in addition, it will report how much error is expected between the value(s) obtained in the sample and what the results would be if the entire population were examined. A measure of confidence in the representativeness of the sample will be reported in terms of a probability or percentage.

In this model certain characteristics common to all problems can be listed: (1) We start with a collection of units called the population; (2) Associated with each unit is information that we desire; (3) To collect this information a measurement has to be made on the unit; (4) If each unit in the population is measured, then a complete census has been made; if only a portion of the units are measured then sample data are obtained; (5) The data are analyzed in order to make an inference about the population from which the units came.

The "statistical-inferential model" (SIM) will be used throughout this text to provide perspective for the course. Each chapter will be introduced with the SIM and the chapter purpose will be related to the appropriate component identified by the cross-hatched region.

Specifically, in this chapter we present common methods for summarizing and describing either population or sample data. Both graphical and numerical methods are introduced along with the appropriate notation. We broadly classify numerical measures into measures of location or "center" and measures of variability, or spread. In the sections that follow, these topics will be organized in that fashion. The SIM for this chapter is presented as Figure 1.2. Note the shaded region for both the population and sample. The techniques of description presented in this chapter apply to either the population or the sample.

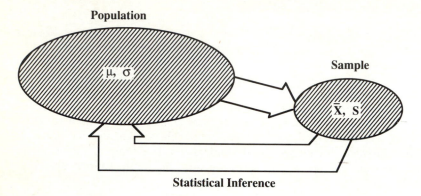

Figure 1.2 The SIM for Chapter 1: describing data numerically and graphically.

The symbols used in Figure 1.2 are common to many statistics texts. The symbol μ denotes the mean of the population and σ denotes the standard deviation, a measure of variability. Greek symbols are commonly used to denote population characteristics. The symbol $\bar{x}$ and s are used to denote the mean and standard deviation of the sample, respectively. We will expand on this notation in later sections.

1.2   THE MEASUREMENT PROBLEM

Introduction

Measurement is the process a scientist goes through to assign numbers to characteristics of the elements of a population or sample under investigation. It is assumed that the numbers assigned reflect useful concepts about the units which in turn makes it easier to manipulate and summarize the characteristics of interest. This is done with the hope that new information and insights will be gained from such an analysis. This manipulation of measurements takes the usual form of ranking or ordering, adding or subtracting, as well as applying the other operations of ordinary arithmetic. These operations will usually be done by a computer and thus have to be amenable to the operations computers are suited to perform.

All of the data we investigate in this course consist of measurements made on some unit in the population or sample. The unit may be a person and the measurements a record of the person's height, weight, marital status, etc. The unit may be a plot of land and the measurements a record of the soil moisture, the pH of the soil, and the density of plant material across the plot. The unit may be a central processing unit (cpu) manufactured by an electronics firm and the measurements the thickness of the electrical contacts, the amperage it draws, the speed with which it performs a given task, and the presence or absence of any visible defects.

In order to choose the correct method of analyzing the data, the data must be classified properly. The method of analysis appropriate for one kind of data may not be appropriate for another. In this section we define the various terms for classifying measurements that produce data and provide examples and practice problems in performing the classification.

Definitions

DEFINITION 1.4: *Quantitative measurement* is a measurement process for which the measurement reflects a quantity or amount.

DEFINITION 1.5: *Qualitative measurement* is a measurement process for which the measurement reflects a quality or characteristic of the unit being measured. It is a nonquantitative measurement.

DEFINITION 1.6: *Discrete measurement* is a measurement process that produces, as possible values, discrete values from the set of real numbers. These values can always be put in a one-to-one correspondence with the set of positive integers. That is, there is a always a first value, a second value, and so forth, in the set.

DEFINITION 1.7: *Continuous measurement* in contrast to discrete measurement is a measurement process that is possible of producing *any* value in an interval from the set of real numbers. This interval may be open or closed, a subset of the real number line, or it may be the entire set of real numbers. The important characteristic is that the set of values in the interval form a continuum of values.

DEFINITION 1.8: The *level of measurement* refers to the level of the mathematical operations that are permissible to use to manipulate the measurements. For instance, with some data it would be appropriate only to indicate whether they are the same or different (the property of equivalence); the operations such as ordering, adding, subtracting and multiplication being inappropriate. The arithmetic operations of addition and multiplication represent a higher order of mathematical manipulation than merely stating whether two things are "the same or different."

DEFINITION 1.9: *Nominal measurement* is the level of measurement where the numerical measurement values, X, have *only* the property of equivalence. That is, if we have two measurement values x and y, all we can say are that $x = y$ or $x \neq y$ —the concept of "greater than" ($>$) doesn't apply.

DEFINITION 1.10: *Ordinal measurement* is the level of measurement where the numerical measurement values, X, possess the property of "greater than" ($>$) in addition to the concept of equivalence. The transitive property of algebra holds. That is, if $x \leq y$ and $y \leq z$, then $x \leq z$.

DEFINITION 1.11: *Interval measurement* is the level of measurement where the numerical measurement values, X, possess the property of differences in addition to all the properties of ordinal measurement . That is, if x is one measurement and y is another, then the *difference* (interval) $x - y$ is interpretable and is subject to the common operations of arithmetic.

DEFINITION 1.12: *Ratio measurement* is the level of measurement where the numerical measurement values, X, possess a true zero point in addition to all the properties of interval measurement. That is, if a reported measurement is a zero, then that implies there is a zero amount of the "quantity" being measured. In addition, if x is one measurement and $y \neq 0$ is another, then the ratio $x \div y$ is interpretable. All of the common operations of arithmetic can be applied to the values X in addition to intervals constructed from the values of X.

DISCUSSION

Suppose a physiologist is collecting information on a sample of subjects for a research study on aerobic conditioning. Some of the measurements taken are heart rate, blood pressure, sex, age, weight, height, eating habits, and previous health history. The data are to be organized and analyzed to see what factors seem correlated with good conditioning. The appropriate methods to analyze one kind of data are not necessarily appropriate to analyze other types of data. In order to make a proper selection of methods of analysis, the characteristics of the measurement processes need to be understood.

Qualitative vs. Quantitative Measurement

One way of classifying data is to determine whether they are quantitative or qualitative. Looking at the definitions above, we can see that they are pretty much self-explanatory and easy to distinguish. Measurements of height, weight, volume, time, blood pressure, heart rate, etc., are common quantitative measurements. Measures of marital status, sex, political preference of persons, hair color, eye color, etc., are qualitative measures. If the characteristic to be measured is a quantity or amount, then the measurement is quantitative. If the measurement simply reports a characteristic, category or "quality" of a unit, then it is a qualitative measurement.

Discrete vs. Continuous Measurement

Another way of classifying data is to note whether it is continuous or discrete. By studying the definitions above, note that discrete measurements are countable. That is, there is a first measurement value, a second measurement value and so on. For instance, if a person reports the number of correct answers on a test, the first possible value is a zero, the second is a one, the third is a two, etc. Thus, you have discrete measurement for the test results. However, if a parent reports the birth weight of their first child, the first *possible* weight may be 4 pounds. Since weight can conceptually take on any value, the next larger weight can't be specified, since by the continuous nature of values, for any number specified as the next possible value, a value between 4 pounds and the value specified can be found. (In reality all common measurement devices have limit accuracy and thus cannot take on any and all values in some range. In deciding whether a measurement process is continuous or not, we must assume the existence of a measurement device of unlimited accuracy.)   Typical examples of continuous measurements are height, weight, volume, and time. Typical discrete measurements are the number of defects in a length of cloth, the number of children in a household, the number of insurance policies a wage earner holds, the brand name of the top selling hand calculator made in the United States.

Nominal Measurement

In order to assess the level of measurement it is assumed all measurements are expressed in numerical terms. For quantitative data this is no problem; but qualitative measurements have to be quantified by some coding scheme. For example, report marital status on a questionnaire, the following coding scheme may be used:

Marital Status	Code
Married	1
Single	2
Widowed	3
Divorced	4

This is an example of nominal data. The numerical values simply distinguish one classification from another but do not imply any concept of quantity or amount. Other examples of nominal measurement are racial classification, social security numbers, political preference, and so forth. (All qualitative data will be nominal measurement.)

Ordinal Measurement

If the numerical measurements allow ranking or ordering from small to large, then the measurement scale is at least ordinal. For example, when a person reports on a scale of: Strongly Agree = 5, Agree = 4, Neutral = 3, Disagree = 2, Strongly Disagree = 1, then we are able to rank or order the different responses that person reports, from small to large, depending upon whether the person reports a lot of agreement or not so much agreement. This same ordering property is present when classifying diseases on some scale of severity, say, from not severe to very severe or when reporting whether a manufactured part displays major, minor or no defects.

Interval Measurement

If the measurements allow a comparison of differences, the measurement scale is interval. For example, the report of the daily high temperature in Celsius degrees allows us to report how many degrees higher today's temperature is than yesterday's. We can not only say whether today's temperature is the same as or different from yesterday's (a nominal classification), whether it is higher or lower than yesterday's (an ordinal classification), but we can say how many degrees more (or less) it is. This is the primary characteristic of interval measure. We can interpret the meaning of a difference or an interval.   The consumer price index and other governmental indices would be regarded as interval measurement processes.

Ratio Measurement

The highest level of measurement is the ratio level. If it makes sense to report that one measurement is twice that of another or only 80% of the size of another, then the measurement level is ratio. Examples of ratio measurement are measurements of height, weight, volume, and so on. (Temperature is an example of interval measurement but not ratio. If the high today is 50° F and the temperature yesterday was 25° F, we cannot say that it is twice as hot today as yesterday, only that it is 25° warmer.)

Classifying the Measurement

To be able to properly classify a measurement, it is necessary to define the basic or elementary unit in a population or sample. This elementary unit is called the *unit of observation* or *experimental unit*. Once the unit is defined, the measurement extracted from this unit can be evaluated according to the definitions of qualitative-quantitative, discrete-continuous, and the level of measurement.

Table 1.1 below shows the relationships between the different classifications of data and their association with the level of measurement. In this table, the empty cells represent measurement combinations that are not possible. If data are nominal only, then they will also be discrete and qualitative. If data are ordinal at most, then they are quantitative *and* discrete, but can't be continuous. If data are interval or ratio they will also be quantitative but they may be either discrete or continuous.

Table 1.1 Measurement classification: empty cells indicate combinations of classification levels that are not possible. Other cells are possible combinations frequently encountered.

Measurement Level	Quantitative		Qualitative	
	Discrete	**Continuous**	**Discrete**	**Continuous**
Nominal			Possible	
Ordinal	Possible			
Interval	Possible	Possible		
Ratio	Possible	Possible		

EXAMPLES

EXAMPLE 1: A voting precinct is to be studied to determine the proportion of registered voters who voted in the last election.

> **Unit of Observation**: A registered voter.

> **Measurement:** Code a 0 if the registered voter didn't vote in the last election. Code a 1 if the registered voter did vote.

> **Classification:** The measurement is qualitative (voting behavior is reported), discrete (measurements are 0 or 1), and nominal (registered voters either voted or did not; there is no concept of "greater than").

EXAMPLE 2: A political scientist studying voting precincts across the state samples 45 precincts randomly from the 385 in the state. Data to be collected are the number of registered voters in the precincts plus other political-social characteristics.

> **Unit of Observation:** A voting precinct

> **Measurement:** The number of voters in a precinct.

> **Classification:** The measurement is quantitative, discrete and ratio. If one precinct reports 300 registered voters and another reports 150 voters, this is obviously a quantitative measurement. The possible set of responses are associated with the set of integers from 0 on up and is thus discrete. There is a true zero point implying that a district has no registered voters, and we can say that the first district has $\frac{300}{150} = 2$ or twice as many voters in it as the latter district. Therefore, the level of measurement is ratio.

EXAMPLE 3: The government plans to open a particular area of forest for logging purposes. The total volume of lumber is to be estimated. A minimum bidding price is established once the volume is estimated. A sample of trees in the area is to be selected and the diameter of each tree is measured at the 2-foot height along with the tree height. Formulas are applied to produce a volume figure for each tree measured based on the diameter and tree height.

> **Unit of Observation:** A particular tree that is part of the sample.
>
> **Measurement:** There are two measurements: the tree diameter and tree height.
>
> **Classification:** The measurements are both quantitative, continuous and ratio. If a tree has a diameter of 3 feet and a height of 40 feet, and another has a diameter of 1 foot and a height of 20 feet, we can say that the first tree's diameter is 3 times bigger than the second and is twice as tall.

EXERCISES

In each of the following exercises identify the population of units under study, decide whether a sample is selected, determine what the unit of observation or experimental unit is, and classify the measurements described.

1. An automotive supplier boxes spark plugs in lots of 200. Before shipment each box is inspected by randomly selecting 20 plugs. The size of the spark plug gap is measured with an appropriate gauge and the ceramic casing is examined for cracks or chips. These are regarded as two separate measurements.

2. An agronomist checks the germination rates of seeds from different suppliers. One hundred seeds from each of 10 suppliers are planted under controlled conditions of planting medium, temperature, and humidity. After 14 days, the number of germinated seeds out of the 100 planted are reported. After 45 days the height of the tallest and shortest plant in the groups from each supplier are measured and the average height of these is reported. (Note: measurements are recorded for each group of 100 seeds and four different numbers are recorded.)

3. Certain minerals found in the soil are taken up by plants, and ultimately find their way into man's bloodstream through the food chain. A nutritionist wanted to know how much of a particular mineral was in the bloodstream of residents in a particular county. These results were to be compared with results from other locations across the country. A random sample of adults was chosen and a blood sample was taken from each adult in the sample. The blood samples were analyzed for the volume (per cubic centimeter) of this particular mineral in the blood of the individuals sampled. (What would be the implication of conducting a "census" in such a setting?)

4. The use of automobile seat belts has been shown to have a definite effect on reducing injuries and fatalities from automobile accidents. However, a large proportion of automobile passengers are "unbelted" when traveling in their automobiles. A county health official wanted to estimate the proportion of passengers using seat belts in the county. A sample of cars passing through an intersection was observed and the number of persons in the front seat wearing safety restraints of any kind was recorded.

5. The manufacturer of hearing aid and watch batteries periodically tests them to measure how long they last until failure. In addition, the tendency of batteries to show corrosion after storage in a humid environment is of interest.

6. In a small community a study is to be conducted to see how many households watched a particular TV show in the preceding week. Households were selected by selecting numbers from a phone book and the person answering the phone was asked if they watched the TV program in question.

7. In a school district the district administrator wanted to determine how many students came from single-parent households. A sample of kindergarten classes was selected and the personal records of the students in each class was examined and the number of single-parent households recorded for each kindergarten class.

8. An experiment was performed to see how many kernels of popcorn out of 20 kernels didn't pop when microwaved for 1.5 minutes. From a bag of popcorn, 20 groups of 20 kernels each were formed and popped. The 400 kernels were presumably selected randomly.

9. Referring to the experiment of exercise 8, the total volume for the popped kernels was approximated by putting all 20 kernels in a glass and the height of the glass at the highest kernel was measured in inches.

10. Blank cassette tapes are tested in a variety of ways before being shipped for sale to retail outlets. Usually a sample of tapes is selected from a production run and examined for cracks, chips and visual defects.

1.3   SUMMARIZING DATA—GRAPHICAL METHODS

Introduction

All applied statistical studies ultimately produce a data set to be analyzed. The data might be a set of test scores on the ACT test from a sample of students who take the test. The data might be the cholesterol values in samples of blood from an overweight group of business managers. It may be the monthly income figures of divorced women in a particular city. Whatever the setting, the individual data needs organizing and summarizing before the important characteristics in the data will begin to show up.

Such organizing and summarizing is often most easily interpreted if the result is displayed in the form of a picture, chart, or diagram. The simplest and most commonly encountered diagrams are pie charts, bar graphs, histograms, scatterplots, etc.

In this section we present guidelines for preparing bar graphs and histograms. These two methods are chosen because they present the data in a form that is analogous to theoretical representations of data presented in later chapters. A few methods such as box plots and stem-and-leaf displays are presented in the appendix. In addition, there are several good books devoted solely to methods of displaying data. These can be consulted if a broader coverage is desired.

Definitions

DEFINITION 1.13: *Empirical data* are data obtained from a sample of a population rather than the entire population itself. Any method used to summarize empirical data will be referred to as an empirical method.

DEFINITION 1.14: A *bar graph* is a two-dimensional graphical technique for displaying discrete data. On one axis, usually the horizontal axis, the possible values or categories associated with the data are marked off. The vertical axis is used to indicate the frequency or percentages of the data taking on each of the possible values or categories. A vertical, rectangular bar is erected at each position on the horizontal axis, with the height of the bar indicating the corresponding frequency or percentage. The bars are usually separated to imply the discreteness of the data.

DEFINITION 1.15: A *histogram* is a two-dimensional graphical technique for displaying continuous data. On one axis, usually the horizontal axis, the possible range of values associated with the data are marked off, usually by marking off a series of intervals. The vertical axis is used to indicate the frequency or percentage of the data taking on values in each of the intervals marked off. A vertical rectangle is erected at each interval on the horizontal axis, with the height of the rectangle corresponding to the frequency or percentage of data values in that interval. The rectangles are usually adjacent to imply the continuity of the data.

DISCUSSION

Suppose the following data were obtained for a group of 30 students. Reported are their class standing and their final exam grades in a statistics course.

Class	Grade	Class	Grade	Class	Grade
So	78	Jr	81	Jr	89
So	94	Sr	100	Sr	78
Jr	57	So	66	So	73
So	71	So	68	So	94
Jr	84	Jr	59	So	67
Jr	81	So	67	Jr	78
So	78	Jr	95	Jr	61
So	89	Sr	73	Sr	80
Jr	69	So	90	So	67
So	82	Jr	76	So	87

First we may wonder about the mix of the students relative to class standing. We then also want to describe the distribution of their scores on the test.

Bar Graphs for Discrete Data

The class standing is a discrete measurement and we want to know what proportion of the 30 students are freshmen, sophomores, juniors and seniors. In Table 1.2 a simple tabulation by hand shows the essential information.

Table 1.2   Frequency table for class standing.

Class Standing	Frequency	Rel. Freq.
Freshmen	0	0÷30 = 0.00
Sophomores	15	15÷30 = 0.50
Juniors	11	11÷30 = 0.37
Seniors	4	4÷30 = 0.13

To display the frequency table graphically a bar graph is the appropriate tool. Figure 1.3 below shows this bar graph. The method of construction is obvious upon examination. In this case the measurement values are placed on the horizontal axis and the frequency or relative frequency is scaled on the vertical axis. Bars are constructed at each measurement value having a height equivalent to the frequency or relative frequency of that value. The bars are separated by a space to indicate the discrete nature of the measurement.

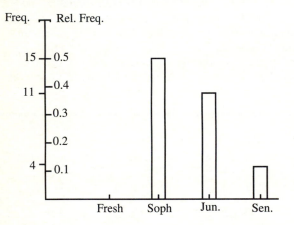

Figure 1.3   Bar graph—frequency and relative frequency by class standing.

It can be seen that the greatest percentage of students are Sophomores followed closely by Juniors, with relatively few Seniors and no Freshman.

Histograms for Continuous Data

A histogram is very similar to a bar graph but is intended for use with continuous data. Using the data consisting of test scores and class standing, we choose the histogram to display the final exam scores. Though there is a strong argument that test score data are discrete, we want to convey the concept that test scores can take on any value between 0 and 100.

In order to construct a histogram we might proceed in several steps. The first is to sort the data from small to large as follows:

57	59	61	66	67	67	67	68	69	71
73	73	76	78	78	78	78	80	81	81
82	84	87	89	89	90	94	94	95	100

Once the data has been sorted then a frequency table can be constructed similar to the one created for the class standing. A frequency table for the test score data is shown in Table 1.3.

Table 1.3 Frequency table for test score data.

Interval	Frequency	%
$50 < x \leq 60$	2	6.67
$60 < x \leq 70$	7	23.33
$70 < x \leq 80$	8	26.67
$80 < x \leq 90$	8	26.67
$90 < x \leq 100$	4	13.33

Figure 1.4 shows the histogram for the test score data using the information from the frequency table. Notice that like the bar graph, the measurement values are scaled on the horizontal axis and the frequency of occurrence is scaled on the vertical axis. However, there is no space between the vertical bars, unlike the bar graph. This is to imply the concept of continuity of measurement values.

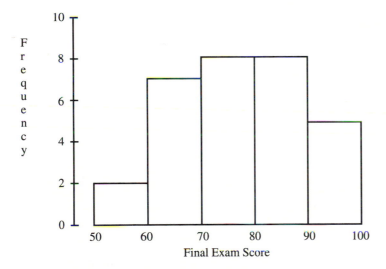

Figure 1.4 Histogram—frequency of final exam scores by class intervals.

The primary decision you face when constructing a histogram is the choice of the number and width of the class intervals. If there are too many class intervals, then you may have too much detail and won't be able to see the main shape of the data. This is illustrated in Figure 1.5.

If you don't have enough class intervals, then the histogram will not show enough detail. This case is shown by Figure 1.6 which has only three class intervals. Obviously, the most extreme case would be a histogram with only one class interval.

It is recommended that you use between 5—12 intervals depending on the number of data points to be summarized—the more data points, the more class intervals you make. Fortunately, there are numerous computer packages available that make histograms with ease. Most of them provide the capability to control the width of the intervals, and thus the number of intervals as well. Therefore, it is easy to make a choice, look at the results, and redo it. The intent is to provide a meaningful visual image of the shape of the distribution of the data. If the display is too sparse (too many intervals) or too gross (too few intervals), then you can choose a different number of intervals.

In doing this, however, you should try to be as objective and "honest" as possible in treating the data. The temptation is always there to show the data in the way that it fits your preconceived notions.

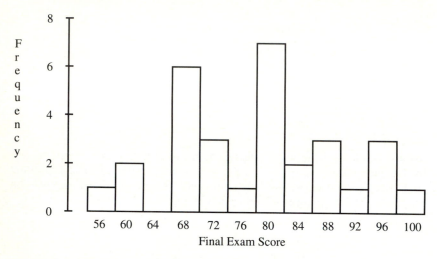

Figure 1.5 Final exam scores—too many class intervals.

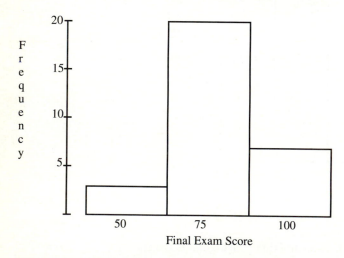

Figure 1.6 Final exam scores—too few intervals.

Symmetry and Skewness

When examining the shape of a histogram, it is useful to look for the presence or absence of symmetry. A histogram shape that is not symmetric will be said to be skewed. However the amount of skewness can be slight or dramatic. If the distribution is skewed and tails off to the right or positive direction we say the distribution is skewed right or skewed positively. If it tails off to the left or in the negative direction, we say the distribution is skewed left or negatively skewed. Figure 1.7 gives some examples of symmetric vs. skewed distributions.

(It may or may not be proper to talk of symmetry relative to bar graphs. The level of measurement must be at least ordinal for the concept of skewness to mean anything.   Data that is only nominal cannot be said to be skewed left or right since the assignment of positions on the horizontal scale is arbitrary.)

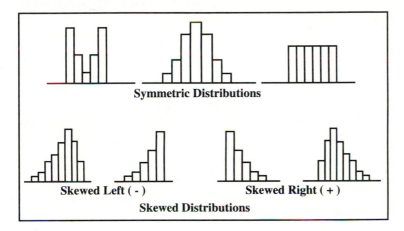

Figure 1.7 Symmetric and skewed histogram shapes.

EXAMPLES

EXAMPLE 1: The manager of a small electronics store keeps a record of the brand name of all the hand calculator that are sold during the month. The following bar graph of Figure 1.8 is a summary of this data. The bar graph shows the percentage of calculators of each brand sold (discrete-nominal data).

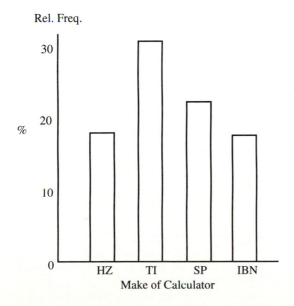

Figure 1.8 Bar graph—percentage of calculators sold by brand code.

In the bar graph of Figure 1.8 you can note that the calculator sold most often was the TI with the SP next. The HZ and IBN sold equally well but not as well as the first two.

EXAMPLE 2: A group of 94 students in a statistics course reported their heights to the nearest whole inch. The following data is presented. The histogram is shown in Figure 1.9.

71	71	69	77	71	71	72	69	65	69	63	72
76	70	67	72	71	66	65	72	70	73	68	72
68	73	69	68	74	74	61	69	69	67	72	73
74	70	66	68	74	73	71	72	73	73	70	73
70	74	74	74	70	72	77	68	63	75	76	67
72	72	72	68	72	70	68	69	66	72	71	69
67	69	70	63	71	69	65	64	70	67	65	62
71	70	67	74	76	58	66	64	74	70		

The histogram of heights shows a fairly clear negative skewness, with most heights being in the upper 60s and lower 70s. The one observation on the left may be atypical. Such an observation is called an outlier–an observation that appears to be rather far away from most of the rest of the data. It would be interesting to note the male-female relationships to height. If the 58 inch category is associated with a female, it may not really be an "outlier" since females tend to be shorter than males.

By looking at the diagram, a "central" number would be expected to be around the low 70s. This would represent a measure of location. The range of the measurements is about 20 units—from 58 to about 78. This reflects a concept of spread or variability in the data.

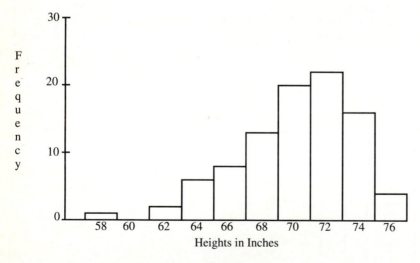

Figure 1.9 Histogram of heights of 94 college students.

EXAMPLE 3: A soft drink company is concerned about the bursting strength of the aluminum cans it uses in the bottling process. Since the contents are under pressure from the carbon dioxide in the drinks, the pull-tab top must withstand the pressures that build up as the temperature of the contents increases. To measure the bursting strength of a particular shipment of cans from a supplier, 20 cans were selected randomly and tested at room temperature with the following results. Make a histogram for the bursting strength data.

265	205	263	307	220
268	260	234	299	215
197	286	274	243	231
267	281	265	214	318

SOLUTION: Figure 1.10 displays a histogram of the 20 observations. It appears that the distribution is skewed somewhat to the left with most of the observations in the 250-275 range. A guess at a central number or measure of location would be about 250, with a range of 325 – 175 = 150 grams. There don't appear to be any obvious, extreme numbers.

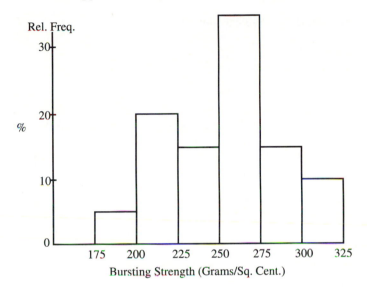

Figure 1.10 Histogram of bursting strength of soft drink containers.

EXERCISES

1.   At the request of a study committee from the state legislature concerning the 55 mph speed limit, the highway patrol set up a radar checkpoint and recorded the speed, in miles per hour, of a random sample of 50 cars that passed the checkpoint during a particular time period. Construct a histogram for the speeds using class intervals 10 units wide. The speeds of the cars were recorded as follows:

74	66	65	55	48	56	50	75	67	65
76	68	50	65	70	65	60	51	68	76
68	77	63	65	52	52	63	65	80	70
65	81	70	63	53	45	65	55	71	64
55	70	64	45	66	64	40	66	55	71

2.   Examine the histogram of exercise 1. Is the histogram symmetric or skewed? If you think it is skewed, decide if it is skewed right or left. By inspection, what would you report as an "average"

mph?   What is the minimum and maximum speeds?

3. Part of the marketing research data collected by a pharmaceutical firm producing a children's vitamin was about the color preferences children have. An equal number of each of five different color-flavor "vitamins" was prepared. (The "vitamins" under study were simple sugar-sweetened capsules the same shape and size of the vitamins to be produced.) A group of 1st and 2nd grade children were asked to select their favorite color from the "vitamins" presented to them with the following results: (Note this is not "raw" data. Rather, it has already been summarized for you as a frequency table.)

Color-Flavor	Number Chosen
Grape	26
Cherry	18
Orange	13
Lemon	23
Lime	20

Construct a bar graph for this data. What conclusions about color preference do you make?

4. Why would it not make sense to refer to skewness or symmetry for the bar graph of exercise 3? What is the unit of observation and the measurement classification for this study?

5. A "weights and measures" inspector for the state is assigned to check public vending machines, periodically. The inspector collected data on soft drink dispensing machines that dispense ice and soft drinks into an eight-ounce cup. The amount of ice dispensed in grams is recorded for 20 fills, along with a note of whether there was any spilling. The following data resulted.

Ice Weight (gms.)	Spilling occurred	Ice Weight (gms.)	Spilling occurred
68	Yes	65	No
59	No	80	Yes
86	Yes	98	Yes
80	No	71	No
70	No	57	No
53	No	82	Yes
71	Yes	67	No
72	Yes	87	Yes
78	Yes	76	No
67	No	85	Yes

Construct a histogram on the ice weight and construct a bar chart on the spilling. Make the vertical axis a "percentage" axis.

6. Using the data in exercise 5, construct a histogram of ice weights for the condition where spilling occurred. Construct another histogram of ice weights where no spilling occurred. Use the same size of class intervals and scaling for both histograms so they can be compared. Do the histograms look the same or different? Are they centered around the same place? What does this suggest about the relationship between the amount of ice and the tendency toward spilling?

7. Using the histogram of ice weights from exercise 6, what is an "average weight" from inspection. Do the two histograms seem to have the same relative spread or dispersion from the center of the histogram?

8. Defective parts found in shipments from different manufacturers of missile parts are reported in the following table.

Manufacturer	Number of Defectives/10,000 inspected
A	22
B	30
C	15
D	10

Convert the number of defectives to a percentage figure and plot an appropriate graph.

9. Looking at the figure obtained in exercise 6, comment on the wisdom of discussing skewness and symmetry properties for the graph.

10. Two persons, examining the same histogram or bar graph may make different conclusions from what they see. Comment on what might be done to make the analysis of a data set less subjective and more objective. What are the advantages and disadvantages of your recommendation?

1.4   SUMMARIZING DATA NUMERICALLY— PRELIMINARIES

Introduction

After you take your first exam in this course, or in any course for that matter, you may wonder how well you did in comparison to the other students in the course. What was the average on the test? What were the high and low scores? Did the test scores vary widely, or were they all pretty much the same? In asking such questions, you expect one or two numbers, such as the average, or the high and low scores, to give you useful information about the entire set of individual scores in the data set. Summarizing a whole set of data into one or two values is the topic of this and the next few sections. Rather than construct a visual picture of the set of numbers, such as a histogram or bar graph, we compute an average or other typical value to represent the whole data set.

In this and the two sections that follow, we introduce several ways of summarizing a set of data numerically. The properties of most interest are the concepts of location and of variability. That is, we want to know what a typical central number is for a set of data and how spread out or dispersed the individual data points are from this central number (how variable the data is). We will address such issues under the headings of "measures of location" and "measures of variability."

In this section, we provide notation and terminology. In section 1.5, we introduce the concepts of "location" and "centrality," while in section 1.6, we present methods of measuring variability.

Definitions

Parameters vs. Statistics

DEFINITION 1.16: A *population parameter* is a number that identifies some characteristic of a population. It is often a function of the set of measurements $X, X_2, \ldots, X_N$ where N denotes the total number of elements in the population.

DEFINITION 1.17: A *sample statistic* is a number that summarizes some characteristic of the measurements associated with a sample. It is a function of the set of measurements X_1, $X_2, \ldots, X_n$ where n denotes the total number of elements in the sample from a population.

Quantiles

DEFINITION 1.18: The qth quantile for a data set is the numerical value such that $q \bullet 100\%$ of the measurements in the data set will fall at that value or below it. For a set of data we define the qth quantile X_q so that it satisfies the following properties: (1) The proportion of X's $< X_q$ is $\leq q$, and (2) The proportion of X's $\leq X_q$ is $\geq q$. [Notation: X_q denotes the population quantile; x_q denotes the sample quantile.]

DEFINITION 1.19: The *percentiles* of the distribution are the quantile values of 0.01, 0.02, ..., 0.99. They are called the 1st, 2nd, ..., 99th, *percentiles* of the distribution.

DEFINITION 1.20: The *deciles* of the distribution are the quantile values of 0.10, 0.20, 0.30, ..., 0.90. They are called the 1st, 2nd, 3rd,..., and 9th *deciles* of the distribution.

DEFINITION 1.21: The *quartiles* of the distribution are the quantile values of 0.25, 0.50, and 0.75 They are called the 1st, 2nd, and 3rd *quartiles* of the distribution sometimes denoted as Q_1, Q_2 and Q_3.

Empirical Moments

DEFINITION 1.22: The *sample mean* also called the *first empirical moment*, m'_1 or x-bar, is the common arithmetic average of the set of measurements. It is denoted by writing

$$\bar{x} = \frac{\Sigma x}{n}.$$

where

    x denotes a general measurement value, and

    n denotes the total number of measurements in the data set.

DEFINITION 1.23: The *second empirical moment, m'_2, or average sum of squares* , is simply the arithmetic average of the squares of all the measurements. It is written symbolically as

$$m'_2 = \frac{\Sigma x^2}{n}$$

DEFINITION 1.24: The *rth empirical (sample) moment, m'_r* for a data set is defined to be

$$m'_r = \frac{\Sigma x^r}{n}$$

where

x	denotes a general measurement value,
r	denotes the power to which x is raised, and
n	denotes the total number of measurements in the data set.

(The definition of the theoretical moments as population parameters will be presented in Chapter 3.)

DEFINITION 1.25: The *cumulative distribution* is a representation of a data set that shows the number or proportion of values in the data set less than or equal to any value on the real number line. In its graphical form it takes the form of a series of steps raising from left to right.

DISCUSSION

Every ten years, by constitutional mandate, a complete census of the population of the United States is conducted. In such a study, every person in the population is to be accounted for. After the data are collected and summarized, "statistics" about the average family size, the proportion of households without a phone, the total number of one-parent families, along with numerous others, are published. By our definition such values are properly called *population parameters,* not statistics. A population parameter, by the definition given above, is a single number that summarizes all the measurements from the entire population relative to some characteristic. Averages, proportions and totals are common population parameters encountered in a census. On the other hand, suppose only a sample of a population is obtained, such as a sample of students taking the ACT test, a sample of tomatoes from all those harvested in a given field, and so on. The average test score of the sample of students taking the ACT test, the proportion of tomatoes with blossom-end rot, or other summarizing values would be regarded as sample statistics. Any numerical summary of a set of sample data is called a *sample statistic.*

By comparing the definitions of a population parameter and a sample statistic, you can note that many population parameters will typically have a sample counterpart which shall be called a sample statistic.

[Note: It is important to realize that a population parameter exists only as a mathematical abstraction. It should not be taken as the "truth" about the population. For example, the federal government may report the average income of employed adults, but at what point in time is that value defined? Obviously, it is a constantly changing value, due to the changing employment status of the citizens. The fact that persons are constantly entering that population as they become of employment age, while others are leaving the population by death and retirement makes it difficult in to really define the population in static terms. Whatever number is published as the "average income" is only useful as a reference value, but does not exist in any real-world sense. However, if a sample is drawn, the average income of the sample can be computed, and then interpreted for what it is worth. The sample average, in this case, is the "average value of the sample."

Such conceptual difficulties are encountered with practically every measurement process. For instance, if we measure a person's weight, do we do it before they inhale, or after exhaling, with or without their clothes on? For a very sensitive weight measuring device, such distinctions can make a difference in the measurements obtained. In addition, how can we be sure that a person has completely filled his lung capacity, or completely exhausted all air. What about the time of day of the measurements? Before a full meal? After eating 1500 calories. Such problems exist in most every measurement setting.

Recognizing these problems, we simplify the world by first doing all we can to minimize their effect. Once this has been done, we shall assume that we have an accurate set of measurements to work with whatever their shortcomings. We then define the various parameters and statistics relative to that obtainable set of measurements, all the while accepting the fact that parameters are only concepts to be estimated.]

Notation

It is necessary to have a concise notation and set of symbols to simplify the method of referring to the various parameters and statistics that will be commonly used in the following pages. In addition, it is extremely helpful to adapt this notation so that the sample statistic can be distinguished from its counterpart, the population parameter. This is usually done by using Greek symbols to denote the various population parameters to be considered, while the regular Roman alphabet will be used for the symbols for the corresponding sample statistics. There are a few exceptions to this policy since it is difficult to obtain a consensus among all who write books and paper in the profession. Therefore, in the following, the notation for both sample and population will be introduced as each term is defined. In subsequent chapters we will try to be consistent in using the Greek-Roman convention for specifying parameters versus sample statistics.

One distinction to make at the outset is the use of upper case "N" to denote the total number of units in a finite population. A lower case "n" will be used to denote the total number of units in a sample from a population.

Quantiles: Using a Frequency Table

One of the more useful ways to summarize a data set is to report various quantile values. As defined, a quantile value is a number reported as having a certain percentage of observations smaller than it. Percentiles are familiar quantile concepts. If on a test it is reported that your score of 75 was at the 85th percentile—the 0.85 quantile—you would know that 85% of the scores were smaller than yours, with only 15% of the scores being larger.

Quartiles are also common quantile values. The quartiles are found roughly by arranging all the scores in ascending order, and then finding the scores that separate the data into four equal size groups. The first division point is called the first quartile (the 0.25 quantile); the second division point is called the second quartile (the 0.50 quantile); and the third division point is the third quartile (the 0.75 quantile). (Percentiles, by the same argument, are the values that divide the data into 100 subdivisions, the deciles are the values that divide the data into 10 equally sized subdivisions.)

To find the qth quantile, X_q, several different approaches can be taken. One way is to construct a table of the cumulative distribution from the data set For example, order the observations from small to large down one column in a table. Then, in an adjacent column record the cumulative relative frequencies associated with each data point. Scan down the cumulative relative frequency column until you find two entries that bracket the value q. The data point corresponding to the larger of the bracketing values will satisfy the definition given above for the qth quantile, X_q.

For instance, consider the following ordered gpa's of 10 students where our objective is to find the first and third quartiles, $X_{0.25}$ and $X_{0.75}$. The data are organized in Table 1.4.

2.69	2.84	3.01	3.33	3.56	3.67	3.72	3.84	3.90	3.95

Table 1.4 Cumulative distribution table of GPA's

Ordered GPA	Rel. Cum. Frequency	Quantile, q
2.69	.10	
2.84	.20	
		0.25
3.01	.30	
3.33	.40	
3.56	.50	
3.67	.60	
3.72	.70	
		0.75
3.84	.80	
3.90	.90	
3.95	1.00	

The first and third quartiles are, respectively, $X_{0.25} = 3.01$ and $X_{0.75} = 3.84$ which we found after bracketing 0.25 between 0.20 and 0.30, and bracketing 0.75 between 0.70 and 0.80. (Obviously, you cannot divide 10 observations into 4 equally sized groups.)   However, the value 3.01 satisfies the requirement in the definition that 25% of the data values are *less than* it—in this case, actually 20%. Also, the value 3.01 satisfies the condition that *at least* 25% of the data values are *less than or equal* to it—in this case actually 30%.

If it should happen that q exactly matches a value in the cumulative relative frequency table, then the corresponding score *and* the next one larger than it (as well as any value between them) will satisfy the definition for X_q. In this case, convention says to average the two scores involved and use the average as X_q.

For instance in Table 1.4, if we want the first decile value, $X_{0.10}$, we find that there is an entry in the cumulative relative frequency column exactly equal to 0.10. The corresponding score is 2.69. The next larger score is 2.84. Both of these values satisfy the definition (check it out). Therefore, we let $X_{0.10} = \dfrac{(2.69 + 2.84)}{2} = 2.765$.

Quantiles: Using a Cumulative Relative Frequency Graph

If the cumulative distribution is graphed, then the following procedure can be used to obtain any of the quantiles. Enter the graph on the vertical axis (which represents the cumulative relative frequency) for the value of q. Then extend a horizontal line until you intersect the step function. If a "plateau" is encountered then average the X values associated with each end point of the horizontal line. If the intersection occurs at a vertical line, the associated X value would be X_q. See Figure 1.11.

From Figure 1.11, you can see that the same values would be obtained for the 0.10, 0.25 and 0.75 quantiles as those we obtained using Table 1.4. That is, $X_{0.10}$ is associated with the first plateau. Therefore, we would average the first two scores. The quantile, $X_{0.25}$, is associated with the third score and $X_{0.75}$ with the eighth score giving us the same results: $X_{0.10} = 2.765$, $X_{0.25} = 3.01$, and $X_{0.75} = 3.84$. The graphical approach has the drawback, however, of accuracy limited to the care and scale with which the graph is drawn.

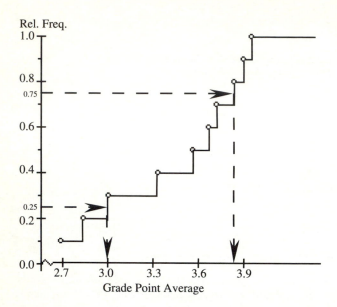

Figure 1.11 Finding the 1st and 3rd quantiles graphically.

Quantiles: Using Order Statistics

Another method, equivalent to but possibly easier to use than the previous two methods is described as follows. To find the qth quantile X_q, order the data set from smallest to largest letting $X_{(1)}, X_{(2)},...,X_{(N)}$ denote this ordered set. (These values are called "order statistics.") Compute the product qN, where N denotes the number of observations in the data set. Let $[qN]$ denote the integer part of the product. If the product, qN, is non-integer valued, then the qth quantile value will be the $[qN]+1$ observation in the ordered set, i.e., $X_q = X_{([qN]+1)}$. If qN is integer valued, then $X_q = \frac{\{X_{([qN])}+X_{([qN]+1)}\}}{2}$, the average of the two values involved. (If the data represents a sample, simply replace N with n in the above description.)

Illustrating this approach with the same set of GPA data, the ordered observations are:

$$X_{(1)} = 2.69, \quad X_{(2)} = 2.84, \quad X_{(3)} = 3.01, \quad X_{(4)} = 3.33, \quad X_{(5)} = 3.56$$

$$X_{(6)} = 3.67, \quad X_{(7)} = 3.72, \quad X_{(8)} = 3.84, \quad X_{(9)} = 3.90, \quad X_{(10)} = 3.95$$

When $q = 0.25$, then $q \bullet n = (0.25) \bullet 10 = 2.5$, therefore, $X_{0.25} = X_{(2+1)} = X_{(3)} = 3.01$.

When $q = 0.75$, then $q \bullet n = (0.75) \bullet 10 = 7.5$, therefore, $X_{0.75} = X_{(7+1)} = X_{(8)} = 3.84$.

When $q = 0.10$, then $q \bullet n = (0.10) \bullet 10 = 1.0$, therefore, $X_{0.10} = \frac{(X_{(1)}+X_{(2)})}{2} = 2.765$.

Empirical Moments

Other numerical values used more directly in subsequent sections of this chapter are called the empirical moments. These are defined at the beginning of this section and as indicated there the moments most likely to be used are the 1st and 2nd moments. The first moment is simply the arithmetic average of the data points; i.e., $m_1' = \bar{x} = \dfrac{\Sigma x}{n}$. This is a concept with which you are already familiar.

The second moment is the arithmetic average of the square of each data point; i.e., $m_2' = \dfrac{\Sigma x^2}{n}$. This number is used in computing what is called a variance. See section 1.6 that follows for the specific computations involved.

To illustrate these computations for the GPA data given above, we get:

$$m_1' = \frac{(2.69 + 2.84 + ... + 3.95)}{10} = 3.451$$

and

$$m_2' = \frac{(2.69^2 + 2.84^2 + ... + 3.95^2)}{10} = 12.09897$$

The first moment, 3.451, as indicated, is the "average" gpa for the ten gpa's reported. The second moment is not easily interpretable other than being the "average" of the squares.

EXAMPLES

EXAMPLE 1: In an engineering class 12 different students were asked to measure, independently, the diameter of the same ball bearing with the same pair of calipers. The difference in measurements would be attributed to what is called measurement error. The data below represent the measurements obtained. What can be said about the distribution of measurement error. What are the 1st and 3rd quartiles of the data? The 1st and 9th deciles?

0.265	0.265	0.266	0.267	0.267	0.267
0.265	0.267	0.265	0.268	0.268	0.265

SOLUTION: Table 1.5 shows the ordered observations, $X_{(1)}, X_{(2)}, ..., X_{(12)}$, and the cumulative relative frequency. To find the 1st quartile we compute $q \bullet n = (0.25)12 = 3$. Therefore,

$$Q_1 = X_{0.25} = \frac{(X_{(3)} + X_{(4)})}{2} = 0.265.$$

Similarly, for the 3rd quartile $q \bullet n = (0.75) \bullet 12 = 9$ so that

$$Q_3 = X_{0.75} = \frac{(X_{(9)} + X_{(10)})}{2} = 0.267.$$

These values are indicated in the table below by double asterisks (**). The 1st and 9th decile are indicated by the triple asterisk (***).

You can verify, using the relative cumulative frequency column, that these values satisfy the basic definition given for the given quantiles. Draw your own cumulative relative frequency graph and verify these values graphically.

The 1st and 2nd empirical moments are:

$$m_1' = \frac{(0.265 + \ldots + 0.268)}{12} = 0.26625, \qquad m_2' = \frac{(0.265^2 + \ldots + 0.268^2)}{12} = 0.07089$$

Table 1.5 Cumulative frequency table of caliper measurements.

$X_{(i)}$		Rel. Cum Freq.
0.265		0.08333
0.265 ***	$X_{.10} = 0.265$	0.16666
0.265 **		0.25000
	$X_{0.25} = 0.265$	
0.265 **		0.33333
0.265		0.41666
0.266		0.50000
0.267		0.58333
0.267		0.66666
0.267 **		0.75000
	$X_{0.75} = 0.267$	
0.267 **		0.83333
0.268 ***	$X_{.90} = 0.268$	0.91666
0.268		1.00000

EXAMPLE 2:   The viscosity of a particular brand of motor oil is measured at 6 different times of the day producing the following values: 25   39   18   50   42.

What is the sample mean for this data and the 50th percentile (2nd quartile).

SOLUTION: First, let's sort the data and then find the sum:

Sorted Data: 18   25   39   42   50
Sum: 174
The sample mean is

$$\frac{174}{5} = 34.80$$

To find the 2nd quartile we compute $qn = 0.5 \cdot 5 = 2.5$. Therefore, $Q_2 = X_{0.50} = X_{(2+1)}$—the third observation in the ordered set. Therefore, $Q_2 = X_{0.50} = X_{(2+1)} = 39$.

EXAMPLE 3: For the following ordered data find the 1st decile, and the 1st and 2nd quartiles.

    5   5   5   5   5   5   6   6   7   8

SOLUTION: First notice that this data is definitely skewed right and that 60% of the observations are equal to the value 5. Therefore, the first decile, which would be halfway between the first and second values in the ordered set is 5; the 1st quartile would be the third observation ($qn = 0.25 \cdot 10 = $

2.5); and the 2nd quartile would be halfway between the 5th and 6th observation is also a 5. In small data sets and with data sets that are skewed, some unusual outcomes may result.

EXERCISES

For each of the data sets given below, determine the quantiles values and the empirical moments requested. Use any procedure described in this section that is easiest for you to use.

1. The following data are the number of chickens produced in 1983 in millions, for 20 different eastern and Midwest states. Compute the 1st, 2nd and 3rd quartiles and the 1st empirical moment. Source: *Statistical Abstracts of the United States: 1985.*

13.9	23.7	3.6	1.0	6.3	15.4	2.8	17.2	8.3	1.4
2.0	4.1	4.1	4.4	8.8	6.6	5.4	6.6	11.3	10.2

2. Using the data of exercise 1, compare the 2nd quartile (called the median) with the 1st empirical moment (called the mean.) These numbers are both "central numbers" in the sense that they are numbers usually used to represent a center point of the data set—an "average" value. If a data set has a perfectly symmetric histogram, these two numbers will be equal. What conclusion would you make about the symmetry of the data? Would you say the data is symmetric, slightly skewed, or extremely skewed.

3. Again using the data and results of exercise 1, what proportion of the data would be found between the 1st and 3rd quartiles? How spread out or how far apart is the inner 50% of the data? (Subtract Q_1 from Q_3.)

4. The number of wells producing natural gas from the years 1970, 1973, and from 1975 to 1982 are given below: Figures are in 1000's. Find the 2nd quartile (also called the median) and the 1st empirical moment. Source: *Statistical Abstracts of the United States: 1985.*

117	124	132	138	148	157	170	182	199	211

5. Using your answers from exercise 4, what is an "arithmetic average" amount of producing wells over the years considered?

6. Per capita income in dollars for 13 Mountain and Pacific states for 1983 are given below. Per capita income for the entire United States the same year is 11,675 dollars. Compute the 2nd quartile (also called the median) and also the 1st and 3rd quartile values. Compute the 1st empirical moment. Compare to the U.S. figure. Source: *Statistical Abstracts of the United States: 1985.*

9999	9342	11969	12580	9560
10719	9031	12516	12051	10920
13239	16820	12101		

7. The median or 2nd quartile value is often used when reporting "average" income rather than the mean or 1st empirical moment. Income data tends to be severely skewed to the right. Can you provide a reason for this preference for skewed data?

8. Using the data of exercise 6, would you think the skewness is severe enough to warrant a choice

of the mean over the median?

9. Using the data of exercise 6 and the values for Q_1, Q_2, and Q_3, look at the difference between each of these values. Also compute the difference between Q_1 and the minimum income figure and the difference between the maximum income figure and Q_3. (You should have computed 4 differences: Q_1-*minimum*, Q_2-Q_1, Q_3-Q_2, and *maximum*-Q_3.) If the distribution of numbers were perfectly symmetric, what would these differences be? Based on what you get for these differences, would you say the data is symmetric? On the real line you might mark the location of these five points: (1) the minimum, (2) the 1st quartile, (3) the 2nd quartile, (4) the 3rd quartile, and (5) the maximum. This way you will see the lack of symmetry these points would suggest.

10. Suppose a set of data had the values:

$$\text{minimum} = 30 \quad Q_1 = 40 \qquad Q_2 = 45 \qquad Q_3 = 50 \qquad \text{maximum} = 60$$

Make a sketch of the general shape you might expect the corresponding histogram to have.

1.5   NUMERICAL METHODS—MEASURES OF LOCATION

Introduction

When students receive test scores back after an examination there are generally two numbers they are vitally interested in–their own personal score and the arithmetic average of all scores. The average denotes a measure of location with which to compare yourself to determine how you are doing compared to the "average" student (if such a student exists.)

A measure of location is a numerical value which identifies a "central number" of a data set. They are sometimes referred to as measures of "central tendency" as well. The common use of the word "average" is another expression for a measure of location. The measures of location, or "averages," that we define in this section are the *mean*, *median*, and *mode*. All these measures of location have graphical interpretations which will be identified as they are defined.

Definitions

DEFINITION 1.26: The *mean* is the ordinary arithmetic average of the values of X in a data set, also called the first empirical moment. It represents the physical center of gravity of a histogram or bar graph of the data. It is commonly denoted by

$$\mu_x$$

when referring to a population, and by

$$\bar{x}$$

(said x-bar) when referring to a sample mean. Its symbolic definition is:

$$\text{Population} \qquad\qquad \text{Sample}$$

$$\mu_x = \sum_{i=1}^{N} \frac{X_i}{N} \qquad\qquad \bar{x} = \sum_{i=1}^{n} \frac{x_i}{n}$$

where

X_i denotes the ith measurement

N is the number of data points in the population, and

n is the number of data points in the sample.

[To refer to an entire set of data we use the symbol X where X_i then denotes the ith measurement in the set. We could just as well denote the set of measurements by Y with its mean denoted by μ_y or $\bar{y}$, or by Z, μ_z, and $\bar{z}$, etc.]

DEFINITION 1.27: The *median* is the same as the 0.50 quantile, the 50th percentile, 5th decile, or 2nd quartile of a data set. It is the value such that roughly 50% of the measurements in the data set are larger than it, and 50% of the values in the data set are smaller than it. Graphically, it is that point on the horizontal axis that divides the area of the histogram into two equal "halves," so that half the "area" in a histogram is to its left and half is to its right. We denote it by:

$$\text{Population} \qquad\qquad \text{Sample}$$
$$M_d \qquad\qquad\qquad m_d$$

DEFINITION 1.28: The *mode* is the value in the data set that occurs most often. Graphically it will represent the value on the histogram or bar graph which has the highest peak. It may or may not occur in the "center" of the distribution and in which case should not be called a true measure of "centrality" or "location." We denote it by:

$$\text{Population} \qquad\qquad \text{Sample}$$
$$M_o \qquad\qquad\qquad m_o$$

DISCUSSION

The concept of an average value may mean different things in different situations. A shoe store stocking various shoe sizes may be interested in the average shoe size in the sense of the size which occurs most often. In this case, looking at the definitions, our "average" is a *mode*. An economist studying household incomes may be interested in the "middle" income value, the equivalent of a *median*. In this case, the economist wants to de-emphasize the impact of the few very high incomes that are typically present in such a data set. Then, again, an instructor is invariably asked what the average score was on the last test. In this case, most questioners are referring to the arithmetic average, or what we define as the *mean*. Each of these terms can properly be called an average and each has its place in terms of its intended use.

One criterion for determining proper use is to look at the level of measurement of the data involved. For nominal data, where numerical size has no meaning, the only measure that makes sense is the mode, the measurement value that occurs most often. For ordinal data, data which lacks interval/ratio properties, the median or mode would serve as an appropriate measure. For interval or ratio data, all three could be alternatives for consideration.

As you look at the definitions of these three measures of location, you should note that in the graphical sense, the mean provides a "center of gravity," the median the "center" from an area point-of-view, and the mode the point associated with maximum height.

For symmetric distributions (histograms), this implies that the mean and the median will be the same. The mode may or may not exist, depending upon the shape of the distribution. These properties can be seen more easily by looking at Figure 1.12.

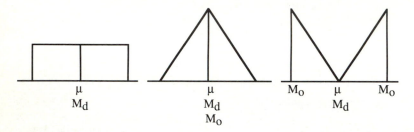

Figure 1.12 Relationship of the mean, median and mode for symmetric distributions.

If the data is highly skewed one way or the other, the median might be more preferred over the mean as a measure of location since the mean is very sensitive to the presence of extremely large or small observations in the data set. (See exercise 4 in this section.)

The position of the mean, median and mode in a positively skewed distribution with a single mode (unimodal) is identified in the histogram of Figure 1.13 below. The mean is the "center of gravity" of the diagram, the median is the middle value that divides the area of the figure into equal "halves," the mode is the value associated with the peak of the figure. The choice of the appropriate measure in a given investigation is not always easy to make. However, if the distribution is definitely skewed, the mean does not tend to represent a "typical" value and thus the median or mode may be a more appropriate selection. When the distribution is symmetric and unimodal, the mean, median and mode will all have the same value. For any symmetric distribution, the mean and median values will coincide.

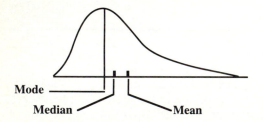

Figure 1.13 Relative position of the mean, median, and mode in a skewed distribution.

While the mean and median will always exist in the cases we examine in this text, the mode may not always exist or represent a meaningful value. In some cases, one value may not be an obvious "most frequent value" but there may be a range of values within which more values fall than any other range of values which we may refer to as a "modal class." In other cases we may have two or three "most frequent values" and thus define a bimodal or trimodal distribution. In other cases, still, there may not be *any* dominating value and thus we may not even cite a value for the mode.

Along with these guidelines, the context and use to be made of the measure of location will also have a bearing on which measure to use.

EXAMPLES

EXAMPLE 1: In an engineering class 12 different students were asked to measure, independently, the diameter of the same ball bearing with the same pair of calipers. The difference in measurements would be attributed to what is called measurement error. The data below represents the measurements obtained. What is the mean, median and mode of this data set?

0.265	0.265	0.266	0.267
0.267	0.267	0.265	0.267
0.265	0.268	0.268	0.265

SOLUTION: The mean is simply calculated (most pocket calculators have special function keys to obtain the mean after entering the individual data points into the calculator). The median and mode can also be obtained by first ordering the data set and applying definitions. The results are as follows:

0.265   0.265   0.265   0.265   0.265   **0.266**   **0.267**   0.267   0.267   0.267   0.268   0.268

The numbers in bold-face are the two middle numbers, the average of which will be the median. It can also be seen that the measurement that occurs most often is 0.265.

Mean	0.26625
Median	0.2665
Mode	0.265

If these values are super-imposed upon a histogram it is easier to evaluate their implications. It can be seen upon such an examination that they tend to validate a conclusion of positive skewness. See Figure 1.14 to confirm this.

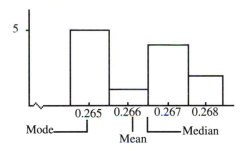

Figure 1.14 Histogram showing the mean, median and mode.

EXAMPLE 2: The systolic blood pressure, in millimeters of mercury, for a group of patients is given below. What are the mean, median and mode for this set of data?

122   110   140   130   140   110   120

SOLUTION: First, order the data to get:

110   110   120   122   130   140   140

The median, as the middle number can easily be seen to be 122. The mean, as the arithmetic average is

$$\frac{872}{7} = 124.57$$

It can be noted that there are two 110's and two 140's. Therefore, on the basis of this very small data set we might be tempted to say the distribution is bimodal (has two modes)—one at 110 and one at 140. However, it would be advisable to get more data and really see whether this pattern persists, or whether this is just a chance characteristic of this data.

EXAMPLE 3: Suppose the first data point in Example 2 was miscoded and reported as 221, rather than 122. What is the effect of this upon the mean and median?

SOLUTION: The data after ordering in this case is

    110   110   120   130   140   140   221

Again it is easy to see that the middle observation is 130, slightly higher than for Example 2. The sum of the 7 numbers is 971 so the mean is

$$\frac{971}{7.00} = 138.71$$

The mean has been affected much more than the median in this example. (Note that if the middle observation had not originally been the 122, the median would possibly have not changed at all.)

EXERCISES

1. An office manager rates the performance of each typist in the typing pool regularly. One aspect of the rating is typing accuracy. To assess this a full page of text is randomly selected from some of their recent work and the number of typing errors per page is recorded. The following data resulted from a recent rating. Compute the mean, median and mode for this set of data.

3	4	3	0	2	4
3	0	1	2	1	2
5	1	3	2	3	4

2. Construct a bar chart for this data set. Locate the mean, median, and mode on this bar chart. Note whether the bar chart is symmetric or skewed right or left.

3. Students in a computer science programming class were assigned a problem in numerical integration. The cpu time was recorded for each student's program to complete a common problem. The data recorded was as follows:

2.33	3.41	1.98	3.03	2.44	2.86	2.23
2.71	2.83	3.10	2.53	2.67	1.89	2.22

4. Consider the following two sets of data, compute both the mean and median for both sets. What do you conclude about the effect of extreme observations on these two measures?

| Set 1: | 1 | 1 | 2 | 2 | 2 | 2 | 2 | 3 | 10 |
| Set 2: | 1 | 1 | 2 | 2 | 2 | 2 | 2 | 3 | 100 |

5. Consider the following set of numbers, compute the mean for this set. Then add 5 to each number in the set and recompute the mean. Next multiply each number by 10 and recompute the mean. Comparing all three "means" what do you conclude about the effect of an additive and multiplication transformation on the mean.

    2   2   4   4   5   5   5   6   6   8   8

6. Using the data from exercise 5, we can make a "bar chart" of the data as follows:

$$
\begin{array}{ccccc}
 & & 5 & & \\
2 & 4 & 5 & 6 & 8 \\
2 & 4 & 5 & 6 & 8
\end{array}
$$

    [If we put this bar chart on its side we get a horizontal bar chart called a "stem-and-leaf" diagram.

                        2 2

                        4 4
                        5 5 5
                        6 6

                        8 8

    Notice that the distribution is symmetric around the value 5. What is the mean, median and mode for this distribution?

7. Using the data from exercises 5 and 6, subtract the mean from each observation. This gives the new set of data:

    − 3   − 3   − 1   − 1   0   0   0   1   1   3   3

    The bar chart for this data looks as follows:

$$
\begin{array}{cccccc}
 & & 0 & & \\
-3 & -1 & 0 & 1 & 3 \\
-3 & -1 & 0 & 1 & 3
\end{array}
$$

    What is the mean, median and mode for this transformed data?

8. Consider a data set that is highly skewed such as the following:

$$
\begin{array}{llll}
5 & & & \\
5 & & & \\
5 & 6 & & \\
5 & 6 & 7 & \\
5 & 6 & 7 & 8 \qquad\qquad\qquad 20
\end{array}
$$

    Find the mean, median and mode for this set of data.

9.   Using the data of exercise 8, create a measure of skewness by using the mean and median.

10. Suppose the data of exercise 8 were:

50				
50				
50	60			
50	60	70		
50	60	70	80	200

Find the mean and median and compute your measure of skewness as created in exercise 9. Does it seem to work, or are there problems that you see?

1.6   NUMERICAL METHODS—MEASURES OF VARIABILITY

Introduction

Another important property of a distribution of measurements is the dispersion, or spread, of the values. Knowing how much variation there is in a set of measurements can be very important. For instance, it is very important for the manufacturer of a medication to not only get the correct dosage in each capsule on the average, but to also ensure that there is little variation around that average. Otherwise, it might be found that sometimes when the patient takes the capsule, there is no observable benefit; while on other occasions there might be the risk of unfortunate side affects due to an unintended overdose. Such a situation could lead either to costly lawsuits from users of the medication, or loss of revenue as customers find the promised results of the medication are not realized. Thus, knowledge of and control over the variation in dosage is very important to the capsule manufacturer.

In other situations, the presence or absence of variation may lead to other conclusions. When an instructor administers an examination to class, the instructor will not only be interested in the average scores but also in the variation in scores. If there is too much variation, it may signal a poor job of preparing the students for the exam. Or it might simply mean there is a large amount of variation in abilities and intellect associated with the students in the class. On the other hand, if there is little variation in the scores, it may mean the exam was too easy and thus can't distinguish between the good and poor students. (If the average is very low along with little variation in scores, it probably means the exam was too difficult, even for the best students.)   In any event, the variation in scores is often as informative as the mean or median score.

In this section, we present several ways of measuring the variation in a data set. Such numbers are called *measures of variability.*

There are a variety of measures that are used. Some use quantile values, such as the *range, interdecile range,* and *interquartile range.* Others use the concept of a *deviation* from a central reference point such as a mean. The measures of dispersion we use are called the *mean deviation,* and the *standard deviation.*

Definitions

DEFINITION 1.29: The *range* (R) of a set of data is the difference between the largest and smallest values in the data set. That is,

$$R = \text{Max}(X) - \text{Min}(X)$$

DEFINITION 1.30: The *interdecile range* (IDR) is the difference between the 9th and 1st decile, i.e., IDR = $D_9 - D_1$ where D_i denotes the *ith* decile.

DEFINITION 1.31: The *interquartile range* (IQR) is the difference between the 3rd and 1st quartile, i.e., *IQR* = $Q_3 - Q_1$ where Q_i denotes the *ith* quartile.

DEFINITION 1.32: The *mean deviation* is the average of the unsigned distances of the measurements from the mean. It is denoted by μ_{Dev} for the population and $\bar{x}_{Dev}$ for the sample. It is computed by taking the absolute value of the deviation of each measurement in the data set from the mean of the data set—$|X_i - \text{mean}|$—and then averaging.   In symbols it is:

<table>
<tr><td align="center">*Population*</td><td align="center">*Sample*</td></tr>
<tr><td align="center">$$\mu_{Dev} = \sum_{i=1}^{N} \frac{|X_i - \mu_x|}{N}$$</td><td align="center">$$\bar{x}_{Dev} = \sum_{i=1}^{n} \frac{|x_i - \bar{x}|}{n}$$</td></tr>
</table>

where

  X_i denotes the *ith* measurement
  N is the number of data points in the population, and
  n is the number of data points in the sample.

DEFINITION 1.33: The *variance* is the arithmetic average of the squares of the deviation of each measurement in the data set from the mean of the data set. The *standard deviation* is the square root of the *variance*. The population standard deviation is denoted by σ and the sample standard deviation is denoted by s. We first compute the deviation of each measurement from the mean and square it, i.e.,

$$(X_i - \text{mean})^2$$

Then compute the mean of these squared deviations. In symbols:

<table>
<tr><td align="center">*Population*</td><td align="center">*Sample*</td></tr>
<tr><td align="center">$$\sigma_x^2 = \sum_{i=1}^{N} \frac{(X_i - \mu_x)^2}{N},$$</td><td align="center">$$s_0^2 = \sum_{i=1}^{n} \frac{(x_i - \bar{x})^2}{n}$$</td></tr>
</table>

The *standard deviation* is:

<table>
<tr><td align="center">*Population*</td><td align="center">*Sample*</td></tr>
<tr><td align="center">$$\sigma_x = \sqrt{\sigma_x^2}$$</td><td align="center">$$s_0 = \sqrt{s_0^2}$$</td></tr>
</table>

Note: We will have reason later on in the course to define the sample variance and standard deviation using a divisor of $n - 1$ rather than n. Either definition serves as a useful measure of variation at this point in the course, and the preference for one over the other is mainly theoretical since in even moderately large samples there will be little practical difference in their numerical values. The reasons for using the a divisor of n–1 will be explained in Chapter 6. To distinguish one computational procedure from the other we will use the following notation:

$$s_0^2 = \sum_{i=1}^{n} \frac{(x_i - \bar{x})^2}{n}$$

$$s^2 = \sum_{i=1}^{n} \frac{(x_i - \bar{x})^2}{n-1}$$

DISCUSSION

Measures Based on the Range

The simplest measure of variation is the range of the measurements. The high and low temperatures that are reported on the evening weather forecast are examples of this. We simply take it one step further and take the difference. For example, if the high is 75° Fahrenheit and the low is 40° Fahrenheit, then we say the range in the daily temperatures is 75 – 40 = 35 degrees Fahrenheit. This gives the idea of spread with a very simple computation. In this case, we use the maximum and minimum values in the data set to report the variation in results.

A simple modification of this same idea is to compute the difference between the ninth and first deciles. This difference is called the interdecile range (*IDR*) and would reflect how spread out the inner 80% of the values are.

Next, is the interquartile range (*IQR*) which is the difference between the third quartile and the first quartile. This difference shows how spread out the inner 50% of the values in the data set are.

The *IDR* and the *IQR* are computed very simply once the data is set up to obtain quantile values. But these these measures do have drawbacks. For example consider Figure 1.15 below. Each figure, representing a distribution of measurements, has a mean and median equal to 50. Similarly, the range in each figure is the same, 60 – 40 = 20. But would we conclude that the data used to produce them have the same amount of variability?

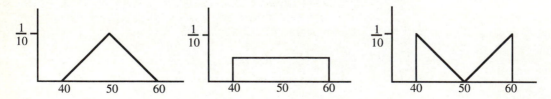

Figure 1.15   Distributions with equal means but unequal variabilities.

Upon examination, the left figure has more observations close to the mean than does the middle figure, which has the values uniformly spread out across the range. The right figure has most observations nearly 10 units away from the mean. This aspect of variation is not reflected in the value of the range. If we reported the interquartile range instead of the range, then we would see a difference of variation among the three figures. However, this would not tell us how spread out the values would be outside of the inner 50% of values. Thus, we desire another measure of variability to take both *deviation* and *frequency* into account.

Measures Based on Deviations

The next measures of variation are all based on the concept of a deviation from a central point in the data set. The case we emphasize uses the mean as the central point though the median could be used just as well. The idea is to compute the deviation of each data point from the mean and to then get an "average deviation." This procedure has some drawbacks that will be explained as we go. We will demonstrate the process with an example.

Suppose the following data represents radar-reported speeds of 10 cars checked on a section of an interstate highway with the purpose of observing compliance with posted speed limits. (The mean of this data is computed to be 59 mph.) Table 1.6 summarizes the computations we need.

54   57   57   58   59   59   59   61   62   64

Table 1.6 Computations for the mean deviation and standard deviation with the mean equal to 59.

	Measurement X	Deviation $X - \text{Mean}$	Absolute Deviation $|X - \text{Mean}|$	Squared Deviation $(X - \text{Mean})^2$
	54	$54 - 59 = -5$	5	25
	57	$57 - 59 = -2$	2	4
	57	-2	2	4
	58	-1	1	1
	59	0	0	0
	59	0	0	0
	59	0	0	0
	61	2	2	4
	62	3	3	9
	64	5	5	25
Sum	590	0	20	72
Mean	59	0	2	7.2

Upon examining this table, first note that the "$X - \text{Mean}$" column sums to zero. This will always be the case since the mean is the "center of gravity" of the distribution of scores. Thus, to average this column would not produce a measure of variability.

Two ways to get around this difficulty would be to take absolute values of the deviations or to square the deviations. These two operations are shown in the third and fourth columns. The third column has the absolute values of the "deviations" in it and the mean of this column produces what is called the "average deviation" or "mean deviation." In this example, the average deviation is $\frac{20}{10} =$ 2. Thus, the average deviation from the mean speed of 59 mph is 2 mph.

The fourth column contains the squares of the deviations, which upon averaging, is called the variance. The square root of the variance, called the "standard deviation," is not strictly an average deviation, but does serve as a "standard" against which comparisons can be made. For the data given, the variance and standard deviation are:

$$s_0^2 = \frac{72}{10} = 7.2 \qquad\qquad s_0 = \sqrt{s_0^2} = \sqrt{7.2} = 2.68 \text{ mph.}$$

You will note that the average deviation and the standard deviation are not equal. But they are of the same relative magnitude. Both provide a useful measure of variability against which other data could be compared. For instance, if 10 cars on a different stretch of highway were also observed, we could compute the standard deviation for those ten and compare with the standard deviation of 2.68 mph to see if there is more variability in speeds in one stretch of highway than another.

Though the average deviation is more easily understood and interpretable than the standard deviation, there are analytical and computational advantages to the standard deviation that have dictated its use in preference to the average deviation. Though this advantage is not so obvious in our computer age, we still use the standard deviation as our most important measure of variability in this and subsequent chapters.

Calculator Formula For A Standard Deviation

After you have practiced computing the standard deviation as illustrated above, and you understand what it means in terms of deviations, there is a more efficient way of computing it with what is called the calculator formula. The calculator formula is obtained by expanding the "squared deviation" algebraically and simplifying the resulting expression. The following formula shows this equivalence.

$$s_0^2 = \sum_{i=1}^{n} \frac{(x_i - \bar{x})^2}{n} = \frac{\sum_{i=1}^{n} x_i^2 - \frac{1}{n}\left(\sum_{i=1}^{n} x_i\right)^2}{n}$$

[When it is obvious that the sum extends across all the observations we will often shorten the notation so that $\sum_{i=1}^{n} x_i$ is denoted by $\sum x$.] This formula is demonstrated on the same highway speed data used above with the aid of Table 1.7.

Table 1.7 Computations for calculator formula: Highway mph data.

	X	X^2
	54	2916
	57	3249
	57	3249
	58	3364
	59	3481
	59	3481
	59	3481
	61	3721
	62	3844
	64	4096
Sum	590	34882

Therefore,

$$s_0^2 = \frac{\sum\limits_{i=1}^{n} x_i^2 - \frac{1}{n}\left(\sum\limits_{i=1}^{n} x_i\right)^2}{n} = \frac{34882 - \frac{1}{10}(590)^2}{10} = 7.2$$

This is the same variance that we obtained before.

[Note: Most "scientific" calculators have the capability of computing a mean and a standard deviation by a single keystroke once the data have been entered. They will often provide the option of using either a denominator of n or $n - 1$. Check your manual to assess the capability of your particular model.]

Eyeballing the Mean and Standard Deviation

It is instructive and often useful to be able to "guess" the approximate value for the mean and standard deviation of a data set before computing them on a calculator. This provides a useful check on your computations. (It is really easy to enter the wrong number on your calculator, especially when you have a large data set to work with.) This will also help you better understand what a standard deviation measures.

A simple approach is to identify the largest and smallest values in the data set and compute their average. This gives the mid-point between the minimum and maximum. If the data set is symmetric this will be right on the mean. If the data set is skewed one way or the other, your guess will be either too high or too low.

Secondly, with most distributions that are encountered in practice, you will find that the largest and smallest observations are within 2 to 3 standard deviations of the mean. Therefore, if you divide the range by 4 or 6, you will have a rough idea of the value of the standard deviation. This is illustrated in Figure 1.16.

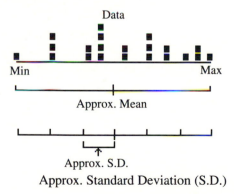

Figure 1.16 Approximating the mean and standard deviation using the smallest and largest data points.

For the example of speeds, the largest and smallest observations were 64 and 54. The average of these two numbers is 59–our guess for the mean. With the range being $64 - 54 = 10$ units, we would expect the standard deviation to be between $\frac{10}{6} = 1.67$ and $\frac{10}{4} = 2.5$ units in size. The actual standard deviation is 2.68 as seen from above, a value of approximately the same size as our guess. The accuracy of this approximation will depend on the shape and skewness present in the data and also the presence of extreme observations in the data. Nonetheless, it does give a useful guideline within which to work.

EXAMPLES

EXAMPLE 1: In an engineering class, 12 different students were asked to measure, independently, the diameter of the same ball bearing with the same pair of calipers. The difference in measurements would be attributed to what is called measurement error. The data below represents the measurements obtained. Compute the various measures of variability and compare them with one another.

0.265	0.265	0.266	0.267	0.267	0.267
0.265	0.267	0.265	0.268	0.268	0.265

SOLUTION: First, make a cumulative relative frequency table so the quantile values can be obtained:

X	Cum.Rel.Freq.
0.265	0.08333
0.265	0.16666
0.265	0.25000
0.265	0.33333
0.265	0.41666
0.266	0.50000
0.267	0.58333
0.267	0.66666
0.267	0.75000
0.267	0.83333
0.268	0.91666
0.268	1.00000

From this table, we obtain $D_1 = 0.265$, $D_9 = 0.268$ (in italics), $Q_1 = 0.265$ and $Q_3 = 0.267$ (bold-face). In addition, a table of deviations would be needed. To find the mean deviation and standard deviation, the mean is used in computing the deviations. These deviations are presented in the next table.

X	|X – Mean|	$(X - \text{Mean})^2$
0.265	0.00125	1.5625×10^{-6}
0.265	0.00125	1.5625×10^{-6}
0.265	0.00125	1.5625×10^{-6}
0.265	0.00125	1.5625×10^{-6}
0.265	0.00125	1.5625×10^{-6}
0.266	0.00025	0.0625×10^{-6}
0.267	0.00075	0.5625×10^{-6}
0.267	0.00075	0.5625×10^{-6}
0.267	0.00075	0.5625×10^{-6}
0.267	0.00075	0.5625×10^{-6}
0.268	0.00175	3.0625×10^{-6}
0.268	0.00175	3.0625×10^{-6}
Sum 3.195	0.01300	16.2500×10^{-6}

From these results, the following measures of variation can be reported.

Range =	0.268 – (0.265) =	0.003
IDR =	0.268 – (0.265) =	0.003
IQR =	0.267 – (0.265) =	0.002
μ_{Dev} =	$\dfrac{0.01300}{12}$ =	0.00108
Std. Dev. =	$\sqrt{\dfrac{16.25 \times 10^{-6}}{12}}$ =	0.00116

The fact that the range and interdecile range are the same indicate that the extreme observations are fairly dense. Also note that the mean deviation and standard deviations are essentially the same in magnitude with only minor differences. This is typical of most cases and suggests that they are measuring essentially the same thing.

EXAMPLE 2:   The systolic blood pressure, in millimeters of mercury, for a group of patients is given below. Estimate the size of the standard deviation by the "eyeball" method described in this section and then compute the standard deviation for this set of data by using both the definition and the calculator formula? (Notice the data has been ordered for us.)

110   110   120   122   130   140   140

SOLUTION: For the "eyeball" estimate we find the range is 140 – 110 = 30 which we then divide by 4. This suggests that the standard deviation ought to be number around 7.5 millimeters of mercury in size. However, because the distribution is high on the ends which produces more variability, this will probably give us an estimate of the standard deviation that is too low.

To compute the standard deviation we form the following table:

X	X – Mean	$(X - \text{Mean})^2$	X^2
110	110-124.57 = 14.56	211.9936	12100
110	110-124.57 = -14.56	211.9936	12100
120	120-124.57 = -4.56	20.7936	14400
122	122-124.57 = -2.56	6.5536	14884
130	130-124.57 = 5.43	29.4849	16900
140	140-124.57 = 15.43	238.0849	19600
140	140-124.57 = 15.43	238.0849	19600
Total 872	0.0500*	956.9891	109584

*This value is not zero because of rounding when computing the mean.

We shall compute the standard deviation by using the divisor $n - 1$.

Definition: St. Dev. = $s = \sqrt{\displaystyle\sum_{i=1}^{n} \frac{(x_i - \bar{x})^2}{n-1}} = \sqrt{\dfrac{956.9891}{6}} = \sqrt{159.4982} = 12.6293$

Calculator Formula : $\text{St. Dev.} = s = \sqrt{\dfrac{\displaystyle\sum_{i=1}^{n} x_i^2 - \dfrac{1}{n}\left(\displaystyle\sum_{i=1}^{n} x_i\right)^2}{(n-1)}} = \sqrt{\dfrac{109584 - \dfrac{872^2}{7}}{6}} = 12.6340$

The difference in these two numbers is due to rounding that takes place when divisions take place and the quotients are rounded. Theoretically the two formulas produce the same numbers. And if we round our answers to two places to the right of the decimal, we get agreement. Also notice our estimate of 7.5 millimeters of mercury was, in fact, low as we anticipated.

EXAMPLE 3: What is the mean deviation for the data of Example 2.

SOLUTION: If the arithmetic average of the absolute value of the deviations in column 2 of the above table is found we get

$$\frac{72.53}{7.00} = 10.3614.$$

This is slightly smaller than the standard deviation obtained in Example 2 above.

EXAMPLE 4: Suppose a set of data has the following summarized computations: $\sum x = 1050$, $\sum x^2 = 315000$, with $n = 10$. What is the mean and standard deviation (use the divisor n in computing the standard deviation.) Can you compute the median from this information?

SOLUTION:

Mean:
$$\sum \frac{x}{n} = \frac{1050}{10} = 105$$

St. Dev. :
$$\sqrt{\dfrac{\displaystyle\sum_{i=1}^{n} x_i^2 - \dfrac{1}{n}\left(\displaystyle\sum_{i=1}^{n} x_i\right)^2}{n}} = \sqrt{\dfrac{315000 - \dfrac{1050^2}{10}}{10}} = \sqrt{20475} = 143.09$$

In order to find the median, more information, in the form of the individual sample points would be needed.

EXERCISES

1. At the request of a study committee from the state legislature concerning the 55 mph speed limit, the highway patrol set up a radar checkpoint and recorded the speed, in miles per hour, of a random sample of 50 cars that passed the checkpoint during a particular time period. The speeds of the cars were recorded as follows:

74	66	65	55	48	56	50	75	67	65
76	68	50	65	70	65	60	51	68	76
68	77	63	65	52	52	63	65	80	70
65	81	70	63	53	45	65	55	71	64
55	70	64	45	66	64	40	66	55	71

Compute the mean, median and mode, the interquartile range, the mean deviation and the standard deviation for this set of data.

2. For the data in exercise 1, use the maximum and minimum values to "guess" the mean and standard deviation. Subjectively compare your guess to the values actually obtained in exercise 1 to see how good your guess would be.

3. Part of the marketing research data collected by a pharmaceutical firm producing a children's vitamin was about the color preferences children have. An equal number of each of five different color-flavor "vitamins" was prepared. (The "vitamins" under study were simple sugar-sweetened capsules the same shape and size of the vitamins to be produced.) A group of 1st and 2nd grade children were asked to select their favorite color from the "vitamins" presented to them with the following results: (Note this is not "raw" data. Rather, it has already been summarized for you as a frequency table.)

Color-Flavor	Number Chosen
Grape	26
Cherry	18
Orange	13
Lemon	23
Lime	20

Can you calculate a measure of variation for this data? What would it mean? (Note that this is nominal data. We could code the responses numerically as: Grape = 1, Cherry = 2, Orange = 3, Lemon = 4, and Lime = 5. How would you interpret the mean for such data? Would the standard deviation have any meaning?)

4. A "weights and measures" inspector for the state is assigned to check public vending machines, periodically. The inspector collected data on soft drink dispensing machines that dispense ice and soft drinks into an eight-ounce cup. The amount of ice dispensed in grams is recorded for 20 fills, along with a note of whether there was any spilling. The following data resulted.

Ice Weight (gms.)	Spilling occurred	Ice Weight (gms.)	Spilling occurred
68	Yes	65	No
59	No	80	Yes
86	Yes	98	Yes
80	No	71	No
70	No	57	No
53	No	82	Yes
71	Yes	67	No
72	Yes	87	Yes
78	Yes	76	No
67	No	85	Yes

Compute the mean and standard deviation for the "ice weight" data.

5. Compute the mean and standard deviation of ice weight for the data of exercise 4 for those cases where spilling occurred and then separately for those case where spilling did not occur. Summarize in the following format for ease of comparison.

Spilling Occurred	Count	Mean	Standard Deviation
Yes			
No			
Overall			

6. Consider the following two sets of data, compute the range, interquartile range and standard deviation for both sets. What do you conclude about the effect of extreme observations on these measures?

Set 1:	1	1	2	2	2	2	2	3	10
Set 2:	1	1	2	2	2	2	2	3	100

7. Consider the following set of numbers, compute the standard deviation for this set. Then add 5 to each number in the set and recompute the standard deviation. Next multiply each number by 10 and recompute the standard deviation. Comparing all three "standard deviations" what do you conclude about the effect of an additive and multiplicative transformation on the mean.

    2   2   4   4   5   5   5   6   6   8   8

8. Using the data from exercises 7, subtract the mean from each observation. This gives the new set of data:

    − 3   − 3   − 1   − 1   0   0   0   1   1   3   3

The bar chart for this data looks as follows:

```
                        0
         − 3    − 1   0   1        3
         − 3    − 1   0   1        3
```

What is the mean and standard deviation for this transformed data?

9. Again starting with the data of exercise 7, transform each value by subtracting the mean and then dividing by the standard deviation. That is if X denotes the original data of exercise 8 and Z is the transformed data, then $Z = \dfrac{X - \mu}{\sigma}$ (or $Y = \dfrac{X - \bar{x}}{s}$ if you regard the set of data as a sample). Such a transformation produces what is called a Z-score. Now compute the mean and standard deviation of these eleven Z-scores.

10. A measure of skewness frequently used for sample data is the ratio of the third moment about the mean to the cube of the standard deviation, i.e.,

$$\text{skewness} = \frac{\sum\limits_{i=1}^{n} \dfrac{(x_i - \bar{x})^3}{n}}{s_0^3}.$$

Apply this formula to the mph data of exercise 1 to determine the direction and amount of skewness. Next apply the formula to the symmetric data of exercise 7.

1.7 GRAPHICAL AND NUMERICAL METHODS FOR BIVARIATE DATA

Introduction

In most studies, there is generally more than one measurement of interest. Often many different kinds of measurements are collected from each unit observed. For example, suppose a sample of students is selected and are asked to report their final exam score along with their overall average, their sex, marital status and current GPA. Thus, four different measurements are obtained from each unit of observation.

In a study of voting behavior, we may collect information about race, sex, who the voter voted for in the last election, the voter's political affiliation, an index of strength of their political affiliations and so on. Thus, several measurements are made on each person.

The purpose for collecting multiple measurements per unit is generally to investigate possible relationships among them. For example, we may wonder how much relationship there is among the scores on the various tests taken in a given class. Does one sex show any tendency to do better on the final exam than the other? Do married students do better or worse than single students? Do males vote differently than females? In other settings, we may want to know whether there is any association between the temperature of the oven and the hardness of the ceramic. Is there a relationship between the stock's closing value and the opening price?

Looking at one pair of scores will not answer the question, but if we examine the entire group, certain patterns may show up that will help us to answer the questions posed.

The simplest way of displaying a set of bivariate data—data collected in pairs—is by using a two-dimensional plot of points. The presence of any obvious patterns will be immediately seen. Less obvious patterns can be detected as you gain experience with the data.

This section explains some methods of examining two variables at a time (bivariate data). The use of scatterplots is explained and the definition of covariance and correlation are given.

Definitions

Graphical Methods

DEFINITION 1.34: A *scatterplot* is a plot on a two-dimensional coordinate system that shows the *x-y* coordinates of the various sets of paired (bivariate) measurements.

DEFINITION 1.35: A *bivariate histogram* is a two-dimensional histogram that displays the frequencies or relative frequencies of pairs of points falling in a particular rectangular region.

DEFINITION 1.36: If the *x*-values in a two-dimensional *x,y* scatterplot are projected to the *x*-axis, then a *marginal distribution* is obtained for the *x*-variable. A histogram can be constructed from this distribution of points which is the same as a simple histogram of the *x* values, ignoring the corresponding *y* values. (The marginal distribution and histogram are obtained for the *y* values by exchanging *y* for *x* in the above definition.)

DEFINITION 1.37: If a "slice" of the bivariate histogram is made, perpendicular to the *x-y* plane, with the slice being parallel to, say, the *x* axis, the resulting figure (after adjustment for a smaller sample size), will be called the *conditional distribution* of *X* given *Y* = *K*, where K is the value of *y* at which point the slice was taken. If the slice is taken parallel to the *y* axis at the point *X* = *K*, then we obtain, after adjustment for a smaller sample size, the conditional distribution of *Y* given *X* = *K*. (By "adjustment for a smaller sample size," we mean that we compute the relative frequencies on the basis of the sample size for the conditional case so that the sum of the relative frequencies totals 100%.)

Numerical Methods

DEFINITION 1.38: The *covariance* of a set of bivariate data measures the pattern of linear association of the two measurements in the data set. It is computed as follows:

$$\text{Population} \qquad\qquad \text{Sample}$$

$$\sigma_{xy} = \frac{\displaystyle\sum_{i=1}^{N}(x-\mu_x)(y-\mu_y)}{N} \qquad\qquad s_{xy} = \frac{\displaystyle\sum_{i=1}^{n}(x-\bar{x})(y-\bar{y})}{n}$$

where σ_{xy} (s_{xy}) denotes the covariance of the population (sample). (If you use the divisor $n-1$ to compute the sample variance, then the same divisor, $n-1$ should be used for the covariance.)

DEFINITION 1.39: The *correlation* is a non-dimensional measure of the linear association between the two measurements in a set of bivariate data. The correlation is defined to be

$$\text{Population} \qquad\qquad \text{Sample}$$

$$\rho_{xy} = \frac{\sigma_{xy}}{\sigma_x\sigma_y} \qquad\qquad r_{xy} = \frac{s_{xy}}{s_x s_y}$$

where ρ_{xy} denotes the population correlation and r_{xy} denotes the correlation of the sample. The correlation is such that $-1 \le \rho \le 1$ with $\rho = 0$ implying no *linear* association.

DISCUSSION

Graphical Methods

Six married couples who had married while both parties were still in high school were randomly selected from a population of such couples. The researcher wanted to study the quality of married life, employment history, marital happiness, etc., for such early marriages. One characteristic examined was their high school GPA at the time of the marriage. The data that resulted are given in Table 1.8. What can be said, if anything, about the relationship of the GPA's of husband and wife? To answer the question, a two-dimensional scatterplot might be constructed to examine the presence of any trend or pattern. If the husband's GPA is plotted on the *X*-axis and the wife's GPA is plotted on the *Y*-axis, the scatterplot of Figure 1.17 is obtained.

TABLE 1.8 High school GPA's of couples married in high school.

GPA	
(X) Husband	(Y) Wife
0.7	1.1
1.2	1.4
1.8	1.6
1.9	2.4
2.3	3.3
2.5	3.2
2.6	3.5

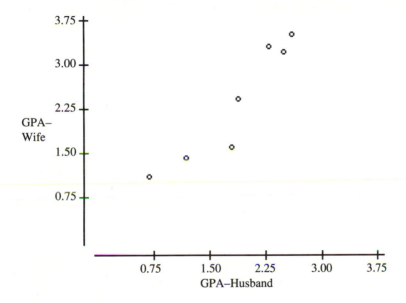

Figure 1.17 Scatterplot of GPA's.

You can very easily see an upward and positive trend in the data as displayed in the scatterplot. This indicates that low x-values tend to be associated to low y-values and high x-values tend to be associated with high y-values. Such an association is said to be positive, since the slope of a straight line plotted through these points is positive. (If the plotted "line" had a negative slope, then we would say the association was negative.)   Another interesting property can noted as well; that is the tendency of the wife's GPA to exceed the husband's. This is observed graphically by extending a 45 degree line through the origin and noting that the majority of points are on the upper side of the line.

Now, to put the same display in a setting that recognizes the frequency of occurrence, the same scatterplot might be laid out as indicated in Figure 1.18 which is intended to convey the three-dimensional aspect of this data set. The horizontal plane is the x-y plane and the vertical axis reflects the frequency or relative frequency.

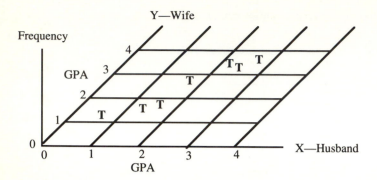

Figure 1.18 Scatterplot of husband/wife high school GPA's.

If we group the data into the intervals $0 \leq x < 1$, $1 \leq x < 2$, $2 \leq x < 3$, and $3 \leq x \leq 4$ for both x and y variables, we can count the number of observations in each group and create a 3-dimensional "histogram" like that of Figure 1.19. If the x values are projected along the x axis, and the y values are projected along the y axis we get the "marginal" histograms of the x values separate from the y-values and vice versa.   These marginal distributions, represented as histograms, are shown in Figure 1.19, below.

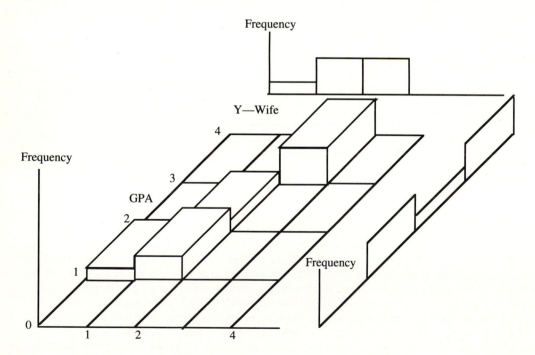

Figure 1.19 Two-dimensional histogram of husband/wife high school GPA's.

Numerical Methods

Scatterplots provide a good visual image of the presence or absence of association or relationship. However, a good quantitative way of expressing the degree of association is also needed. Such a measure is based on a number, called the covariance of X and Y.

The covariance of a set of data reflects the general nature of the trend of the (x,y) pairs, relative to their respective means. By examination of the formulas given in the definitions, either for the population or the sample, the covariance will be positive whenever the sum of the products of the deviations from the mean is positive. The equation for the covariance is:

$$\sigma_{xy} = \frac{\displaystyle\sum_{i=1}^{N} (x - \mu_x)(y-\mu_y)}{N}$$

for a population, and for a sample,

$$s_{xy} = \frac{\displaystyle\sum_{i=1}^{n} (x - \bar{x})(y - \bar{y})}{n}$$

The individual product in the covariance will be positive whenever the deviation of the x value from its mean pairs up with a y value whose deviation from the mean is of the same sign. The sum of these products will be positive whenever the sum of the positive products exceeds the sum of the negative products. It can be seen from Figure 1.20 below, the quadrants where such pairing creates positive products and where it creates negative products. By examination of this figure it can be seen that data with a positive covariance tend to show a positive slope when plotted on the xy-plane. Data with a negative covariance tend to show a negative slope when plotted on the xy-plane.

Quadrant	Signs of Deviations: $[(X - \mu_x), (Y - \mu_y)]$	Sign of Product: $(X - \mu_x)(Y - \mu_y)$
I	(+,+)	+
II	(−,+)	−
III	(−,−)	+
IV	(+,−)	−

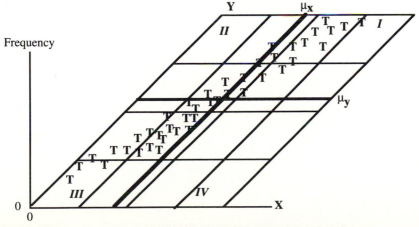

Figure 1.20   Bivariate data: Determining the sign of the covariance.

However, the magnitude of the covariance is difficult to interpret since it depends on the size of the data values from which it is computed. A more meaningful measure that is easier to interpret is the correlation. The correlation, as given in the definition section, is the ratio of the covariance to the product of the respective standard deviations. This provides a number, scaled between -1 and $+1$ where the closer the number is to ±1, the more linear the trend between X and Y. A correlation close to zero would imply no strong *linear* association.

Consider the GPA's of the seven married couples we have been examining. Their GPA's are given here again along with the deviations from their respective means as Table 1.9. The five columns of the table contain (1) the GPA of the husband, (2) the deviation from the mean for the husband GPA's, (3) the GPA of the wife, (4) the deviation from the mean of the wife's GPA's, and (5) the product of the deviations (column 2 times column 4).

Table 1.9 Computations for the covariance of husband/wife GPA's.

			GPA		
	Husband (X)	Deviation (X–mean)	Wife (Y)	Deviation (Y–mean)	Product-of-Deviations
	0.7	(0.7 – 1.86)	1.1	(1.1 – 2.36)	1.4616
	1.2	(1.2 – 1.86)	1.4	(1.4 – 2.36)	0.6336
	1.8	(1.8 – 1.86)	1.6	(1.6 – 2.36)	0.0456
	1.9	(1.9 – 1.86)	2.4	(2.4 – 2.36)	0.0016
	2.3	(2.3 – 1.86)	3.3	(3.3 – 2.36)	0.4136
	2.5	(2.5 – 1.86)	3.2	(3.2 – 2.36)	0.5376
	2.6	(2.6 – 1.86)	3.5	(3.5 – 2.36)	0.8436
Sum	13.0	0.00	16.5	0.00	3.9372

The standard deviations for X and Y (using n for the denominator) are

$$s_x = 0.648 \qquad s_y = 0.924$$

The covariance and correlation are

$$s_{xy} = \frac{3.9372}{7} = 0.5625, \quad r_{xy} = \frac{s_{xy}}{s_x s_y} = \frac{0.5625}{(0.648) \bullet (0.924)} = 0.939$$

Since the covariance and thus the correlation are positive it confirms the upward, linear trend exhibited by the data in the scatterplot. In addition the correlation is quite close to 1, indicating a fairly strong linear trend in the data. A scatterplot for this data was shown in Figure 1.17.

A calculator form for the covariance is also available which may be a more efficient computational approach than that given above. This formula is given as:

$$\sigma_{xy} = \frac{\sum_{i=1}^{N} (x - \mu_x)(y - \mu_y)}{N} = \frac{\sum_{i=1}^{N} xy - \frac{1}{N} \left(\sum_{i=1}^{N} x \right) \left(\sum_{i=1}^{N} y \right)}{N} = \frac{\sum_{i=1}^{N} xy}{N} - \mu_x \bullet \mu_y$$

Thus, after some simplification, the correlation coefficient can be written as

$$\rho_{xy} = \frac{N\left(\sum_{i=1}^{N} x_i y_i\right) - \left(\sum_{i=1}^{N} x_i\right)\left(\sum_{i=1}^{N} y_i\right)}{\sqrt{N\left(\sum_{i=1}^{N} x_i^2\right) - \left(\sum_{i=1}^{N} x_i\right)^2} \sqrt{N\left(\sum_{i=1}^{N} y_i^2\right) - \left(\sum_{i=1}^{N} y_i\right)^2}}$$

The sample correlation can be obtained using the same computational format, replacing N with n. bivariate data

EXAMPLES

EXAMPLE 1: The following data reports the systolic and diastolic blood pressure readings for 10 individuals. Form a scatterplot for this data and then compute the correlation between the two measurements.

Systolic	141.8	140.2	131.8	132.5	135.7	141.2	143.9	140.2	140.8	131.7
Diastolic	89.7	74.4	83.5	77.8	85.8	86.5	89.7	89.3	88.8	82.2

SOLUTION: First, we show a scatterplot of the data. Looking at this table, a general upward, positive linear trend can be noted. However, there is one observation that seems out of place. (See in bold face in table below.) You might wonder how dropping that observation from the data set would affect the computed correlation.

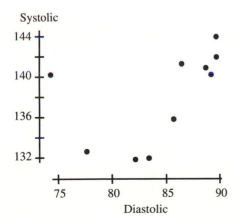

The bottom row of the following table contains the various numbers that are needed to compute the correlation.

	Systolic X	Diastolic Y	X^2	Y^2	$X \cdot Y$
	141.8	89.7	20107.240	8046.0900	12719.460
	140.2	**74.4**	19656.040	5535.3600	10430.880
	131.8	83.5	17371.240	6972.2500	11005.300
	132.5	77.8	17556.250	6052.8400	10308.500
	135.7	85.8	18414.490	7361.6400	11643.060
	141.2	86.5	19937.440	7482.2500	12213.800
	143.9	89.7	20707.210	8046.0900	12907.830
	140.2	89.3	19656.040	7974.4900	12519.860
	140.8	88.8	19824.640	7885.4400	12503.040
	131.7	82.2	17344.890	6756.8400	10825.740
Total	1379.8	847.7	190575.480	72113.2900	117077.470

$$\frac{n\left(\sum_{i=1}^{n} x_i y_i\right) - \left(\sum_{i=1}^{n} x_i\right)\left(\sum_{i=1}^{n} y_i\right)}{\sqrt{n\left(\sum_{i=1}^{n} x_i^2\right) - \left(\sum_{i=1}^{n} x_i\right)^2}\sqrt{n\left(\sum_{i=1}^{n} y_i^2\right) - \left(\sum_{i=1}^{n} y_i\right)^2}}$$

$$= \frac{10 \cdot 117077.470 - 1379.8 \cdot 847.7}{\sqrt{10 \cdot 190575.48 - 1379.8^2} \cdot \sqrt{10 \cdot 72113.29 - 847.7^2}} = 0.508$$

(If the row containing the bold-faced data points is deleted and the computation is redone, the correlation becomes 0.871 instead of 0.508. Verify it for yourself.)

EXAMPLE 2: An investigation of the traffic capacity of rotary intersections in England was investigated to determine the reliability of the design formula in current use. The actual passenger car units per hour were recorded along with the theoretical flow obtained from the formula. What is the correlation between the actual and theoretical values. The data is shown below.

Measured(Y)	Theoretical(X)
2290	3060
2100	2520
1830	2260
3290	3350
3130	3440
3400	3460

SOLUTION: A scatterplot of the data is shown first. A positive, linear trend can be seen in this scatterplot. This would suggest a positive correlation between the measured and the theoretical passenger cars per hour.

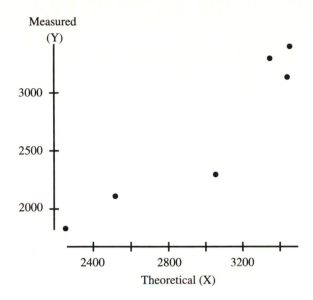

Secondly, the following table contains the numbers needed to calculate the correlation which is seen to be 0.937, a fairly high correlation.

	Measured(Y)	Theoretical(X)	Y^2	X^2	$Y \cdot X$
	2290	3060	5244100	9363600	7007400
	2100	2520	4410000	6350400	5292000
	1830	2260	3348900	5107600	4135800
	3290	3350	10824100	11222500	11021500
	3130	3440	9796900	11833600	10767200
	3400	3460	11560000	11971600	11764000
Total	16040	18090	45184000	55849300	49987900

$$r_{xy} = \frac{6 \cdot 49987900 - 16040 \cdot 18090}{\sqrt{6 \cdot 45184000 - 16040^2} \cdot \sqrt{6 \cdot 55849300 - 18090^2}} = 0.937$$

EXAMPLE 3: For the following three data sets, make scatterplots and compute the correlation for each set.

Set 1		Set 2		Set 3	
Y	X	Y	X	Y	X
1.2	1.2	1.2	1.2	2.4	1.2
1.2	1.6	1.2	1.6	1.9	1.5
1.8	2.0	1.8	2.0	1.4	1.9
2.1	2.4	2.1	2.4	1.2	2.2
1.3	2.4	1.3	2.4	1.1	2.5
2.0	2.8	2.0	2.8	1.2	2.8
2.6	2.8	2.6	2.8	1.4	3.1
2.9	3.0	2.9	3.0	1.9	3.5

(Data continued.)

Set 1		Set 2		Set 3	
3.0	3.2	3.0	3.2	2.4	3.8
3.5	3.8	3.5	3.8	3.0	4.1
4.0	4.2	4.0	4.2	3.8	4.4
4.8	4.8	4.8	4.8	4.7	4.8
2.3	1.8	1.4	3.0		
2.9	2.2	1.3	3.8		
3.5	2.9	1.5	4.6		
4.0	3.4	2.1	4.6		
3.0	4.0	2.9	4.7		
2.4	3.7	3.4	4.5		
2.6	3.8	3.5	4.6		
3.3	2.4	3.7	4.6		
3.5	4.9	3.5	3.9		
4.5	3.8	2.8	3.2		
1.8	1.4	2.0	3.8		
3.7	2.8	3.0	3.4		
2.6	1.7	2.8	3.9		
2.1	3.2	4.2	4.5		
3.3	4.4	3.8	4.0		
3.1	1.9	2.5	3.0		
1.5	2.8	2.1	2.5		
4.0	4.9	2.3	3.1		
		3.2	3.5		
		4.5	4.5		
		1.5	1.6		
		1.6	1.4		
		4.6	4.5		

SOLUTION: The scatterplots are as follows:

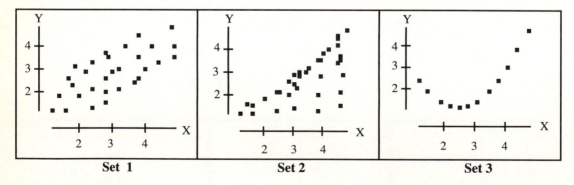

Using a statistical package to compute the correlations it is found that the correlation for set 1 is 0.70, for set 2 it is 0.70 and for set 3 it is 0.69. The implication of this outcome is that it is very unwise to examine correlations by themselves. A scatterplot should always be made so that you will better understand the relationships the data are showing. In the above, the correlations as all approximately the same, but the patterns of relationship is very different.

EXERCISES

1. An electronics engineer, trying to find clues to the reason for failure for a particular electronic part, suspected that the operating temperature may have something to do with it. She was able to collect data on length of life (y), and average operating temperature (x), for a random sample of 10 components with the following data resulting.

Hours (y)	Temp (x)
673	175
699	185
739	210
719	236
696	198
704	192
754	233
769	216
754	229
743	241

Use the data to make a scatterplot. Compute the covariance and correlation. After examining the scatterplot and the correlation coefficient, would you conclude there is a relationship between operating temperature and "time to failure?"

2. An insurance company researcher is studying relationships between the type of policy and the characteristics of the policy holder. One issue examined was the dollar amount of life insurance (y) held and family income (x). Six different policy holders were randomly selected with the following data resulting. (Data is in $1000's.) Plot the data and compute the correlation coefficient. What conclusions do you make for this data?

Income (x)	Life Ins. (y)
45	70
20	50
40	60
40	50
30	55
55	105

3. In a fuel economy test, 5 different drivers tested two different compact cars from different manufacturers. The following miles per gallon (mpg) data were obtained.

Driver	Car 1	Car 2
1	26.2	27.3
2	26.5	28.0
3	27.9	28.7
4	29.1	29.9
5	28.3	29.9

Would you conclude there is a correlation in these measurements across the various drivers? Or do you conclude that the measurements are independent regardless of the driver? (Make a scatterplot to help you see what is going on and then compute a correlation coefficient.)

4. Using the data of exercise 3, each mpg figure is replaced with its rank or ordinal value. Make a scatterplot and compute the correlation coefficient for this data and compare to your results for exercise 3. (Note: the average of ranks is given for driver 4 and 5 under Car 2.)

Driver	Car 1	Car 2
1	1	1
2	2	2
3	3	3
4	4	4.5
5	5	4.5

5. Ten different judges rated two popular cola drinks on a scale from 1 to 10. Each judge operated independently from the others in a completely "blind" experiment so that one judges had no knowledge of another judge's "ratings," and the cola's were unmarked and randomly given to the judge for ranking. Are the cola ratings correlated across the judges?   The data follow.

Judge	Cola 1	Cola 2
1	2	8
2	4	3
3	2	5
4	1	3
5	9	7
6	9	8
7	2	4
8	9	8
9	7	9
10	2	3

6. In exercise 5, what would a high positive correlation mean concerning the way the judges rated the two cola's? What would a high negative correlation mean?

7. Compute the covariance and correlation for the following paired numbers. Then add 5 to each number in the pair and redo your computations. Next, multiply each number in the pair by 10 and again redo your computations. What do you conclude about the effect of such transformations of the data?

Pair	X	Y
1	3	7
2	4	9
3	1	8
4	2	6
5	5	10

8. Using the data in Example 1, drop the data point in bold face and determine the correlation without that observation in the data set.

9. Using the data in Example 3, compute the correlation for each data set and confirm the values that are listed in the example.

10. It is claimed that the correlation between the births in a particular country is highly correlated with the suicide rate over time. Comment on this observation from a cause and effect point of view. What other variables may be highly correlated with both?

1.8   MORE GRAPHICAL METHODS

Introduction

In preparing bar graphs and histograms, there are obvious guidelines to follow, but there is great flexibility allowed as well. It is impossible to say there is a correct or incorrect way to do things; there are simply good choices or bad choices. In addition, there are other useful displays for describing a set of data. Two methods are presented here in a brief and simplified presentation—the "stem-and-leaf" and the "boxplot."

Boxplots and the stem-and-leaf display are of fairly recent origin having been introduced in the statistical literature in the 1970s and are finding more acceptance among the statistical community for their flexibility and usefulness. More complete discussions are available in John W. Tukey's book, *Exploratory Data Analysis,* published in 1977 by Addison-Wesley.

Definitions

DEFINITION 1.40: A *stem-and-leaf* is a display, similar to a histogram. In it simplest form, it shows the number of observations associated with class intervals 10 units wide. These intervals cover the range from 0-9, 10-19, 20-29, etc. If a set of data consists of 2-digit numbers between 00 and 99, the ten's digit will be called the stem and the unit's digit the leaf. The stems are ordered from small to large and placed in a column. The various leaves are placed to the right of each stem, forming a row. When completed, the display looks like a histogram laid on its side. For 3–digit numbers or numbers with decimals, a different choice of stems and leaves would have to be made, with your own skill and intuition being your best guide. (See Table 1.10 for an example of a stem-and-leaf display.)

DEFINITION 1.41: A *boxplot* is a graphical display of a set of data that divides the data into fourths. The values displayed are the minimum and maximum*, the 1st, 2nd (median), and 3rd quartiles. (See section 1.3 for the definition of a quartile.) If a vertical orientation is used, a vertical axis is erected showing the scale of the observations. A rectangular box is then drawn, extending from the 1st quartile to the 3rd quartile. A vertical line is drawn that extends from the mid-point of the box starting at the 1st quartile and ending at the minimum. Another line extends from the mid-point of the box starting at the 3rd quartile, ending at the maximum. These lines have been given the name: "whiskers." A horizontal line across the width of the box is drawn at the location of the median. (This definition will become immediately obvious upon inspection of Figure 1.21.) Properties of location, variability, skewness and symmetry can be easily seen in a boxplot.

* The minimum and maximum values are not the formally defined end points of the boxplot but are used to simplify our presentation here.

DISCUSSION

Consider the final exam and total point data of the chapter exercise. We want to understand the properties of the distribution of the scores involved. Histograms and scatterplots have already been used to display the shape and "distribution" of those numbers. Here we present two additional techniques that have been developed recently to convey as much useful information about a set of numbers as succinctly as possible.

Stem-and-leaf Displays

The following table, Table 1.10, presents stem-and-leaf displays of the test data. These displays were prepared using Minitab, a widely used statistical software program. Notice the information included in these two stem-and-leaf displays are: (1) the number of data points, (2) the median, and (3) the 1st and 3rd quartiles. For these particular data sets the "stems" are the 10's digits, the "leaves" are the "unit" digits. To obtain more detail from the data, each stem is listed twice to allow for class interval widths of 5 units wide. For instance, the first interval goes from 20 to 24, the second interval goes from 25 to 29, and so forth. The first 4 entries for the final exam scores of Table 1.10 represent scores of 20, 27, 30, 30, 35, etc. A close examination of the table will reveal the basic idea of the "stem-and-leaf" display.

Table 1.10 Stem-and-leaf display—final exam scores and "% of Total" scores.

Final Exam Scores	% of Total Scores
N = 102 Median = 75 Quartiles = 64, 86	N = 102 Median = 79 Quartiles = 70, 89
Decimal point is 1 place to the right of the colon	Decimal point is 1 place to the right of the colon
	Low: 26
2 : 0	
2 : 7	
3 : 00	
3 : 5	
4 : 22233	
4 : 77	
5 : 0344	4 : 6
5 : 555699	5 : 4
6 : 13444	5 : 5899
6 : 55667789	6 : 01234
7 : 001111112233444	6 : 55556778899
7 : 5566777889999	7 : 00000111123333
8 : 000334444444	7 : 555555667889999
8 : 56799	8 : 000111122233333444
9 : 0002444444	8 : 55677799
9 : 6667999999	9 : 0011111222334444
10 : 00	9 : 6677889
	10 : 0

One advantage the stem-and-leaf has over the histogram is that the individual data set values are still preserved in the stem-and-leaf but are lost in the histogram. In comparing the two sets of scores you can see that the "total scores" produced a more compact distribution with a somewhat higher average than did the "final exam scores." Both distributions are skewed left. Other similarities and differences can be noted as you compare the two tables more carefully.

Boxplots

A boxplot, as defined earlier, is a graphical alternative to the histogram and stem-and-leaf. In Figure 1.21 a boxplot is presented for each set of scores. The end points of the straight line projecting from each end of the box identify the practical maximum and minimum of the respective data set. The top and bottom of each box are located at the first and third quartile respectively and the horizontal line through the middle of each box is located at the median. Upon examination, it shows differences in spread, location, symmetry or lack of it, and so forth.

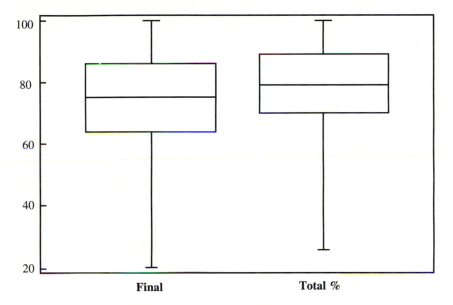

Figure 1.21 Boxplots for final and total scores.

Looking at Figure 1.21 you can see that both sets of scores show a negative skewness. The median of the "total %" is higher along with the 1st and 3rd quartiles than the "final exam" scores. Thus the final exam data seems to be spread out a little more with a lower average than the "total %" scores.

EXAMPLES

EXAMPLE 1: A stem-and-leaf and boxplot of the height data of Example 1 in section 1.3 is presented here. Comparing the stem-and-leaf to the histogram of section 1.3 some of the same characteristics can be seen. The same tendency for right-skewness, seen in the stem-and-leaf, can also be seen in the boxplot which has a shorter tail at the top. The boxplot, as drawn by a computer package, also indicates there are two "outliers" on the low end of the heights. These may be important or not; your experience has to come to your rescue.

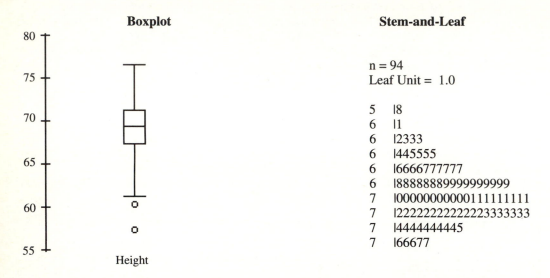

Boxplot   **Stem-and-Leaf**

n = 94
Leaf Unit = 1.0

```
5   |8
6   |1
6   |2333
6   |445555
6   |6666777777
6   |888888889999999999
7   |000000000000111111111
7   |222222222222223333333
7   |4444444445
7   |66677
```

Height

EXAMPLE 2: A major soft drink company is concerned about the bursting strength of the aluminum cans it uses in the bottling process. Since most contents are under pressure from the carbon dioxide in the drinks, the pull-tab top has to be able to withstand the pressures that build up especially as the temperature of the contents increases. To measure the bursting strength of a particular shipment of cans from a supplier, 20 cans were selected randomly and subjected to test at room temperature. The following data resulted as reported in grams/square centimeter of pressure.

265	205	263	307	220
268	260	234	299	215
197	286	274	243	231
267	281	265	214	318

Make a stem-and-leaf and a boxplot for the bursting strength data above.

SOLUTION: A boxplot and stem-and-leaf are presented side-by-side on the following page. The stem-and-leaf clearly shows that mode of the observations is in the 260+ interval. The median is approximately this size as seen by the horizontal line in the box of the boxplot. When the data set is not very large, the boxplot may not be as useful as a stem-and-leaf would be in summarizing the characteristics of the data.

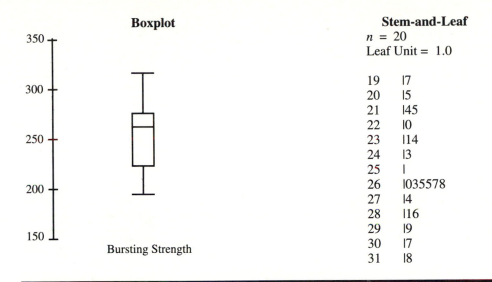

Boxplot

Stem-and-Leaf

$n = 20$

Leaf Unit = 1.0

19	l7
20	l5
21	l45
22	l0
23	l14
24	l3
25	l
26	l035578
27	l4
28	l16
29	l9
30	l7
31	l8

EXERCISES

1. Construct a stem-and-leaf display and side-by-side boxplots for the following data taken from cola ratings data of exercise 5, section 1.7. Describe characteristics of symmetry vs. skewness in the data along with comparisons of location and variability.

Judge	Cola 1	Cola 2
1	2	8
2	4	3
3	2	5
4	1	3
5	9	7
6	9	8
7	2	4
8	9	8
9	7	9
10	2	3

2. At the request of a study committee from the state legislature concerning the 55 mph speed limit, the highway patrol set up a radar checkpoint and recorded the speed, in miles per hour, of a random sample of 50 cars that passed the checkpoint during a particular time period. The speeds of the cars were recorded below. Construct a stem-and-leaf display and a boxplot for this data.

74	66	65	55	48	56	50	75	67	65
76	68	50	65	70	65	60	51	68	76
68	77	63	65	52	52	63	65	80	70
65	81	70	63	53	45	65	55	71	64
55	70	64	45	66	64	40	66	55	71

3. The following data represents various quantiles for the height and age of 19 secondary school

students. Use this information to produce a boxplot of both height and age. Decide whether there is a strong tendency towards skewness or not for each boxplot and whether it is positive or negative.

	Height Percentile	Q_p		Age Percentile	Q_p
maximum	100.0%	72.000	maximum	100.0%	16.000
	99.5%	72.000		99.5%	16.000
	97.5%	72.000		97.5%	16.000
	90.0%	69.000		90.0%	15.000
quartile	75.0%	66.500	quartile	75.0%	15.000
median	50.0%	62.800	median	50.0%	13.000
quartile	25.0%	57.500	quartile	25.0%	12.000
	10.0%	56.300		10.0%	11.000
	2.5%	51.300		2.5%	11.000
	0.5%	51.300		0.5%	11.000
minimum	0.0%	51.300	minimum	0.0%	11.000

4. The following histogram and boxplot was produced using the statistical package JMP-IN created by SAS Institute. It summarizes the births per 1000 of population for 74 countries. From this graphic, what are the minimum and maximum birth rates, the median and the IQR? (The diamond shape in the middle of the box is a 95% confidence interval on the mean birth rate. See Chapter 7 for more information.)

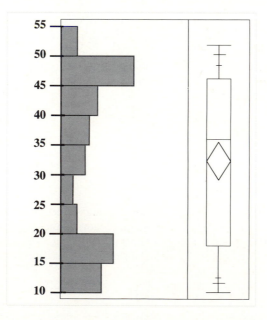

5. Looking at the histogram of exercise 4, what would a stem-and-leaf display look like for the same data? What is the advantage of a stem-and-leaf over the above histogram?

6. Using the quantiles and boxplots of exercise 3, sketch the general shape for a histogram of both

height and age.

1.9   ANALYSIS OF THE CHAPTER EXERCISE

In the Chapter Exercise presented at the beginning of this chapter, you were given a bivariate data set consisting of "final exam" and "total percentage" scores for a group of students taking a course in mathematical statistics. In this section, the techniques of this chapter are demonstrated for you, using that data set. You should verify numbers and operations where ever possible for practice.

The data is presented again here for your use. Figures and calculations have been obtained using common statistical analyses packages available on many computer systems.

The following data occur in pairs: A total of 96 student scores on the final (F) and their corresponding "total %" score (T) are reported below. For instance student 1, shown in bold type, had a 77 "total %" score with a score on the final exam of 71, etc.

T	F	T	F	T	F	T	F	T	F	T	F
77	**71**	91	99	79	73	81	70	65	69	82	64
61	42	79	65	85	75	87	94	75	63	66	59
55	77	59	50	83	74	26	20	90	92	92	94
71	61	70	55	65	27	79	73	85	84	80	71
99	100	73	71	96	96	86	84	73	80	93	94
91	72	78	85	68	74	64	72	75	64	70	65
71	64	67	53	91	100	97	99	70	59	58	55
93	90	75	71	94	90	82	84	75	80	75	43
82	89	72	42	71	47	91	83	69	47	67	54
63	67	94	94	94	94	73	66	73	66	68	77
90	84	83	87	98	99	100	99	46	30	83	78
97	94	87	84	75	71	60	54	76	56	71	55
76	74	65	43	54	30	98	99	81	71	80	79
89	84	80	75	96	96	79	76	83	89	65	42
83	83	59	79	70	68	70	67	81	79	89	79
92	97	69	35	78	76	91	96	94	99	87	84

Questions of interest in examining this data are: What is the average score on the final exam? What is the average total percentage score? How do they compare with one another? What kind of variation in scores is there for each set of scores? Of greater interest in examining a set of data such as this is the question of association between final exam score and overall score on which the grade is based. Is there strong evidence that a person doing well on the final exam also does well overall? If so how strong is the evidence? What kinds of tools can be used to assess such issues? How many students who do poorly on the final are able to compensate for that by a steady, consistent overall performance? Is there any evidence that would suggest that students can make up a semester of inferior work by "pulling it out on the final"?

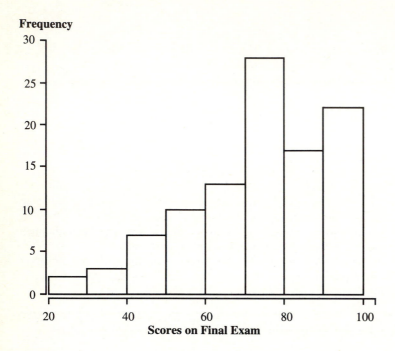

Figure 1.22   Histogram of final exam scores.

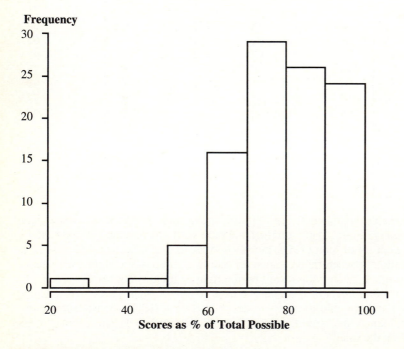

Figure 1.23 Histogram of total % scores.

Graphical Methods

Figures 1.22 and 1.23 present "histograms" for the two sets of scores. These histograms are easy to compare visually and you can see that the distribution of "total points" is slightly less variable with a slightly higher measure of location than the distribution for the "final exam" scores. Both distributions are skewed negatively.

To examine both variables together, a two-dimensional scatterplot of pairs of points is provided in Figure 1.24. Each point on the graph corresponds to a particular student's Final and Total Point average score for the semester. From this scatterplot, you can see that there is a strong upward (positive) trend in the data. This implies a tendency for those scoring high on the final to also get a high overall average, though the scatter of points indicates the postivie association is not perfect.

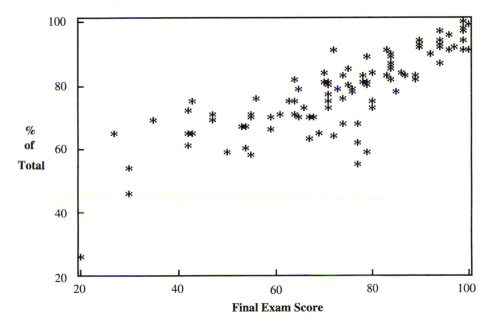

Figure 1.24 Scatterplot of total points vs. final exam scores.

Another observation is that most of the points are above a 45° line passing through the origin. This suggests that the overall average score tends to be higher than the final exam score for most of the students.

[**Note:** If you think of the practical connection between the overall average score and the final exam score, you recognize that there will be an association which would be expected to be positive. For example, suppose that there was only one exam in the course—a final exam. In that case, the total points and the final exam values would be the same number and you would end up with a set of points falling exactly on a 45° line through the origin. Since the set of points shown in Figure 1.24 don't fall on a straight line, it suggests that the other exams plus other contributing scores on homework and so on, account for the deviations from the line. This is an indication of inconsistency of performance by the student—or an inconsistency between the other exams and the final. It is impossible to say which, without further information.]

Numerical Measures

Conclusions based on the graphs and plots presented previously are subjective in nature. To put a little more objectivity in our descriptions of the data, we provide some numerical measures of locations and spread. First, we prepare Table 1.11 below to enable us to obtain any of the quantiles of interest. This table is a cumulative relative frequency table for the data we are examining. Typically, each unique score (a few are shown in bold face at the beginning of the table) would be listed with its associated percentage which represents the proportion of scores that size or smaller. All the scores are listed in the table below with their associated relative frequencies. Whenever there are repeat scores, the last score and its relative frequency is the only one that should be interpreted. Thus we would say about 1.9% of the observations are 27 or smaller, 7.8% are 42 or smaller, and so forth.

Table 1.11 Cumulative relative frequency distribution.

F	T	Rel. Cum. Freq.	F	T	Rel. Cum. Freq.	F	T	Rel. Cum. Freq.
20	26	0.0098039	69	73	0.3431373	84	84	0.6764706
27	46	0.0196078	70	73	0.3529412	84	84	0.6862745
30	54	0.0294118	70	73	0.3627451	84	85	0.6960784
30	55	0.0392157	71	75	0.3725490	84	85	0.7058824
35	58	0.0490196	71	75	0.3823530	84	86	0.7156863
42	59	0.0588235	71	75	0.3921569	84	87	0.7254902
42	59	0.0686275	71	75	0.4019608	84	87	0.7352941
42	60	0.0784313	71	75	0.4117647	85	87	0.7450981
43	61	0.0882353	71	75	0.4215686	86	89	0.7549020
43	62	0.0980392	72	76	0.4313725	87	89	0.7647059
47	63	0.1078431	72	76	0.4411765	89	90	0.7745098
47	64	0.1176470	73	77	0.4509804	89	90	0.7843138
50	65	0.1274509	73	78	0.4607843	90	91	0.7941176
53	65	0.1372549	74	78	0.4705883	90	91	0.8039216
54	65	0.1470588	74	79	0.4803922	90	91	0.8137255
54	65	0.1568627	74	79	0.4901961	92	91	0.8235295
55	66	0.1666666	75	79	0.5000000	94	91	0.8333333
55	67	0.1764705	75	79	0.5098040	94	92	0.8431373
55	67	0.1862745	76	80	0.5196079	94	92	0.8529412
56	68	0.1960784	76	80	0.5294118	94	92	0.8627451
59	68	0.2058823	77	80	0.5392157	94	93	0.8725490
59	69	0.2156862	77	81	0.5490197	94	93	0.8823530
61	69	0.2254901	77	81	0.5588235	96	94	0.8921568
63	70	0.2352941	78	81	0.5686275	96	94	0.9019608
64	70	0.2450980	78	81	0.5784314	96	94	0.9117647
64	70	0.2549019	79	82	0.5882353	97	94	0.9215686
64	70	0.2647058	79	82	0.5980392	99	96	0.9313725
65	70	0.2745098	79	82	0.6078432	99	96	0.9411765
65	71	0.2843137	79	83	0.6176471	99	97	0.9509804
66	71	0.2941176	80	83	0.6274510	99	97	0.9607843
66	71	0.3039215	80	83	0.6372549	99	98	0.9705883
67	71	0.3137254	80	83	0.6470589	99	98	0.9803922
67	72	0.3235294	83	83	0.6568627	100	99	0.9901961
68	73	0.3333333	83	84	0.6666667	100	100	1.0000000

Figure 1.25 below displays a graph of the relative cumulative frequency distribution for the Final Exam Scores. The horizontal position for each point on the graph represents the relative cumulative frequency for the associated Final Exam Score. This graph provides an easy, approximate method for finding quantile scores. To do this, you simply identify the q-value on the vertical (rel. cum. freq.) axis; project a horizontal line to the right until it intersects the graph, and then read off the corresponding quantile score X_q, on the Final Exam axis.

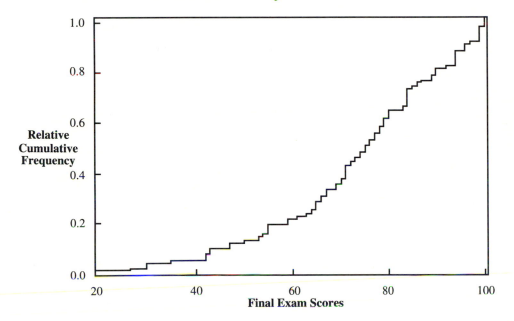

Figure 1.25 Relative cumulative frequency graph—final exam scores.

In the next table, Table 1.12, a summary is presented of all the various measures of location and variability that we have presented in this chapter. In addition, the covariance and correlation is provided. As you examine these numbers, turn back to the histograms and the scatterplot to relate what was seen there to the numerical measures that are computed here.

Looking at Table 1.12, you can see that the final exam scores are lower on the average than the total point score. All the measures of variation support the conclusion that the final exam scores are more variable, more greatly dispersed than the overall total-point scores. (You should also note the similarity of the standard deviation, mean deviation and median deviation values within each column. Though they are different, each provides the same basic information—that the "average" deviation from the center of the distribution of scores is about 12-14 points.) The interquartile range indicates that the middle 50% of scores is spread across a range of about 20 points for both distributions. However, looking at the interdecile range the we see that the difference between 49 and 31 indicates that the final exam scores tend to tail off more than the total-point scores; i.e., there are more observations in the extreme tails of the final exam and the distribution would be flatter in comparison to the total-point scores. In addition, the smaller mean score compared to the median supports a conclusion of negative skewness. (The graphical displays confirm this.)

The measures of covariance and correlation confirm the positive upward trend seen in the data, but the fact that the correlation is only 83% indicates the presence of variation about the upward trend line.

Table 1.12   Summary of Numerical Measures.

	Final Exam	Total %
Measures of Location		
Mean	73.19	78.20
Median	75.00	79.00
Measures of Variation		
Range	80.00	74.00
Interdecile Range	49.00	31.00
Interquartile Range	22.00	19.00
Variance	339.70	164.85
Standard Deviation	18.43	12.84
Mean Deviation	14.45	10.18
Bivariate Measures		
Covariance	196.4087	
Correlation	0.8300	

In terms of overall conclusions, we could say that those students who do well overall also tend to do well on the final, though there are some exceptions that the scatterplot of Figure 1.24 shows. There are some points on the scatterplot showing final exam scores in the 30s and 40s with a corresponding overall score in the 60s and 70s. Also there are a few points showing final exam scores of about 80 with overall scores in the 50s. The fact that the overall average is higher than the final exam indicates that students tend to do better on the other sources of points during the semester when compared to the final—the final exam tends to be lower than their overall average score for most students.

CHAPTER SUMMARY

In this chapter you have learned how to summarize data, whether it is population or sample data. Remember that the methods used for discrete data are slightly different from the methods used for continuous data. The methods used for nominal data are different and of necessity more crude than those used for interval or ratio data. Table 1.13 and Table 1.14 summarize this for both graphical and numerical summaries. Note especially that the interval and ratio rows have been combined since the techniques used on one are appropriately used on the other.

As you examine these tables, note that the operations that we ordinarily perform on data—adding, subtracting, multiplying, dividing—are reserved for data that are interval or ratio data. However, much of the data commonly encountered are ordinal or nominal. Special statistical techniques have been developed to handle such data and will be presented in later chapters.

In addition, it is important to remember that a primary purpose in summarizing data is to become better acquainted with the information the data contains. Often, simply ordering the data will reveal many important insights about the data you are studying. The other techniques of this chapter are additional aids for extracting information from data for decisions that you are trying to make.

Finally, although we have dwelt extensively in this chapter on the computational aspects of summarizing data, the important issue is the concept of the number behind the computation performed. Make sure that you understand these concepts before going on to other topics.

Table 1.13 Methods for summarizing data graphically by data classification.

Measurement Level	Quantitative		Qualitative	
	Discrete	**Continuous**	**Discrete**	**Continuous**
Nominal			Bar Chart	
Ordinal	Bar Chart			
Interval Ratio	Bar Chart Boxplot Stem & Leaf	Histogram Boxplot Stem & Leaf		

Table 1.14 Methods for summarizing data numerically by data classification.

Measurement Level	Quantitative		Qualitative	
	Discrete	**Continuous**	**Discrete**	**Continuous**
Nominal — Location			Mode	
Nominal — Variability			None	
Ordinal — Location	Mode Median			
Ordinal — Variability	Max vs. Min Corr.			
Interval/Ratio — Location	Mode Median Mean	Mode Median Mean		
Interval/Ratio — Variability	Range, IQR, Stand. Dev. Cov. Corr.	Range, IQR, Stand. Dev. Cov. Corr.		

CHAPTER EXERCISES

1. The IQ scores for a sample of 24 students entering their first year of high school are given. Using the techniques of this chapter, summarize as completely as you can the basic

characteristics of this distribution of scores. Use both graphical and numerical procedures. Write a short paragraph summarizing these characteristics.

115	119	119	134	121	128
128	152	97	108	98	130
108	110	111	122	106	142
143	140	141	151	125	126

2. The following data was obtained from the responses of 42 students. They were asked to pick a random digit between 0 and 9, then they were asked to report on what they regarded as the ideal number of children in a family. They also reported the number of brothers and sisters they had, the number of children in their father's family and the number of children in their mother's family.

What can you say about the ability of individuals to generate "random numbers?" What should the histogram look like theoretically. Compare to the histogram you get. (In theory a 0 should occur with a frequency of one-tenth, etc. What is the theoretical mean and variance of such a distribution. That is, if we have 10 numbers from 0 to 9 with a frequency of one each, what is the mean and variance/standard deviation for such a set of numbers?)

Random #	Idl Fam Size	# of Br/Sis	Fam. Size/ Father	Fam. Size/ Mother	Random #	Idl Fam Size	# of Br/Sis	Fam. Size/ Father	Fam. Size/ Mother
1	6	7	9	7	3	4	8	3	3
7	5	6	3	3	4	3	6	6	5
6	4	4	2	2	3	7	4	4	5
9	10	9	2	10	4	5	5	6	3
8	3	3	1	3	4	5	6	8	9
9	4	3	8	9	3	6	4	6	8
2	5	5	5	1	3	5	2	7	4
4	4	4	4	2	7	2	4	8	5
3	5	5	4	3	3	3	5	2	4
4	5	5	6	4	4	6	7	3	5
9	6	5	10	7	7	5	3	2	5
8	6	5	10	7	6	4	3	2	7
7	6	5	6	5	5	6	3	2	4
7	5	1	8	2	6	4	2	2	10
8	6	4	4	11	8	3	3	4	6
6	4	5	2	4	0	8	4	7	8
6	5	2	9	11	4	4	3	4	6
8	6	7	6	5	3	4	5	4	5
1	5	5	7	2	7	5	1	8	1
5	5	3	3	3	5	4	2	6	2
7	4	2	2	4	7	2	6	5	6

3. What is the correlation between the a student's "ideal family size" and the actual family size with which they are associated? Use the data of exercise 2.

4. What is the correlation between a person's actual family size and the family size of the person's father? the person's mother?

5. A person's family size is one larger than the number of brothers and sisters they report since they have to count themselves. Does adding one to one column of data change the correlation with another column of data?

6. Make four boxplots, all scaled to the same axis, for the family size data in exercise 2. Make one each for: Ideal family size, Number of Brothers/Sisters, Family size of Father, and Family size of Mother. Describe similarities or differences you observe in the plots.

7. The following data represents various quantiles for the height and age of 19 secondary school students. Provide an approximation for the mean and standard deviation for the data that produced these numbers.

	Height Percentile	Q_p		Age Percentile	Q_p
maximum	100.0%	72.000	maximum	100.0%	16.000
	99.5%	72.000		99.5%	16.000
	97.5%	72.000		97.5%	16.000
	90.0%	69.000		90.0%	15.000
quartile	75.0%	66.500	quartile	75.0%	15.000
median	50.0%	62.800	median	50.0%	13.000
quartile	25.0%	57.500	quartile	25.0%	12.000
	10.0%	56.300		10.0%	11.000
	2.5%	51.300		2.5%	11.000
	0.5%	51.300		0.5%	11.000
minimum	0.0%	51.300	minimum	0.0%	11.000

8. Provide a rough sketch for the histograms for both height and age using the information in exercise 7.

9. Can anything be said about the size of the correlation of height and age using the data from exercise 7? Give an explanation for your answer.

10. A histogram and boxplot was produced using the statistical package JMP-IN created by SAS Institute. This is shown in exercise 4 of section 1.8. It summarizes the births per 1000 of population for 74 countries. From this graph estimate a value for the mean and the standard deviation of the birth rate.

2

INTRODUCTION TO PROBABILITY

OBJECTIVES

When you complete this chapter you should be able to solve simple probability problems involving (1) the basic probability postulates, (2) the addition property, (3) the multiplication property, and (4) conditional probability. You should understand and be able to use the principle of independence in solving probability problems involving the intersection of events. You should know the connection between set theory and the definition of of events in the probability context. However, you will not be expected to deal with any of the complex operations of union, intersections, and complementation of events. That is reserved for another course.

INTRODUCTION

The foundations of probability theory were first established as mathematicians tried to develop a logical basis for games of chance. Probability, as developed from this perspective, assumed equal likelihood of outcomes. For example, if a die was balanced, then it was very natural to assume the probability associated with each outcome to be 1 in 6. However, the assumption of equal likelihood did not apply to many other situations of practical importance. (What is the probability of an outcome if the die is *not* balanced?) To handle this predicament, the frequency theory of probability was assumed as an alternative explanation of the meaning of a probability.

The frequency theory of probability operates by simply replicating a basic operation a large number of times. As we do this we observe the frequency that a particular observation turns up. The ratio of this frequency to the total number of replicates is an approximation to the theoretical probability, with the approximation improving as the number of replicates increases. For instance, with a loaded die, to find the probability of a 1 occurring, we roll the die a large number of times (a 1000 times would satisfy most people) and note the number of times in the thousand rolls that a 1 occurred. This we would accept as a fairly good approximation to the theoretical probability of a 1 occurring.

This still does not adequately treat all situations, for there are some situations that simply can't be repeated time after time. In such a case, we may have to make a subjective assignment of probability for those situations where replication of experiments is not feasible.

In this chapter we present the basic definitions and theorems of elementary probability theory. However, we will limit our treatment to those concepts which have particular application in this course. No pretense is made that this single chapter will provide a complete coverage of introductory probability. A more complete coverage would properly take place in an elementary probability course. There are numerous texts available that would provide such an introduction.

Our approach will tend toward the frequentist approach, though the properties and postulates to be stated apply equally to any of the assumptions of the other philosophies upon which probability is based.

Where We Are Going in This Chapter

Recall the statistical–inferential model introduced in Chapter 1 reproduced here as Figure 2.1. In Chapter 1 the perspective given was that we wanted to make inferences about the population based on information gained from the sample. In such an inferential process, we would expect the sample information to differ from the true population since the sample would not be expected to mirror, exactly, the population from which it came. If chance was used to select the sample, then the deviation of the sample from the population from which it was drawn could be attributed to the sampling process rather than to other sources. Therefore, with an understanding of chance and probability, we would hope to be able to predict how much difference (error) there may be between the sample and the population. We would also expect the size of the difference or error to be related to the amount of variation in the population. If there is very little variation in the population, then we would not expect the sample to differ very much from the population. If there is a lot of variation in the population then we would have to allow that the sample could differ quite a bit from the population. (We can use the size of the population standard deviation to indicate the amount of variation present in the population.) Therefore, the more variation there is in the population, the more likely it would be to find a large difference between the sample and the population. Conversely, intuition suggests that for larger sample sizes, the difference between sample and population would be smaller.

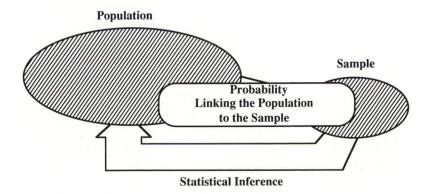

Figure 2.1 How probability fits into the SIM.

The connection between the error in the sample, the sample size, and the population variation is formally spelled out in Chapter 7 and applied in the following chapters of this book. This connection is depicted by the shaded region in Figure 2.1 above. However, this connection is rooted firmly in the principles of probability. Therefore, it is essential that we introduce the principles of probability as a prelude to the important concepts to follow from it. In fact, the principles of probability are the only bases developed thus far that allow us to measure the errors in prediction. Therefore, any course in statistics, by necessity, requires a review and study of the basic properties and postulates of probability in order to be able to assess the likelihoods and magnitude of errors.

Chapter Exercise

In a manufacturing process it has been determined from experience that, as each new shift begins work, the process machinery is either producing at the 1% defective level, an acceptable level; at the 3% defective level, with minor adjustments needed; or at the 5% defective level with major adjustment required. Historically it has been found that 90% of the time as they start a new shift, the process is at the 1% level, 9% of the time it will be at the 3% level, and only 1% of the time will it be at the 5% level.

At the beginning of each shift the first two products are inspected for quality. If one defective is found, what is the probability that the process is at the 1% defective level?

2.1   TERMINOLOGY APPLIED TO PROBABILITY AND A REVIEW OF SET THEORY

Introduction

Most formal presentations of probability usually use the language and terminology of sets and their properties. We assume that a student coming into a course with a calculus prerequisite will possess the necessary background in set theory. However, as a review, we present the following short section to indicate the specific terms we will use and the concepts that are of particular interest in this course.

Definitions

DEFINITION 2.1: An *experiment* is any repeatable operation whose outcomes are attributed to chance.

DEFINITION 2.2: A *sample space* is a list of all the possible outcomes that could occur when an experiment is conducted. Such a list is a set to be denoted by *S*.

DEFINITION 2.3: Each element in the sample space *S* is called a *sample point* or *sample outcome*.

DEFINITION 2.4: Any subset *A, B, ….* of the sample space *S* is called an *event*, with ϕ denoting the *empty set*.

DEFINITION 2.5: The *complement* of the event A with respect to S, denoted by A' is the set of points in S not belonging to the subset A.

DEFINITION 2.6: The *intersection* of two events A and B, denoted by A & B, is the set of points in the sample space S that belong to A and also to B.

DEFINITION 2.7: The events A and B are said to be *mutually exclusive* if their intersection is empty; i.e. A & $B = \phi$.

DEFINITION 2.8: The *union* of two events A and B denoted by A *or* B is the set containing the points in the sample space S that belong to either A or B or both.

DISCUSSION

The concept of probability occurs as we consider "future" events whose specific outcomes are unpredictable. Tomorrows weather is uncertain; a missile may or may not successfuly hit the target; we roll a die not knowing what outcome will occur but at least knowing what the possible outcomes are. To formalize the language in such a discussion we introduce some new terms: an *experiment* is any operation that produces unpredictable outcomes. A list of the possible outcomes is called the *sample space* and each individual possible outcome is called a *sample point* or *sample outcome*.

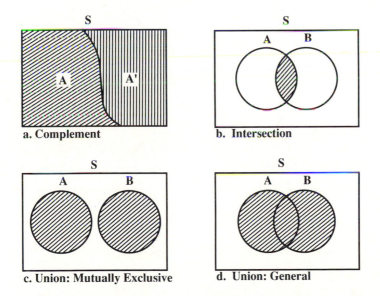

a. Complement   b. Intersection

c. Union: Mutually Exclusive   d. Union: General

Figure 2.2 Set properties: complementation, intersection and unions.

The events about whose probability we are concerned are the same thing as sets or subsets of the sample space S in the context of the theory of sets. Therefore it is useful to utilize the properties of set theory to present and discuss the properties of probability. The event "A or B" refers to the union of events (sets) A and B; the event "A & B" refers to the intersection of A and B, and so on. The formal definition of most common terms in set theory are given above and are most easily visualized by using a Venn Diagram. Figure 2.2 demonstrates these concepts graphically. Note that in the context of probability ideas, it will be true that for the sample space S, $P(S) = 1$, and for the null set ϕ, $P(\phi) = 0$. That is if an experiment occurs, then it is certain that some outcome in the sample space will occur.

In Figure 2.2 the shaded areas represent the areas of interest. Diagram (a) shows a set and its complement; in diagram (b) the shaded area represents the intersection of the events A and B. In diagram (c) and (d), the shaded area shows the union of the two events A and B for mutually exclusive and non-mutually exclusive events, respectively.

EXAMPLE 1: Consider a sample space that results from rolling a die twice. The sample space S is:

$$S = \begin{Bmatrix} (1,1) & (1,2) & (1,3) & (1,4) & (1,5) & (1,6) \\ (2,1) & (2,2) & (2,3) & (2,4) & (2,5) & (2,6) \\ (3,1) & (3,2) & (3,3) & (3,4) & (3,5) & (3,6) \\ (4,1) & (4,2) & (4,3) & (4,4) & (4,5) & (4,6) \\ (5,1) & (5,2) & (5,3) & (5,4) & (5,5) & (5,6) \\ (6,1) & (6,2) & (6,3) & (6,4) & (6,5) & (6,6) \end{Bmatrix}$$

Suppose the set A is "a 1 occurs" and the set B is "a 6 occurs." The union of A and B is the set of points around the perimeter of the array, shown in italics below. The intersection of A and B is shown in bold-face.

$$S = \begin{Bmatrix} (1,1) & (1,2) & (1,3) & (1,4) & (1,5) & \mathbf{(1,6)} \\ (2,1) & (2,2) & (2,3) & (2,4) & (2,5) & (2,6) \\ (3,1) & (3,2) & (3,3) & (3,4) & (3,5) & (3,6) \\ (4,1) & (4,2) & (4,3) & (4,4) & (4,5) & (4,6) \\ (5,1) & (5,2) & (5,3) & (5,4) & (5,5) & (5,6) \\ \mathbf{(6,1)} & (6,2) & (6,3) & (6,4) & (6,5) & (6,6) \end{Bmatrix}$$

The complement of A, those sample points not containing a 1, is shown as the boldface sample points in the next set below. It is easy to note that A and A' are mutually exclusive.

$$S = \begin{cases} (1,1) & (1,2) & (1,3) & (1,4) & (1,5) & (1,6) \\ (2,1) & \mathbf{(2,2)} & \mathbf{(2,3)} & \mathbf{(2,4)} & \mathbf{(2,5)} & \mathbf{(2,6)} \\ (3,1) & \mathbf{(3,2)} & \mathbf{(3,3)} & \mathbf{(3,4)} & \mathbf{(3,5)} & \mathbf{(3,6)} \\ (4,1) & \mathbf{(4,2)} & \mathbf{(4,3)} & \mathbf{(4,4)} & \mathbf{(4,5)} & \mathbf{(4,6)} \\ (5,1) & \mathbf{(5,2)} & \mathbf{(5,3)} & \mathbf{(5,4)} & \mathbf{(5,5)} & \mathbf{(5,6)} \\ (6,1) & \mathbf{(6,2)} & \mathbf{(6,3)} & \mathbf{(6,4)} & \mathbf{(6,5)} & \mathbf{(6, 6)} \end{cases}$$

EXAMPLE 2: List the sample space associated with the experiment of guessing on a 3-question true-false test.

SOLUTION: If we record all the possible ways of answering the questions using T for true and F for false, we get:

$$S = \{TTT,\ TTF,\ TFT,\ FTT,\ TFF,\ FTF,\ FFT,\ FFF\}$$

EXAMPLE 3: For the experiment described in Example 2 and its sample space identify the events:

$$A = \{Exactly\ 2\ questions\ that\ are\ answered\ true\},$$

$$B = \{At\ least\ 2\ questions\ that\ are\ answered\ true\}.$$

SOLUTION: The events A and B are

$$A = \{TTF,\ TFT,\ FTT\} \qquad\qquad B = \{TTT,\ TTF,\ TFT,\ FTT\}$$

EXERCISES

For each of the experiments described indicate the sample space by listing the sample points and identify the sample points in the events described.

1. Experiment: A card is selected from a standard deck of cards.
    Event: A card is a king.

2. Experiment: Three coins as tossed.
    Event: Exactly 1 head appears.

3. Experiment: A child takes one of a penny, dime or quarter out of a pocket.
    Event: The amount is less than an quarter.

4. Experiment: A child takes 2 coins from the pocket described in exercise 3.
    Event: The sum of the money is less than $.20.

5. Experiment: A red and black ball are distributed into two numbered cells.
    Event: A cell contains 2 balls.

6. Experiment: Two red balls are distributed into two numbered cells.
    Event: A cell contains 2 balls.

7. Experiment: A survey of families with 3 children is made and the sexes of the children (in order
        of age, oldest first) are recorded.
    Event: A family has at most 2 females.

8. Experiment: From six club members, *A*, *B*, *C*, *D*, *E*, and *F*, three are chosen for a committee.
    Event: Club member *A* is a member of the committee.

9. Experiment: A die is rolled and then a coin is tossed.
    Event: An even number occurs with a head.

10. Experiment: Box 1 contains a red and black ball; Box 2 contains a blue and white ball. A ball is
    selected from Box 1 and placed in Box 2 and then a ball is selected from Box 2.
    Event: A red ball is selected from Box 2.

2.2   THE PROBABILITY POSTULATES

Introduction

 Like other courses in mathematics, probability is based upon mathematical logic and on deductive
proofs.   Therefore, to begin a formal discussion of probability we need to establish the basic
definitions and postulates upon which probability is based. Once these postulates are understood then
various theorems and properties will follow as a matter of deductive proof.
        This section contains the basic postulates that are commonly used in defining an algebra of
probabilities. Each postulate stated, though quite intuitive, is important in setting up the framework
from which other theorems can be proved.
        You should be aware that probability is a complete course of study in itself.   Therefore, in this
short chapter we will only scratch the surface. Our intent is to present only the essentials that will be
needed in subsequent chapters for an understanding of statistical inference, the main objective of this
course.
        Remember that there are at least 3 philosophies that have been argued to relate probability to
real-world phenomena. They are the frequency point of view, the equal likelihood argument, and the
subjective point-of-view. In most situations each perspective fills a need and has its attractions.
However, because of the purpose to which we put probability, we will discuss and demonstrate the
postulates of probability mainly from the frequentist philosophy.   Regardless of the application or
perspective, the postulates presented in this chapter are valid for any of the philosophies used to
understand and interpret probabilities.

Postulates

If an event, A , is of interest then the probability of that event, denoted by $P(A)$, satisfies the following properties:

1.   The probability of A is greater than or equal to 0; i.e.,
$$P(A) \geq 0$$

2.   If $A_1, A_2, \ldots$, is a sequence of mutually exclusive events (See Definitions: Section 1) then

$$P(A_1 \text{ or } A_2 \text{ or } \ldots) = P(A_1) + P(A_2) + \ldots$$

3.   If the event A is the entire sample space and ϕ is the empty or null set, (See Definitions: Section 2.1) then

$$P(A) = 1, \; \& \; P(\phi) = 0.$$

DISCUSSION

From your experience in playing various board games where rolling a die or spinning spinners is an integral part, you undoubtedly have developed an intuitive feeling for probability. From the die rolling experience, it is easy to conclude that for a balanced die, the probability of any face turning up is $\frac{1}{6}$, a number greater than or equal to 0 (postulate 1). Likewise, it seems intuitive that the probability of a 1 or 2 or 3 occurring would be $\frac{1}{6} + \frac{1}{6} + \frac{1}{6} = \frac{3}{6}$ (postulate 2). Also, if each face of the die is red, the probability of a red face occurring is 1 and the probability of a black face is 0, the probabilities of a certain and impossible event respectively (postulate 3). In this simple example, we have illustrated all three postulates of probability for a case with which most of us are familiar.

The above example takes the perspective of equal likelihood of events. However, the above postulates don't require an assumption of equal likelihood or even for that matter that the probabilities make sense—only that they satisfy the 3 postulates.

For example, suppose we assigned the probabilities to the 6 sides of a die as shown in Table 2.1 that follows. Note that: (1) each probability is greater than or equal to 0, (2) mutually exclusive events have probabilities that can be summed together, and (3) that the probabilities of certain and impossible events are 1 and 0. Thus, all postulates are satisfied. But the die is also very unusual.

Though this example satisfies the postulates, it doesn't relate to what we would expect to happen if we were to roll a die that is balanced or nearly balanced. From the strict mathematical view of a study of probability, that shouldn't disturb us too much. However, we want probability to serve as a tool to help us understand real-world occurrences. Thus, we want the probability assignment to reflect experience as much as possible. Thus to accommodate this desire, and to have as much flexibility as possible, we take the relative frequency approach to assigning probabilities as described in the next paragraph. This way, probabilities would reflect experience as nearly as possible. This view is formalized in the next statement.

Table 2.1 Probability assignment to the six sides of a loaded die.

Face	Probability
1	$\dfrac{1}{12}$
2	$\dfrac{1}{12}$
3	$\dfrac{1}{24}$
4	$\dfrac{1}{24}$
5	$\dfrac{9}{12}$
6	0
Sum	$\dfrac{12}{12} = 1$

Frequency and Probability

The frequentist view of probability is to say that if in *'n'* repetitions of a given experiment or process, a particular event A were to occur *'a'* times, then the probability of that event occurring is the limit of the ratio $\dfrac{a}{n}$ as n goes to infinity; i.e.,

$$P(A) = \lim_{n \to \infty} \frac{a}{n}$$

This view of probability allows the conclusion that the ratio a/n is an estimate of the probability of the event, and as such can serve in its place. Thus, from a practical point-of-view, we can work with relative frequencies as if they were probabilities, interchanging the concepts without making an issue of which one we are dealing with.

In addition, the mathematical properties of relative frequencies are directly interchangeable with the postulates of probability, so the rules and theorems we state for probabilities will carry directly across to any discussion involving relative frequencies. That is, relative frequencies are numbers greater than or equal to 0, are bounded by 1, with the additive property of mutually exclusive events.

Probability comes into play in this course, typically, in either of two ways. One way is seen when we sample randomly from some population whose outcomes are, therefore, dependent upon chance. The other way occurs when we conduct a basic experiment which is repeatable but whose outcomes are unpredictable. In either of the above two situations, we define some event "A" whose probability of occurrence we would like to predict. We may envision repeating this sampling operation time after time. The relative frequency of occurrence of some event we assume will approach a number that, in the limit, we will call the probability of that event.

EXAMPLES

EXAMPLE 1: Suppose a biologist is studying the occurrence of a rare eye disorder of newborn infants. She has found that in 500 observed births, the rare eye disorder was encountered only 10 times for a relative frequency of 0.02. Suppose that the event A is that "an observed birth will produce an infant with the eye disorder." The biologist then claims that the chance of a particular infant being born with this eye disorder is 0.02, that is, the probability of the event A is 0.02; i.e., $P(A) = 0.02$. Notice that this assignment of probability satisfies postulate 1, and since A is not the entire sample space, it agrees with postulate 3 that $P(A) < 1$.

A further observation made by the biologist is that all 10 infants with the eye-disorder were Caucasian (white). Also of the 500 infants observed, 330 were white, the remaining being non-white. Therefore, the biologist reports that the chance of a non-white showing the disorder is $\frac{0}{170} = 0$, while the chance of a "Caucasian" infant showing the disorder is $\frac{10}{330} = 0.0303$. Again these "probabilities" satisfy the probability postulates. This sample space is shown as Table 2.2 below.

Table 2.2 Sample space for 500 new-born infants and eye-disorder.

Race	No Eye disorder	Eye disorder	Total
White	320	10	330
Black	170	0	170
Total	490	10	500

EXAMPLE 2: A medical researcher working for Medicare was studying hospital records, recording information relative to length of stay, type of illness, cost of prescribed drugs, sex, age, among other things. He selected the records for study randomly since time did not allow an examination of all records on file. There are 10,000 records on file from which the records are selected. What is the probability of any particular record being selected? If 55% of the patient records are male, what is the probability that the first record selected will be for a male patient?

SOLUTION: The implication of a random selection procedure using random numbers or any other equivalent procedure is that any particular member of the population has the same chance of being selected as any other. This being the case we would assign our probabilities as follows:

$$P(A \; particular \; record) = \frac{1}{10000} \qquad P(Male \; patient \; record) = 0.55$$

(**Note:** As stated previously the probability postulates do not require the probabilities assigned to be reasonable reflections of reality, only that they satisfy the 3 conditions as stated. Therefore, there are infinitely many ways of assigning probabilities, consistent with the postulates, but out of harmony with what would happen from a frequentist point of view. The probabilities given above should reflect what would happen in the long run.)

EXAMPLE 3: A manufacturer of golf balls produces one with a defective logo imprinted upon it only 1 in every 3000 balls produced. In an inspection program a ball is randomly chosen after imprints are made. What probability would you assign for the likelihood of finding a defective imprint?

SOLUTION: There are many ways of assigning probabilities, but only one that would seem reasonable under a random sampling procedure. The probability we assign is

$$P(Defective \; Imprint) = \frac{1}{3000}$$

EXERCISES

In the following problems make a "reasonable" assignment of probabilities to the events described.

1. A student guesses on a true-false question on a test. What outcomes could occur and what assignment of probability do you make to them?

2. A student guesses on two true-false questions on a test. What outcomes could occur on the two guesses made? [Hint: The student may guess correctly on both questions. We might denote that outcome by C_1C_2. Then again the student may get the first one correct and be incorrect on the second one. We might denote this by C_1I_2. List the other possibilities that are suggested by this approach.] What probabilities would you assign to each item you have listed? What is the sum of these probabilities? Confirm the applicability of all three postulates.

3. An ordinary six-sided die is to be rolled. It is believed that the die is completely balanced. What outcomes could occur and what assignment of probability do you make to them?

4. A six-sided die is to be rolled. However, it is known that the die is loaded so that a 1 and 6 are equally likely and are three times as likely to occur as any of the other sides. What are the probabilities for the associated sides that satisfy the postulates and the conditions described above?

5. It has been observed that when a particular part's diameter has been found out of specifications, that it is 3 times as likely to be due to being too large as it is to being too small. If we examine a randomly chosen part that is out of specification, what outcomes are possible and what assignment of probability do you make to them?

6. The employees of a certain company are twice as likely to call in sick on a Friday and Monday than on any other day of the week. What are the probabilities you would assign to the probabilities of absence due to sickness for each day of the week? (Consider only Monday through Friday.)

7. From experience, the quality inspector for a manufacturer of plastic washers knows that 0.5% of the washers will fail a symmetry test, 1% will have plastic ridges that have to be removed and 0.75% will have both problems. Let these three conditions be represented by events *A, B,* and *C*. What are the probabilities you would assign to these three events? How would you describe the relationship between events *C* and the events *A* and *B*? (See Appendix A for a discussion of set relationships.)

8. A spinner used in psychological experimentation has 10 numbered divisions around its circumference, each of equal size. Three of the divisions are red, four are green, and three are blue. If someone spins the spinner, what is the probability of *division 1* appearing? A red division?

2.3   ADDITION RULE OF PROBABILITY—PROBABILITY OF THE UNION

Introduction

One of the most useful properties of probability is summarized in this section and is referred to as the additive rule of probability. In the context of set theory it is the same as the probability of the union of two events. This rule handles events or sets which are either mutually exclusive or not mutually exclusive. It can also be extended in a logical fashion to the union of more than two events. The following theorems state these properties.

Theorems

THEOREM 2.1: The probability of the event "A or B" (or both) occurring, denoted by $P(A \text{ or } B)$ is:

$$P(A \text{ or } B) = P(A) + P(B) - P(A \text{ and } B)$$

Proof Let "a" denote the set of points belonging only to the event A, excluding any points that also might be in B. Let "b" denote the set of points belonging exclusively to "A and B" the intersection of A and B.   And lastly, let "c" denote the set of points belonging exclusively to B, excluding any points belonging also to A.   These sets are identified by the three different shadings in the Venn diagram of Figure 2.3 below. (See Section 2.1 for a review of set ideas: unions, intersections, complements, mutually exclusive events, etc.)

Sample Space, S

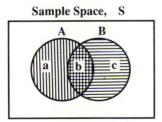

Figure 2.3 Venn diagram showing the probability of the union.

By definition the sets "a", "b", and "c" are pairwise mutually exclusive with no points in common, so the following applies.

1. $P(A \text{ or } B) = P(a) + P(b) + P(c)$ by Postulate 2

2. $P(A) = P(a) + P(b)$ "   "

3. $P(B) = P(c) + P(b)$ "   "

Therefore, by adding and subtracting $P(b)$ to equation 1 above we get:

$$P(A \ or \ B) = [P(a) + P(b)] + [P(c) + P(b)] - P(b)$$

$$= P(A) + P(B) - P(A \ and \ B)$$

COROLLARY 2.1: If A and B are mutually exclusive then

$$P(A \ or \ B) = P(A) + P(B)$$

PROOF: (See Exercises)

COROLLARY 2.2: If A and B are complementary, i.e., $B = A'$, A' denoting the complement of A, then

$$P(A \ or \ A') = P(A) + P(A') = P(S) = 1$$

Therefore,

$$P(A) = 1 - P(A')$$

PROOF: (See Exercises)

DISCUSSION

In many situations, we are interested in the occurrence of either or both of two events A and B. For example, suppose a shipment of parts is sampled and examined for defective parts. We may wonder whether 2 or 3 defectives turn up. If the event A is "2 defectives" and the event B is "3 defectives" then the event "2 or 3 defectives" is the union of event A and event B.

Suppose two marksmen shoot at a target. We want to know if either of them hit the target. If A is the event that the first marksman hits the target and B the event that the second marksman hits the target, then the event that either hits the target is the union of the events, A and B.

To evaluate the probability of "$A \ or \ B$," according to the theorem, we must find the probability of event A, the probability of event B, the probability of "$A \ and \ B$" (the intersection), and then apply the theorem above.

EXAMPLES

EXAMPLE 1: Two production line inspectors working on the same product line are observed to reject items, requiring an aesthetic evaluation of defect, 3% of the time and 1% of the time, respectively. It is also observed that without them knowing they were observed, that they reject the same items about .5% of the time. What is the probability that a given item will be rejected by one or the other of the inspectors?

SOLUTION: Let the event that an item will be rejected by inspector A be denoted by A, and similarly for inspector B. Therefore, $P(A) = 0.03$, $P(B) = 0.01$, and $P(A \ \& \ B) = 0.005$. To find $P(A \ or \ B)$ we get

$$P(A \ or \ B) = 0.03 + 0.01 - 0.005 = 0.035$$

Therefore, the proportion of parts observed by the two inspectors that will be classified as defective is 3.5%.

EXAMPLE 2: A sociologist sent out 800 questionnaires and after repeated follow-ups received 500 back. A statistical clerk was responsible for coding the data and it has been observed from past experience that the clerk wrongly codes an item 1 time in 100. What is the probability that a given item will be coded correctly?

SOLUTION: Let the event A denote "an item is wrongly coded." Therefore, the event A' is that "an item is correctly coded," the complement of A. Therefore,

$$P(A') = 1 - P(A) = 1 - 0.01 = 0.99$$

EXAMPLE 3: In test-firing a ballistic missile on a missile firing range two targets X and Y are set up with a "kill" circumference defined for each. The probability that it will fall within the "kill" circumference of target X is the same as for target Y and is 0.95. The circumferences overlap and the probability of a simultaneous "kill" is 0.9025. What is the probability that one or the other target will be hit?

SOLUTION: Let A denote the event that "target X is hit" and B the event that "target Y is hit." We are interested in $P(A \text{ or } B)$ which by the addition theorem is

$$P(A \text{ or } B) = P(A) + P(B) - P(A \text{ and } B) = 0.95 + 0.95 - 0.9025 = 0.9975.$$

Remember that this is the probability that one or the other or both targets are hit.

Example 4: Suppose a die is rolled twice. Suppose the set A is "a 1 occurs" and the set B is " a 6 occurs." What is the probability that $A \text{ or } B$ occurs?

SOLUTION: Using the sample space as written in Section 1, the italicized pairs are the events A and B with the bold-faced points denoting the intersection:

$$S = \left\{ \begin{array}{cccccc} (1,1) & (1,2) & (1,3) & (1,4) & (1,5) & \mathbf{(1,6)} \\ (2,1) & (2,2) & (2,3) & (2,4) & (2,5) & (2,6) \\ (3,1) & (3,2) & (3,3) & (3,4) & (3,5) & (3,6) \\ (4,1) & (4,2) & (4,3) & (4,4) & (4,5) & (4,6) \\ (5,1) & (5,2) & (5,3) & (5,4) & (5,5) & (5,6) \\ \mathbf{(6,1)} & (6,2) & (6,3) & (6,4) & (6,5) & (6,6) \end{array} \right\}$$

Therefore,

$$P(A) = \frac{11}{36}, \qquad P(B) = \frac{11}{36}, \qquad P(A\&B) = \frac{2}{36}, \qquad \text{so that} \quad P(A \text{ or } B) = \frac{11}{36} + \frac{11}{36} - \frac{2}{36} = \frac{20}{36}.$$

EXERCISES

1. Consider an ordinary 6-sided die that is assumed to be balanced. Let the event A denote that an even number is thrown on the die, the event B denote that a number larger than 3 is thrown. What is the probability of A or B occurring; that is, what is $P(A$ or $B)$?

2. Consider the case of the statistical clerk described in Example 2 above who has a 1 in 100 chance of wrongly coding an entry. Suppose in 2 entries chosen randomly the probability of at least one error is 0.030. What is the probability of no errors showing up in the 2 examined?

3. In a local community served by two daily newspapers, information provided by the papers indicate that 40% of households in the community take the Daily Bugle, and 35% take the Clarion Express. From a study done by a research firm, it was found that 15% subscribe to both newspapers. What proportion of the community have at least one newspaper coming into their home?

4. Using the information from exercise 3, what proportion of the households in the community is not receiving any local newspaper in their homes?

5. From experience an engineer has determined that the probability that a given manufactured part fails to meet specifications is p. Using principles of probability he has been able to determine that in 10 parts randomly selected from the assembly line, the probability of finding only one of the parts "out of specs" is p_1, of finding only 2 "out of specs" is p_2, ... of finding only i parts out of specs is p_i. Using this general notation what is the probability of finding "at least 1 part out of specs" in the ten examined? What is the probability of finding no parts out of specs?

6. From the context of exercise 5, suppose the engineer has found the probability of no parts out of ten examined being out of specification is 0.0202. What is the probability that at least one part will be out of specification?

7. A regular six-sided die is being used to randomly select a household from a cluster of six households in a neighborhood. The household selected will be part of a nationwide survey on crime. If three of the households are single-parent households, what is the probability that a single-parent household will be selected? Use the principle of unions to come to your solution.

8. From experience, the quality inspector for a manufacturer of plastic washers knows that 0.5% of the washers will fail a symmetry test, 1% will have plastic ridges that have to be removed and 0.75% will have both problems. Let these three conditions be represented by events A, B, and C. What are the probabilities you would assign to these three events? How would you describe the relationship between events C and the events A and B?

9. Prove that if A and B are mutually exclusive then $P(A$ or $B) = P(A) + P(B)$.

10. Prove that if A' is the complement of the event A, then $P(A') = 1 - P(A)$.

11. Using the information in Example 3, draw a Venn diagram to represent the problem and insert the probabilities for events A, B and $A\&B$. What is the probability that the missile will hit target X outside the joint "kill" range? (This event is denoted by $A\&B'$ and is the intersection of A with the complement of B.)

12. Using a Venn diagram similar to that created in exercise 11, argue that the following probability statement is true.

$$P[(A\&B') \text{ or } (A'\&B)] = P(A) + P(B) - 2 \cdot P(A\&B).$$

2.4   CONDITIONAL PROBABILITY, THE MULTIPLICATION RULE AND INDEPENDENCE

Introduction

Often we are interested in the probability of event A occurring knowing that event B has occurred. This is referred to as conditional probability. For example, suppose we want to know the probability that a person is married given that the person is a male over 30 years of age. Or suppose that a pair of dice rolled in a game of chance produces a sum of 9 and we want to know the probability of one of the dice being a 5. Both of these situations are examples of conditional probability.

In the following section we formally define what is meant by "conditional probability." Once we have defined conditional probability, the probability of the intersection of events follows naturally. Next, it is natural to ask whether the knowledge of the event B's occurring has any influence on the probability of A's occurring. Thus, we ask if A and B are related or whether they behave independently in a probability sense.

These are the topics to be developed in this section: conditional probability, the probability of the intersection, and independence of events.

Definitions

DEFINITION 2.9: The conditional probability of the event A given that the event B has occurred, denoted by $P(A \mid B)$ is defined to be:

$$P(A \mid B) = \frac{P(A \text{ and } B)}{P(B)}$$

THEOREM 2.2: The probability of the simultaneous occurrence of the event A and B, denoted by $P(A \text{ and } B)$, is

$$P(A \text{ and } B) = P(A) \cdot P(B \mid A) \qquad \text{or} \qquad P(A \text{ and } B) = P(B) \cdot P(A \mid B)$$

(The simultaneous occurrence of the event A and B is also called the intersection of events A and B.)

PROOF: This is a simple algebraic rearrangement of the expression defining the conditional probability of B *given* A, or A *given* B.

DEFINITION 2.10: Two events, A and B, are said to be independent if either of the following conditions hold

1. $P(A \mid B) = P(A)$ or $P(B \mid A) = P(B)$
2. $P(A \ and \ B) = P(A) \bullet P(B)$

DISCUSSION

Determining a conditional probability is intuitive in many cases. For instance, suppose 2 cans of soup are selected randomly from a store shelf which contains ten cans, two of which are dented and potentially damaged. If the first can selected is not dented, we may ask, "What is the probability that the second can selected will be dented?"

The answer to this question is fairly simple. The shelf originally had 10 cans on it, 2 of which were dented; but a nondented can was removed from the shelf, leaving 9 cans, 2 of which are dented. Therefore, the conditional probability of interest is:

$$P(dented \ can \ on \ 2nd \ selection \ given \ that \ the \ first \ selection \ was \ not \ dented) = \frac{2}{9}.$$

This probability was obtained by a very simple logic, but not all cases will be solved quite so easily.   Therefore, to handle all cases, a more formal definition is needed. This definition was given above in the Definitions Section.

To be convinced this definition makes sense and is consistent with our other perceptions of probability, consider another obvious case. Let S denote the entire sample space, so that by definition, $P(S)=1$. If we then write $P(A \mid S)$, we can evaluate its probability by applying the definition in the following way:

$$P(A \mid S) = \frac{P(A \ and \ S)}{P(S)} = \frac{P(A)}{1} = P(A)$$

However, all we are saying here is that the probability of the event A relative to its sample space is simply what we call the probability of A, $P(A)$.

Now consider the event, $P(B \mid B)$. This obviously must equal 1, that is, the probability of the event B occurring given that event B has occurred is obviously a certainty.

Applying the definition to this case produces the following result.

$$P(B \mid B) = \frac{P(B \ \& \ B)}{P(B)} = \frac{P(B)}{P(B)} = 1.$$

From examination of the above outcomes, we can think of the conditional event as the event that defines the sample space of reference.   Thus, when we write $P(A \mid B)$ we can think of B as the "new" sample space that controls the assignment of probabilities, so they will be consistent with our three postulates of probability. From this point of view, the event A can occur only if A has an intersection with the event B, the "new" sample space. Thus we need to consider the intersection of A and B and its probability, $P(A \ \& \ B)$. However, to constrain the conditional probability so that the probability of the "new" or redefined sample space has a probability of 1, we must divide by $P(B)$. These ideas can be visualized by examining Figure 2.4. The first diagram frame in Figure 2.4 shows that if the conditional event is the sample space S, then $P(A \mid S) = P(A)$. The second and third frames show the relationship of the conditional probability of A *given* B when (1) A and B have a partial intersection and (2) when *A* and *B* are the same sets and intersect completely.

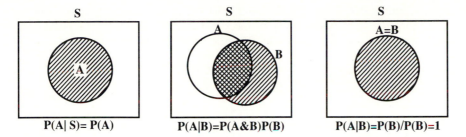

Figure 2.4 Venn diagrams and conditional probability—shaded region denotes the conditional event.

For instance, in a previous example a study was described that indicated that a biologist was studying the occurrence of a rare eye disorder of newborn infants. She found that in 500 observed births, the rare eye disorder was encountered only 10 times for a relative frequency of 0.02.

A further observation made by the biologist was that all 10 infants with the eye disorder were Caucasian (white). Also of the 500 infants observed, 330 were white; the remaining being nonwhite. These results can be presented in a two-way table, Table 2.3 below. Suppose we accept these frequencies to be typical of the entire population.

Table 2.3 Eye disorder by race.

		Race		
		White	Non-White	Total
Eye Disorder	Present	10	0	10
	Absent	320	170	490
	Total	330	170	500

Summarizing the results of this table we can say:

$$P(Eye\ disorder) = \frac{10}{500}$$

$$P(Eye\ disorder \mid White) = \frac{10}{330}$$

$$P(Eye\ disorder \mid Nonwhite) = \frac{0}{170}$$

Note that the first probability above is based on the entire sample space of 500, the latter two probabilities are based on the given conditions of 330 whites, and 170 nonwhites, each of which becomes the new sample space under the given conditions.

The Multiplication Rule of Probability—Probability of the Intersection

Following directly from the definition of conditional probability is the multiplication rule of probability. If we define the conditional probability of the event B given A and then solve for the probability of the intersection of A and B found in the definition, we get the multiplication rule as shown below:

$$P(B \mid A) = \frac{P(A\&B)}{P(A)} = \frac{P(B\&A)}{P(A)}$$

Therefore,

$$P(A\&B) = P(A)\bullet P(B \mid A).$$

The multiplication rule provides a very useful tool for evaluating the probability of the intersection of events and is a very intuitive process especially in the setting of "sampling without replacement." Coupled with the following definition of independence, it makes it possible to solve many types of probability problems.

For example, suppose from a classroom of 20 people, 15 of which are male and 5 of which are female, 2 are to be chosen randomly for a physical stress test. Suppose we want to know the likelihood of our selection resulting exclusively in females. The desired probability can be derived in the following fashion. (Note that the sample is drawn *without replacement.*)

$$P(2\ Females) = P(Female\ on\ 1st\ selection\ \&\ Female\ on\ 2nd\ selection)$$

$$= P(Female\ on\ 1st)\bullet P(Female\ on\ 2nd \mid Female\ on\ the\ 1st)$$
$$= (\frac{5}{20})\bullet(\frac{4}{19}) = \frac{1}{19}$$

Independent Events

Suppose that a situation were to be found where the conditional probability of A given B was the same as the unconditional probability, i.e., $P(A|B) = P(A)$. What would such a situation imply about the connection of the two events A and B? By examination of the equality, we would conclude that the occurrence of the event B has no influence on the likelihood of A's occurrence, and thus A seems to be "independent" of B. This leads to a natural definition of independence that if either $P(A \mid B) = P(A)$ or $P(B \mid A) = P(B)$ we will claim that A and B are independent events. Then following directly from this in the context of the multiplication rule is the result that if $P(A\&B) = P(A)\bullet P(B)$, we will say that A and B are independent. All of these conditions are "if and only if " statements.

Many experimental situations involve operations that can be considered to operate independently in a probability sense. That is, one operation has no influence on the probability of outcomes of the other operation. Also, in the following chapters we frequently assume that measurements to be analyzed resulted by a sampling process that guarantees independence of measurements. This ultimately comes down to a situation where we assume that we have sampled units from a population in an independent fashion. Such an assumption is crucial to the analysis methods to be used which are often based on the assumption of independence in the sampling of observations.

A note of caution: There is a temptation for students first encountering the concept of independence to confuse the concepts of *mutually exclusive events* and *independent events*. This can lead to errors in understanding since mutually exclusive events describes a set theoretic idea—events having no intersection—while independence is based on the way the probabilities are assigned. Since the assignment of probability does not necessarily have to take into account set ideas as long as the assignment is consistent with basic postulates, independence represents a different concept from that of mutually exclusiveness.

For example, consider two mutually exclusive events A and B. It is true, therefore, that $P(A \& B) = 0$. If A and B are to be independent then

$$P(A \& B) = P(A) \cdot P(B)$$

However, the left side of this equation is zero, therefore implying that either $P(A)$ or $P(B)$ or both must be zero, a possible but uninteresting condition.

This argument leads to the following conclusion: if A and B are independent and both have positive probability, they cannot be mutually exclusive. Or if A and B are mutually exclusive, each having nonzero probability, they are necessarily dependent events. The only conditions under which mutually exclusive events will be independent are if either or both events have a zero probability.

EXAMPLES

EXAMPLE 1: The probability that a certain electrical component will work is 0.9. A machine contains two such components. The machine will operate as long as at least one of these components works.

    a.   Regarding the working or nonworking status of both components, what are the possible outcomes and their respective probabilities? (What assumption do you have to make?)

    b.   What is the probability that the machine will operate?

SOLUTION: a. Let W_i denote the event: *Component i works.* Let W_i' denote the event: *Component i does not work.* With this notation we have the following four possibilities:

W_1W_2:	Both components work
$W_1'W_2$:	Component 1 fails, 2 works
W_1W_2':	Component 1 works, 2 fails
$W_1'W_2'$:	Both components fail

If we assume that the components operate independently; i.e., the working or nonworking status of one has no effect on the probability that the other will work, then we have the following probabilities:

$$P(W_1W_2) = (0.9)(0.9) = 0.81$$

$$P(W_1'W_2) = (0.1)(0.9) = 0.09$$

$$P(W_1W_2') = (0.9)(0.1) = 0.09$$

$$P(W_1'W_2') = (0.1)(0.1) = 0.01$$

b. The machine works if either of the following conditions apply: (a) Component 1 works or component 2 works; (b) At least one works. To calculate these probabilities we have:

$$P(W_1 \text{ } or \text{ } W_2) = P(W_1) + P(W_2) - P(W_1 W_2)$$

$$= 0.9 + 0.9 - 0.81 = 0.99$$

$$P(At \text{ } least \text{ } one \text{ } works) = 1 - P(None \text{ } works) = 1 - P(Both \text{ } fail)$$

$$= 1 - .01 = 0.99$$

$$P(At \text{ } least \text{ } one \text{ } works) = 0.81 + 0.09 + 0.09 = 0.99$$

Note that all answers agree though approached in a slightly different manner.

EXAMPLE 2: Suppose the sum on a pair of dice is 9. What is the probability that one of the dice is a 5?

SOLUTION: Let the event A be: *A five is obtained on the roll of a pair of dice*. And let the event B be: *The sum of the dice is 9*. To apply the definition we need $P(A\&B)$ and $P(B)$. First, we assume the die are balanced. Then, the easiest way to get these probabilities is to list them and count. The event "A" is:

$$A = \{(1,5) \text{ } (2,5) \text{ } (3,5) \text{ } (4,5) \text{ } (5,5) \text{ } (6,5) \text{ } (5,1) \text{ } (5,2) \text{ } (5,3) \text{ } (5,4) \text{ } (5,6)\}.$$

Next, we list the event B:

$$B = \{(3,6) \text{ } (4,5) \text{ } (5,4) \text{ } (6,3)\}$$

The event, $A\&B$, consists of those points common to both A and B and is:

$$A\&B = \{(4,5) \text{ } (5,4)\}$$

Since there are 36 equally likely events, we have that

$$P(A) = \frac{11}{36}$$

$$P(B) = \frac{4}{36}$$

$$P(A\&B) = \frac{2}{36}$$

Therefore, by the definition

$$P(A|B) = \frac{P(A\&B)}{P(B)} = \frac{\frac{2}{36}}{\frac{4}{36}} = \frac{1}{2}.$$

If we were to ask, "*Are A and B independent?*" we simply note that

$$P(A) = \frac{11}{36} \neq P(A \mid B) = \frac{1}{2}.$$

Since these two probabilities are unequal, the events A and B are not independent.

EXAMPLE 3: A box of 10 tubes of construction adhesive has two tubes with defective seals. Three tubes are selected from the box by a customer. If it is assumed the tubes are randomly arranged in the box, what is the probability that the customer will obtain no defective tubes? Given that the customer finds the first tube selected as defective, what is the probability that the customer will also have the other defective tube?

SOLUTION: Let the three events N_1, N_2 and N_3 denote the outcomes that "*no defective is obtained on the 1st selection*," "*no defective is obtained on the second selection*" and "*no defective is obtained on the third selection*," respectively. Thus,

$$P(No\ defectives) = P(N_1 \& N_2 \& N_3) = P(N_1) \bullet P(N_2 \mid N_1) \bullet P(N_3 \mid N_1 \& N_2) = \frac{8}{10} \bullet \frac{6}{9} \bullet \frac{5}{8} = \frac{7}{24} = 0.29167$$

$$P(A\ second\ defective \mid 1st\ selection\ is\ defective) = \frac{P(A\ second\ defective\ \&\ 1st\ selection\ is\ defective)}{P(1st\ selection\ is\ defective)}$$

$$= \frac{P(N_1' \& N_2'\ or\ N_1' \& N_2 \& N_3')}{P(N_1' \&\ anything\ else\ on\ 2nd\ and\ 3rd\ selections)}$$

$$= \frac{\dfrac{2}{10} \bullet \dfrac{1}{9} \bullet 1 + \dfrac{2}{10} \bullet \dfrac{8}{9} \bullet \dfrac{1}{8}}{\dfrac{2}{10} \bullet 1 \bullet 1} = \frac{2}{9} = 0.2222$$

Thus, the chance of getting no defectives is just about 30% and the chance of getting the second defective given that you got the first defective on the first selection is about 22%.

EXERCISES

In the following exercises, apply the preceding theorems and definitions to determine the probability asked for. In many cases, it will be helpful to list some if not all the possible outcomes that could result in order to direct your thinking to a logical conclusion.

1. Batteries for use in ordinary flashlights are packaged with six to a package. Two batteries the research department had developed having longer life than the regular batteries are accidentally mixed with 4 regular batteries in a package. Suppose a battery is selected at random for use in a flashlight. What is the probability that the one selected is one of the "longer-life" batteries? Suppose a second battery is selected after the first battery is identified as a long-life battery. What is the probability that the second selection will turn up the other long-life battery? A regular battery on the second selection?

2. Under the conditions of exercise 1 calculate the following probabilities: (A) The probability that both long-life batteries are selected in the two selections made. (b) The probability that two regular batteries are selected in the two selections made. (c) The probability that the first battery is long-life and the second battery is regular. (d) The probability that the first battery is regular

and the second is long-life.

3. Suppose in exercise 1 three batteries are randomly selected. What is the probability that no long-life batteries are selected? At least 1 long-life battery is selected?

4. The following table summarizes the typical characteristics of one hundred computer jobs submitted to a main frame computer installation.

Type of Job	Programming Bugs	No Bugs Encountered	Total
Fortran Job	20	50	70
Non-Fortran Job	20	10	30
Total	40	60	100

What proportion (probability) of jobs have programming bugs? Non-Fortran jobs? What proportion of Fortran jobs have programming "bugs"? What proportion of non-Fortran jobs have programming "bugs"? Based on your answers to these questions, would you say the type of job is independent of the presence or absence of programming bugs? (Hint: Is $P("Bugs") = P("Bugs" | Fortran Job)$?)

5. Using the table of exercise 4, assume that a "job" is randomly selected from the 100 jobs described. What is $P(A\&B)$ if A is the event *a Fortran Job is selected,* and B is the event that *programming bugs are encountered.* What is $P(A)$? Use these answers to calculate $P(B|A) = \dfrac{P(A\&B)}{P(A)}$. Compare this answer to the answer obtained in exercise 4.

6. The records of the two randomly selected single births in a hospital are examined. What is the common probability that the sex of the first child is male? What is the probability that the sex of the second child is male? What does this imply about the two events: $A =$ *"the first child is male"* and $B =$ *"the second child is male?"* What is the probability that both babies are male? Only 1 is male? Both are female? Do these three probabilities sum to one? Should they?

7. A student, being completely unprepared, takes a 5-question true-false test. As a result he guesses on each question. What is the probability that he misses every question? What is the probability that he gets only 1 question correct?

8. The unfortunate student of exercise 7 failed to learn from experience and encounters another pop quiz, again being unprepared. However, the test is a 5-part multiple-choice test with 5 questions again. What is the probability that he misses every question? What is the probability that he gets only 1 question correct?

9. In an engineering context, redundancy (a backup system) is often built into systems to improve their overall reliability. Suppose a critical component in a fire alarm has a probability of .05 of failing at any point in time. Two backup components, operating independently of the original, with the same failure probabilities, are installed in the alarm. If all other components in the alarm are essentially 100% reliable, what is the probability the alarm will work properly?

10. A physiologist is conducting an experiment with 10 human subjects, five of which are to receive a steroid supplement, the other 5 serving as controls. The subjects are numbered from 1 to 10 and the numbers placed in a bowl. The first 5 numbers to be selected will be the control group,

the remaining 5 will be the treatment group. It turns out that of the 10 subjects, 5 are male and 5 are female. What is the probability that the control group will all be of the same sex? (This is called sampling without replacement.)

11. Suppose 2 students are to be selected randomly from a group of 10 students of which you are a member. What is the probability you will be selected? [Hint: Note that you will be selected if you are the first person selected and anyone else is selected, or if anyone else is selected first and you are selected secondly.)

12. Use the hint and approach of exercise 11 to determine the probability of a particular person being selected if n persons are to be selected from a group of N persons.

2.5   BAYES' THEOREM

Introduction

Suppose a sample space is partitioned into k events; that is, it is divided up into mutually exclusive, and collectively exhaustive events. The events are denoted by B_1, B_2, ..., B_k, and we are able to assign probabilities to these events, $P(B_1)$, $P(B_2)$, ..., $P(B_k)$. (These probabilities are called the "a priori" probabilities.)

Subsequent to these events, an event, A occurs, often associated with the collection of data, experimental evidence, and so forth. For each of the prior events we are able to determine the conditional probabilities:

$$P(A \mid B_i) \quad i = 1, ..., k$$

The question of primary interest is how knowledge that the event A has occurred may cause us to revise our probabilities of the prior events B_i. That is, does the occurrence of A have an affect on the probability of B_i occurring? Will $P(B_i)$ be the same or different from $P(B_i \mid A)$, (called the "posteriori" or "revised" probability)? Bayes' Theorem, presented below, provides the method of evaluating these revised probabilities using the given conditions.

Theorem

BAYES THEOREM: If the events B_1, B_2, ..., B_k, constitute a partition of the sample space S such that $P(B_i) \neq 0$ for i = 1, 2, ..., k, then for any event A such that $P(A) \neq 0$,

$$P(B_i \mid A) = \frac{[P(B_i) \cdot P(A \mid B_i)]}{\sum_{j=1}^{k} [P(B_j) \cdot P(A \mid B_j)]}$$

PROOF: Using the concept of conditional probability we want to evaluate the "a posteriori" probabilities, $P(B_i \mid A)$. By definition of conditional probability, we may write the above expression as

$$P(B_i \mid A) = \frac{P(A \text{ and } B_i)}{P(A)}$$

By examining Figure 2.5 below it can be seen that the event A is the union of the k mutually exclusive events, $A \& B_1, A \& B_2, ..., A \& B_k$.

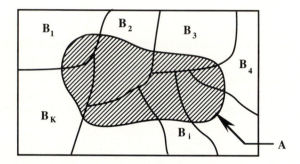

Figure 2.5   Venn diagram—Bayes' Theorem: Event A is oblong and event B_i a part of a partition of the sample space.

Therefore, using the basic postulates and the multiplication rule we can write:

$$P(A) \quad = P\{(A \& B_1) \text{ or ... or } (A \& B_K)\}$$

$$= P(A \& B_1) + \ ... \ + \ P(A \& B_K)$$

$$= P(B_1) \cdot P(A \mid B_1) + \ ... \ + \ P(B_k) \cdot P(A \mid B_k)$$

Therefore, the "a posteriori" or "revised" probabilities are

$$P(B_i \mid A) = \frac{P(A \text{ and } B_i)}{P(A)} = \frac{[P(B_i) \cdot P(A \mid B_i)]}{\displaystyle\sum_{j=1}^{k} [P(B_j) \cdot P(A \mid B_j)]}$$

The posteriori probabilities show the effect of revising the prior probabilities on the basis of the experimental evidence, denoted by A.

DISCUSSION

A statistics instructor keeps three different boxes for demonstration purposes. Box B_1 contains 20 red and 30 black balls; box B_2 contains 30 red and 20 black balls, and box B_3 contains 40 red and 10 black balls. During the evening, as the custodian cleans the instructor's office, one of the boxes is bumped and a ball is lost. Later the custodian finds the lost ball which turns out to be black, but the three boxes appear identical and the custodian can't remember which box it was that was bumped. He writes a note explaining what happened and puts it and the black ball on the instructors desk. The instructor, finding the note the next day, hasn't time to check each box and so decides to present the problem to the class as a probability problem and asks: "*What is the probability that the black ball came from box B_1? Box B_2? Box B_3?*"

The solution to this problem is determined by Bayes' Theorem after some simplifying assumptions are made. First, we assume that the chance of a box being bumped is the same for each of the three boxes, and secondly, that the ball that pops out is a random selection from that box. With these two assumptions made then we can solve the problem as follows:

Let B_i = "*Box B_i is selected*," and A = "*A black ball is selected*."

Therefore,

$$P(B_1) = P(B_2) = P(B_3) = \frac{1}{3},$$

the "prior" probabilities, and

$$P(A \mid B_1) = \frac{30}{50} = \frac{3}{5}, \qquad P(A \mid B_2) = \frac{20}{50} = \frac{2}{5}, \qquad \text{and} \qquad P(A \mid B_3) = \frac{10}{50} = \frac{1}{5}$$

The computations involved in applying Bayes' Theorem are easily summarized in Table 2.4 that follows. In this table, column 2 contains the "prior" probabilities, column 3 the conditional probabilities. Column 4 is the row product of columns 2 and 3 which is then summed. The Bayesian probabilities in column 5 are obtained by dividing each entry in column 4 by the column total, 6/15.

Table 2.4 Computations for posteriori probabilities of Bayes' Theorem.

	Prior Event	Prior Prob.	Conditional Probability	Joint Probability	Post. Prob.
Column #	1 B_i	2 $P(B_i)$	3 $P(A \mid B_i)$	$4 = 2 \times 3$ $P(B_i)P(A \mid B_i)$	5 $P(B_i \mid A)$
	B_1	$\frac{1}{3}$	$\frac{3}{5}$	$\frac{3}{15}$	$\frac{\frac{3}{15}}{\frac{6}{15}} = \frac{3}{6}$
	B_2	$\frac{1}{3}$	$\frac{2}{5}$	$\frac{2}{15}$	$\frac{\frac{2}{15}}{\frac{6}{15}} = \frac{2}{6}$
	B_3	$\frac{1}{3}$	$\frac{1}{5}$	$\frac{1}{15}$	$\frac{\frac{1}{15}}{\frac{6}{15}} = \frac{1}{6}$
		1.0		$\frac{6}{15}$	1.0

Comparing the entries in column 2 with those of column 5 shows the way the probabilities have been revised on the basis of having observed a black ball from the "bumped" box. The probability has gone from 1/3 to 1/2 for box 1, stayed the same for box 2 and has gone from 1/3 to 1/6 for box 3. Restating it another way; having observed the ball is black, the probability that it came from box 1 is 1/2 and only 1/6 for box 3.

EXAMPLES

EXAMPLE 1: In a manufacturing process it has been determined from experience that, as each new shift begins work, the process machinery is either producing at the 1% defective level, an acceptable level; at the 3% defective level, with minor adjustments needed; or at the 5% defective level with major adjustment required. Historically it has been found that 90% of the time as they start a shift, the process is at the 1% level, 9% of the time it will be at the 3% level, and only 1% of the time will it be at the 5% level.

At the beginning of each shift the last product from the preceding shift's production is sampled and examined for defects. If it is found to be defective, an administrative decision needs to be made concerning the defective level of the whole process. What is the probability that the process is at the 1% defective level given that a defective item is found?

SOLUTION: Let the events B_i represent the three production levels of 1%, 3%, and 5% defective. The event A is the outcome of sampling the last item of the previous shift which produced a defective. Thus, $P(Defective \mid 5\% \ are \ defective) = 0.05$; $P(Defective \mid 3\% \ defective) = 0.03$, and $P(Defective \mid 1\% \ defective) = 0.01$. These numbers are entered into column 3 of Table 2.5 as the probabilities that a defective is obtained, given a specific proportion are defective from a production run. The prior probabilities for column 2 of the table are obtained from the problem description which indicated that 90% of the time the machinery is at the 1% level, 9% of the time it is at the 3% level and only 1% of the time is it at 5% level. The other columns of the table are obtained by computations indicated in the table itself, but which are defined in Bayes' theorem.

Table 2.5 Calculations for Bayes' Theorem.

Column #	Prior Event	Prior Prob.	Conditional Probability	Joint Probability	Revised Prob. or Post. Prob.
	1	2	3	4 = 2x3	5
	B_i	$P(B_i)$	$P(A \mid B_i)$	$P(B_i)P(A \mid B_i)$	$P(B_i \mid A)$
	B_1	0.90	0.01	0.0090	$\dfrac{0.009}{0.0122} = 0.738$
	B_2	0.09	0.03	0.0027	$\dfrac{0.0027}{0.0122} = 0.221$
	B_3	0.01	0.05	0.0005	$\dfrac{0.0005}{0.0122} = 0.041$
		1.00		0.0122	1.000

From examination of column 5 of the above table, it can be seen that the probability that the process is at the 1% defective level given that a defective item is found is 0.738, revised down from the 0.90 "a priori" probability. Likewise, the probability of being at the 3% level, given a defective, is revised up to 0.221 from 0.09; and the probability of being at the 5% level, given a defective, is revised up to 0.041 from 0.01. Thus, when a defective is found, it revises our perceptions about the true status of the machinery. Bayes' theorem, gives us an objective measure of the state of the machinery in terms of a "revised" or posteriori probability.

EXAMPLE 2: Based on many geological surveys, an energy company has classified geological formations beneath potential oil wells as types 1, 2, and 3. For a particular site in which the company is considering drilling for oil, the probabilities of 0.35, 0.40, and 0.25 are assigned to the three types of formations, respectively. It is known from experience that oil is discovered in 40 percent of type 1 formations, in 20 percent of type 2, and in 30 percent of type 3 formations. If oil is not discovered at this site, determine the probabilities of it being of a type 1 formation; type 2; and type 3.

SOLUTION: Let the events B_i represent the various type of sites, and the event A is that oil is *not* discovered. From the information given above we can construct Table 2.6 to tabulate the computations needed.

Table 2.6 Calculations for Bayes' Theorem.

Column #	Prior Event	Prior Prob.	Conditional Probability	Joint Probability	Post. Prob.
	1	2	3	4 = 2x3	5
	B_i	$P(B_i)$	$P(A \mid B_i)$	$P(B_i)P(A \mid B_{ii})$	$P(B_i \mid A)$
	Type I	0.35	0.6	0.210	$\dfrac{0.210}{0.705} = 0.298$
	Type II	0.40	0.8	0.320	$\dfrac{0.320}{0.705} = 0.454$
	Type III	0.25	0.7	0.175	$\dfrac{0.175}{0.705} = 0.248$
		1.00		0.705	**1.000**

Examining the probabilities in column 5 of Table 2.6 as compared to the prior probabilities in column 2 we see that the failure to discover oil tends to enhance the likelihood of a type II site which has increased, with a type I site probability decreasing somewhat and and type III site hardly dropping at all.

EXERCISES

In the following exercises carefully identify the events B_i and the subsequent event A. Describe each symbol in words to clarify them in your mind. Then from the information given fill in a table similar to Tables 2.4 through 2.6 to help summarize your calculations.

1. The federal government discovered serious contaminants in the tuna fish packed by Seaside Packing, requiring a recall of the suspected cases of tuna. Before the discovery was made they had produced a total of 10,000 cases of which 7000 came from an early run and the remaining 3000 came from a later run. It is believed that 30% of the cans in the early run have been contaminated, and 60% of the cans in the later run have been contaminated. A laboratory has a case of unlabeled tuna that is to be inspected for the contaminant, and it is impossible to determine which run the case came from. A can is chosen randomly from a case to be inspected. What is the probability that the can will be contaminated?

2. Continuing with the conditions of exercise 1, suppose the can is found to be contaminated. What is the probability the case came from the early run? What is the probability the case came from the later run?

3. Suppose another can from the case under investigation in exercise 1 is tested and found to be contaminated. (A total of two contaminated cans have been found.) Use the "revised" probabilities" of exercise 2 as the new "prior" probabilities for this problem to determine a new set of "revised" probabilities that the case came from the early run or the later run.

4. A psychologist conducted an experiment with a 5-year-old child. The child is shown three indistinguishable boxes and told that the contents are as follows (assume the child can't see the contents of a specific box):

	Box 1	Box 2	Box 3
Marbles	2	2	4
Hard Candy	3	3	2
Candy Coated Bubble Gum	5	5	1

    The child is told to select one box and select one item out of the box, after which the child is allowed to repeat the process (with replacement). The question of interest is to see which box the child selects the second time after observing the outcome of the first selection. Note that boxes 1 and 2 have identical contents. Suppose the child obtains a piece of candy. Calculate the Bayesian "revised" probabilities for each box, given that the selection resulted in a piece of hard candy. Compare to the prior probabilities. What decision would you make? Would you select from the same box or choose another? (**Note:** the revised probabilities you compute serve as a the new "priors" for the second selection.)

5. Consider the example given at the beginning of this section wherein a statistics instructor keeps three different boxes for demonstration purposes. Box B_1 contains 20 red and 30 black balls; box B_2 contains 30 red and 20 black balls, and box B_3 contains 40 red and 10 black balls. During the evening, as the custodian cleans the instructor's office, one of the boxes is bumped and a ball is lost. Later the custodian finds the lost ball which turns out to be black, but the three boxes appear identical and the custodian can't remember which box it was that was bumped. He writes a note explaining what happened and puts it and the black ball on the instructors desk. The instructor, finding the note the next day, hasn't time to check each box and so decides to present the problem to the class as a probability problem and asks: "What is the probability that the black ball came from box B_1? Box B_2? Box B_3?" Assume that box 1 and box 2 are on the edge of the table and are considered to have equal probability of being bumped but box 3 is farther from the edge and has only half the probability of being bumped as box 1 and 2.

6. Supposed two black balls are knocked to the floor in the situation described in exercise 5. What are the "revised" probabilities for this case?

7. A large university houses guests and potential donors at one of three local motels. Over the last few years 60% of guests have been assigned to Sleepway Village, 30% to Roadway Motel, and 10% to Village Green Motor Inn. After the visit, the guests are asked to fill out a short questionnaire about the comfort of their stay, responding to questions about food service and quality, friendliness of motel personnel, etc. Of those guests staying at the Sleepway Village, 80% give it a composite rating of 10 or above, while only 60% of guests at Roadway Motel, and 50% of guests at Village Green Motor Inn score them with at least a 10.

    University personnel received a card from a guest who scored the motel a 6 on the composite score, and then proceeded to describe a very unpleasant experience that they had at the motel. They obliterated their name and failed to indicate which motel it was. On the basis of the information given what are the Bayesian probabilities for each motel, and which motel manager do you recommend calling to discuss the alleged incident?

8. The number of persons having the HIV virus (AIDS) is unknown and the only way to estimate it unbiasedly is to take a random sample of people and test them for the virus. Suppose the best current estimate of the number of persons having AIDS is 6 out of a thousand, i.e., $P(AIDS) = 0.006$. In addition, the test of AIDS can produce errors—a small percentage of the time it will declare non-AIDS carriers to be carriers and a small percentage of the time it will fail to detect true carriers of the virus. Suppose a "+" denotes the test is positive for AIDS and $P(+ | AIDS) = 0.999$ and $P(- | AIDS') = 0.99$. Use Bayes' Theorem to calculate the probability of a person being a true carrier of the AIDS virus given that the test is positive, i.e., $P(AIDS | +)$.

9. Suppose a second independent test of each person is performed in the situation described in exercise 8. Assume the probabilities of testing positive are the same as those given in exercise 8. Suppose a person tests positive on both tests. What is the probability a person has AIDS given two positive tests? That is, what is $P(AIDS | ++)$? Compare your answer to that obtained in exercise 8.

10. Suppose we have 3 identical cards in shape and size. However, one card is black on both sides, one card is red on both sides, and the third card is black on one side and red on the other. The 3 cards are mixed in a hat and 1 card is randomly selected and placed on a table. If the upper side of the chosen card is black, what is the probability that the other side is red?

2.6   ANALYSIS OF THE CHAPTER EXERCISE

In a manufacturing process it has been determined from experience that, as each new shift begins work, the process machinery is either producing at the 1% defective level, an acceptable level; at the 3% defective level, with minor adjustments needed; or at the 5% defective level with major adjustment required. Historically, it has been found that 90% of the time as they start a shift, the process is at the 1% level, 9% of the time it will be at the 3% level, and only 1% of the time will it be at the 5% level.

At the beginning of each shift the last two products from the preceding shift are inspected for quality. Depending on the number of defectives found, an administrative decision needs to be made concerning the defective level of the whole process.

(a)   If two defectives are found, what is the probability that the process is at the 1% defective level? the 3% defective level? the 5% defective level?

(b)   If only one defective is found, what is the probability that the process is at the 1% defective level? the 3% defective level? the 5% defective level?

SOLUTION: The set-up of this problem is similar to Example 1. The difference is that an intermediate calculation of the probabilities $P(A \mid B_i)$ is necessary, the probabilities not being immediately obvious. We will carefully do this for the case $P(A|B_1)$ and leave you to do the others, following the method presented.

(a)   First recall that we will let the events B_i represent the conditions that the machinery is operating at the 1%, 3% and 5% level, and the event A is that 2 defectives are encountered from the preceding shift.

$P(A \mid B_1) = P(2 \text{ defectives} \mid \text{ the process produces 1% defective}) = P(D_1 \& D_2 \mid 1\% \text{ defectives})$

$= P(D_1) \cdot P(D_2 \mid D_1) = (0.01)(0.01) = 0.0001$

where D_i is the event that "*the ith item selected in the sample is defective.*" Similarly,

$P(A \mid B_2) = (0.03)(0.03) = 0.0009$

$P(A \mid B_3) = (0.05)(0.05) = 0.0025$

The calculations can be summarized in the following table when observing 2 defectives in the sample of two items.

Column #	Prior Event	**Prior Prob.**	Conditional Probability	Joint Probability	**Post. Prob.**
	1	**2**	3	4 = 2x3	5
	B_i	$P(B_i)$	$P(A \mid B_i)$	$P(B_i)P(A \mid B_i)$	$P(B_i \mid A)$
	1% Defective	0.90	0.0001	0.000090	0.459
	3% Defective	0.09	0.0009	0.000081	0.413
	5% Defective	0.01	0.0025	0.000025	0.128
		1.00		0.000196	**1.000**

Again compare the post (revised) probabilities with the prior probabilities to see how the sample changed the probabilities. That is, observing two defectives made it more likely that the process is at the 3% or 5% level than at the 1% level.

(b)   If the sample were to produce only 1 defective item in the two sampled we would evaluate the probabilities in the following way:

Let G denote a good item, and B denote a bad item with a subscript to indicate whether it is obtained on the first selection or second. That is,

$P(A \mid B_1) = P(1 \text{ item is defective \& one is not defective } | the \text{ process produces } 1\% \text{ defective})$

$\qquad = P(G_1B_2 \text{ or } B_1G_2) = P(G_1B_2) + P(B_1G_2)$

$\qquad = P(G_1) \cdot P(B_2 \mid G_1) + P(B_1) \cdot P(G_2 \mid B_1) = (0.01)(0.99) + (0.99)(0.01)$

$\qquad = 2(0.01)(0.99) = 0.0198$

Similarly,

$$P(A \mid B_2) = 2(0.03)(0.97) = 0.0582$$

$$P(A \mid B_3) = 2(0.05)(0.95) = 0.0950$$

The calculations can be summarized in the following table for the case of observing 1 defective in the sample of two items.

	Prior Event	Prior Prob.	Conditional Probability	Joint Probability	Post. Prob.
Column #	1	2	3	4 = 2x3	5
	B_i	$P(B_i)$	$P(A \mid B_i)$	$P(B_i)P(A \mid B_i)$	$P(B_i \mid A)$
1% Defective		0.90	0.0198	0.017820	0.742
3% Defective		0.09	0.0582	0.005238	0.218
5% Defective		0.01	0.0950	0.000950	0.040
		1.00		0.024008	**1.000**

Again comparing the revised probabilities with the prior probabilities you see that observing only 1 defective does not revise the probabilities toward the 3% and 5% levels nearly as much as observing 2 defectives.

The same line of reasoning used in this exercise could be extended to sampling 20 items or 100 items. The application of Bayes' theorem would follow along just as we have shown it, it would simply require more extensive calculations to find $P(A \mid B_i)$.

CHAPTER SUMMARY

In this chapter you have a very brief exposure to some of the basic principles and theorems of probability. You should be aware that some textbooks are often devoted solely to the study of this subject and there is much more for you to learn if you have the interest. However, for the purposes of this text, the principles you have learned here will allow you to perform the probability tasks encountered in the following chapters with the level of understanding expected.

In review, you should remember the use of the "addition rule for the union of two events, A and B." That is:

$$P(A \ or \ B) = P(A) + P(B) - P(A \ and \ B)$$

This relationship simplifies when A and B are mutually exclusive, and the result when A and B are complementary is also useful for simplifying problems.

Next, you encountered the use of the "multiplication rule for the intersection of events, A and B." That is:

$$P(A \ \& \ B) = P(A) \cdot P(B \mid A)$$

This relationship derived from the basic definition of conditional probability. And when A and B are assumed to be independent, the probability of the intersection becomes the product of the separate probabilities. That is:

$$P(A \ \& \ B) = P(A) \cdot P(B)$$

Lastly, Bayes' Theorem was presented along with examples showing how to apply it.

CHAPTER EXERCISES

In the following exercises you will have to apply many of the preceding definitions and theorems to obtain a solution. Work methodically, carefully defining events in words and phrases, identifying events that constitute the union or intersection of other events. Once you obtain a solution, stop and examine your answer for consistency with other results and for any insights it gives about the more meaningful issues involved in the problem.

1. It has been observed historically that the proportion of defective items in a manufacturing process is about 0.5% (0.005). Three items are randomly selected from the production process and examined for the defect in question. In advance of the examination, what is the likelihood of finding no defects? one defect? two defects? all three with defects?

2. Engineers have estimated that on any given space shuttle launching the chances of having a lift-off without unexpected delays is about 0.95. What is the probability that in the next 5 lift-offs, there will be no unexpected delays?

    Suppose that the next 2 go off without a hitch? What is the probability that the next 3 will not suffer any delays? What assumption do you make?

3. A large university has found that about 80% of applying Freshmen meet admissions requirements. Of the 80% meeting admissions requirements, 10% will not show up in the fall. What is the probability that a randomly selected admissions application will be associated with a student who meets requirements and actually will show up on campus the following fall? (Thus with 2500 applications being submitted, what is the approximate number that you would expect to show up on campus the following fall?)

4. A large manufacturer of electronic calculators requires a specific integrated circuit in its configuration. To meet their own demand they order parts from 3 different semi-conductor firms, 50% coming from firm A, 30% from firm B and 20% coming from firm C. From experience it has been found that the probability of a defective circuit from firm A is 0.002, from firm B is 0.005, and from firm C is 0.006. What proportion of integrated circuits will be found to be defective? If a calculator fails due to a failure of this specific circuit, what is the probability the circuit came from firm A? firm B? firm C?

5. A state inspector regularly visits grocery stores to take samples of their dairy products. Suppose that of 20 quarts in the dairy case, 2 are sour. He chooses 2 to test. What is the chance of getting neither sour quart? one of the sour quarts? both sour quarts?

6. Under the conditions of exercise 5, it has been found that 90% of all milk cartons are over a day old. Of the 2 cartons of milk that are sour in exercise 5 above, suppose both are over a day old. The inspector decides on a new strategy of first inspecting the milk for dates, and then takes 1 carton at random from the cartons over a day old and the second carton from the fresh milk. What is the probability of getting at least 1 carton of sour milk?

7. Suppose a population in a small community of 500 registered voters is evenly split between Mayoral Candidate Smoothly and Mayoral Candidate Cleverly. A very small random sample of 6 registered voters is taken from the list of 500 registered voters. (Note that the sampling is without replacement.) What is the probability that all 6 voters report themselves as favoring Candidate Smoothly? [Questions like this arise in many situations as we deal with larger populations and a larger sample size and ask "How likely is it that the sample will deviate in some fashion from the population, and provide a misleading conclusion about who is ahead?"]

8. Suppose the population described in exercise 7 is sampled with replacement. What is the probability that all 6 will report themselves as favoring Candidate Smoothly? Compare the difference of this number with the answer obtained in exercise 7. Would this serve as an acceptable "approximation" to the answer of exercise 7?

9. Suppose that each child born to a couple is equally likely to be a boy or a girl. For a family having 5 children, compute the probability of the following events.
    (a) All children are of the same sex.
    (b) The 3 eldest are boys and the others are girls.
    (c) Exactly 3 boys.
    (d) At least 3 boys.

10. Suppose that there is a cancer diagnostic test that is 90 percent accurate both on those that do and do not have the disease. If the proportion in the population having this particular kind of cancer is 0.005, compute the probability that a tested person has cancer, given that the test result indicates cancer.

11. Consider two boxes having red and black balls in them. Box 1 has 2 red and 8 black and Box 2 has 7 red and 3 black. A ball is randomly selected from Box 1 and placed in Box 2. Then a ball is randomly selected from Box 2. What is the probability that the ball selected from Box 1 is red and the ball selected from Box 2 is red?

12. For the boxes described in exercise 11, what is the probability that the ball selected from Box 1 is black and the ball selected from Box 2 is red?

13. Using the results of exercises 11 and 12 what is the probability that a red ball will be selected from Box 2?

14. Suppose *two* balls are selected from Box 2 in the problem description given in exercise 11. What is the probability that both are red? one is red and one is black? both are black?

APPENDIX A—SOME BASIC COUNTING RULES

Introduction

In solving many probability problems, we assume the outcomes are equally likely to occur. Such is the case for rolling a balanced die, tossing a balanced coin, selecting a sample randomly, and so forth. Under the assumption of equal likelihood, evaluating probabilities is essentially a counting problem. If the sample space consists of N equally likely outcomes and n of them have a particular characteristic in common, then the probability of that characteristic is simply $\frac{n}{N}$, a result obtained by counting the total possibilities and then counting how many of those possibilities have the common characteristic. For instance, a balanced die has six sides, three of which are even numbers, then the probability of an even number is $\frac{3}{6} = \frac{1}{2}$. (This implies that the relative frequency of even numbers to total rolls of the die should settle down to $\frac{1}{2}$.) Similarly, if a population consists of 4000 females and 6000 males, then a randomly selected person has a 40% chance of being female.

In these two illustrations, the counting process was fairly simple. However, if we roll the die 4 times or sample 400 persons, questions of probability using the counting approach are much more difficult to treat. Therefore, in such situations, it is useful to be familiar with some of the basic counting rules that we rely on in these contexts.

This section presents some of these rules. We limit our discussion to those which will have specific application later in the course and we will keep things as simple as possible.

Definitions

DEFINITION 2.11: The *fundamental rule of counting* states that if task 1 can be performed in k_1 different possible ways, task 2 can be performed in k_2, different possible ways, ..., and task m can be performed in k_m different possible ways, then the sequence of the tasks could be performed in any of $k_1 \cdot k_2 \cdot ... \cdot k_m$ different possible ways.

DEFINITION 2.12: A *permutation* of n objects is a specific ordering or arrangement of those n items.

THEOREM 2.3: The number of permutations of n items of which n_1 are identical and of one kind, n_2 which are identical of another kind, ..., and n_k which are identical and of the kth kind, is

$$\frac{n!}{(n_1)!\bullet(n_2)!\bullet\ldots\bullet(n_k)!}$$

THEOREM 2.4: The total number of permutations of *r* items selected without replacement from *n* items, denoted by $_nP_r$, is evaluated by

$$_nP_r = n(n-1)(n-2)\ldots(n-r+1) = \frac{n!}{(n-r)!}$$

DEFINITION 2.13: A *combination* is a list or cluster of items considered as a group, without regard to order.

THEOREM 2.5: The number of combinations that could be formed by selecting *r* items without replacement from a group of *n* items is denoted by $\binom{n}{r}$ and evaluated as:

$$\binom{n}{r} = \frac{n!}{r!(n-r)!}$$

DISCUSSION

Fundamental Rule of Counting

All the results presented in this section are derived from the fundamental rule of counting. However, our purpose here is not to prove the results but to illustrate their use and application. Therefore, this section will mainly consist of examples.

First, consider two simple situations that use the fundamental rule of counting:

1.   A penny, nickel, and dime are tossed. How many different ways can they turn up?
       Let each coin tossed represent a different task. Therefore task 1, tossing the penny,
       has 2 possible outcomes. Task 2, tossing the nickel, has 2 possible outcomes. And
       task 3, tossing the dime, has 2 possible outcomes. Therefore the sequence of 3
       tasks has $2\bullet2\bullet2 = 8$ possible outcomes.   (You might list the sample space to identify
       the 8 possible outcomes.   Your list would look like: $S = \{HHH, HHT, ..., TTT\}$.)

2.   Suppose we have 10 people from which we select 4 at random to serve as a
       chairman, 1st assistant, 2nd assistant and secretary for a special subcommittee.
       How many different committee arrangements are possible? In this situation, let
       each person selected for a position be a task. Therefore task 1 has 10 possibilities,
       task 2 has 9, task 3 has 8 and task 4 has 7. Therefore, there is a total of $10\bullet9\bullet8\bullet7 =$
       5040 different committee arrangements possible.

Permutations

Suppose in a taste testing experiment comparing 3 different cola drinks a researcher wants to know how many different arrangements of the three cups can be made. In this case there are $n = 3$ items, and we are selecting all three, so that $r = 3$. Thus there are:

$$_3P_3 = \frac{3!}{(3-3)!} = 6 \text{ possible arrangements.}$$

Note that in order to provide a sensible result we accept 0! to equal 1.

As another example suppose that a ward in a hospital consisting of 6 beds has 2 cases where the patient has broken bones, 1 appendectomy, and 3 cases of pneumonia. If the patients are assigned to the beds in a random fashion, how many arrangements or permutations of the different illness types are possible?

In this case we are permuting 3 different sets of indistinguishable objects, 2 broken bone cases, 1 appendectomy, and 3 cases of pneumonia. This is done in 60 different ways since

$$\frac{6!}{2! \cdot 1! \cdot 3!} = 60$$

Combinations

Suppose we are sampling randomly from some population having 10,000 elements in it. If we were to try to list all of the possible samples of size r that could be taken, how many would we have to list? This is a particular application of the combination rule of counting where we need to evaluate

$$\binom{n}{r} = \frac{n!}{r!(n-r)!}$$

Specifically, if $n = 10,000$ and $r = 400$, then the total number of samples of size 400 are

$$\frac{(10,000!)}{[(10,000-400)! \cdot 400!]} = \frac{(10,000!)}{[(600)! \cdot 400!]},$$

a very large number.

EXAMPLES

EXAMPLE 1: A student takes a 5-question, true-false test. The test is not handed back but she is told that she missed 2. She wonders which questions she got right and which ones she got wrong. How many different possibilities should she consider?

SOLUTION: There are a total of $n = 5$ answers on the test of which 3 are correct and 2 are wrong. Therefore, there is a total of

$$\frac{5!}{[3!2!]} = 10$$

different orders of right and wrong arrangements to consider. (The permutations would be listed as: *RRRWW, RRWRW, RRWWR,* etc.)

EXAMPLE 2: A sociologist is studying birth order effects in families of size 5. If the sex of the children is to be taken into account, how many arrangements of 1 male and 4 females are possible in the various birth order positions?

SOLUTION: In this case, we have $n = 5$ children of which 4 females are of one kind and 1 male is of another kind. Therefore, there is a total of $\dfrac{5!}{4!1!} = 5$ different orderings when the sex is considered. They could be listed as:

$$\{MFFFF, FMFFF, FFMFF, FFFMF, FFFFM\}$$

EXAMPLE 3 Suppose out of a class of 8 graduate students, 4 are to be randomly selected to give oral presentations. Of the 8 graduate students 3 are non-English-speaking students. What is the chance that of the 4 selected all will be English-speaking?

SOLUTION: The solution to this problem will illustrate both the rules of counting when selecting various combinations, as well as how these rules of counting allow us to calculate probabilities under an assumption of equally likely selection, which the assumption of "random selection" implies.

First, the total number of possible student combinations (possible samples) that could occur is:

$$\frac{8!}{[4!4!]} = 70.$$

Of these 70 combinations we want to know how many of them consist of 4 English-speaking and 0 non-English-speaking students. To count this number we consider two tasks:

1. Select 4 English-speaking students from the 5 in the group.
    This can be done in

$$\frac{5!}{[4!1!]} = 5 \text{ ways.}$$

2. Select 0 of the non-English-speaking students from the 3.
    This can be done in

$$\frac{3!}{[3!0!]} = 1 \text{ ways.}$$

Performing these two tasks in sequence will produce a particular group of 4 English-speaking and 0 non-English-speaking students. Therefore, there is a total of $5 \cdot 1 = 5$ ways this sequence of two tasks could be performed.

The probability of such an outcome resulting by random selection is, therefore,

$$\frac{5}{70} = \frac{1}{14},$$

or 1 chance in 14.

EXERCISES

1. Evaluate the expressions $\binom{n}{n}$, $\binom{n}{n-1}$, $\binom{n}{n-2}$, $\binom{n}{2}$, $\binom{n}{1}$, and $\binom{n}{0}$. What does this imply about the combinatorial expressions, in general?

2. On a particular exam containing 10 questions students could either get each question correct, partially correct or incorrect. In how many different ways could a student get 5 correct, 2 partially correct and 3 incorrect answers?

3. A taste test is being done on three different colas. A test subject is to be given a cup of each cola in one of three different colored cups. How many different ways could the colas be put in the different colored cups?

4. A license plate can accommodate 6 characters. They can be either numerical or alphabetical. How many unique license plates could be manufactured if there where to be 4 numerical and 2 alphabetical characters? Three numerical and 3 alphabetical?

5. In assigning telephone numbers to new hookups the last 4 digits are to be filled with a new common prefix to be used. Use the counting rules to determine the total number of phone numbers that could be allocated to this "bank" of numbers?

6. From a group of 10 people, 5 are to be chosen randomly to serve as a "control" group in an experiment on learning how to use a word processor. How many different combinations are there for the control group?

7. Suppose 5 of these 10 people described in exercise 6 have good typing skills. How many different ways are there for all 5 of them to get selected for the control group?

8. What is the probability of getting all 5 skilled typists in the control group as described in exercises 6 and 7?

9. What is the probability of getting 4 skilled and 1 unskilled typist in the control group described in exercises 6 and 7? (The general pattern of probability represented here is called the hypergeometric probability distribution.) This distribution is discussed in more detail in Chapter 4 and has the expression:

$$\frac{\binom{k}{r} \cdot \binom{N-k}{n-r}}{\binom{N}{n}}$$

where

N = the total group size,
n = sample size,
k = total number in the subgroup, and
r = number in the subgroup selected.

When applied to our problem, $N = 10$ persons, $n = 5$ persons selected, $k = 5$ skilled and $r = 4$, the number of skilled typists in the control group.)

10. A forest contains 20 moose, of which 7 are captured, tagged and then released again. Some time later, 4 of the 20 moose are captured. What is the probability that 1 of the 4 are tagged? Apply the formula of exercise 9 to solve this problem.

11. A room of applicants for a position contains 20 persons. Four of them are randomly selected to take an aptitude test. Of the 20 applicants, 10 are humanities majors, 7 are physical science majors and 3 are biological science majors. What is the probability that 2 of the 4 selected for the aptitude test are physical science majors? [Hint: Use the formula of exercise 9.]

12. Using the information given in exercise 11, what is the probability that you will end up with 2 humanities majors, 1 physical science major and 1 biological science major among the four chosen for the aptitude test?

13. Suppose a box of marbles contains N elements of which k are red, m are black and $N-k-m$ are white. A random sample of n marbles is selected from the box. Write a general expression similar in form to that given in exercise 9 to calculate the probability of getting r red, s black and $n-r-s$ white marbles in the sample.

14. Apply the formula of exercise 13 to solve the problem proposed in exercise 12.

15. Suppose Box 1 contains a red and black ball; Box 2 contains a blue and white ball. A ball is selected from Box 1 and placed in Box 2 and then a ball is selected from Box 2. What is the probability of a red ball being selected from Box 2?

3

RANDOM VARIABLES, PROBABILITY FUNCTIONS, AND DISTRIBUTION FUNCTIONS

OBJECTIVES

Upon completing this chapter you should understand the concept of a probability model, and be able to define a random variable on that probability model. You should know the difference between a continuous and discrete random variable, and know how to verify that their probability functions are legitimate probability functions. For simple continuous and discrete models, you should be able to compute probabilities, the mean, the variance and other measures of location, variability and skewness.

You should be able to sketch the graph of simple probability functions and their distribution functions. You should be able to: (1) plot the joint probability function, $f(x,y)$; (2) obtain the marginal distributions; (3) find conditional distributions along with their means and variances; and (4) find the covariance and correlation coefficient for the joint distribution. For the multivariate case, you should be able to find the joint distribution for independent and identically distributed random variables, and find the mean and variance of a linear combination of a sequence of random variables.

INTRODUCTION

Probability Modeling

The challenge many scientists face as they study various types of phenomena is to find a fairly simple mathematical model or expression that explains the various properties under investigation whether they be physical, chemical, electrical, biological or whatever. If an adequate mathematical model can be found, then you can express the various inter-relationships present in the system under investigation using the principles of mathematical analysis, or using simulation on a computer.

For instance, the physical relationship between force, mass, and acceleration of a body has been expressed by the equation, $F = m \cdot a$. Thus the force of an object of constant mass, can be expected to increase in a linear fashion as the acceleration of the body increases. This simple conclusion is readily apparent by examining the above equation. Mathematical analysis reveals this quickly and easily. If the model is accurate, these conclusions are much more simply obtained than the expensive laboratory experimentation required in the absence of such a model.

Similarly, if a population is to be studied by taking a sample from it and random selection is to be used to select the units in the sample, then we can better understand the whole sampling process if we can write an expression that models the probabilities of getting certain kinds of sample outcomes.

For a very simple example, suppose a person is to be selected at random from a room containing 10 men and 5 women. If we let $Y = 1$ if a woman is chosen, and $Y = 0$ if a man is chosen, then we can express the probability model for this sampling process by the expression:

$$f(y) = (\frac{5}{15})^{y}(\frac{10}{15})^{1-y} \quad y = 0,1$$

In this case, if $Y = 1$ (a woman is selected), then $f(y) = \frac{5}{15} = \frac{1}{3}$, the probability of selecting a woman. If $Y = 0$, (a man is selected), then $f(y) = \frac{10}{15} = \frac{2}{3}$. Thus this probability "model" describes the probability properties of the sampling experiment defined on the population.

In this chapter we present the notation, terminology, and theory for developing probability models. In we do so, you should make note of the parallel between the concepts and terminology defined here with those introduced in chapter one.

Where We Are Going in This Chapter

In Chapter 1 we presented both graphical and numerical methods of summarizing data, regardless of whether the data were the entire population, or merely a sample of a population. We also introduced the concept of a *distribution* of a data set as we referred to a histogram and relative frequency table. (In the two-dimensional scatter-plot, we also presented the idea of a bivariate distribution or joint distribution of a set of bivariate data.) In Chapter 2, the relationship between probability and relative frequency was made so that we could then treat the probability characteristics of a data set. (Probability enters the picture as we take a random sample from a population, or as we assume that a random process has generated the data to be analyzed.)

This chapter formalizes the concept of a probability description of a set of data. Figure 3.1 shows the Statistical-Inference Model (SIM) for this chapter. We define a probability density function (pdf) $f(x;\omega)$ for the population that describes the probability characteristics of getting sample values x when the sample is chosen randomly. This probability model is a function of the measurement process, X, and certain population parameters, ω. For the present we are concerned strictly with the process of predicting the probability of obtaining a certain sample, given specific assumptions and characteristics of the population and assuming a random sampling process as controlled by the probability density function $f(x;\omega)$.

As you go through this chapter, you should particularly note the similarity between Chapter 1 ideas and related ideas presented in this chapter. For example: (a) the *histogram* (Chapter 1) and a *probability distribution* (Chapter 3) are similar concepts, as are (b) the *cumulative relative frequency distribution* (Chapter 1) and the *theoretical cumulative distribution function* (Chapter 3), and (c) the *empirical moments* (Chapter 1) and the *theoretical moments*. (Chapter 3). Other parallels will be noted as they arise.

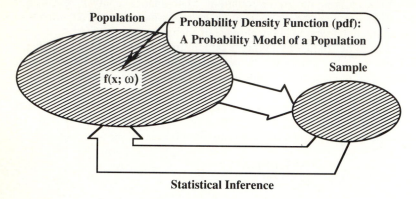

Figure 3.1 The Statistical-Inferential Model—Chapter 3.

Chapter Exercise

Suppose the price P of a certain commodity (in dollars), and S, the total sales (in 10,000 units), are random variables whose distribution can be approximated with the joint probability density function:

$$f(p,s) = \begin{cases} 5pe^{-ps} & \text{for } 0.2 < p < 0.4 \text{ and } s > 0 \\ 0 & \text{elsewhere} \end{cases}$$

Find:

     (a)   The marginal distribution of p, its mean, variance and 1st and 3rd quartiles.

     (b)   The effect of price on sales. Predict sales for a given price.

     (c)   The probability that sales will exceed 20,000 units if the price is $0.35.

3.1   THE PROBABILITY DENSITY FUNCTION AND THE CUMULATIVE DISTRIBUTION FUNCTION

Introduction

Suppose a sample is drawn from a population under investigation. From each element in the sample a series of quantitative measurements are taken. If X denotes one of the measurements made, then X is called a random variable. If the same measurement were to be made on each and every element in the population, then the chance of obtaining a particular set of measurement values in the sample selected becomes important to us. A probability model is desired to enable us to predict the probability of a set of sample outcomes occurring. This probability model will be denoted by $f(x)$ where x is associated with the measurement referred to above.

For instance, suppose we want to study the potential drug abuse problem in a population of teenagers. We decide to collect a variety of information, one of which is the age, X, at which a particular teenager first used a particular drug. Then, $f(x)$ would be the probability model that would allow us to predict the probability of getting a certain distribution of ages in a sample of 100 teenagers selected from the population of all teenagers under investigation.

In this section we introduce the formal concept of a mathematical-probability model, $f(x)$. This mathematical function describes the probability properties of a measurement process X, which we call a random variable. In addition, we present some of the logical properties associated with the function $f(x)$, under the restrictions imposed by the probability postulates of chapter 2. We will do this for both continuous and discrete measurement processes. The logical similarities to histograms, and other graphical methods will be noted. You should refer back to chapter 1 and note the concepts and terminology used there to help better understand the concepts and terminology of this chapter.

Definitions

DEFINITION 3.1: A *random variable (RV)*, X, is a real-valued function defined on a sample space, S, such that X transforms each outcome in S into a point on the real line.

DEFINITION 3.2: A random variable X, is said to be *discrete* if the values it can assume is countable (either finite or infinite), that is, if the values that the measurement process produces can be arranged in a sequence corresponding, one-to-one, to the set of positive integers.

DEFINITION 3.3: A random variable X, is said to be *continuous* if the possible values it can take on consist of one or more intervals on the real line.

DEFINITION 3.4: The *probability density function* of a discrete random variable X, is obtained when we associate with each possible numerical value x of X, its corresponding probability. For notation we will write:

$$P(X = x) = f(x).$$

That is, $f(x)$ will symbolically stand for the probability density function, (pdf), of the random variable X and will mean "the probability that X takes on the value x". The probability density function will have the following properties:

1. $f(x) \geq 0$ for all values x of X;

2. $\sum f(x) = 1$

DEFINITION 3.5: The *cumulative distribution function* of the discrete random variable X, denoted by $F(x)$ is defined as the probability that X will take on a value less than or equal to x; i.e.,

$$F(x) = P(X \leq x) = \sum f(t) \text{ for all } t \leq x$$

(The cumulative distribution function is so named because as we move from left to right on the horizontal axis, $F(x)$, the probability of being less than or equal to x, accumulates as x gets larger. The cumulative distribution function is often simply called the "distribution function" for short.)

DEFINITION 3.6: The probability density function (pdf) of a *continuous random variable* X, is a continuous function $f(x)$ with the following properties:

1. $f(x) \geq 0$ for all x such that $\infty < x \leq +\infty$.

2. $\displaystyle\int_{-\infty}^{\infty} f(x)dx = 1$

3. $P(a < x \leq b) = \displaystyle\int_{a}^{b} f(x)dx$ for any $a \leq b$.

DEFINITION 3.7: The *cumulative distribution function $F(x)$, of the continuous random variable X*, or simply the distribution function is:

$$F(x) = P(X \leq x) = \int_{-\infty}^{x} f(t)dt$$

which has the following properties:

1. $F(x)$ will be a smooth non-decreasing function of x
2. $F(-\infty) = 0$
3. $F(+\infty) = 1$
4. $P(a < X \leq b) = F(b) - F(a)$
5. $\dfrac{dF(x)}{dx} = f(x)$

DISCUSSION

Random Variables

Whenever data are collected to be analyzed, it is a natural desire to quantify the measurement process so that the usual arithmetic and algebraic processes can be used upon them. However, not all data are quantitative by nature. Thus some system of quantification is desired to deal with such cases. In addition, it may be important to look at some other transformation of a measurement than that originally collected. To accommodate such a desire, the concept of a random variable is introduced. A formal definition of a random variable is given in the Definitions above. More simplistically, a random variable, usually denoted by X, Y, Z, etc., is a "rule" that, when applied, quantifies a measurement made on a unit of observation. This can be a very straight forward process when measuring a person's height, weight, number of brothers and sisters, and age. In these cases the measurement process produces a numerical value naturally. Thus the rule and the measurement process are the same. This will typically be the case for measurements producing interval or ratio measurements. However, no such natural numerical value is produced when measuring a person's race, political affiliation, marital status, and so forth.

For ordinal data, the random variable generally describes a rule that associates ranks or ordered values with various measurement values.   For nominal data, the random variable is the rule that "codes" the nominal categories into numerical form. For instance, suppose a survey questionnaire asks, among other things, about a person's marital status, their height in inches, weight in kilograms, and the number of brothers and sisters they have. For the question on marital status, we could define the random variable (rule), Y, in the following fashion:

$$Y(\text{single}) = 1$$
$$Y(\text{married}) = 2$$
$$Y(\text{widowed}) = 3$$
$$Y(\text{divorced}) = 4.$$

Note that this is simply an arbitrary coding convention. Other coded values could have been used instead.

For measurements which are quantitative by nature, the random variable might be to define specifically the measurement device to be used in the following way, $X(u) = u$. For instance, suppose we let the random variable V be associated with the process of measuring a person's height in inches, X a person's weight in kilograms, and Z the number of brothers and sisters a person has. Thus for a specific person who reports that they are widowed, 72 inches tall, 82 kilograms with 2 brothers and sisters, we get:

$$Y(\text{widowed}) = 3$$
$$V(72) = 72 \text{ inches,}$$
$$X(82) = 82 \text{ kilograms,}$$
$$Z(2) = 2.$$

In this case, the reported values are assigned to be the values the random variables take on. However, we might have defined $X(u) = u^2$, that is we square the measurement u once we have obtained it. Or we might let $Z(t) = \dfrac{(t - \mu)}{\sigma}$, called a Z–score transformation. There are many possibilities to choose from.

Discrete Vs. Continuous Random Variables

The next distinction we make is the contrast between discrete and continuous random variables. To distinguish between discrete and continuous random variables we apply the same principles here as those covered in chapter one. In chapter one, we contrasted discrete and continuous *measurements*. Here we use the phrase random variables instead of measurements but the concept is the same. A measurement process that is continuous produces a random variable that is continuous; a measurement process that is discrete produces a random variable that is discrete. Therefore, if you were successful in distinguishing between a continuous and discrete measurement in chapter one, you can distinguish between a continuous and discrete random variable in this chapter.

For instance, if the random variable Y refers to the rule to "record" the family size of a household, then Y is a discrete random variable because it takes on the discrete values, 1,2,3,4, and so forth.   If the random variable Z refers to the time it takes to process a computer job, then the random variable Z is continuous because it can take on any value in the interval $[0,\infty]$. (The computer job could get into an "infinite loop.")

The Probability (Density) Function: Discrete Random Variables

A random variable, by definition, is connected to a sample space which will have some probability "measure" associated with it. Under such a situation, the random variable will also have a probability "measure" associated with it. The way the probability is assigned to the various values of the random variable is denoted by a function $f(x)$ called a *probability density function* or *pdf*. This connection is illustrated for a discrete random variable by Figure 3.2. In this figure, assume that the sample points e_2, e_6, and e_8 all map onto the same value x on the real line through the random variable X, and that they are the only values in the sample space that do so. Therefore, we can write:

$$f(x) = P(X = x) = P(e_2) + P(e_6) + P(e_8).$$

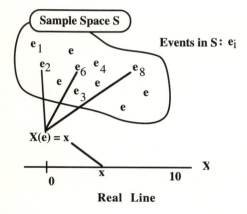

Figure 3.2 Mapping simple events in S to the real line.

If we continue this process for all the sample points in the sample space S, mapping them onto their respective values on the real line, we will have defined the complete probability density function, $f(x)$, which shows how the probability is distributed. Thus $f(x)$ is often also referred to as a probability distribution. In this case, it becomes obvious that:

1. $f(x) \geq 0$

2. $\displaystyle\sum_{all\ x_i} f(x_i) = 1$

If we display the probability density function graphically, it might appear as shown in Figure 3.3, (a) analogous to a bar graph or as (b) similar to a histogram. Note in Figure 3.3 (a) that probability is represented by a "spike" located at a specific value of the random variable and the height of the spike is equal to or proportional to the probability of getting that particular value of the random variable. In Figure 3.3 (b) probability is represented by a bar of width one and height equal to the probability so that "area" of the bar equals the "probability of occurrence."

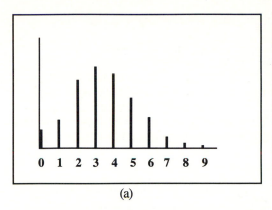 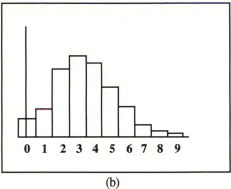

(a)   (b)

Figure 3.3 Graph of a discrete probability density function.

(Some authors will refer to $f(x)$ as simply a probability function whenever the random variable X is discrete. They will reserve the phrase "probability density function" for the continuous case. Since this difference in wording adds no crucial insight, we will adopt the simpler approach of calling $ff(x)$ the probability density function regardless of whether X is discrete or continuous.)

The cumulative distribution function $F(x)$ for the probability function given in Figure 3.3 might look like the graph of Figure 3.4.

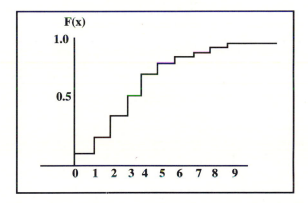

Figure 3.4 Graph of a cumulative distribution function. (See figure 1.2).

As an example, suppose that in an investigation of drunk driving arrests, the sex of the offender is noted. In the next five records, let X denote the number of females. In this case X can take on any integer value between 0 and 5. If we assume that the sex of the offender is as likely to be a male as a female then we can write,

$$f(x) = \begin{cases} \dfrac{5!}{x! \cdot (5-x)!} \left(\dfrac{1}{32}\right) & x = 0,1,2,3,4,5 \\ \\ 0 & \text{elsewhere} \end{cases}$$

You can verify these results by listing the sample space. The sample space will consist of 32 equally likely elements and would appear as:

$$S = \{MMMMM, \; MMMMF, \; MMMFM, \; ..., \; FFFFF\}$$

Though it might be time consuming to make the list, the elements of the sample space are easy to specify. The probability of all males ($X = 0$) or all females ($X = 5$) yields a probability of 1/32 each; of 4 males and 1 females ($X = 1$) or 1 male and 4 females ($X = 4$) is 5/32 each; etc. The probability density function $f(x)$ and cumulative distribution $F(x)$ are summarized in Table 3.1 below.

Table 3.1 The probability density function and the cumulative distribution function.

x	$f(x)$	$F(x)$
0	$\dfrac{1}{32}$	$\dfrac{1}{32}$
1	$\dfrac{5}{32}$	$\dfrac{6}{32}$
2	$\dfrac{10}{32}$	$\dfrac{16}{32}$
3	$\dfrac{10}{32}$	$\dfrac{26}{32}$
4	$\dfrac{5}{32}$	$\dfrac{31}{32}$
5	$\dfrac{1}{32}$	$\dfrac{32}{32}$

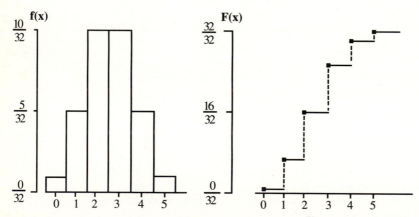

Figure 3.5 Graphs of $f(x)$, and $F(x)$ where $f(x) = \dfrac{5!}{x! \bullet (5 - x)!} (\dfrac{1}{32})$.

The graph of both $f(x)$ and $F(x)$ is shown in Figure 3.5. Note especially the scale of the vertical axes are different. $F(x)$ is also called a step function, and that although we have added vertical dotted lines to provide a more continuous looking appearance, they would not be included in a precise mathematical sense. Also note that in the graph of $f(x)$ that instead of drawing "spikes" showing the probability at each of the points from 0 to 5, rectangles are drawn, centered on the integer values involved, each interval being of width one. This feature guarantees that the area of each separate rectangle equals the probability.

The connection between area and probability, though a little contrived in this situation, is nonetheless a very important connection that is essential to make when discussing continuous random variables. We will more formally define this connection in the next paragraphs that follow.

The Probability (Density) Function: Continuous Random Variables

Consider a simple spinner used for a board game. Usually, the circumference is marked off into segments of different colors or with numbered segments.   When the spinner is spun, the color on which it stops or the number assigned to the segment in which it lands is noted. Suppose, instead, that we consider an idealized extension of this operation. Let the circumference be marked off in degrees from 0 to 360, and allow the spinner to conceptually fall at any position on the interval 0 – 360 considering the interval to represent a continuum of values, not just the discrete integer values from 0 to 360 degrees.

There are obviously an infinite number of positions upon which the spinner might land. The probability of it landing on any single position, specified in advance, would be assigned the probability of $\frac{1}{\infty} = 0$. However, the probability that it will land between 0 and 90 degrees would logically take on the value of $\frac{1}{4}$, since $\frac{1}{4}$ of the possible values it could take on are in that interval. If we think of the circle having a total area equal to one, then the area of the circle associated with the values of the circumference from 0 to 90 degrees would be 1/4. Thus, we can equate the concepts of area and probability. Specifically, for continuous random variables, we equate the area under the curve of $f(x)$ to the probability that the random variable will take on a value in that region or interval.

For this example, we could graph the probability density function as a rectangle of height $\frac{1}{360}$ and width equal to 360. Thus the area between 0 and 90 degrees would be $\frac{90}{360} = \frac{1}{4}$. If we let $f(x) = \frac{1}{360}$ for $0 \le x \le 360$, then the probability becomes equal to the integral of $f(x)$ over the range from 0 to 90. See Figure 3.6.

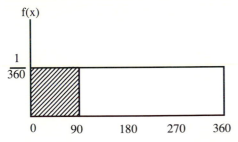

Figure 3.6 Graph of probability density function for $f(x) = \frac{1}{360}$. Shaded area is $P(0 < x \le 90)$.

That is,

$$P(0 < x \le 90) = \int_0^{90} \frac{1}{360}\, dx \; = \frac{x}{360}\Big|_0^{90} = \frac{90}{360} = \frac{1}{4}.$$

Generalizing from this example for a continuous random variable, first, we equate area under the continuous curve of $f(x)$ to probability and when doing so we equate probability to the integral over an appropriate range of values of X. Thus we may conclude that:

1. $$P(X = x) = \int_{x}^{x} f(x)dx = 0$$

2.   The function $f(x)$ always plots above or asymptotic to the horizontal axis in the usual x,y plane.

3. $P(a < X < b) = P(a \leq X < b) = P(a < X \leq b) = P(a \leq X \leq b)$ since the "area" at the point a or b is zero and equals:

$$P(a < X \leq b) = \int_{a}^{b} f(x)dx$$

A graph of a hypothetical pdf, $f(x)$ is displayed in Figure 3.7 below. The probability, $P(a < X \leq b)$, is indicated by the shaded area under the curve.

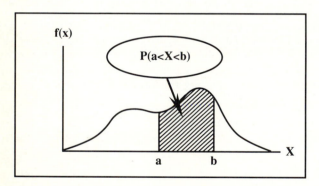

Figure 3.7 Graph of an arbitrary pdf: the shaded area is $P(a < x \leq b)$.

The cumulative distribution function, $F(x)$, is defined initially the same as for a discrete random variable, but will represent the area under the curve of $f(x)$ to the left of the specified value of x in $F(x)$. As the definition section states,

$$F(x) = P(X \leq x) = \int_{-\infty}^{x} f(t)dt$$

which by the nature of the continuity of the random variable and the probability properties it possesses, will have the following properties.

• $F(x)$ will be a smooth non-decreasing function of x

• $F(-\infty) = 0$ and $F(+\infty) = 1$

- $P(a < X < b) = F(b) - F(a)$

- $\dfrac{dF(x)}{dx} = f(x)$

A graph of an arbitrary distribution function $F(x)$ is given in Figure 3.8 below. Note that it approaches the value 1.0 asymptotically from the right. Here, *vertical points* on the curve $F(x)$ correspond to *areas under the curve* of $f(x)$. The areas under $f(x)$ corresponding to $F(a)$ and $F(b)$ are shown in the small inset.

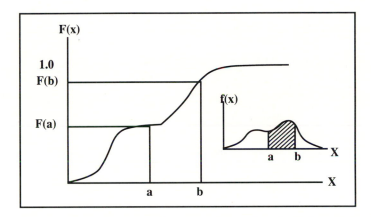

Figure 3.8 Graph of a continuous distribution function showing the relationship of $f(x)$ and $F(x)$.

EXAMPLES

EXAMPLE 1: For the following probability density function (pdf) determine $P(1 < x \le 3)$, find $F(x)$, and then graph both $f(x)$ and $F(x)$.

$$f(x) = \begin{cases} \dfrac{1}{2}e^{-\frac{x}{2}} & x > 0 \\ 0 & \text{elsewhere} \end{cases}$$

SOLUTION: To find $P(1 < X < 3)$ we perform the following integration.   (See Figure 3.9 below.) The probability requested will be obtained by integrating the probability density function across the limits 1 to 3 as indicated.

$$P(1 < X \le 3) = \int_{1}^{3} \frac{1}{2}e^{-\frac{x}{2}}\,dx = -e^{-\frac{x}{2}}\Big|_{1}^{3} = -e^{-\frac{3}{2}} + e^{-\frac{1}{2}} = 0.3834.$$

To find $F(x)$ for this pdf, we use the definition, $F(x) = \int_{-\infty}^{x} f(t)dt$. When $x \le 0$, $f(x) = 0$ therefore $F(x) = 0$. For any $x > 0$ then $F(x)$ is equal to

$$F(x) = \int_{-\infty}^{x} f(t)dt = \int_{0}^{x} \frac{1}{2} e^{-\frac{t}{2}} dt = 1 - e^{-\frac{t}{2}}$$

Therefore, summarizing the results we get,

$$F(x) = \begin{cases} 0 & x \le 0 \\ 1 - e^{-\frac{x}{2}} & x > 0 \end{cases}$$

The graph of $f(x)$ is given in Figure 3.9 below and $F(x)$ is in Figure 3.10 which follows.

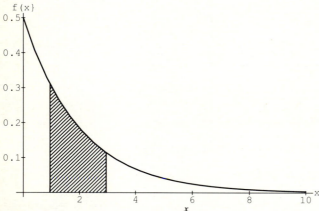

Figure 3.9 Graph of $f(x) = \frac{1}{2} e^{-\frac{x}{2}}$.

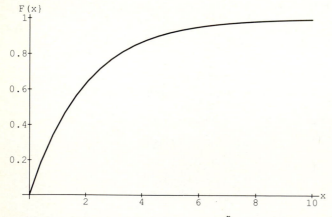

Figure 3.10 Graph of $F(x) = 1 - e^{-\frac{x}{2}}$ associated with $f(x) = \frac{1}{2} e^{-\frac{x}{2}}$.

We can use $F(x)$ to find $P(1 < x \le 3)$ by evaluating $F(3) - F(1) = (1 - e^{-3/2}) - (1 - e^{-1/2}) = e^{-1/2} - e^{-3/2} = 0.3834$, the same value we obtained previously.

EXAMPLE 2: For the following probability density function, determine the value of k that will ensure that the area under the curve is equal to 1. Find the cumulative distribution function, $F(x)$ as well, carefully defining it over the various regions of X on the real line. Plot both $f(x)$ and $F(x)$ on separate graphs. However scale the horizontal axis the same on both graphs.

$$f(x) = \begin{cases} kx^2 & -2 < x < 2 \\ 0 & \text{elsewhere} \end{cases}$$

SOLUTION: To ensure that this is a legitimate probability density function we require that:

$$1 = \int_{-\infty}^{\infty} f(x)dx = \int_{-2}^{2} kx^2 dx = k\frac{x^3}{3}\Big|_{-2}^{2} = k[(\frac{8}{3}) - (\frac{-8}{3})] = k\frac{16}{3}$$

Therefore $k = \dfrac{3}{16}$. As a result we can write $f(x)$ as:

$$f(x) = \begin{cases} \dfrac{3}{16}x^2 & -2 < x < 2 \\ 0 & \text{elsewhere} \end{cases}$$

The cumulative distribution function is given as:

$$F(x) = \begin{cases} 0 & x \le 2 \\ \displaystyle\int_{-2}^{x} (\frac{3}{16})t^2 dt = (\frac{1}{16})(x^3 + 8) & -2 < x < 2 \\ 1 & x \ge 2 \end{cases}$$

The graphical representations of the pdf, $f(x)$, and the distribution function, $F(x)$ are given in Figure 3.11 that follows.

EXAMPLE 3: For the following function, find the value of k to make it a legitimate probability density function. Graph the function. Secondly determine the expression for $F(x)$ and also graph it.

$$f(x) = \begin{cases} kx^3(1 - x) & 0 \le x \le 1 \\ 0 & \text{elsewhere} \end{cases}$$

SOLUTION: We first determine the value of k to make this a legitimate probability distribution. Therefore, the integral from $-\infty$ to ∞ must equal 1. However, $f(x)$ is zero except for $0 \le x \le 1$. Thus the following integral is set up and solved for k:

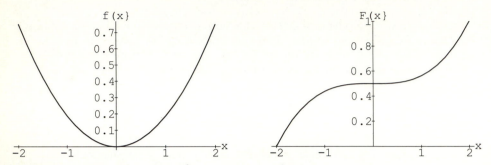

Figure 3.11 Graph of $f(x) = \frac{3}{16}x^2$, and $F(x) = (\frac{1}{16})(x^3 + 8)$.

$$1 = \int_0^1 kx^3(1-x)dx = k(\frac{x^4}{4} - \frac{x^5}{5})\,\Big|_0^1$$

$$= k(\frac{5}{20} - \frac{4}{20}) = \frac{k}{20}.$$

Therefore, $k = 20$ and $f(x)$ becomes

$$f(x) = 20x^3(1-x) \qquad 0 \le x \le 1$$

Graphically, this is shown in Figure 3.12 below.

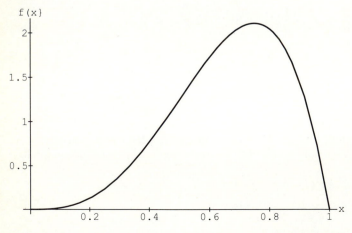

Figure 3.12 Graph of $f(x) = 20x^3(1-x)$ over the interval $0 \le x \le 1$.

To determine the expression for $F(x)$ first recognize that there are three sections on the real line for which $f(x)$ is defined: (1) when $x < 0$, (2) when $0 \leq x \leq 1$, and (3) when $x > 1$. $F(x)$ will have different forms over each of those three sections. Also, we can interpret $F(x)$ to represent the area under $f(x)$ to the left of the value x. Therefore, the area under $f(x)$ when $x < 0$ is zero; the area under $f(x)$ when $x > 1$ is 1, and so we have only to solve for $F(x)$ when $0 \leq x \leq 1$. To determine the expression we form the following integral.

$$F(x) = \int_0^x 20t^3(1-t)dt$$

$$= 20(\frac{t^4}{4} - \frac{t^5}{5})|_0^x \qquad 0 \leq x \leq 1$$

$$= 5x^4 - 4x^5$$

From this we can summarize these results as follows:

$$F(x) = \begin{cases} 0 & x < 0 \\ 5x^4 - 4x^5 & 0 \leq x \leq 1 \\ 1 & x > 1 \end{cases}$$

The graph of $F(x)$ is shown below as Figure 3.13.

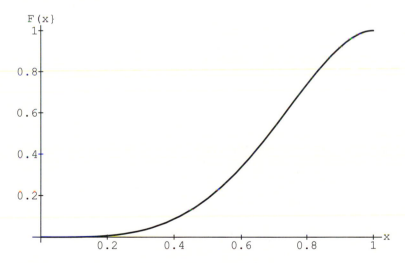

Figure 3.13 Graph of $F(x)$ for Example 3.

EXERCISES

In the following exercises confirm wherever necessary that the given function is indeed a probability density function. In addition, plot or graph the various functions in order to have a better visual understanding of the properties of the function given.

1.   Suppose a salesperson in an electronic store sells between 0 and 6 TV's per week according to the

probability function

$$f(x) = \frac{(6-x)}{k} \quad \text{for } x = 0, 1, ..., 6$$

What is the value of k to make this a legitimate probability density function, i.e. the sum of the probabilities is 1.0? Using this function what is the probability of selling at least 2 TV's in a given week? More than 4?

2. Evaluate the expression $F(x)$ from exercise 1. Make a plot of $f(x)$ and $F(x)$. Evaluate the probability of selling at least 2 TV's in a given week by using the expression for $F(x)$.

3. Using the expression for $F(x)$ of exercises 1 and 2 above, find:
   a. $P(2 < X < 4)$ c. $P(2 \le X < 4)$
   b. $P(2 < X \le 4)$ d. $P(2 \le X \le 4)$

4. A biologist wants to model the proportion of mutant genes in a given insect after irradiation, by the function

$$f(x) = \begin{cases} kx(1-x) & 0 < x < 1 \\ 0 & \text{elsewhere} \end{cases}$$

Determine the value of k to make this a probability density function (pdf).

5. For the pdf of exercise 4 evaluate $F(x)$ over the entire real line, graph $f(x)$ and $F(x)$.

6. Using the pdf of exercises 4 and $F(x)$ of exercise 5, compute the following probabilities.
   a. $P(X < 0.5)$ c. $P(X > 0.25)$
   b. $P(0.25 < X < 0.75)$ d. $P(X < 0.75)$

7. Out of curiosity the biologist of exercise 4 tried the function $f(x) = kx^2(1-x)$ for $0 < x < 1$ instead of the expression $f(x) = kx(1-x)$. Determine the value of k to make $f(x) = kx^2(1-x)$ for $0 < x < 1$ a probability density function (pdf). Determine the formula for $F(x)$ for this pdf.

8. Graph the pdf of exercise 7 along with $F(x)$ using the same scale used in exercise 5 so that you can visually compare the two sets of graphs.

9. Using the pdf and $F(x)$ of exercise 7, compute the following probabilities.
   a. $P(X < 0.5)$ c. $P(X > 0.25)$
   b. $P(0.25 < X < 0.75)$ d. $P(X < 0.75)$

10. The number of phone calls arriving at a brokerage office during the lunch hour averages 2 per quarter hour, and is thought to have the following probability density function

$$f(x) = \frac{e^{-2} 2^x}{x!} \quad\quad\quad \text{for } x = 0, 1, 2, 3, ...$$

Evaluate this function for the values of x from 0 to 4. What is the probability that more than 4 calls will arrive in a particular quarter hour time period?

11. Suppose the density function of a random variable Z is given by

$$f(z) = \begin{cases} -kz & -1 < z < 0 \\ kz & 0 \le z < 1 \\ 0 & \text{elsewhere} \end{cases}$$

Find the value of k and $F(z)$. Sketch the graph of $f(z)$ and $F(z)$.

12. Sketch the graph of $f(u)$ and $F(u)$ for a random variable U which has a pdf of

$$f(u) = \begin{cases} ku^2 & -2 < u < 2 \\ 0 & \text{elsewhere} \end{cases}$$

3.2   "EXPECTED VALUES"—THE MEAN AND VARIANCE OF A PROBABILITY DENSITY FUNCTION

Introduction

Since the primary application of a probability density function, in this course, is in modelling natural phenomena, the same characteristics that we observe in data sets that we collect from nature should be found in the modelling distribution. Thus, a pdf should have a mean, variance, quantiles of any value, and so forth. Of primary interest is the mean and the variance (or standard deviation). In this section we formally define how to obtain a mean and variance of a probability density function. We do this by introducing the concept of the "expected value" of a random variable, X. Other measures that may be of interest will be discussed in the next section.

Remember that the mean and variance that we compute for a probability density function are only theoretical values, or parameters of the population being modelled and will be indicative of the truth only insofar as the model is an accurate description of the population of interest. They are only an indication of what may be real; they should not be viewed as the ultimate "truth" of things. If the model fails, then the parameters of that model will not be realistic.

Definitions

DEFINITION 3.8: The *expected value* of the random variable X, denoted by $E(X)$, is defined as:

$$E(X) = \begin{cases} \displaystyle\sum_{\text{all } x} xf(x) & \text{if } X \text{ is discrete} \\ \displaystyle\int_{-\infty}^{\infty} xf(x)dx & \text{if } X \text{ is continuous} \end{cases}$$

The expected value of a *function* of a random variable X, say g(X), is:

$$E(g(X)) = \begin{cases} \displaystyle\sum_{\text{all } x} g(x)f(x) & \text{if } X \text{ is discrete} \\ \displaystyle\int_{-\infty}^{\infty} g(x)f(x)dx & \text{if } X \text{ is continuous} \end{cases}$$

PROPERTIES OF EXPECTATION OPERATOR:

1. $E(a) = a$ where a is a constant

2. $E(aX + b) = aE(X) + b$, where a and b are constants.

DEFINITION 3.9: The mean of a probability density function denoted by μ_x is the expected value of X, $E(X)$. That is,

$$\mu_x = \begin{cases} E(X) = \displaystyle\sum_{\text{all } x} xf(x) & X \text{ is discrete} \\ E(X) = \displaystyle\int_{-\infty}^{\infty} xf(x)dx & X \text{ is continuous} \end{cases}$$

DEFINITION 3.10: The variance σ_x^2 of a random variable X (also denoted by V(X)) having a probability density function or pdf $f(x)$ is the expected value of the squared error, that is,

$$\sigma_x^2 = V(X) = E(x - \mu)^2 = \begin{cases} \displaystyle\sum_{\text{all } x}(x - \mu_x)^2 \bullet f(x) & X \text{ is discrete} \\ \displaystyle\int_{-\infty}^{\infty} (x - \mu_x)^2 \bullet f(x)dx & X \text{ is continuous} \end{cases}$$

By expanding $(x - \mu)^2$ and then simplifying, we get an expression easier to evaluate, i.e.,

$$\sigma_x^2 = E(X^2) - \mu_x^2$$

PROPERTIES OF VARIANCE OPERATOR:

1. $V(constant) = 0$

2. $V(aX + b) = a^2 V(X)$ where a is a constant

DISCUSSION

Suppose that we have an jar as shown in Figure 3.14 with 10 balls in it, one of which is numbered with a 1, 2 have a 2, 3 have a 3 and 4 have a 4.

Figure 3.14 Urn: $\mu = 3$, $\sigma = 1$.

The mean of this population would be found by the following computation:

$$\frac{1 + 2 + 2 + 3 + 3 + 3 + 4 + 4 + 4 + 4}{10} = \frac{1 + 2\bullet 2 + 3\bullet 3 + 4\bullet 4}{10}$$

$$= 1\bullet\frac{1}{10} + 2\bullet\frac{2}{10} + 3\bullet\frac{3}{10} + 4\bullet\frac{4}{10} = 3$$

Looking at this expression, we can interpret it as the sum of the product of the value on the ball *multiplied by* the relative frequency of that ball occurring. That is, the ball with a *one* on it occurs with frequency 1 out of 10, the *2* with frequency 2 out of 10, and so forth. Thus we get:

$$1\bullet\frac{1}{10} + 2\bullet\frac{2}{10} + 3\bullet\frac{3}{10} + 4\bullet\frac{4}{10} = 3.0$$

Another way of stating the same thing is that the number we *expect* on the ball is a 1 with probability 1/10, or a 2 with probability 2/10, and so on. Thus, if a ball is randomly selected from the jar, the *expected value* of the number on the ball selected is:

$$1\bullet\frac{1}{10} + 2\bullet\frac{2}{10} + 3\bullet\frac{3}{10} + 4\bullet\frac{4}{10} = 3.0$$

Or you may imagine that you are playing a game of chance. You randomly select a ball from the jar, and your payoff in dollars is equal to the number on the ball. Thus you would earn $1 with probability 1/10, $2 with probability 2/10 etc. Thus, the long run, average value of your payoff or *expected payoff* would be $3.00.

We have demonstrated with this example that if X denotes the random variable which takes on the values x with probabilities $f(x)$, then the *expected value of X*, denoted by E(X), is the mean of the distribution, which is computed by the formula:

$$\mu_x = E(X) = \sum_{\text{all } x} xf(x) \quad \text{if } X \text{ is discrete}$$

A similar argument provides a definition for the variance. For instance to find the variance of the numbers in the jar, we would use the formula in Chapter 1 to get:

$$\sigma_x^2 = \frac{(1-3)^2 + (2-3)^2\bullet2 + (3-3)^2\bullet3 + (4-3)^2\bullet4}{10} = 1.0$$

$$= \sum_{i=1}^{4} (x_i - 3)^2 f(x_i)$$

where $f(x_i)$ represents the probability of getting x_i and $x_i = 1,2,3,4$.

This gives the analogous definition for a variance of a discrete random variable as:

$$V(X) = \sigma_x^2 = \sum_{\text{all } x} (x - \mu_x)^2 \bullet f(x) \quad \text{for discrete } X.$$

If X is continuous, the definition of the mean and variance are a natural extension of the concepts of an integral as the limit of a sum as the distance between the x's get closer and closer together. That is, in the discrete case, $f(x)$ is the probability at the point x. In the continuous case, we have to consider an interval of width dx and of height $f(x)$ so that the probability at the point x is approximately $f(x)dx$. Therefore, the definition of a the expected value E(X) for a continuous random variable becomes in the limit:

$$E(X) = \mu_x = \int_{-\infty}^{\infty} xf(x)dx$$

which is interpreted as the mean. The variance is developed by a similar argument and becomes

$$V(X) = \sigma_x^2 = \int_{-\infty}^{\infty} (x - \mu_x)^2 f(x)dx.$$

The mean defined as above is also called the first central moment and is the center of gravity of the graphical representation defined by $f(x)$. Thus, it serves as a measure of location in the same sense as the mean did in Chapter 1. The variance $V(X)$ of a pdf measures the variability about the mean for the set of values the pdf takes on. (Note that it takes into account the frequency or density of values at a given point, as well.) It thus serves as a legitimate measure of variability or dispersion associated with a probability distribution of values in the same sense as the variance defined in Chapter 1.

The expected value operator E( ) has some useful properties that are important to remember and fairly easy to prove. (You will be asked to prove some of these results in the exercises of this section. In addition, in exercise 8 you are asked to demonstrate them for a particular probability density function.) First, if a and b are constants, then

$$E(a) = a$$

$$E(aX + b) = aE(X) + b.$$

Here, we are claiming that the expected value of a constant is itself. And secondly, that if a random variable is multiplied by a constant a and then is shifted by the amount b, the mean of the resulting transformation will be affected in the same way.

Similarly, we can write

$$V(a) = 0$$

$$V(aX + b) = a^2 V(x)$$

$$V(X) = E(X^2) - [E(X)]^2 = E(X^2) - \mu_x^2.$$

These properties are intuitive and very easy to show using the definition of a variance in terms of expected values. Obviously a constant doesn't vary, therefore its variance will be zero. In addition, if a constant b is added to a set of numbers, it will not change how the numbers vary, but multiplying the set of numbers by a constant $a \neq 1$ will affect their variance.

Finally, again using the properties of "expected values" it is straight forward to show that the variance of the random variable X equals the expected value of the square of the random variable from which is subtracted the square of the expected value of the random variable. This proof comes by first squaring the expression $(X - \mu_x)^2$ and then taking the expectation. That is, $E[(X - E(X))^2] = E[X^2 - 2X \cdot E(X) + E(X)^2]$. The formal proofs of these three properties will be left for the exercises.

EXAMPLES

EXAMPLE 1: Find the mean and variance for the pdf previously defined as

$$f(x) = \begin{cases} \dfrac{1}{2} e^{-\frac{x}{2}} & x > 0 \\ \\ 0 & \text{elsewhere} \end{cases}$$

SOLUTION: By definition

$$E(X) = \mu_x = \int_{-\infty}^{\infty} xf(x)dx = \int_{0}^{\infty} x\left(\frac{1}{2}e^{-\frac{x}{2}}\right)dx = \int_{0}^{\infty} \frac{x}{2}e^{-\frac{x}{2}}dx$$

By a change of variable to $u = \dfrac{x}{2}$, we get $dx = 2du$ so that

$$E(X) = \int_{0}^{\infty} 2ue^{-u}du$$

The above integral can be integrated by parts with the result that $E(X) = 2$. However, instead of integrating by parts, a general property that will be useful to us here and in many problems in the future is presented next.

Special Note: The above integral is a particular case of an integral called the gamma function. The gamma function is a definite integral with the general form:

$$\Gamma(n) = \int_0^\infty u^{n-1}e^{-u}du \text{ where } u > 0.$$

This integral has several useful properties obtained by successive "integrations by parts" (to be proved in the Exercises):

    1. $\Gamma(n) = (n-1) \cdot \Gamma(n-1)$

    2. $\Gamma(n) = (n-1)!$ if n is integer

    3. $\Gamma(1/2) = \sqrt{\pi}$

For other values of n, tables have been prepared to evaluate the gamma function. However we do not include that table in this text.

To determine the variance we use the property that

$$V(X) = E(X^2) - \mu_x^2$$

Let us first evaluate $E(X^2)$ again using the change of variable to $u = \dfrac{x}{2}$ with $dx = 2du$.

$$E(X^2) = \int_0^\infty (x)^2(\frac{1}{2}e^{-\frac{x}{2}})dx = \int_0^\infty (\frac{(x)^2}{2}e^{-\frac{x}{2}})d(x) = \int_0^\infty 2(\frac{(x)}{2})^2 e^{-\frac{x}{2}}dx = 2\int_0^\infty u^{3-1}e^{-u}2du = 4 \cdot \Gamma(3) = 8.$$

With this result we have that

$$V(X) = E(X^2) - \mu_x^2 = 8 - (2^2) = 4.$$

Therefore, the above random variable has a theoretical mean of 2 and a variance of 4 (a standard deviation of 2).

EXAMPLE 2: Consider the example of Section 3.1 concerning an investigation of drunk driving arrests where the sex of the offender is noted. In that example, X denoted the number of females which could take on any integer value between 0 and 5. We assumed that the sex of the offender was as likely to be a male as a female so that we could write,

$$f(x) = \begin{cases} \dfrac{5!}{x! \cdot (5-x)!}(\dfrac{1}{32}) & x = 0,1,2,3,4,5 \\[2ex] = 0 & \text{elsewhere} \end{cases}$$

The probability density function and cumulative distribution are summarized in Table 3.2 that follows.

Table 3.2 The probability density function and the cumulative distribution function for $f(x) = \dfrac{5!}{x!\bullet(5-x)!}(\dfrac{1}{32})$.

x	f(x)	F(x)
0	$\dfrac{1}{32}$	$\dfrac{1}{32}$
1	$\dfrac{5}{32}$	$\dfrac{6}{32}$
2	$\dfrac{10}{32}$	$\dfrac{16}{32}$
3	$\dfrac{10}{32}$	$\dfrac{26}{32}$
4	$\dfrac{5}{32}$	$\dfrac{31}{32}$
5	$\dfrac{1}{32}$	$\dfrac{32}{32}$

Find the mean and standard deviation of $f(x)$.

SOLUTION: Let us first find $E(X)$ and $E(X^2)$:

$$E(X) = 0\bullet\frac{1}{32} + 1\bullet\frac{5}{32} + 2\bullet\frac{10}{32} + 3\bullet\frac{10}{32} + 4\bullet\frac{5}{32} + 5\bullet\frac{1}{32} = \frac{80}{32} = 2.5$$

$$E(X^2) = 0^2\bullet\frac{1}{32} + 1^2\bullet\frac{5}{32} + 2^2\bullet\frac{10}{32} + 3^2\bullet\frac{10}{32} + 4^2\bullet\frac{5}{32} + 5^2\bullet\frac{1}{32} = \frac{240}{32}$$

Thus the mean is 2.5 females with a standard deviation of

$$\sigma = \sqrt{\sigma^2} = \sqrt{E(X^2) - E(X)^2} = \sqrt{\frac{240}{32} - (\frac{80}{32})^2} = 1.11803$$

EXAMPLE 3: Find the mean and variance for the following pdf (see Example 3 of Section 3.1 of this chapter):

$$f(x) = \begin{cases} 20x^3(1-x) & 0 \le x \le 1 \\ 0 & \text{elsewhere} \end{cases}$$

SOLUTION: We will find $E(X)$ and $E(X^2)$ and use these results to give us the mean and variance of the distribution. We first find $E(X)$

$$E(X) = \int_0^1 x \cdot [20x^3(1-x)]dx$$

$$= 20 \int_0^1 (x^4 - x^5)dx$$

$$= 20 \left(\frac{x^5}{5} - \frac{x^6}{6}\right)\Big|_0^1 = \frac{20}{30} = 66.67\%.$$

Next, we find $E(X^2)$:

$$E(X^2) = \int_0^1 x^2 \cdot [20x^3(1-x)]dx$$

$$= 20 \int_0^1 (x^5 - x^6)dx$$

$$= 20\left(\frac{x^6}{6} - \frac{x^7}{7}\right)\Big|_0^1 = \frac{20}{42}$$

The mean of the distribution equals $E(X)$ and the variance is found by the relationship $\sigma^2 = E(X^2) - [E(X)]^2$. Thus the mean and variance are:

$$\mu = E(X) = \frac{2}{3} = 0.6667$$

and

$$\sigma^2 = E(X^2) - [E(X)]^2 = \frac{20}{42} - [\frac{2}{3}]^2 = \frac{12}{378}$$

(The standard deviation is: $\sigma = \sqrt{\frac{12}{378}} = 0.1782$.)

EXERCISES

In the following determine the mean and variance of the pdf's given. In doing so, make sure you understand how to interpret them in the context of the problem setting.

1. Suppose a student guesses on a True-False question. If the random variable Y takes the value 0 for an incorrect answer and 1 for a correct answer, determine what $f(y)$ is and find its mean and variance.

2. Simulate the experience described in exercise 1 by tossing a coin 30 times, recording a 0 for a tail (incorrect answer), and a 1 for a head (correct answer). Compute the sample mean and variance of these 30 numbers using the formulas of chapter 1. Compare to the theoretical results obtained in exercise 1.

3. Suppose a salesperson in an electronic store sells between 0 and 6 TV's per week with the probability function

$$f(x) = \frac{(6 - x)}{k} \quad \text{for } x = 0, 1, ..., 6$$

Using the value of k you obtained in Section 3.1, exercise 1 determine the expected number of TV's sold per week along with the variance. That is, obtain E(X), and V(X) for the above random variable.

4. A biologist wants to model the proportion of mutant genes in a given insect after irradiation, by the function

$$f(x) = \begin{cases} kx(1 - x) & 0 < x < 1 \\ 0 & \text{elsewhere} \end{cases}$$

You obtained a value for k to make this a legitimate pdf in Section 3.1. Use those results and find the theoretical mean and standard deviation for this random variable. Interpret what these numbers mean in the context of a biological study.

5. Prove that E($aX + b$) = aE(X) + b. Assume that the random variable is continuous, and thus

$$E(aX + b) = \int_{-\infty}^{\infty} (ax + b)f(x)dx.$$ Remember in simplifying this expression that $\int_{-\infty}^{\infty} f(x)dx = 1.$

(The same arguments and approach will hold for summation signs when X is discrete.)

6. Prove that V(X) = E(X^2) − μ^2. (Hint: Start with E($X - \mu$)^2, square the quadratic term, and then take its expectation using the properties of expectation previously cited.)

7. Prove that V($aX + b$) = a^2V(X). (Hint: Start with the result, V($aX + b$) = E[($aX + b$) − E($aX + b$)]^2. Simplify the terms within the brackets using the properties of expectation, then recognize the definition of the term you have remaining.)

8. Demonstrate the properties of exercises 5, 6, and 7 on the probability density function given below. (This is a discrete pdf.)

$$f(x) = \frac{1}{4} \quad x = 1, 2, 3, 4$$

Let a = 3 and b = 5. (Hint: If $Y = aX + b$, then when $X = 2$, $Y = (3 \cdot 2) + 5 = 11$ with probability $f(y) = f(x) = 1/4$.)

9. Let $Y = \frac{(X - \mu)}{\sigma}$, prove that E(Y) = 0 and V(Y) = 1. (Note this transformation from X to Y is called "standardizing the random variable X.")

10. Demonstrate the property of exercise 9 by verifying the result with the distribution $f(x) = \frac{1}{4}$, for x = 1, 2, 3, 4. Do this by finding the mean μ, and the standard deviation σ. Transform each x to a y and then find the mean and variance of the new random variable Y.

11. Suppose the running time in hours or fractions thereof to discharge of a rechargeable battery follows the probability density function:

$$
f(x) = \begin{cases} \dfrac{1}{4}\, e^{-\frac{x}{4}} & x > 0 \\[2mm] 0 & \text{elsewhere} \end{cases}
$$

Find the mean "time to discharge" for the random variable X.

12. For the random variable of exercise 11, find the standard deviation and interpret it in the context of the problem.

13. Prove that $\Gamma(n) = (n-1) \bullet \Gamma(n-1)$ for the gamma function defined as $\Gamma(n) = \displaystyle\int_{0}^{\infty} u^{n-1} e^{-u} du$.

14. Show that $\Gamma(1/2) = \sqrt{\pi}$ where the gamma function is defined as in exercise 13.

15. Prove that if n is integer-valued that $\Gamma(n) = (n-1)!$.

16. Find the mean and variance for the random variable U whose pdf is:

$$
f(u) = \begin{cases} ku^2 & -2 < u < 2 \\[2mm] 0 & \text{elsewhere} \end{cases}
$$

3.3   OTHER MEASURES OF LOCATION AND VARIABILITY

Introduction

As we mentioned at the beginning of this chapter, we introduce a probability density function as a model for a real-world population from which we have a set of data. In doing this, we want to define such measures of location and variability that provide useful summaries of the data in helping us to understand the population from which the data came. In Section 3.2, we defined the mean and variance, but other measures are also desired, such as the various quantiles and concepts related to them such as the median and interquartile range; the median deviation might also be of interest along with measures of skewness. In short we want to be able to define, for a probability density function, any numerical value that might serve a descriptive purpose for a set of measurements. This section catalogs a few of those definitions.

Definitions

Measures of Location

DEFINITION 3.11: The *qth quantile* x_q for any pdf, $0 < q < 1$ is the value of the random variable X such that

$$P(X < x_q) \leq q \text{ and } P(X \leq x_q) \geq q \qquad \text{if } X \text{ is discrete}$$

$$P(X \leq x_q) = q \qquad \qquad \text{if } X \text{ is continuous}$$

DEFINITION 3.12: The *median* of a pdf $x_{0.5}$, is the value of the random variable such that the following probability statements are true.

$$P(X < x_{0.5}) \leq 0.5 \text{ and } P(X \leq x_{0.5}) \geq 0.5 \qquad \text{if } X \text{ is discrete}$$

$$P(X \leq x_{0.5}) = F(x_{0.5}) = 0.5 \qquad \text{if } X \text{ is continuous}$$

DEFINITION 3.13: The *mode* of a pdf is defined to be the value of the random variable X where the pdf is a maximum. In other words,

Mode = Value of x such that $f(x)$ is a maximum.

Measures of Variability

DEFINITION 3.14: The *interquartile range* is:

$$x_{0.75} - x_{0.25} = Q_3 - Q_1$$

DEFINITION 3.15: The *mean deviation* of any random variable X is $E|X - \mu|$ for X continuous or discrete.

Other Moments

DEFINITION 3.16: The *rth central moment* or *moment about the origin*, denoted by μ_r', is defined to be:

$$\mu_r' = E(X^r)$$

Of particular interest is $\mu_1' = \mu = E(X)$ and $\mu_2' = E(X^2)$.

DEFINITION 3.17: The *rth moment about the mean*, denoted by μ_r, is defined to be:

$$\mu_r = E(X - \mu)^r$$

Of particular interest is $\mu_2 = E(X - \mu)^2 = V(X)$, the variance of X, and $\mu_3 = E(X - \mu)^3$, useful in providing a measure of skewness or symmetry. This measure of skewness is the standardized third moment, the ratio of the third moment to the cube of the standard deviation, called the *coefficient of skewness,* defined as

$$\alpha_3^* = \frac{\mu_3}{(\sigma)^3}$$

DISCUSSION

The mean and the standard deviation serve as our primary measures of location and variability. However, there will be occasions when the median, the interquartile range, the mean deviation, etc., may provide useful information about a hypothetical population. In those instances, we want to be able to determine their values using the probability density function that is used to model the data. The interpretations of these values will be the same as the interpretations given in Chapter 1. However, their definitions are given in the context of a probability density function $f(x)$ rather than from a "data" perspective.

For instance, for a continuous random variable, the median is the center of "area"—that location on the X–axis that has half the area to its left. Thus, $x_{0.5}$ is the value of x such that $F(x) = 0.5$. The 1st and 3rd quartiles are defined similarly: Q_1 is the value of x such that $F(x) = 0.25$, and Q_3 is the value of x such that $F(x) = 0.75$. The mode is the value of x associated with the maximum probability. For a continuous random variable the mode would be that value of x where $f(x)$ is a maximum. The principles of calculus—finding the value of x where the first derivative equals zero will usually work in the continuous case to find the mode.

The mean deviation has a definition analogous to the definition given in Chapter 1:

$$\text{Mean Deviation} = E|X - \mu|$$

which computes the "mean" of the absolute values of the deviations between the random variable and the mean of the random variable. The "expectation" results in a sum of terms if X is discrete, and a integral if X is continuous. The interquartile range is defined as would be expected: $Q_3 - Q_1$.

There are also some useful mathematical properties, called moments that are important to note. First, we give the definition and notation for *central moments* or moments about the origin followed by *moments about the mean* :

Central Moments

$$\mu_1' = E(X) = \mu$$
$$\mu_2' = E(X^2)$$
$$\mu_3' = E(X^3)$$
$$\cdots$$
$$\mu_r' = E(X^r)$$

Moments About The Mean

$$\mu_1 = E(X - \mu)^1 = 0$$
$$\mu_2 = E(X - \mu)^2 = \sigma^2$$
$$\mu_3 = E(X - \mu)^3$$
$$\cdots$$
$$\mu_r = E(X - \mu)^r$$

We can also express moments about the mean in terms of a linear combination of the central moments. This will be demonstrated for two cases; other cases follow the same development.

$$\mu_2 = E(X - \mu)^2 = E(X^2 - 2X\mu + \mu^2) = E(X^2) - 2E(X)^2 + \mu^2 = E(X^2) - [E(X)]^2 = \mu_2' - \mu^2$$

Likewise,

$$E(X - \mu)^3 = E(X^3 - 3X^2\mu + 3X\mu^2 - \mu^3) = E(X^3) - 3E(X^2)E(X) + 2E(X)^3$$

$$= \mu_3' - 3\,\mu_2'\,\mu + 2\mu^3$$

Since it is often much easier to obtain the central moments, we can use the above relationships to obtain the moments about the mean of any order.

One application for the third moment about the mean, $\mu_3 = E(X - \mu)^3 = \mu_3 - 3\,\mu_2\mu + 2\mu^3$, is to provide a measure of skewness. This skewness measure is defined as:

$$\alpha_3^* = \frac{\mu_3}{(\sigma)^3}$$

which will be negative for negatively skewed distributions and positive for positively skewed distributions.

EXAMPLES

EXAMPLE 1: Find the median and mode for the distribution previously given by

$$f(x) = \begin{cases} \dfrac{1}{2}\,e^{-\frac{x}{2}} & x > 0 \\[2mm] 0 & \text{elsewhere} \end{cases}$$

In addition, find the interdecile, interquartile and mean deviation. Locate the mean, median and mode on the graph of Figure 3.15 below.

SOLUTION:
The median: To find the median $x_{0.5}$, we must find the value of x such that the area to the left of it is 50% of the total area. That is, $F(x_{0.5}) = 0.5 = P(X \leq x_{0.5})$. We have previously found $F(x)$ to be (see Section 3.1 of this chapter):

$$F(x) = 1 - e^{-\frac{x}{2}} \qquad x > 0$$

Writing the equation yields:

$$0.5 = 1 - e^{-\frac{x_{0.5}}{2}}$$

and solving for $x_{0.5}$

$$x_{0.5} = 2 \cdot \ln\left(\frac{1}{(1 - .5)}\right) = 1.386.$$

The mode: By examination it can be seen that the pdf approaches a maximum as you approach zero from the right. Therefore, we would say the mode is 0 in the limit. (Taking derivatives and setting to zero will not help in this case.)

The interquartile range: The interquartile range, *IQR*, is defined to be $x_{0.75} - x_{0.25}$ where the 1st and 3rd quartiles can be obtained by the same process that produced the median since they involve using quantiles. Therefore,

$$x_{0.75} = 2 \cdot \ln\left(\frac{1}{(1 - .75)}\right) = 2.773, \text{ and } x_{0.25} = 2 \cdot \ln\left(\frac{1}{(1 - .25)}\right) = 0.575$$

So that, $IQR = 2.773 - 0.575 = 2.198$.

The mean deviation: The mean deviation will be obtained by the following integration (note that $|x - \mu| = -(x - \mu)$ when $x < \mu$ and equals $(x - \mu)$ when $x > \mu$):

$$\text{Mean Deviation} = E|X - \mu| = \int_{-\infty}^{\mu} -(x - \mu)f(x)dx + \int_{\mu}^{\infty} (x - \mu)f(x)dx$$

$$= \int_{0}^{2} -(x - 2)(\frac{1}{2}e^{-\frac{x}{2}})dx + \int_{2}^{\infty} (x - 2)(\frac{1}{2}e^{-\frac{x}{2}})dx.$$

This integral, though not impossible to integrate, is quite involved, much more so than the procedure to obtain the variance and standard deviation. When finished, the integration produces the value

$$\text{Mean Deviation} = 1.47$$

(Remember, the standard deviation was equal to 2.) A graph of the probability density function is given in Figure 3.15 with the location of the mode, median and mean shown respectively.

EXAMPLE 2: For the probability function of Example 1, determine the coefficient of skewness α_3^*.

SOLUTION: We need the first three central moments, $E(X)$, $E(X^2)$ and $E(X^3)$. These are found respectively by performing the following integrations. (The first two have previously been found.)

$$E(X) = \int_{0}^{\infty} x(\frac{1}{2}e^{-\frac{x}{2}})dx = 2 \cdot \Gamma(2) = 2$$

$$E(X^2) = \int_{0}^{\infty} x^2(\frac{1}{2}e^{-\frac{x}{2}})dx = 2^2 \cdot \Gamma(3) = 8$$

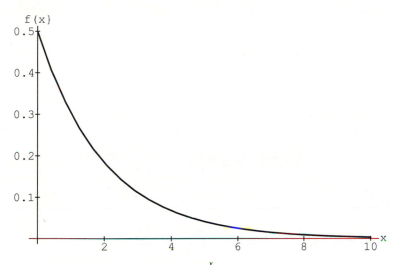

Figure 3.15   Graph of $f(x) = \dfrac{1}{2} e^{-\frac{x}{2}}$.

$$E(X^3) = \int_0^\infty x^3 \left(\frac{1}{2} e^{-\frac{x}{2}}\right) dx = 2^3 \cdot \Gamma(4) = 48$$

Therefore, using the relationship that

$$E(X - \mu)^3 = \mu_3' - 3\,\mu_2'\mu + 2\mu^3$$

we get

$$\mu_3 = 48 - 3 \cdot 8 \cdot 2 + 2 \cdot (2)^3 = 16.$$

Also, from Section 3.2 we found that $\sigma = 2$, so that

$$\alpha_3^* = \frac{\mu_3}{(\sigma)^3} = \frac{16}{2^3} = 2.0,$$

a fairly strong indication of positive skewness.

EXAMPLE 3: For the following discrete probability distribution, find the mean deviation and the coefficient of skewness α_3^*.

x	2	3	4	5
f(x)	$\dfrac{1}{8}$	$\dfrac{3}{8}$	$\dfrac{3}{8}$	$\dfrac{1}{8}$

SOLUTION: First let us find $E(X)$, $E(X^2)$ and $E(X^3)$. By symmetry we can see that $E(X) = 3.5$, but we shall verify it as well as calculate the other values:

$$E(X) = 2 \bullet \frac{1}{8} + 3 \bullet \frac{3}{8} + 4 \bullet \frac{3}{8} + 5 \bullet \frac{1}{8} = \frac{28}{8} = 3.5$$

$$E(X^2) = 2^2 \bullet \frac{1}{8} + 3^2 \bullet \frac{3}{8} + 4^2 \bullet \frac{3}{8} + 5^2 \bullet \frac{1}{8} = \frac{104}{8}$$

$$E(X^3) = 2^3 \bullet \frac{1}{8} + 3^3 \bullet \frac{3}{8} + 4^3 \bullet \frac{3}{8} + 5^3 \bullet \frac{1}{8} = \frac{406}{8}$$

The mean deviation is:

$$E|X - \mu| = E|X - 3.5| = |2-3.5| \bullet \frac{1}{8} + |3-3.5| \bullet \frac{3}{8} + |4-3.5| \bullet \frac{3}{8} + |5-3.5| \bullet \frac{1}{8} = \frac{6}{8} = 0.75$$

The third moment about the origin is:

$$E(X - \mu)^3 = \mu_3' - 3\,\mu_2'\mu + 2\mu^3 = \frac{406}{8} - 3 \bullet \frac{104}{8} \bullet \frac{28}{8} + 2 \bullet \left(\frac{28}{8}\right)^3 = 0$$

Therefore, the coefficient of skewness $\alpha_3^* = \frac{\mu_3}{(\sigma)^3}$ equals zero which upon inspection is to be expected by noting that the distribution is symmetric around 3.5.

EXERCISES

In the following exercises, the objective is to find other descriptive measures for a pdf such as quantiles, quartiles, and so on which represent the theoretical counterpart of the same measures as introduced in Chapter 1. As you do the following exercises, compare them with the more familiar measures, such as the mean and standard deviation, in order to be sure that you know what kind of information they provide you about a pdf.

1. For the pdf $f(x) = \frac{3x^2}{8}$ for X in the domain $0 \le x \le 2$, and 0 elsewhere, determine the mean, median, mode and note how similar or dissimilar they may be. Plot the pdf.

2. For the pdf of exercise 1 compute the standard deviation, interquartile range, mean deviation, and coefficient of skewness, α_3^*.

3. Let a pdf $f(x) = \frac{1}{4}x^3$ for X in the domain $0 \le x \le 2$. Determine the mean, median, and mode for this distribution. Plot the pdf on the same axis as a plot of the pdf for exercise 1. Note the differences and similarities of the two plots.

4. For the pdf of exercise 3, compute the standard deviation, interquartile range, mean deviation, and

coefficient of skewness, α_3^*. Compare these results to exercise 2.

5. Suppose the proportion of the time that there is a queue at a printer is modeled by the pdf $f(x) = 6x(1 - x)$ for $0 \le x \le 1$, $f(x) = 0$ elsewhere. Plot the pdf and compute the mean, median and mode.

6. For the pdf of exercise 5, compute the standard deviation, mean deviation, and coefficient of skewness, α_3^*.

7. Functions such as $f(x) = kx^{\alpha}(1 - x)^{\beta}$ for $0 \le x \le 1$ are used to model the behavior of percentages. Determine the value of k for $\alpha = 2$ and $\beta = 1$, also for $\alpha = 1$ and $\beta = 2$. For each case find the mean and mode. Plot the two pdf's and compare them for similarities and dissimilarities. (What problems would you have in finding the median?)

8. For the two pdf's of exercise 7 which would you use to model the percentages on a test having a mean of about 70%?

9. Suppose $f(x) = 1$, $0 < x < 1$ and is 0 elsewhere. (This pdf is referred to as the uniform distribution.) Plot the pdf and then determine the mean, median, mode, standard deviation, interquartile range, mean deviation, and coefficient of skewness, α_3^*.

10. The distribution of exercise 9 is the theoretical distribution from which random numbers are obtained. Go to a random number table (see Table G of the tables at the end of the book) and choose 25, 3-digit random numbers from the table. (If the random numbers are without decimals, simply place a decimal in front of each 3-digit number so that all numbers are between 0 and 1.) Determine the mean, median, mode, standard deviation, and interquartile range for this sample and compare to the theoretical results of exercise 9.

11. The random variable U has the following pdf.

$$f(u) = \begin{cases} ku^2 & -2 < u < 2 \\ 0 & \text{elsewhere} \end{cases}$$

Graph this function and determine the mode(s), the mean and then compute the mean deviation.

3.4 CHEBYSHEV'S INEQUALITY

Introduction

One has to ask the question at this point: "The mean I know, the standard deviation I know. So what?" Just *how much* do these numbers tell you about a distribution of measurements? As you know the mean conveys the concept of "location"—of a typical or middle number for the set of measurements. The standard deviation provides some information about how spread out a population is in an "average" sense. Can any other conclusions be made about the entire distribution of measurements if the mean and standard deviation of the measurements are known? The answer is yes.

Chebyshev's inequality provides a rough rule by which to grasp a feeling for the entire distribution of measurements. However, because of its generality and application to *any* distribution, discrete or continuous, unimodal, bimodal and so forth, it provides only a rough guideline. Nonetheless, it is useful and provides insights that should help you.

Theorem

THEOREM 3.1: Let X be a random variable with probability density function $f(x)$ having a finite mean and variance. Then

$$P(|X - \mu| \le k\sigma) \ge 1 - (\frac{1}{k^2})$$

for any constant $k \ge 1$. (The proof of this theorem, called Chebyshev's inequality, is found in Appendix A of this chapter.)

DISCUSSION

Suppose we know or have a strong guess as to the mean and standard deviation for a distribution $f(x)$ but we don't know its particular functional form $f(x)$. Chebyshev's inequality provides a way of placing bounds on the proportion of the distribution's values to be found within a particular distance of the population mean. The inequality is commonly written:

$$P(|X - \mu| \le k\sigma) \ge 1 - (\frac{1}{k^2}).$$

The interpretation of this expression becomes clearer if we remove the absolute value signs and rewrite the statement of the theorem as:

$$P(\mu - k\sigma < X < \mu + k\sigma) \ge 1 - (\frac{1}{k^2})$$

That is, the proportion of the population within k standard deviations of the mean is at least $1 - (\frac{1}{k^2})$. This relationship can be visualized by looking at Figure 3.16 below. The shaded area is the proportion of the area within "k" standard deviations of the mean, and is at least $(1 - \frac{1}{k^2}) \times 100\%$ of the area. The proportion of the population beyond the shaded area is at most $(\frac{1}{k^2})$.

For example, if we set $k = 2$, we can say that at least 0.75 or 75% of the population is within 2 standard deviations of the mean (that is: $P(\mu - 2\sigma < X < \mu + 2\sigma) \ge 1 - \frac{1}{4^2} = 0.75$); at least 89 % of the population will lie within 3 standard deviations of the mean (that is: $P(\mu - 3\sigma < X < \mu + 3\sigma) \ge 1 - \frac{1}{3^2} = 0.89$); at least 94% of the population will lie within 4 standard deviations, and so forth. (Note that k doesn't have to be integer valued as the previous examples may imply. We could let k equal 1.5 or 2.6 and apply the theorem without difficulty.)

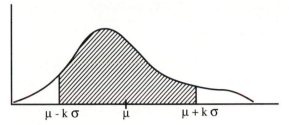

Figure 3.16 Shaded area is "at least $1 - (\frac{1}{k^2})$." Unshaded area is "at most $(\frac{1}{k^2})$."

A Z-score or Standardized Score

A useful transformation of a random variable X, referred to as a Z-score or standardized score, is obtained by subtracting the mean μ from the random variable X and then dividing the result by the standard deviation, i.e.

$$Z = \frac{X - \mu}{\sigma}.$$

Notice that the numerator represents the deviation from X to μ. This deviation is then "standardized" by dividing by the standard deviation.

The interpretation of the value that is produced by this transformation is "Z is the number of standard deviations that X is from the mean μ." This is essentially the same numerical value conceptually that is represented by k in Chebyshev's inequality. Therefore, to refer to the proportion of the population that is within k standard deviations of the mean is to refer to the proportion of the population that is within Z standard deviations of the mean. For instance, if the number X is 2 standard deviations to the right of the mean then $X = \mu + 2\sigma$. Standardizing this value we get:

$$Z = \frac{X - \mu}{\sigma} = \frac{(\mu + 2\sigma) - \mu}{\sigma} = 2 = k.$$

An Application of Chebyshev's Inequality and Rare Events

The concept of a "rare event" is an important concept in statistics. It is a fundamental piece in the logic that we weave to make decisions. For instance, if a "balanced coin" is tossed 20 times and 20 heads occur we all agree that a "rare event" has occurred and our reaction is to question whether the coin was really balanced. In this sense a "rare event" is an event that occurs where the odds are against its occurrence; that is, the probability of the event occurring is very small. It is intuitive that the odds are against the occurrence of 20 heads in 20 tosses.

The situation becomes less obvious if the 20 tosses produced 14 heads. This outcome is obviously not as "rare" as 20 heads. The question is: Is it rare enough that it would still cause us to question the balance of the coin? The principles of probability allow us to calculate the probability of getting 14 or more heads (which is 0.0577), and you have to decide whether the probability is sufficiently small to convince you that it is more reasonable to conclude that the coin is unbalanced rather than balanced.

Convention and tradition suggest using a probability of 0.05 or less as denoting a "rare" probability. Using this convention the probability of 20 heads in 20 tosses is 0.0000 up to 4 decimal places, definitely a rare event. The probability of 14 or more heads is 0.0577, a borderline case, especially when the choice of 0.05 as the cutoff point is chosen subjectively.

Chebyshev's inequality can serve as a useful tool for assessing "rare event" probabilities under very general assumptions. By examination of the theorem, we see that the probability of getting an observation as far away from the mean as k standard deviations is *at most* $\frac{1}{k^2}$. If $k = 4$, this translates into $\frac{1}{16} = 0.0625$, fairly close to the 0.05 convention. If $k = 5$, then we have $\frac{1}{25} = 0.04$. Thus, if an observation is found to be at least 4 to 5 standard deviations from the mean, we might call that observation a rare event. Note that $k = 4.5$ implies that at most $\frac{1}{4.5^2} = 0.0494$, a borderline value for a 5% rare event definition.

[Remember that Chebyshev's Inequality applies to any distribution of any shape and because of its generality, it may not be very precise. For instance, if it is known that the distribution is somewhat bell-shaped, observations more than 2 standard deviations from the mean occur with an approximate probability less than 0.05. Thus any observation from a bell-shaped distribution falling outside 2 standard deviations might be called a "rare event."]

EXAMPLES

EXAMPLE 1: In tossing a balanced coin 20 times, the mean is known to be 10 heads with a standard deviation of $\sqrt{5} = 2.24$ heads. (See Section 4.2 of Chapter 4.) Using to Chebyshev's inequality, what is the probability of getting 16 or more, or 4 or fewer heads?

SOLUTION: The outcome, sixteen or four heads, is 2.68 standard deviations from the mean, that is:

$$\left(\frac{16 - 10}{\sqrt{5}}\right) = 2.68.$$

Therefore, the probability of an outcome that far or farther from the mean is *at most*

$$\frac{1}{2.68^2} = 0.1392,$$

not a rare event using the Chebyshev's approach.

EXAMPLE 2: The amount of cpu time required for several thousands of computer jobs was recorded over several weeks. The mean was 10.3 seconds with a standard deviation of 3.7 seconds. What can be said about the distribution of cpu times? Would a job that required 23 seconds be considered typical? Or unusual?

SOLUTION: At least 75% of the times are between $10.3 \pm 2 \cdot 3.7$ or from 2.9 to 17.7 seconds; at least 89% of the cpu times are between $10.3 \pm 3 \cdot 3.7$ or from 0 to 21.4 seconds. (Obviously, a time cannot be negative.) A job that requires 23 cpu seconds to complete falls at $\frac{23 - 10.3}{3.7} = 3.43$ standard deviations from the mean, not something we would call a rare event.

EXAMPLE 3: An archaeologist working at a site knows from experience that the mean measurement across the forehead from temple to temple of the skulls found in the area was 8.3 inches with a standard deviation of 0.4 inches. A skull from a slightly different location measures 9.4 inches. Would you conclude that the skull is typical or atypical of the other skulls found in the site.

SOLUTION: A measurement of 9.4 inches compared to a mean of 8.3 inches is 2.75 standard deviations from the mean ($\frac{9.4-8.3}{0.4} = 2.75$).   If we are willing to make no more assumptions about the distribution of measurements than knowing the mean and standard deviation, then by Chebyshev's inequality this is not a rare event.

EXERCISES

1. From experience, it is known that the mean height of college-age students is about 70 inches with a standard deviation of about 3 inches. Using this information, what proportion of college-age students will be between 64 and 76 inches in height? What proportion of college-age students will be taller than 79 inches or shorter than 61 inches?

2. The final exam in an introductory statistics course had a mean score of 65 with a standard deviation of 12. What proportion, at most, of A's and E's would be given if those scoring 2 standard deviations away from the mean were to be given and A's and E's?

3. Suppose the following pdf describes the time to failure in hours of a particular integrated circuit when subjected to high temperatures.

$$f(x) = \frac{1}{20}\,e^{-\frac{x}{20}} \qquad x > 0$$

Apply Chebyshev's inequality for $k = 1.5$, and then determine the actual probability using the above model. That is, what is the probability of an observation falling within 1.5 standard deviations of the mean? (Hint: You must first find the mean and standard deviation for this distribution. Then you apply the inequality.)

4. The random variable X has the pdf $f(x) = \frac{1}{5}$ for $1 < X < 6$. Apply Chebyshev's inequality for $k = 2$ and $k = 3$. Find the mean and standard deviation for this pdf and then determine the actual probability of falling within 2 and 3 standard deviations of the mean using the above model.

5. The following data are number of errors per page for 20 beginning type students. Apply Chebyshev's inequality for $k = 2$ and then determine the actual proportion within 2 standard deviations using the data. Data: 2, 8, 10, 13, 14, 12, 3, 0, 14, 0, 2, 2, 13, 2, 9, 8, 8, 10, 14, 20.   [Hint: find the mean and standard deviation of these 20 numbers. Predict the proportion that will be within 2 standard deviations of the mean, and then actually count the number of observations within 2 standard deviations.]

6. An instructor has given the same examination to his students over the course of 6 semesters. Because of the demand for his course and his enrollment policies, the same number of students took the course each semester. He found that the mean over the period of semesters was about 65. He computed the standard deviation of the 6 means and found the standard deviation of the "means" to be 1.2. The students in his most recent semester (the 7th semester) had a mean of 69. Is this a "rare event" or would this represent a fairly typical outcome based on the experience?

7. Short steel rods used as an axle in manufacturing bicycles are to have diameters of 3/8 inches

(0.375"). For quality control purposes the steel rods are measured with calipers at each end and in the middle before the ends are threaded. These three measurements are "averaged" and recorded. A set of 50 of the latest averages is examined and are summarized: The mean of the set of 50 averages is found to be 0.3763 inches with a standard deviation of 0.0003 inches. An inspector measuring another axle off the assembly line obtained measurements of 0.3748, 0.3746, 0.3751 inches. Would this set of measurements be regarded as a "rare event"?

3.5   THE MOMENT-GENERATING FUNCTION

Introduction

The central moments for any pdf are of interest to us with the first moment, the mean, being the most common measure of location that we use. The variance, as we've shown, is a function of both the first and second moments. Higher moments provide information about skewness versus symmetry and so on. In this section we present a useful mathematical result that provides a method of generating the central moments for the pdf's we'll study. This is done by differentiating an appropriate function. This function is referred to as the moment-generating function, (mgf). In addition, it serves another useful purpose besides generating the moments, as will be seen in Chapter 5 that follows.

In this text, we will not expect you to derive the moment-generating function except for fairly simple cases, since the integration or summation involved can become very difficult. However, you should be able to obtain whatever moments are asked for in the specific cases cited.

Definitions

DEFINITION 3.18: The *moment-generating function* (mgf) of a random variable X with its associated pdf is defined to be:

$$M_x(t) = E(e^{tX})$$

THEOREM 3.2: If the mgf function exists for t in the interval $-c < t < c$ for some c then the rth derivative of $M_x(t)$ relative to t which is then evaluated at $t = 0$ will yield the rth moment about the origin μ'_r ; i.e.

$$\mu'_r = d^r \frac{M_x(t)}{dt^r} |_{t=0}$$

Proof: Note that the power series expansion of $e^{(tX)}$ can be written as:

$$e^{(tX)} = 1 + tX + \frac{t^2X^2}{2!} + \dots + \frac{t^r X^r}{r!} + \dots$$

Taking the expected value of each side yields

$$M_x(t) = E(e^{(tX)}) = 1 + tE(X) + \frac{t^2 E(X^2)}{2!} + \cdots$$

Taking the derivative of mgf with respect to t and setting $t = 0$ yields

$$d\frac{M_x(t)}{dt}\Big|_{t=0} = E(X) = \mu$$

Taking the second derivative of mgf and setting $t = 0$ yields

$$d^2\frac{M_x(t)}{dt^2}\Big|_{t=0} = E(X^2) = \mu'_2$$

Taking the rth derivative of mgf and setting $t = 0$ yields

$$d\frac{M_x(t)}{dt^r}\Big|_{t=0} = E(X^r) = \mu'_r$$

DISCUSSION

A moment-generating function (mgf) is a mathematical function defined for most probability density functions by taking an "expected value." As can be seen by the theorem above, using the power series representation of e^x, the moment-generating function can be expanded into an expression involving the various central moments for that probability density function. By taking derivatives, then setting $t = 0$, we produce the various moments of the random variable and thus can find the mean, variance or any other function of the moments of a distribution.

It is also possible to prove mathematically that the moment-generating function for a given random variable X and its associated pdf, is unique to that particular pdf. That is, if two random variables have the same mgf then they also have the same probability density function. This means that a pdf can be uniquely represented by its mgf.

In the next chapter as we examine some of the standard and commonly encountered probability density functions; we will also provide the expression for their moment-generating functions when they exist and are tractable.

EXAMPLE

EXAMPLE 1: Suppose the random variable X has the mgf defined to be $M_x(t) = (1 - p + pe^t)$. Find the first two moments for this random variable.

SOLUTION: The first derivative is $d\frac{M_x(t)}{dt} = pe^t$. Therefore, setting $t = 0$ we get $\mu = p$. The second derivative is also pe^t. Again setting $t = 0$, we get $\mu'_2 = p$. Thus, the first two moments (as well as any other moment for this mgf) is the value p. Therefore, if we want to know the variance we find that $\sigma^2 = p - p^2 = p(1 - p)$. Note that this is the same mean and variance you would get for the pdf $f(x) = p^x(1 - p)^{1-x}$, $x = 0,1$, called the Bernoulli probability density function. You might verify this result.

EXAMPLE 2: Suppose a random variable X has the pdf $f(x) = \frac{1}{3}$ for $x = 0,1,2$ and is 0 elsewhere. Find the moment-generating function for this pdf.

SOLUTION: The moment-generating function is the $E(e^t x)$ which is derived as follows:

$$M_x(t) = E(e^{tx}) = e^{t0} \cdot \frac{1}{3} + e^{t1}\frac{1}{3} + e^{t2}\frac{1}{3} = \frac{1}{3} + \frac{1}{3} \cdot (e^t + e^{2t})$$

EXAMPLE 3: Using the mgf derived in Example 3, find the mean and variance for the random variable X.

SOLUTION: To find $E(X)$ and $E(X^2)$ we take the first two derivatives of the moment-generating function, setting $t = 0$. This gives:

$$\mu = E(X) = d\frac{M_x(t)}{dt}\Big|_{t=0} = \frac{1}{3} \cdot (e^t + 2e^{2t})\Big|_{t=0} = \frac{1}{3} \cdot (1 + 2) = 1$$

$$E(X^2) = d^2\frac{M_x(t)}{dt^2}\Big|_{t=0} = \frac{1}{3} \cdot (e^t + 4e^{2t})\Big|_{t=0} = \frac{1}{3} \cdot (1 + 4) = \frac{5}{3}$$

Therefore, the variance equals:

$$V(X) = E(X^2) - [E(X)]^2 = \frac{5}{3} - 1^2 = \frac{2}{3} .$$

EXERCISES

1. The probability density function for the random variable X has the following form: $f(x) = e^{-x}$ for $x > 0$. This has the moment-generating function $\frac{1}{(1-t)}$. Find the mean and variance for this random variable using the mgf.

2. For the probability density function defined in exercise 1 above find $E(X)$ and $E(X^2)$ by integration using the definition of "expected values."

3. The random variable X has the following mgf: $M_x(t) = \frac{2e^t}{(3 - e^t)}$. Find the first moment using this moment-generating function.

4. Suppose a random variable Y takes on the values 1 and 2 with probabilities of $\frac{1}{3}$ and $\frac{2}{3}$ respectively. Find the moment-generating function for this random variable. [Hint: In the discrete case the moment-generating function would be defined to be: $E(e^{tX}) = \sum_{\text{all } x} e^{tx} \cdot f(x)$.]

5. For exercise 4, use the moment-generating function to find the mean of the random variable Y.

6. For exercise 4, find the mean of the random variable Y using the concept of expectation.

7. Suppose the moment-generating function of the random variable Y is $(\frac{1}{3} + \frac{2et}{3})^3$. Use the moment-generating function to find the mean and variance for the random variable. This random variable has the pdf $f(y) = \frac{3!}{y! \cdot (3-y)!}(\frac{2}{3})^y(\frac{1}{3})^{3-y}$, $y = 0,1,2,3$. Verify the mean and variance for this random variable by finding $E(X)$ and $E(X^2)$.

8. The random variable Z has the mgf $e^{2(e^t - 1)}$. Find the mean and variance of Z.

9. For the probability density function defined in Example 2, the mean and variance were shown to be 1 and $\frac{2}{3}$, respectively using the moment-generating function. (See Example 3.) Verify the mean and variance values for that distribution by taking expected values. Remember this is a discrete distribution when taking expected values.

10. Suppose a discrete random variable Y has the probability density function $f(y) = p^y(1-p)^{1-y}$, $y = 0,1$, and is 0 elsewhere. Find the moment-generating function $M_Y(t)$ and use it to confirm that the mean and variance is p and $p(1-p)$ as shown in Example 1.

3.6 JOINT PROBABILITY DENSITY FUNCTIONS

Introduction

In chapter 1 we introduced the concept of bivariate data. We demonstrated bivariate relationships by using a scatterplot, and then defined the covariance and correlation for a set of bivariate data. We want to pursue the same ideas in the context of a probability model that relates two random variables X and Y together. Such a probability model will be called a joint probability distribution or joint probability density function.

In this section, we define these concepts for the discrete case as well as the continuous case, though we pursue the continuous case more completely than the discrete case.

Definitions

DEFINITION 3.19: Let X and Y be discrete random variables. The joint probability that $X = x$ and $Y = y$ is denoted by

$$f(x,y) = P(X = x, Y = y)$$

and is called the joint probability function or the bivariate probability density function.

Similarly, the cumulative distribution function is denoted by $F(x,y)$ and is defined to be:

$$F(x,y) = P(X \le x, Y \le y) = \sum_{s \le x} \sum_{t \le y} f(s,t).$$

DEFINITION 3.20: Let X and Y be two continuous random variables. Suppose there exists a function $f(x,y)$ such that the joint probability

$$P(a < X < b, c < Y < d) = \int_a^b \int_c^d f(x,y)\,dy\,dx$$

for any $a \le b$, and $c \le d$ and where $f(x,y) \ge 0$, and

$$\int_{-\infty}^{\infty} \int_{-\infty}^{\infty} f(x,y)\,dy\,dx = 1.$$

Then $f(x,y)$ is a bivariate probability density function of X and Y. Also, the bivariate cumulative distribution function of X and Y is denoted by $F(x,y)$ and is defined to be

$$F(x,y) = P(X \le x, Y \le y) = \int_{-\infty}^{x} \int_{-\infty}^{y} f(s,t)\,dt\,ds.$$

DISCUSSION

Suppose we are interested in modeling the joint probability behavior of two measurements, such as wind speed and humidity associated with a cold weather front moving through the Central Plains, or water temperature and suds level associated with a particular detergent. Questions of interest are the association between the two variables of interest, the behavior of one variable for given levels of the other variable and so forth. Such questions arise in the context of a joint probability density function as we look at covariances, correlations, conditional distributions.

In this and the next few sections we examine such joint probability properties. But first we need to get the fundamental notations and concepts defined.

EXAMPLES

EXAMPLE 1: As a very simple first case, where both measurements or random variables are discrete suppose we have two three-sided die that we roll where the numbers on the sides are 1, 2 and 3. Upon a particular roll of the pair of dice we can get as the possible ordered pairs in our sample space:

$$S = \{(1,1), (1,2), (1,3), (2,1), (2,2), (2,3), (3,1), (3,2), (3,3)\}$$

Under the assumption that each die is balanced, then the probability of each ordered pair is equally likely. As a result we could define the joint probability density function to be

$$f(x,y) = \frac{1}{9} \quad x = 1,2,3; \text{ and } y = 1,2,3$$

This probability density function is represented graphically in 3-dimensional space by Figure 3.17.

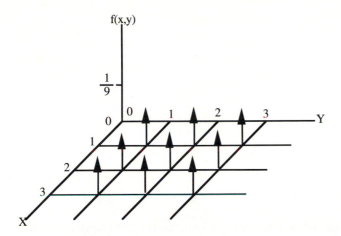

Figure 3.17 An example of a graph of a discrete probability function.

From the definition and by examining Figure 3.17, it can be noted that

$$f(2,2) = \frac{1}{9} = f(1,3) = \dots = f(x,y).$$

However, the cumulative distribution function $F(x,y)$ is a sum of probabilities. Therefore, it can be seen by again closely examining Figure 3.17, that

$$F(2,2) = f(1,1) + f(1,2) + f(2,1) + f(2,2) = \frac{4}{9}.$$

Also note, by the same property, that $F(2,3) = \frac{6}{9}$ and $F(3,3) = \frac{9}{9}$.

If the random variables X and Y are continuous rather than discrete, then the joint probability density function has properties that utilize integration rather than summation to get probabilities and so forth. Likewise, in a logical extension from the two-dimensional setting of X and $f(x)$ where "area" was "probability," in the bivariate, 3-dimensional space involving X, Y and $f(x,y)$, then "volume" and "probability" become analogous concepts. To obtain probabilities or "volumes" under a surface requires the use of double integration. This is illustrated in Example 2 below. (Note: the Appendix of this chapter explains how to evaluate a double integral if you are unfamiliar with the procedure.)

EXAMPLE 2: Consider the following joint probability function:

$$f(x,y) = \begin{cases} x + y & 0 \le x \le 1,\, 0 \le y \le 1 \\ 0 & \text{elsewhere} \end{cases}$$

Find the probability: $P(0 < x < \frac{1}{2},\, 0 < y < \frac{1}{2}) = F(\frac{1}{2}, \frac{1}{2})$

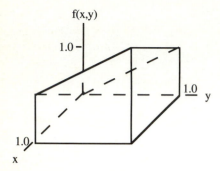

Figure 3.18   Graph of $f(x,y) = x + y$ when $0 \le x \le 1$, $0 \le y \le 1$.

SOLUTION: The graph of Figure 3.18 identifies the shape of $f(x,y)$, and the corresponding "solid" whose volume is to be obtained is shown in Figure 3.19 (the block within the block).

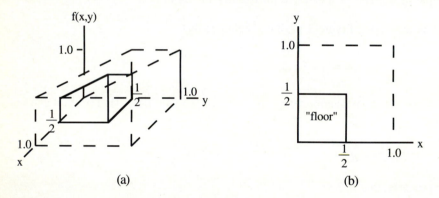

(a)   (b)

Figure 3.19   Graph of $f(x,y) = x + y$ when $0 \le x \le 1$, $0 \le y \le 1$:"Solid" denoting $P(0 < x < \frac{1}{2}, 0 < y < \frac{1}{2})$.

Double integration is simply repeated single integration so that to find $P(0 < x < \frac{1}{2}, 0 < y < \frac{1}{2})$, we would perform the following integration:

$$P(0 < x < \frac{1}{2}, 0 < y < \frac{1}{2}) = \int_0^{1/2} \left\{ \int_0^{1/2} (x + y)dx \right\} dy$$

After integrating on x first (holding y constant) we get

$$F(\frac{1}{2}, \frac{1}{2}) = P(0 < x < \frac{1}{2}, 0 < y < \frac{1}{2}) = \int_0^{1/2} (\frac{1}{8} + \frac{y}{2})dy = \frac{1}{16} + \frac{1}{16} = \frac{1}{8}$$

EXAMPLE 3: For the same joint probability function found in Example 1 find $P(Y < X)$.

SOLUTION: The appropriate region of integration in the x,y *plane* is identified in Figure 3.20 with the "walls and floor" surrounding the proper region being shaded. Note that on the horizontal x,y plane, the set of points where $y < x$ is the triangular shaded region. All the points in this region satisfy the inequality: $y < x$. The resulting integration will produce the volume under the $x + y$ surface for that triangular region. By inspection, it can be concluded that the volume will equal 1/2, as the following integration confirms.

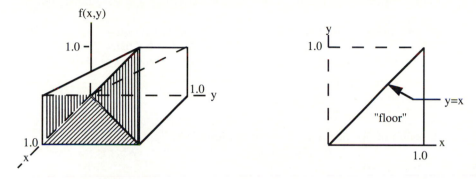

Figure 3.20   Graph of $f(x,y) = x + y$ when $0 \le x \le 1$, $0 \le y \le 1$:   Volume of shaded region is $P(y < x)$.

The probability is obtained by integrating as follows.

$$P(y < x) = \int_0^1 \left\{ \int_y^1 (x + y)dx \right\} dy.$$

$$P(y < x) = \int_0^1 (\frac{1}{2} + y - \frac{3y^2}{2})dy = \{ \frac{y}{2} + \frac{y^2}{2} - \frac{y^3}{2} \}|_0^1 = \frac{1}{2}$$

EXERCISES

1. A janitor, changing the fluorescent light tubes in a large room, inadvertently mixed two previously used but still-working tubes with 5 new tubes. Each fixture requires 3 fluorescent tubes. The janitor picks 3 tubes for the next fixture. Let X denote the number of fresh tubes and Y the number of used tubes out of the three chosen for the light fixture. In a table list the joint probabilities for the various possible values of the two random variables, X and Y.

2. Let X and Y have the joint pdf $f(x,y) = k \cdot \frac{1}{4}$, $0 \le X \le 2, 0 \le Y \le 2$.   On a three-dimensional coordinate system, plot $f(x,y)$ and integrate over the appropriate range to verify that it is legitimate pdf.

3. For the function of exercise 2, calculate $P(X \ge 1, Y \le \frac{1}{2})$. On your plot from exercise 2 identify the volume of the figure you have drawn that represents this probability.

4. Let X and Y have the joint pdf $f(x,y) = k \cdot 2$ for $0 \le x \le y \le 1$. On a three-dimensional coordinate system, plot $f(x,y)$ and integrate over the appropriate range to verify that it is a legitimate pdf.

5. For the function of exercise 3 find $P(0 \le X \le \frac{1}{2}; 0 \le Y \le \frac{1}{2})$. On a three-dimensional coordinate system, plot $f(x,y)$ and integrate over the appropriate range to verify that it is a legitimate pdf.

6. Suppose a joint probability density function (pdf) is defined by the expression $f(x,y) = kx^2y$ for $0 \le x \le 2$, $y \le x$. Find the appropriate region of integration for x and y pairs that satisfy this function and then determine the value of k that will make the entire volume equal unity.

7. For the function of exercise 6 above find $P(y \le \frac{x}{2})$.

8. Suppose the length of life of two microbes exposed to ultra-violet light has the following joint pdf: $f(x,y) = e^{-x-y}$ for $x > 0$ and $y > 0$. Plot the graph of this function. What is the probability that both will "die" before 2 hours?

9. Create a joint pdf for X and Y of the form $f(x,y) = kx(2 - y)$ over an appropriate domain. Make sure you choose the domain such that $f(x,y) \ge 0$. Find the value of k to make this a legitimate probability density function.

10. For the function of exercise 9 find $P(X > 2Y)$. Draw this region carefully on the domain you defined in exercise 9 and then integrate over the correct limits.

3.7 MARGINAL AND CONDITIONAL DISTRIBUTIONS AND THEIR PROPERTIES

Introduction

In section 3.6 the joint probability behavior of two random variables, X and Y, and their basic properties were examined. In this section we continue our examination of joint distributions by first examining the probability behavior of each random variable separately, essentially examining the behavior of X regardless of Y and then of Y regardless of X.

Such an examination produces what we call the marginal distributions of X and Y respectively. You will see that the marginal distributions have properties that are identical to those of univariate distributions discussed in Section 3. 2. The concepts of measures of location and variation, moments, etc., as discussed in Sections 3.3 through 3.5 apply without any conceptual difference other than a notational change. Thus, other than defining what we mean by a marginal distribution, when our starting point is the joint distribution, $f(x,y)$, there are really no new concepts introduced here. Means, variances, quantiles, and so forth are obtained as shown earlier in the chapter.

Once having explored the properties of each random variable unconditionally, we next investigate the conditional behavior of X, given that Y is a certain value (or vice versa). That is, we want to be able to describe the probability behavior of X knowing that Y takes on a specific value. For instance, given that we restrict our study to all college students whose high school GPA was 3.7, we would like to study their college GPA. Such an examination would be called the conditional distribution of "college GPA" given a "high school GPA equal to 3.7." Such an examination produces a distribution known as the conditional distribution.

Once a conditional distribution is obtained, then we can obtain the mean, variance and other measures of interest for that distribution.

Definitions

Marginal Distributions

DEFINITION 3.21: Let X and Y be two continuous random variables having the joint probability density function $f(x,y)$. The marginal pdf's of X and Y respectively are defined to be

$$f_X(x) = \int_{-\infty}^{\infty} f(x,y)dy$$

$$f_Y(y) = \int_{-\infty}^{\infty} f(x,y)dx$$

DEFINITION 3.22: The mean and variance of the marginal distributions are given as:

$$\mu_x = E(X) = \int_{-\infty}^{\infty} x{\cdot}f_X(x)dx, \qquad \sigma_x^2 = V(X) = \int_{-\infty}^{\infty} x^2{\cdot}f_X(x)dx) - \mu_x^2$$

$$\mu_y = E(Y) = \int_{-\infty}^{\infty} y{\cdot}f_Y(y)dy, \qquad \sigma_y^2 = V(Y) = \int_{-\infty}^{\infty} y^2{\cdot}f_Y(y)dy - \mu_y^2$$

Other moments of interest can be defined in the usual way if they exist and are of interest.

Conditional Distributions

DEFINITION 3.23: Given that we have the continuous random variables X and Y and their associated joint (bivariate) probability density function, the conditional distribution of X given Y, and the conditional distribution of Y given X are given, respectively, by:

$$f(x \mid y) = \frac{f(x,y)}{f_Y(y)}$$

$$f(y \mid x) = \frac{f(x,y)}{f_X(x)}.$$

That is, the conditional distributions represent the joint distribution divided by the marginal distribution of the conditioning random variable.

DEFINITION 3.24: The mean of the conditional distribution of Y given X, also called the regression of Y on X, and variance of the conditional distribution of Y given X are defined for continuous X and Y to be:

$$\mu_{y|x} = E(Y|X) = \int_{-\infty}^{\infty} y \bullet f(y|x)dy$$

$$\sigma_{y|x}^2 = V(Y|X) = \int_{-\infty}^{\infty} y^2 \bullet f(y|x)dy - (\mu_{y|x})^2$$

NOTE: Other moments of interest are defined in the same way for conditional distributions as for marginal distributions as for univariate distributions associated with a single random variable.

DISCUSSION

Marginal Distributions

The concepts of the section can be more easily understood and be more intuitively clear by using an example involving two discrete random variables, X and Y. For example, suppose that an experiment is being conducted to determine whether a college freshman has extra-sensory perception (ESP). Eight cards are placed on a table face down. Of the eight cards 3 are Kings, 1 is a Queen and 4 are Jacks. The freshman is told to select two cards one of which is to be a King and the other a Queen. Let X denote the number of Kings selected and Y the number of Queens selected by the freshman. Let us assume that the freshman has no ESP skills and therefore chance controls the selection process. The joint probability function of X and Y are displayed in Table 3.3. (The counting techniques of Chapter 2 were used to obtain the probabilities found in this table. Recognize that X can be either 0, 1 or 2 and Y can be only a 0 or 1.*) In Table 3.3 note that

$$f(1,1) = P(X = 1, Y = 1) = P(1 \text{ King, 1 Queen, and 0 Jacks}) = \frac{3}{28}.$$

Therefore, the probability of chance producing the desired King and Queen is only 3 chances out of 28.

Next note row and column totals in the "margins" of the table. These are called the "marginal probabilities" and show the probabilities associated with one variable at a time. For instance, the "marginal total" associated with the three rows represent the marginal distribution of X and are interpreted in the following way:

$$P(0 \text{ Kings}) = f_X(0) = f(0,0) + f(0,1) = \frac{6}{28} + \frac{4}{28} = \frac{10}{28}$$

$$P(1 \text{ King }) = f_X(1) = f(1,0) + f(1,1) = \frac{12}{28} + \frac{3}{28} = \frac{15}{28}$$

$$P(2 \text{ Kings}) = f_X(2) = f(2,0) + f(2,1) = \frac{3}{28} + 0 = \frac{3}{28}$$

Table 3.3   The joint and marginal probability distributions of X = # of kings, Y = # of queens.

| | | | Y Number of Queens | | |
			0	1	Total
	Number	0	$\frac{6}{28}$	$\frac{4}{28}$	$\frac{10}{28}$
X	of	1	$\frac{12}{28}$	$\frac{3}{28}$	$\frac{15}{28}$
	Kings	2	$\frac{3}{28}$	0	$\frac{3}{28}$
	Total		$\frac{21}{28}$	$\frac{7}{28}$	1.0

Notice that the probability concerning a particular value of X is obtained by summing over all Y values for that value of X. This marginal distribution would be the very same set of probabilities if we were to pose the problem of determining the number of Kings selected when two cards are selected from eight cards containing 3 Kings and 5 non-Kings.

Looking at the column totals, by summing across the values of X, we have the marginal distribution of Y. That is:

$$f_Y(0) = P(0 \text{ Queens}) = \frac{21}{28}$$

$$f_Y(1) = P(1 \text{ Queen }) = \frac{7}{28}$$

For each of the marginal distributions we can determine a mean and variance in a very straightforward way using the principle of expected values. For the random variable X we have:

$$\mu_x = E(X) = 0 \bullet \frac{10}{28} + 1 \bullet \frac{15}{28} + 2 \bullet \frac{3}{28} = \frac{21}{28} = \frac{3}{4}$$

$$\sigma_x^2 = V(X) = E(X^2) - [E(X)]^2 \; [0^2 \bullet \frac{10}{28} + 1^2 \bullet \frac{15}{28} + 2^2 \bullet \frac{3}{28}\,] - (\frac{3}{4})^2 = \frac{45}{112} = 0.40179$$

For the random variable Y we have:

$$\mu_y = 0 \bullet \frac{21}{28} + 1 \bullet \frac{7}{28} = \frac{7}{28} = \frac{1}{4}$$

$$\sigma_y^2 = [0^2 \bullet \frac{21}{28} + 1^2 \bullet \frac{7}{28}\,] - (\frac{1}{4})^2 = \frac{1}{4} \bullet (1 - \frac{1}{4}) = \frac{3}{16} = 0.1875$$

Thus the "expected number of Kings" to be drawn is $\frac{3}{4}$ and the "expected number of Queens" to be drawn is $\frac{1}{4}$. That is, if this experiment were repeated an infinite number of times, the average number of Kings and the average number of Queens drawn out of the infinite number of draws would work out to be $\frac{3}{4}$ and $\frac{1}{4}$, respectively.

In summary, from this example we can see that for the discrete case, the marginal distribution for the random variable X is obtained by summing across the various values of Y for each specific value of X. Similarly, the marginal distribution for Y is obtained by summing across the various values of X. In addition, once the marginal distribution is obtained, the mean and variance are obtained in a straightforward fashion regardless of whether you are looking at the marginal distribution of X, or the marginal distribution of Y.

However, in this section, we are more concerned about the continuous case than the discrete case. We have used this example to develop the logic presented in the definitions above. Pursuing this idea for a pair of continuous random variables, we can conclude that the marginal distribution associated with the random variable X (say), would be obtained by integrating (summing) across the random variable Y. Once this marginal distribution is obtained, it is a legitimate probability density function whose mean, variance and other summary properties can be obtained by the definitions given in earlier sections of this chapter.

To obtain the marginal distribution of Y, we would integrate the joint probability density function with respect to X. Then we would obtain the mean and variance in the logical way. (See the definitions above.)

From a geometric point of view, a marginal distribution becomes a 2-dimensional projection of the 3-dimensional solid that represents the joint probability density function. This projection is an "average" shape of the three-dimensional solid on the $X, f_X(x)$ plane for the marginal distribution of X, or on the $Y, f_Y(y)$ plane for the marginal distribution of Y. These ideas and procedures are shown in the following as Example 1.

Conditional Distributions

The conditional distribution of Y given X represents the probability behavior of Y for a given value or level of X, unlike the marginal distribution that represents the probability behavior of Y regardless of the value of X. (The conditional distribution of X given Y has the same interpretation as given in the preceding sentence by simply interchanging the X and Y.) Again, refer to Table 3.3, listed here as Table 3.4 with an addition to be explained as we go.

For instance, in the ESP experiment described earlier in this section, suppose that we know that no Kings have been selected. And suppose that we want to know the probability of having 0 or 1 Queens selected. This distribution of probability would represent the probability of Y given $X = 0$. This probability, shown in boldface in Table 3.4, is 0.6 and 0.4 respectively. That is, the probability of having, by chance, no Queens given that no Kings have been selected is $\frac{6}{10}$ and the probability of having 1 Queen given that no Kings have been selected is $\frac{4}{10}$. (These probabilities can also be obtained using the conditional probability argument of Chapter 2.)

As you look at these results and the computations involved, you can see that the conditional probability of Y given X is obtained by taking the joint probability $f(x,y)$ and dividing it by the marginal probability of $X, f_X(x)$. This is in agreement with the definition given above.

The mean and variance are again obtained by the same procedure used for obtaining a mean from any distribution. That is:

$$\mu_{y|x=0} = 0 \cdot \frac{6}{10} + 1 \cdot \frac{4}{10} = 0.4$$

$$\sigma_{y|x = 0}^2 = 0^2 \cdot \frac{6}{10} + 1^2 \cdot \frac{4}{10} - (\frac{4}{10})^2 = \frac{4}{10}\frac{6}{10} = 0.24$$

Table 3.4 The joint, and marginal probability distribution of X = # of kings, Y = # of queens and and the conditional probability distribution of Y given X..

			Y Number of Queens		
			0	1	Total
	Number	0	$\frac{6}{28}$	$\frac{4}{28}$	$\frac{10}{28}$
X	of	1	$\frac{12}{28}$	$\frac{3}{28}$	$\frac{15}{28}$
	Kings	2	$\frac{3}{28}$	0	$\frac{3}{28}$
	Total		$\frac{21}{28}$	$\frac{7}{28}$	1.0

Conditional Distribution of Y given X

			Y Number of Queens		
			0	1	Total
	Number	0	$\frac{\frac{6}{28}}{\frac{10}{28}} = \frac{6}{10}$	$\frac{\frac{4}{28}}{\frac{10}{28}} = \frac{4}{10}$	**1.0**
X	of	1	$\frac{12}{15}$	$\frac{3}{15}$	1.0
	Kings	2	1.0	0.0	1.0

Thus, the expected number of Queens, given that no Kings are drawn is 0.4 with a variance of 0.24.

In summary, we can see that the conditional distribution of Y given X describes the probability behavior of Y at the given value of X. This distribution is obtained by dividing the joint distribution by the marginal distribution of X. This division properly scales the probabilities in the conditional setting so that the total probability accumulates to 1.

This argument for obtaining the conditional distribution carries across logically to the case of continuous X and Y. The mean and variance are defined using integration rather than a summation.

(Graphically, the conditional distribution of Y given X can be thought of as representing the surface that a slice of the three dimensional solid would produce if the slice were taken perpendicular to the x,y plane and done parallel to the y axis at the point $X = x$. The two dimensional figure that results from taking the slice described would have to be adjusted to ensure that it accounted for a total probability of 1. See Figure 3.22 for an example of such an operation.)

The conditional distribution again is a legitimate pdf and thus possesses all the characteristics in which we have previously expressed interest: measures of location, variability, skewness and symmetry. This is demonstrated for the continuous case in the following example.

EXAMPLES

EXAMPLE 1: Suppose a joint probability density function is given as follows:

$$f(x,y) = \begin{cases} x + y & 0 \le x \le 1, \ 0 \le y \le 1 \\ 0 & \text{elsewhere} \end{cases}$$

Find the marginal distributions of X and Y respectively.

SOLUTION: This joint distribution is examined in Section 3.6 of this chapter and some of its probability properties are demonstrated there. The marginal distribution of X is:

$$f_X(x) = \begin{cases} \displaystyle\int_0^1 (x + y)dy = [xy + \frac{y^2}{2}]\Big|_0^1 = x + \frac{1}{2} & 0 < x < 1 \\ \\ 0 & \text{elsewhere} \end{cases}$$

The marginal distribution of Y, by symmetry of integration, is:

$$f_Y(y) = \begin{cases} y + \frac{1}{2} & 0 < y < 1 \\ 0 & \text{elsewhere} \end{cases}$$

These marginal distributions are depicted by the shaded areas in Figure 3.21 below. They show up as trapezoids on the X and $f(x,y)$ axis, and the Y and $f(x,y)$ axis.

It is possible to show that the area of the trapezoid equals 1 and satisfies all other requirements for it to be a legitimate probability density function.

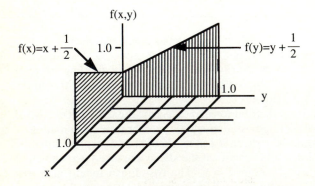

Figure 3.21 Marginal distributions for X and Y shown projected to X, Y, and $f(x,y)$ planes.

The mean of the random variable X is

$$\mu_x = \int\limits_0^1 x(x + \frac{1}{2})dx = \frac{7}{12}$$

and again by symmetry of the function in terms of x and y, the mean of the random variable Y is also

$$\mu_y = \frac{7}{12}$$

The variances of X and Y are

$$V(X) = \int\limits_0^1 x^2(x + \frac{1}{2})dx - (\frac{7}{12})^2 = \frac{11}{144},$$

and again by symmetry

$$V(Y) = \int\limits_0^1 y^2(y + \frac{1}{2})dy - (\frac{7}{12})^2 = \frac{11}{144}.$$

EXAMPLE 2: Suppose that the joint pdf of X and Y is:

$$f(x,y) = \begin{cases} x + y & 0 \le x \le 1, \ 0 \le y \le 1 \\ 0 & \text{elsewhere} \end{cases}$$

Find the conditional distribution of y given x; i.e., $f_{Y|X}(y)$; find the mean of y given x, plot it on the x,y plane as the regression of y on x.

SOLUTION: By definition

$$f_{Y|X}(y) = \frac{f(x,y)}{f_X(x)}$$

which requires the derivation of the marginal distribution of the random variable X which previously has been found to be $f_X(x) = x + \frac{1}{2}$. Therefore, the conditional distribution of y given x is:

$$f_{Y|X}(y) = \frac{(x + y)}{(x + \frac{1}{2})} \qquad 0 < x < 1, \ 0 < y < 1$$

This distribution is shown in Figure 3.22 below for the specific cases where $x = 0$ and $x = 0.5$.

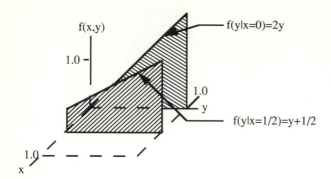

Figure 3.22 Conditional distributions of y given $x = 0$ and $x = 1/2$ shown as shaded areas.

$$\mu_{y|x} = \int_{-\infty}^{\infty} yf(y \mid x)dy) = \int_{0}^{1} \frac{y(x + y)}{(x + (1/2))}dy = \frac{(3x + 2)}{3(2x + 1)} \quad \text{for } 0 \le x \le 1.$$

For the two cases considered in Example 2 where $x = 0$ and $x = \frac{1}{2}$, we have $\mu_{y|x = 0} = \frac{2}{3}$, and $\mu_{y|x = 0.5}$ $= \frac{7}{12}$. These values, as well as the general regression equation, $\mu_{y|x} = \frac{(3x + 2)}{3(2x + 1)}$, are plotted in Figure 3.23 below.

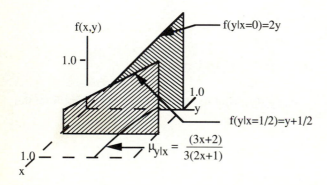

Figure 3.23 Curve of the regression of y on x shown as a curved line on the X,Y plane.

Exercises

Several of the exercises from Section 3.6 are reproduced here. In each case you are to determine the marginal distributions, their means and variance as requested. You also may be requested to find the conditional distribution of Y on X and develop its properties.   Use your work from Section 3.6 to help in your understanding of the concepts of this section.

1.   A janitor, changing the fluorescent light tubes in a large room, inadvertently mixed two previously used but still-working tubes with 5 new tubes. Each fixture requires 3 fluorescent

tubes. The janitor picks 3 tubes for the next fixture. Let X denote the number of fresh tubes and Y the number of used tubes out of the three chosen for the light fixture. Using the table of joint probabilities previously prepared to find the marginal distributions for both X and Y.

2. For the joint probability distribution described in exercise 1 prepare another table showing the conditional distribution of X given Y. Note that if X denotes the rows of the table and Y denotes the columns of the table, the columns will each sum to 1 for the conditional distribution table.

3. Find the means and variances for the conditional distributions of exercise 2 above. Are the means the same for each value of X? What does this imply about the relationship between X and Y?

4. Let X and Y have the joint pdf $f(x,y) = \dfrac{1}{4},\ 0 \le X \le 2, 0 \le Y \le 2$.   Find the marginal distributions $f_X(x)$ and $f_Y(y)$ along with their means and variances.

5. For the joint distribution described in exercise 4 find the conditional distribution of Y given X, and the regression of Y on X, $\mu_{Y\backslash X}$. What does the regression of Y on X imply about the relationship of X and Y? That is, if $X = 0.5$ what would you predict the mean of the Y's to be in contrast to when $X = 1$?

6. Let X and Y have the joint pdf $f(x,y) = 2$ for $0 \le x \le y \le 1$.   Find the marginal distributions $f_X(x)$ and $f_Y(y)$ along with their means and variances.

7. For the joint distribution described in exercise 6 find the conditional distribution of Y given X, and the regression of Y on X, $\mu_{Y\backslash X}$. Plot the regression of Y on X on graph. What does the regression of Y on X imply about the relationship of X and Y?

8. Suppose a joint probability density function (pdf) is defined by the expression $f(x,y) = kx^2y$ for $0 \le x \le 2, y \le x$. Find the marginal distributions $f_X(x)$ and $f_Y(y)$ along with their means and variances.

9. For the joint distribution described in exercise 8 find the conditional distribution of Y given X, and the regression of Y on X, $\mu_{Y\backslash X}$. Plot the regression of Y on X on graph. What does the regression of Y on X imply about the relationship of X and Y?

10. Suppose the length of life of two microbes exposed to ultra-violet light has the following joint pdf: $f(x,y) = e^{-x-y}$ for $x > 0$ and $y > 0$. Find the marginal distributions $f_X(x)$ and $f_Y(y)$ along with their means and variances.

11. For the joint distribution described in exercise 10 find the conditional distribution of Y given X, and the regression of Y on X, $\mu_{Y\backslash X}$. Plot the regression of Y on X on graph. What does the regression of Y on X imply about the relationship of X and Y?

12. Create a joint pdf for X and Y of the form $f(x,y) = kx(2 - y)$ over an appropriate domain. Find the marginal distributions $f_X(x)$ and $f_Y(y)$ along with their means and variances.

13. For the joint distribution described in exercise 12 find the conditional distribution of Y given X, and the regression of Y on X, $\mu_{y\backslash x}$. Plot the regression of Y on X on graph. What does the regression of Y on X imply about the relationship of X and Y?

3.8   EXPECTATIONS AND MOMENTS FOR BIVARIATE DISTRIBUTIONS

Introduction

We have previously defined what is meant by the joint distribution of the random variables X and Y (see Section 3.6). However, no numerical summary measures were presented in that discussion. In this section, we re-examine the joint distribution and define several measures associated with the joint distribution. Of particular interest is the centroid μ_{XY} and the covariance and correlation. The covariance and correlation measures the relationship between the random variables X and Y. If these random variables are independent of one another, the covariance and correlation will be zero. If the random variables are not independent, then in most cases, the covariance and correlation will be non-zero. (There are situations that exist or can be constructed where dependent random variables have a zero covariance and correlation, but in most cases this will not be so.)

The formulas for obtaining these moments and their interpretation will be presented in this section.

Definitions

DEFINITION 3.25: The r-sth product moment for the joint pdf of the continuous random variables X and Y is:

$$E(X^r \bullet Y^s) = \int_{-\infty}^{\infty} \int_{-\infty}^{\infty} x^r \bullet y^s f(x,y) dx dy)$$

Of particular interest is the case of $r = s = 1$, so that we get

$$\mu_{XY} = E(X \bullet Y) = \int_{-\infty}^{\infty} \int_{-\infty}^{\infty} xy f(x,y) dx dy$$

DEFINITION 3.26: The r-sth product moment about the marginal means for continuous random variables is defined to be:

$$E\{(X - \mu_x)^r (Y - \mu_y)^s\} = \int_{-\infty}^{\infty} \int_{-\infty}^{\infty} (X - \mu_x)^r (Y - \mu_y)^s f(x,y) dx dy$$

Of particular interest is the case where $r = s = 1$, so that we get what is called the covariance of the random variables X and Y; i.e.,

$$COV(X,Y) = E\{(X - \mu_x) \bullet (Y - \mu_y)\} = \sigma_{XY}$$

(It is true that $COV(X,Y) = E(XY) - \mu_x \mu_y$. You will be asked to prove it in the exercises of this section.)

DEFINITION 3.27: The correlation coefficient of the joint pdf for the random variables X and Y is:

$$\rho = \frac{\sigma_{XY}}{\sigma_X \sigma_Y} \quad \text{or} \quad \rho = \frac{COV(X,Y)}{\sqrt{V(X) \cdot V(Y)}}$$

DISCUSSION

The covariance and correlation were introduced in Chapter 1 as measures of association between two measurements or random variables. We re-introduce these concepts in this section without modification other than that we define them using the "expected value" notation found in this chapter.

We define the covariance as the expected value of the product of deviations from the mean, i.e. $E(X - \mu_x)(Y - \mu_y)$. As such, the covariance primarily picks up a linear association between X and Y. For instance, if the association between X and Y is approximately linear with a positive slope, then the product of the deviations will tend to be positive rather than negative, and the covariance will turn out to be positive. If the association between X and Y is approximately linear with a negative slope, then the product of the deviations will tend to be negative rather than positive. If the association is curvilinear, then the covariance could be positive, negative or zero, for that matter and it is not possible to say which without specifying the nature of the curvilinearity.

The correlation, ρ, is simply a scaled value of the covariance, so that it ranges between ± 1. Thus, it provides a more interpretable measure of association than does the covariance. If $\rho = 1$, then the association between X and Y is perfectly linear with a positive slope. If $\rho = -1$, then the association between X and Y is perfectly linear with a negative slope. A correlation of zero may mean no *linear* association but doesn't rule out the possibility of other types of association.

In order to define the covariance and correlation precisely, it is useful to introduce a more general concept of "expected value" in the context of a joint distribution. Thus, we define the r–s th product moment, both about the origin and about the marginal means involving the random variables X and Y. These definitions are presented formally in the Definitions above. However, the case of particular interest to us is when r = 1 and s = 1. In that case, the following properties hold true:

$$\sigma_{XY} = COV(X,Y) = E\{(X - \mu_x) \cdot (Y - \mu_y)\}$$

$$= E(XY) - \mu_x \mu_y$$

$$= \mu_{xy} - \mu_x \mu_y$$

In addition, the correlation is such that

$$-1 \le \rho = \frac{\sigma_{XY}}{\sigma_X \sigma_Y} \le +1$$

To illustrate these ideas, we present Example 1.

EXAMPLE

EXAMPLE 1: Suppose a joint probability density function is given as follows:

$$f(x,y) = \begin{cases} x + y & 0 \leq x \leq 1, \ 0 \leq y \leq 1 \\ 0 & \text{elsewhere} \end{cases}$$

Find the marginal means, variances, the covariance and correlation of the random variables X and Y.

SOLUTION: The marginal distributions have been found previously to be

$$f_X(x) = x + \frac{1}{2} \quad 0 < x < 1$$

and

$$f_Y(y) = y + \frac{1}{2} \quad 0 < y < 1.$$

The mean of the random variable X and Y has previously been show to be:

$$\mu_x = \int_0^1 x(x + \frac{1}{2})dx = \frac{7}{12} = \mu_y.$$

The variance of X and Y is

$$V(X) = \int_0^1 x^2(x + \frac{1}{2})dx - (\frac{7}{12})^2 = \frac{11}{144} = V(Y).$$

Applying the definition of the covariance, we get

$$COV(X,Y) = \int_0^1 \int_0^1 xy(x + y)dxdy - (\frac{7}{12})(\frac{7}{12})$$

$$= \frac{1}{6} + \frac{1}{6} - \frac{49}{144} = -\frac{1}{144}.$$

The correlation is therefore computed to be:

$$\rho = -\frac{\frac{1}{144}}{\frac{11}{144}} = -\frac{1}{11}.$$

Note that both the covariance and correlation are negative. This pattern of association would indicate that there is a tendency for X values larger than μ_x to show a slight tendency to pair up with Y values that are below their mean, μ_y and vice versa, i.e. large X's tend to pair up with small Y's and large Y's tend to pair up with small X's. This tendency can be understood by looking at the shape of the pdf and conjecturing about the pattern of x,y pairs that would most likely result from such a distribution. This is demonstrated in Figure 3.24 below. The dots on the horizontal, x,y plane show some of the more likely x,y points to occur. The pattern of points displayed would tend to produce a line with a negative slope. (The exact line that may be plotted through these points can be obtained but is not presented until Chapter 10.)

EXAMPLE 2:   Production of a particular item in a manufacturing plant comes from three different production lines that are undergoing experimentation with the sequencing of operations. Ten items were pulled from the days output with 5 of them coming from line 1, 3 from line 2 and 2 from line 3. Two items are to be randomly selected from the ten for extensive testing. Let X = the number of items selected from line 1, and Y = the number of items selected from line 2. Find the covariance between X and Y knowing that $X + Y \le 2$.

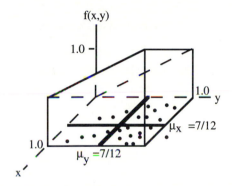

Figure 3.24 Illustration of a random sample of x,y pairs from the pdf $f(x,y) = x + y$.

SOLUTION: The joint distribution of X and Y along with their marginal distributions are shown in the following table.

	$f(x,y)$	Y 0	1	2	$f_X(x)$
	0	$\frac{1}{45}$	$\frac{6}{45}$	$\frac{3}{45}$	$\frac{10}{45}$
X	1	$\frac{10}{45}$	$\frac{15}{45}$	0	$\frac{25}{45}$
	2	$\frac{10}{45}$	0	0	$\frac{10}{45}$
$f_Y(y)$		$\frac{21}{45}$	$\frac{21}{45}$	$\frac{3}{45}$	1.0

The covariance is obtained by the following computations (note μ_x and μ_y have to be found from the marginal distributions to complete the computation and are found to be 1 and 0.6 respectively):

$$\sigma_{xy} = \text{Cov}(X,Y) = E[(X - \mu_x)(Y - \mu_y)] = \{\Sigma\Sigma x \cdot y f(x,y)\} - \mu_x \mu_y$$

$$= 0 \bullet 0 (\frac{1}{45}) + 0 \bullet 1 (\frac{6}{45}) + \dots + 1 \bullet 1 (\frac{15}{45}) + 0 - 1 \bullet (0.6) = -0.2667$$

EXAMPLE 3: Find the correlation of X and Y in Example 2 and interpret it.

SOLUTION: The computation of the correlation requires that we know σ_x and σ_y which are found to be 0.67 and 0.61, respectively. Therefore,

$$\rho = \frac{\sigma_{xy}}{\sigma_x \, \sigma_y} = \frac{-0.2667}{(0.67)(0.61)} = -0.6547$$

The negative correlation should be expected since if X is large, Y must be small, and vice versa.

EXERCISES

Some exercises from Sections 3.6 and 3.7 are reproduced here. Use the results from those exercises to answer the questions asked here.

1. A janitor, changing the fluorescent light tubes in a large room, inadvertently mixed two previously used but still-working tubes with 5 new tubes. Each fixture requires 3 fluorescent tubes. The janitor picks 3 tubes for the next fixture. Let X denote the number of fresh tubes and Y the number of used tubes out of the three chosen for the light fixture. From a table of the joint probabilities calculate the covariance and correlation of X and Y.

2. In exercise 3 of Section 3.7 you were asked to assess whether there was a relationship between X and Y based on examining the means of the conditional distributions. Was your conclusion about a relationship there the same as it is here?

3. Let X and Y have the joint pdf $f(x,y) = \frac{1}{4}$, $0 \le X \le 2, 0 \le Y \le 2$. Find the covariance and correlation between X and Y.

4. Let X and Y have the joint pdf $f(x,y) = 2$ for $0 \le x \le y \le 1$. Find the covariance and correlation between X and Y.

5. Suppose a joint probability density function (pdf) is defined by the expression $f(x,y) = kx^2y$ for $0 \le x \le 2, y \le x$. Find the covariance and correlation between X and Y.

6. Suppose the length of life of two microbes exposed to ultra-violet light has the following joint pdf: $f(x,y) = e^{-x-y}$ for $x > 0$ and $y > 0$. Find the covariance and correlation between X and Y.

7. Create a joint pdf for X and Y of the form $f(x,y) = kx(2 - y)$ over an appropriate domain. Make sure you choose the domain such that $f(x,y) \ge 0$. Find the value of k to make this a legitimate pdf. Find the covariance and correlation between X and Y.

8. For the function of exercise 7 find $P(X > 2Y)$. Draw this region carefully on the domain you defined in exercise 8 and then integrate over the correct limits.

9. Create a function $f(x,y)$ that will have a negative covariance and correlation. Verify that it does

have a negative correlation.   Suppose a random sample of x,y pairs were selected from such a distribution. What would the plot of those x,y pairs look like?

10. Examine the shape of the three dimensional plots of exercises 3, 4, 5, 6, 7, 8 and 9. Are there any general conclusions that would lead you to conclude what kind of a shape leads to a positive correlation, negative correlation or no correlation?

3.9   STATISTICAL INDEPENDENCE

Introduction

Statistical independence is a vital condition in the assumptions we make to validate most of the statistical methods that are used in statistical practice. (See Chapters 7-10.) Therefore, it is essential to have a clear understanding of what this phrase means.

The concept of independence of random variables carries over naturally from our discussion of independent events as presented in Chapter 2 on probability. There, as you recall, independence was defined as a property of probability referring to the effect one event and its occurrence has on the probability of another event taking place. If the occurrence of one event has no effect on the probability of the other event's occurrence, then we say that the two events are independent. In this section we extend this idea to apply to random variables, where the "events" that take place are associated with the set of values that the random variables of interest take on. The formal definition follows.

Definitions

DEFINITION 3.28:   The random variables X and Y are said to be independent if any of the following conditions hold:

1.   The conditional distribution of X given Y is the same as the marginal distribution of X; i.e.,

$$f(x|y) = f_X(x)$$

2.   The conditional distribution of Y given X is the same as the marginal distribution of Y; i.e.,

$$f(y|x) = f_Y(y)$$

3.   The joint distribution of X and Y is equal to the product of the marginal distributions; i.e.,

$$f(x,y) = f_X(x) \bullet f_Y(y)$$

(Recall the definition of independence for events A and B in the Chapter 2 on probability and note the similarities.)

DEFINITION 3.29: Suppose $f(x_1, x_2, ..., x_n)$ is the joint probability density of the n random variables. $X_1, X_2, ..., X_n$ and $f_i(x_i)$ is the marginal distribution of X_i. The random variables $X_1, X_2, ..., X_n$ are independent if and only if

$$f(x_1, x_2, ..., x_n) = f_1(x_1) \bullet f_2(x_2) \bullet \; ... \; \bullet f_n(x_n)$$

for all $(x_1, x_2, ..., x_n)$ within their range.

THEOREM 3.3: Suppose X and Y are independent. Then,

$E(XY) = E(X) \bullet E(Y)$.

PROOF: If X and Y are independent then $f(x,y) = f_X(x) \bullet f_Y(y)$. For continuous X and Y we then have

$$E(XY) = \int_{-\infty}^{\infty} \int_{-\infty}^{\infty} xyf(x,y)dxdy$$

$$= \int_{-\infty}^{\infty} \int_{-\infty}^{\infty} xyf_X(x) \bullet f_Y(y)dxdy$$

$$= \left[\int_{-\infty}^{\infty} xf_X(x)dx \right]\left[\int_{-\infty}^{\infty} yf_Y(y)dy \right]$$

$$= E(X) \bullet E(Y).$$

DISCUSSION

Suppose we have a random variable X which has a pdf of $f(x)$ from which a random sample of n observations is taken. If we let $X_1, X_2, ..., X_n$ denote the values of the random variable X that is associated with each of the n sample observations, and $f(x_i)$ denotes the common "marginal" distribution of the random variable X_i, then by the assumption of random sampling, we will call the random variables $X_1, X_2, ..., X_n$ a set of *independent and identically distributed* (IID) random variables. By the independence condition, and Definition 3.29, their joint distribution will be the product of the marginal distributions. That is:

$$f(x_1, x_2, ..., x_n) = f(x_1) \bullet f(x_2) \bullet \; ... \; \bullet f(x_n)$$

This is a very useful result because it provides a means of obtaining the joint distribution of a set of random variables from knowledge of the marginal distributions when the random variables are independent. By knowing the joint distribution, many of the statistical properties we will study later, probabilities, etc., are derivable, which otherwise might not be the case.

Many of the data sets we study as we go through this text meet the above conditions since we assume that the data are a random sample of observations from some population modelled by $f(x)$. The assumption of a random sample is essentially an assumption of independence. That is, we assume that the measurement obtained from the first observation in a sample has no affect on the outcome of the 2nd sample observation, etc. Thus, if the marginal distribution is known or can be assumed, the joint distribution can be determined. This is one of the important implications of independence.

Next consider three examples applying the above results.

EXAMPLES

EXAMPLE 1: Suppose a population of measurements is modeled by the pdf,

$$f(x) = \frac{1}{2} e^{-\frac{x}{2}} \qquad x > 0$$

from which a random sample of 3 observations is taken, denoted by X_1, X_2, X_3. What is the joint distribution of the random variables?

SOLUTION: By Definition 3.29 above we can write the joint distribution as the product of the marginal distributions. That is,

$$f(x_1, x_2, x_3) = \prod_{i=1}^{3} \frac{1}{2} e^{-\frac{x_i}{2}} = (\frac{1}{2})^3 e^{-(\frac{x_1}{2} + \frac{x_2}{2} + \frac{x_3}{2})}$$

EXAMPLE 2: Suppose a joint probability density function is given as follows:

$$f(x,y) = \begin{cases} x + y & 0 \le x \le 1, \ 0 \le y \le 1 \\ 0 & \text{elsewhere} \end{cases}$$

Determine if the random variables X and Y are independent.

SOLUTION: The marginal distributions have been found previously to be

$$f_X(x) = x + \frac{1}{2} \qquad 0 < x < 1$$

and

$$f_Y(y) = y + \frac{1}{2} \qquad 0 < y < 1$$

If X and Y are independent, then the product of their marginal distributions must necessarily equal the joint pdf. However this is obviously not the case, i.e.

$$f(x,y) = x + y \ne (x + \frac{1}{2})(y + \frac{1}{2}) = f_X(x) f_Y(y).$$

EXAMPLE 3: Suppose a joint probability density function is given as follows:

$$f(x,y) = \begin{cases} xy & 0 \leq x \leq 1,\ 0 \leq y \leq 2 \\ 0 & \text{elsewhere} \end{cases}$$

Determine if the random variables X and Y are independent.

SOLUTION: The marginal distributions are found to be:

$$f_X(x) = \int_0^2 xy\, dy = 2x \ \text{ for } 0 \leq x \leq 1$$

and

$$f_Y(y) = \int_0^2 xy\, dy = \frac{y}{2} \ \text{ for } 0 \leq y \leq 2.$$

For X and Y to be independent the joint distribution must equal the product of the marginals. It is obvious that $f(x,y) = xy = 2x \cdot \dfrac{y}{2} = f_X(x) \cdot f_Y(y)$. Therefore, X and Y are independent random variables.

EXERCISES

Some exercises from Section 3.6 are reproduced here. Determine whether the two random variables are independent using marginal or conditional distribution arguments.

1. A janitor, changing the fluorescent light tubes in a large room, inadvertently mixed two previously used but still-working tubes with 5 new tubes. Each fixture requires 3 fluorescent tubes. The janitor picks 3 tubes for the next fixture. Let X denote the number of fresh tubes and Y the number of used tubes out of the three chosen for the light fixture. Using the marginal distributions for X and Y to determine whether they are independent.

2. For the joint probability distribution described in exercise 1 prepare another table showing the conditional distribution of X given Y. Compare the conditional distribution of X given Y to the marginal distribution of X to prove the dependence between X and Y.

3. Let X and Y have the joint pdf $f(x,y) = \dfrac{1}{4}$, $0 \leq X \leq 2$, $0 \leq Y \leq 2$. Find the marginal distributions $f_X(x)$ and $f_Y(y)$. Does $f(x,y) = f_X(x)f_Y(y)$ for all values for X and Y?

4. Let X and Y have the joint pdf $f(x,y) = 2$ for $0 \leq x \leq y \leq 1$. Find the marginal distributions $f_X(x)$ and $f_Y(y)$. Does $f(x,y) = f_X(x)f_Y(y)$ for all values for X and Y?

5. For the joint distribution described in exercise 4 find the conditional distribution of Y given X. Compare $f_Y(y)$ to $f_{Y|X}(y)$. What does this imply about the relationship of X and Y?

6. Suppose a joint probability density function (pdf) is defined by the expression $f(x,y) = kx^2y$ for $0 \leq x \leq 2$, $y \leq x$. Find the marginal distributions $f_X(x)$ and $f_Y(y)$. Does $f(x,y) = f_X(x)f_Y(y)$?

7. For the joint distribution described in exercise 6 find the conditional distribution of Y given X. Compare $f_Y(y)$ to $f_{Y|X}(y)$. What does this imply about the relationship of X and Y?

8. Suppose the length of life of two microbes exposed to ultra-violet light has the following joint pdf: $f(x,y) = e^{-x-y}$ for $x > 0$ and $y > 0$. Find the marginal distributions $f_X(x)$ and $f_Y(y)$. Does $f(x,y) = f_X(x)f_Y(y)$?

9. For the joint distribution described in exercise 8 find the conditional distribution of Y given X. Compare $f_Y(y)$ to $f_{Y|X}(y)$. What does this imply about the relationship of X and Y?

10. Create a joint pdf for X and Y of the form $f(x,y) = kx(2 - y)$ over an appropriate domain. Find the marginal distributions $f_X(x)$ and $f_Y(y)$. Does $f(x,y) = f_X(x)f_Y(y)$?

11. Suppose the random variables X and Y have the joint pdf given by

$$f(x,y) = \begin{cases} 24y(1-x-y) & x > 0, y > 0, x + y < 1 \\ 0 & \text{elsewhere} \end{cases}$$

Determine whether the two random variables are independent.

3.10 SOME USEFUL RELATIONSHIPS OF LINEAR COMBINATIONS OF RANDOM VARIABLES

Introduction

Suppose you are asked to analyze a set of sample data from a data set found in Chapter one. One of the first things you might do is to find the mean of the data. If x_1 denotes the first observation in the data set, x_2, the second, and so forth, then the sample mean would be denoted by

$$\bar{x} = \sum_{i=1}^{n} \frac{x_i}{n}.$$

Mathematically, this expression is a linear combination of the x's comprising the data set.

The analog of such an operation in this chapter would amount to referring to a set of n random variables $X_1, X_2, ..., X_n$, sampled from a pdf $f(x)$, from which we compute a mean. We might denote this by

$$\bar{X} = \sum_{i=1}^{n} \frac{X_i}{n}.$$

In this case, $\overline{X}$ is a random variable also, being a linear combination of random variables. As such we may be interested in its expected value, $E(\overline{X})$ as well as its variance, $V(\overline{X})$.

Similarly, in many of the situations where we analyze a set of sample data, we often look at *sums* or *means* or other linear combinations of the sample observations. In this section we present a few theorems that provide some useful relationships concerning the expectation and variance of the most common linear combinations that we will investigate.

Theorems

THEOREM 3.4: Let $X_1, X_2, ..., X_n$ be a set of random variables having means $\mu_1, \mu_2, ..., \mu_n$, and variances $V(X_1), V(X_2), ..., V(X_n)$. Then, if $Y = a_1X_1 + a_2X_2 + ... + a_nX_n$

$$E(Y) = a_1E(X_1) + a_2E(X_2) + ... + a_nE(X_n)$$

$$= a_1\mu_1 + a_2\mu_2 ... + a_n\mu_n$$

and

$$V(Y) = \sum_{i=1}^{n} a_i^2 V(X_i) + \sum_{i=1}^{n} \sum_{\substack{j=1 \\ i \neq j}}^{n} (a_i a_j) \text{Cov}(X_i, X_j)$$

COROLLARY 3.1: If the random variables $X_1 X_2, ..., X_n$ are *independent* and we for the sum, $Y = \Sigma a_i X_i$, then

$$V(Y) = \sum_{i=1}^{n} a_i^2 V(X_i)$$

THEOREM 3.5: Let $X_1, X_2, ..., X_n$ be a set of random variables as described in the above theorem, then if

$$Y = a_1X_1 + ... + a_nX_n$$

and

$$Z = b_1X_1 + ... + b_nX_n$$

where the a's and b's are constants, then

$$\text{Cov}(Y,Z) = \sum_{i=1}^{n} a_i b_i V(X_i) + \sum_{i=1}^{n} \sum_{\substack{j=1 \\ i < j}}^{n} (a_i b_j + a_j b_i) \text{Cov}(X_i, X_j)$$

COROLLARY 3.2: Let $X_1, X_2, ..., X_n$ be independent and let Y and Z be defined as above, then

$$\text{Cov}(Y,Z) = \sum_{i=1}^{n} a_i b_i V(X_i)$$

THEOREM 3.6: Let $X_1, X_2, ..., X_n$ and $Y_1, Y_2, ..., Y_m$ be a set of random variables where we let

$$U = a_1 X_1 + a_2 X_2 + ... + a_n X_n$$

and

$$V = b_1 Y_1 + b_2 Y_2 + ... + b_m Y_m$$

Then

$$\text{Cov}(U,V) = \sum_{i=1}^{n} \sum_{j=1}^{m} (a_i b_j) \text{Cov}(X_i, Y_j)$$

DISCUSSION

In many of the practical applications of statistics, we will select a random sample of observations from a population having $f(x)$ as a population model. Many of the computations we make using the sample data can be expressed as a linear combination of the sample values. In order to perform various statistical operations on the sample data it is necessary for us to know the expected value and variance of the particular linear combination we have formed. The above theorems show how to find the mean and variance of various linear combinations.

EXAMPLES

EXAMPLE 1: The sample mean is a linear combination of the X's with each $a_i = \dfrac{1}{n}$; i.e.

$$\bar{X} = \frac{X_1}{n} + \frac{X_2}{n} + ... + \frac{X_n}{n} = \sum_{i=1}^{n} \frac{X_i}{n}$$

Use the above relationships to find $E(\bar{X})$. and $V(\bar{X})$ as follows:

SOLUTION: Let $X_1, X_2, ..., X_n$ be a random sample from the same pdf $f(x)$. Thus, they are a set of random variables which are independent and identically distributed (IID) having a common mean μ and variance σ^2. Applying Theorem 3.4 we get

$$E(\bar{X}) = E(\frac{X_1}{n} + \frac{X_2}{n} + ... + \frac{X_n}{n})$$

$$= \frac{\mu}{n} + \frac{\mu}{n} + ... + \frac{\mu}{n} = \frac{n\mu}{n} = \mu$$

EXAMPLE 2: Use the theorems of this section and the assumptions made in doing Example 1 to find $V(\bar{X})$.

SOLUTION: Applying Corollary 1a for the variance of a linear combination of independent random variables we get

$$V(\bar{X}) = \sum_{i=1}^{n} (\frac{1}{n})^2 V(X_i)$$

$$= \frac{n\sigma^2}{n^2} = \frac{\sigma^2}{n} .$$

Thus, in summarizing Examples 1 and 2, the expected value of the sample mean is the mean of the population, i.e.,

$$E(\bar{X}) = \mu ;$$

the variance of the sample mean is the variance of the population divided by the sample size, n; so that we can write,

$$V(\bar{X}) = \frac{\sigma^2}{n} .$$

These two results express very important properties. The applications to which we will apply them is presented formally in Chapter 6.

EXAMPLE 3: A student takes a mid-term and a final examination in a college biology course. From experience the mean of the mid-term is known to be 72 with a standard deviation of 12 points. For the final examination the mean is known to be 68 with a standard deviation of 10 points, and students generally have a correlation between the mid-term and final exams of 0.8. What is the expected value and variance of the total number of points a student will earn?

SOLUTION: Let Total Points $= X_1 + X_2$, a linear expression. Thus

$$E(\text{Total Points}) = E(X_1 + X_2) = E(X_1) + E(X_2) = 72 + 68 = 140 \text{ points}$$

$$V(\text{Total Points}) = V(X_1 + X_2) = V(X_1) + V(X_2) + 2 \, \text{Cov}(X_1,X_2)$$

$$= 12^2 + 10^2 + 2\bullet(0.8)\bullet12\bullet10 = 436$$

Note that $\rho = \dfrac{\sigma_{xy}}{\sigma_x \bullet \sigma_y}$ and therefore, $\sigma_{xy} = \rho\bullet\sigma_x\sigma_y$, which is used to evaluate the covariance part of the above expression.

$$\text{Standard Deviation (Total Points)} = \sqrt{436} = 20.88 \text{ points}$$

EXAMPLE 4: Consider two linear combinations involving a set of independent and identical random variables X_1, X_2, and X_3 having the common mean μ and variance σ^2.

$$Y = X_1 + 0{\bullet}X_2 - X_3 = X_1 - X_3$$

$$Z = X_1 - 2X_2 + X_3.$$

What are the expected values, variances and covariance of Y and Z?

SOLUTION: Using Theorem 3.4, Theorem 3.5 and its corollary 3.2 for independent random variables, we have

$$E(Y) = E(X_1 - X_3) = E(X_1) - E(X_3) = \mu - \mu = 0$$

$$V(Y) = V(X_1 - X_3) = V(X_1) + (-1)^2 V(X_3) = \sigma^2 + \sigma^2 = 2\sigma^2$$

$$E(Z) = E(X_1 - 2X_2 + X_3) = E(X_1) - 2E(X_2) + E(X_3) = \mu - 2\mu + \mu = 0$$

$$V(Z) = V(X_1 - 2X_2 + X_3) = V(X_1) + (-2)^2{\bullet}V(X_2) + V(X_3) = \sigma^2 + 4\sigma^2 + \sigma^2 = 6\sigma^2$$

$$\text{Cov}(Y,Z) = 1{\bullet}1{\bullet}V(X_1) + 0{\bullet}(-2)V(X_2) + (-1){\bullet}(1)V(X_3) = \sigma^2 - \sigma^2 = 0.$$

[Note: Because the covariance of Y and Z is zero, Y and Z are said to be orthogonal. Also because the coefficients of the X's in Y sum to zero as do the coefficients in Z, Y and Z are each said to represent linear contrasts.]

EXERCISES

1. Let $U = X + Y$, and $V = X - Y$. Find $E(U)$ and $V(U)$, $E(V)$ and $V(V)$. Assume $E(X) = \mu_x$, $V(X) = \sigma_x^2$, $E(Y) = \mu_y$, $V(Y) = \sigma_y^2$, and $\text{Cov}(X,Y) = \sigma_{xy}$).

2. Let U and V be defined as in exercise 1. Find $\text{Cov}(U,V)$.

3. Suppose in exercise 1, X and Y are identically distributed so that $E(X) = \mu$, $V(X) = \sigma^2$, $E(Y) = \mu$, $V(Y) = \sigma^2$, and $\text{Cov}(X,Y) = \sigma_{xy}$.. Find $E(U)$ and $V(U)$, $E(V)$ and $V(V)$.

4. Suppose ρ denotes the correlation between X and Y. Let X and Y be as defined in exercise 3. Simplify the expressions for $V(X + Y)$ and $V(X - Y)$ in terms of σ^2 and ρ.

5. Let $X_1, ..., X_n$ and $Y_1, ... , Y_m$ represent a set of independent random variables. Assume $E(X) = \mu_x$, $V(X) = \sigma_x^2$, $E(Y) = \mu_y$, $V(Y) = \sigma_y^2$, and $\text{Cov}(X,Y) = 0$. Knowing that $E(\bar{X}) = \mu_x$, $V(\bar{X}) = \dfrac{\sigma_x^2}{n}$, $E(\bar{Y}) = \mu_y$ and $V(\bar{Y}) = \dfrac{\sigma_y^2}{m}$, let $T = \bar{X} - \bar{Y}$. Then find $E(T)$ and $V(T)$.

6. Suppose that $X_1, ..., X_n$ and $Y_1, ..., Y_m$ represent two independent random samples from the same population. Therefore, $E(X) = \mu$, $V(X) = \sigma^2$, $E(Y) = \mu$, $V(Y) = \sigma^2$, and $\text{Cov}(X,Y) = \sigma_{xy} = \rho\sigma^2$. Use the results of exercises 3, 4 and 5 to simplify expressions for the random variable $E(T) = E(\bar{X} - \bar{Y})$ and $V(T) = V(\bar{X} - \bar{Y})$.

7. Given a set of bivariate observations X_i and Y_i let $D_i = X_i - Y_i$. Find $E(D_i)$ and $V(D_i)$. Next let $T = \sum_{i=1}^{n} \frac{D_i}{n}$. Find $E(T)$ and $V(T)$. (Assume $E(X) = \mu_x$, $V(X) = \sigma_x^2$, $E(Y) = \mu_y$, $V(Y) = \sigma_y^2$, and $\text{Cov}(X,Y) = \sigma_{xy}$).

8. Under the conditions of exercise 7, assume the random variables X and Y have identical means and variances and correlation ρ. Find $E(T)$ and $V(T)$.

9. Referring to the random variable T in problems 6, 7 and 8, let $Z = \frac{[T - E(T)]}{\sqrt{V(T)}}$. Form the Z-score expression for each of those three different cases.

10. A weighted mean for calculating a grade from two tests X_1 and X_2, and a final X_3 is:

$$\bar{X} = \frac{X_1 + X_2 + 2 \cdot X_3}{4}.$$

What is the mean and variance of $\bar{X}$ ($E(\bar{X})$ and $V(\bar{X})$) if the test scores are independent and identically distributed with a mean and variance of 70 and 100, respectively?

11. Using the expression for the weighted mean of exercise 10, suppose the X's are identically distributed but are not independent, having a correlation of 0.80. What is $E(\bar{X})$ and $V(\bar{X})$ under these conditions. (Again, assume a mean and variance of 70 and 100 for each of the three random variables.)

3.11 ANALYSIS OF THE CHAPTER EXERCISE

Let P denote the price of a certain commodity (in dollars), and S denote the total sales (in 10,000 units). Also assume these are random variables whose distribution can be approximated with the joint probability density

$$f(p,s) = \begin{cases} 5pe^{-ps} & \text{for } 0.2 < p < 0.4 \text{ and } s > 0 \\ 0 & \text{elsewhere} \end{cases}.$$

Find:

(a) The marginal distribution of p, its mean, variance and 1st and 3rd quartiles.

(b) The effect of price on sales. Predict sales for a given price.

(c) The probability that sales will exceed 20,000 units if the price is $0.35.

SOLUTION: (a) The marginal distribution of p is obtained by integrating as follows:

$$f_P(p) = \int_0^\infty 5pe^{-ps}\, ds = 5 \quad \text{for } .2 < p < .4$$

The mean and variance are:

$$\mu_P = \int_{.2}^{.4} p(5)\,dp = 0.3$$

$$\sigma_P^2 = \int_{.2}^{.4} p^2(5)\,dp - (0.3)^2 = 0.005$$

To find the quantiles we first determine $F_P(p)$, the distribution function for the random variable P. We then form the equation, $q = F_P(p_q)$ from which we can find any quantile of interest. For this specific case we have:

$$F_P(p) = \int_{.2}^{p} 5\,dt = 5(p - 0.2)$$

from which the quantiles can be obtained. Specifically for $q = 0.25$ and $q = 0.75$ we get

$$0.25 = 5(p_{0.25} - 0.2)$$

so that
$$Q_1 = p_{0.25} = 0.25,$$
and

$$0.75 = 5(p_{0.75} - 0.2)$$

so that

$$Q_3 = p_{0.75} = 0.35.$$

Thus, the mean price for the commodity is $.30, with price being distributed "uniformly" over the interval from $.20 to $.40. The interior 50% range is from $.25 to $.35.

(b) To assess the effect of price on sales, we examine the conditional distribution of sales, for a given price. This conditional distribution, applying the definition, is:

$$f_{S|P}(s) = \frac{5pe^{-ps}}{5} = pe^{-sp} \quad \text{for } .2 < p < .4 \text{ and } s > 0$$

The conditional mean of S given P (the regression of sales on price) is:

$$\mu_{S|P} = E(S|P) = \int_{0}^{\infty} s(pe^{-ps})ds = \frac{1}{p} \text{ for } .2 < P < .4.$$

This shows an inverse relationship between sales and price. Therefore, as the price increases, sales decrease. For instance, for a price, say, of \$.25, the expected sales would be $\frac{1}{.25} = 4$ or 40,000 units.

If the price were to increase to $\$0.33\frac{1}{3}$, then sales would decrease to $\frac{1}{0.33\frac{1}{3}} = 3$ or 30,000 units.

(c) To compute the probability of selling more than 20,000 units if the price is \$.35 we evaluate the conditional pdf of S given P, $f_{S|P}(s) = pe^{-sp}$:

$$P(20,000 \text{ or more units}) = P(S \geq 2| \, p = .35) = 1 - F_{S|P = 0.35}(2)$$

$$= 1 - \int_{0}^{2} (.35)e^{-0.35s}\, ds = 1 - (1 - e^{-0.35(2)}) = 0.4966,$$

or about a 50% chance.

CHAPTER SUMMARY

In this chapter we have presented the basic theoretical properties and vocabulary of probability density functions for one and two random variables. You should be able to verify that a pdf $f(x)$ is a valid pdf and find its cumulative distribution function $F(x)$, its mean and variance, and its various quantiles. For a joint probability density function you should be able to find the marginal distributions, their means and variances, etc., the conditional distribution of Y given X or X given Y, the conditional mean, called the regression of Y on X (or X on Y), and the covariance and correlation of Y on X.

In addition you have been shown Chebyshev's Inequality and some of its implications for "rare event" applications. You should be able to generate the various moments using the moment-generating function for a random variable.

For multivariate distributions the implications of independence have been presented and the properties of linear combinations of random variables has been given as it relates to the mean, variance, and covariance of those linear combinations.

A useful property of linear combinations can be summarized as follows: *The expected value of a sum is the sum of the expected values.* In addition, t*he variance of a sum is the sum of the variances **if** the random variables are independent.* If the random variables are not independent then the variance of the sum needs to include the various covariance terms that would appear in such an expression as indicated in the theorems of section 3.10.

This chapter forms the groundwork for Chapter 4 which follows in which various commonly used probability density functions are presented along with their properties and tables of probabilities.

CHAPTER EXERCISES

In the following exercises, you will be asked to utilize many of the concepts covered in this chapter. If you get stuck, recheck definitions and the exercises in the preceding sections for help. Most exercise solutions will follow naturally if you use basic definitions and/or the appropriate theorems.

1. Suppose the time to the next phone call X and the length of the phone call Y are jointly distributed according to the following joint pdf.

$$f(x,y) = \begin{cases} xye^{-(x^2 + y^2)} & \text{for } x > 0 \text{ and } y > 0 \\ 0 & \text{elsewhere} \end{cases}$$

What can you say about the independence of these two random variables? What is $E(X)$ and $E(Y)$ and $V(X)$ and $V(Y)$? What is the probability that the length of the phone call will exceed the length of the wait for the phone call?

2. A local branch bank has both a drive-up teller and a walk-up window. Let X and Y denote the proportions of the time of each business day that the two windows, respectively, are in use. The joint probability density function of the two random variables is given by

$$f(x,y) = \begin{cases} \dfrac{2(x + 2y)}{3} & \text{for } 0 \le x \le 1 \text{ and } 0 \le y \le 1 \\ 0 & \text{elsewhere} \end{cases}$$

What are the means and variances associated with the two random variables and interpret their values in the context of the problem? Are the two random variables independent? If not, what is their correlation? If the drive-up window is busy 50% of the time, what is the probability that the walk-up window will be busy more than 50% of the time?

3. A propane gas supplier keeps a large 1000-liter tank for customer sales. The owner is trying to determine the characteristics of supply and demand during a particular winter month. The amount of propane at the beginning of a given day is denoted by Y and the amount sold to customers is X where obviously, $X \le Y$. The pdf for these two random variables is given by

$$f(x,y) = \begin{cases} 2 & \text{for } 0 < x < y \text{ and } 0 < y < 1 \\ 0 & \text{elsewhere} \end{cases}$$

Are X and Y independent? If not, what is their correlation? What is the probability that more than half the supply will be demanded on a given day? What is the regression of x on y, the demand on supply? If the tank is full, what is the expected demand? If the tank is full what is the probability that demand will exceed 57%?

4. Suppose a set of test scores historically has had a mean of 70 and a standard deviation of 12. A random sample of 64 scores from this population is selected and the mean of these 64 scores is $\bar{x} = 79$. Theoretically what is $E(\bar{X})$ and $V(\bar{X})$?

5. Using the values for $E(\bar{X})$ and $V(\bar{X})$ from exercise 4, compute the Z-score

$$Z = \frac{\bar{x} - E(\bar{X})}{\sqrt{V(\bar{X})}}$$

Use Chebyshev's Equality to decide whether a sample mean of 79 is an unusual occurrence when sampling 64 observations from a distribution having a mean of 70 and a standard deviation of 12.

6. A rechargeable battery is designed to lose very little life from a initial charge and a second charge. A new process is being investigated for the manufacture of such batteries. The average number of hours of standard use under test has been found to be 6 hours with a standard deviation of 1 hour. The correlation from charge to charge of battery life until discharge is believed to be 0.90. If 36 of the new batteries are to be tested for an initial and follow-up charge and X denotes the time to discharge for the initial charge and Y is the time to discharge for the follow-up charge, and $D = X - Y$. What is $E(D)$, and $V(D)$.

7. Suppose in exercise 6 we compute $\bar{D} = \bar{X} - \bar{Y}$, the difference in the average test lifes between charges for the 36 test batteries. What is $E(\bar{D}) = E(\bar{X} - \bar{Y})$ and $V(\bar{D}) = E(\bar{X} - \bar{Y})$?

8. Suppose in the test of the 36 rechargeable batteries described in exercise 6 results in $\bar{D} = 0.40$ hours. Using the results of exercise 7, is the difference in average life an indication that the new process is different than the old process. (Assume that under the old process $E(\bar{D}) = 0.00$.) Use Chebyshev's Inequality to decide whether $\bar{D} = 0.40$ hours is an unusual occurrence compared to an expected value of 0.00.

9. A self-employed individual reimburses himself for gasoline according to the amount spent on gasoline for business travel. The amount of the reimbursement is never more than the expense but may be as little as half the expense. Suppose that the amount spent and the amount of the reimbursement are both random variables that follow a joint probability density function where $X = \$$ spent on gasoline for business and $Y = \$$ for reimbursement. Suppose the joint pdf is

$$f(x,y) = \begin{cases} \dfrac{1}{25}\left(\dfrac{(20 - x)}{x}\right) & 10 < x < 20, \ \dfrac{x}{2} < y < x \\ \\ 0 & \text{elsewhere} \end{cases}$$

Plot the joint pdf and find the marginal distributions for both X and Y and determine their means and variances.

10. For the pdf of exercise 9, determine the correlation of X and Y.

11. For the pdf of exercise 9, determine the conditional distribution of Y given X, and interpret the expression obtained in terms of dollars and cents implications for the individual.

12. Using the pattern of proof in Appendix A, prove Chebyshev's inequality for a discrete random variable X.

13. Suppose a random variable X has the following probability density function:

$$f(x) = \begin{cases} e^{-x} & x > 0 \\ 0 & \text{elsewhere} \end{cases}$$

A random sample is selected from this pdf producing three independent random variables X_1, X_2, and X_3 having the pdf's

$$f(x_i) = \begin{cases} e^{-x_i} & x_i > 0 \\ 0 & \text{elsewhere} \end{cases} \quad i = 1,2,3$$

Find $E(X_i)$, $V(X_i)$, the joint distribution of the X's and $E(X_1 \cdot X_2 \cdot X_3)$, the expectation of the product. Also find the covariance of X_1 with X_2 and generalize to the other covariances possible.

APPENDIX A: PROOF OF CHEBYSHEV'S INEQUALITY

THEOREM 3.1: Let X be a random variable with probability density function $f(x)$ having a finite mean and variance. Then

$$P(|X - \mu| \leq k\sigma) \geq 1 - (\frac{1}{k^2})$$

for any constant $k \geq 1$.

PROOF: The following proof assumes the random variable X is continuous. A similar proof follows for a discrete random variable X. See exercise 12 in the Chapter Exercises.

By definition we can write

$$\sigma^2 = \int_{-\infty}^{\infty} (x - \mu_x)^2 f(x)dx = \int_{-\infty}^{\mu-k\sigma} (x - \mu_x)^2 f(x)dx + \int_{\mu-k\sigma}^{\mu+k\sigma} (x - \mu_x)^2 f(x)dx + \int_{\mu+k\sigma}^{\infty} (x - \mu_x)^2 f(x)dx$$

where the three integrals on the right are defined over the three regions shown in Figure 3.25 below.

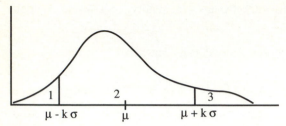

Figure 3.25 Three regions used to prove Chebyshev's theorem.

Since the integrand $(x - \mu_x)^2 f(x)dx$ is non-negative then we can drop out the middle term of the three expressions above and write the inequality

$$\sigma^2 \geq \int_{-\infty}^{\mu-k\sigma} (x - \mu_x)^2 f(x)dx \quad + \quad \int_{\mu+k\sigma}^{\infty} (x - \mu_x)^2 f(x)dx$$

Looking at the second integral it is true that for any x in the region $\mu + k\sigma \leq x < \infty$ then $x \geq \mu + k\sigma$ so that $x - \mu \geq k\sigma$ and thus $(x - \mu)^2 \geq k^2\sigma^2$. Similarly, for the first integral it is true that $x \leq \mu - k\sigma$ and therefore $|x - \mu| \geq k\sigma$ so that $(x - \mu)^2 \geq k^2\sigma^2$. It follows by substitution that

$$\sigma^2 \geq \int_{-\infty}^{\mu-k\sigma} k^2\sigma^2 f(x)dx \quad + \quad \int_{\mu+k\sigma}^{\infty} k^2\sigma^2 f(x)dx \; = k^2\sigma^2 \left[\int_{-\infty}^{\mu-k\sigma} f(x)dx \quad + \quad \int_{\mu+k\sigma}^{\infty} f(x)dx \right]$$

and hence, after cancelling the common term σ^2, that

$$\frac{1}{k^2} \geq \left[\int_{-\infty}^{\mu-k\sigma} f(x)dx \quad + \quad \int_{\mu+k\sigma}^{\infty} f(x)dx \right]$$

However, the two terms in the brackets are nothing more than

$$P(X \leq \mu-k\sigma) + P(X \geq \mu + k\sigma) = P(|X| \geq \mu + k\sigma) = P(|X - \mu| \geq k\sigma)$$

Therefore it is true that

$$P(|X - \mu| \geq k\sigma) \leq \frac{1}{k^2} \; .$$

It also follows that

$$P(|X - \mu| \leq k\sigma) \geq 1 - \frac{1}{k^2} \; .$$

APPENDIX B: DOUBLE INTEGRATION

The evaluation of a double integral requires an integration over a region S which gives the volume under the surface defined by $f(x,y)$. In this course, we find a double integral in Section 3.7 when we work with a bivariate or joint probability density function and are evaluating a probability jointly for X and Y. The volume under the surface formed by $f(x,y)$ represents the probability over that region S. To evaluate the double integral, the principle of repeated integration is used. For example, if the region S is defined as

$$S = \{ a < x < b, \ c < y < d \}$$

then

$$\iint_S f(x,y)dxdy = \begin{cases} \int_c^d [\int_a^b f(x,y)dx]dy & (1) \\[2em] \int_a^b [\int_c^d f(x,y)dy]dx & (2) \end{cases}$$

A repeated integral takes place in two steps. For the repeated integral identified as (1) above, the two steps are:

1. Integrate the expression enclosed in brackets relative to x, regarding y as a constant as you integrate on x.

2. Integrate the result obtained by step 1 with respect to y, doing this over the defined limits of integration.

(The integral defined as (2) above would be evaluated by the same two steps, simply reversing the order of integration.) The outcome of such a two-step integration produces the desired volume which for our situation is the joint probability desired.

4

DISCRETE AND CONTINUOUS PROBABILITY MODELS

OBJECTIVES

When you complete this chapter you should be able to graph, evaluate probabilities or obtain tabled probabilities for any of the special discrete or continuous probability density functions defined here. You should be able to list several real-world situations that could be modelled by each pdf and know the assumptions upon which each pdf is based. You should be able to find the means and variances for each pdf either by direct summing or integrating, or by using the moment-generating function.

Introduction

In the sections that follow, various typical probability density functions are presented. You should note, by examining the examples and exercises in a particular section, that their probability properties are similar though the settings of the problems are different. Thus, it becomes possible for one probability density function to serve as a model for a variety of different but similar real-world populations under investigations.Thus, the biologist, the economist, the engineer or sociologist may use the same probability model to serve them in modelling data from widely different areas of application.

The format of each section will follow the same pattern. The basic assumptions of the model will be given, and the form for the pdf, $f(x)$, will be provided. The mean and variance will follow next along with moment-generating function, $M_x(t)$. In the discussion section the model will be elaborated upon with some common applications being suggested. Graphs of the pdf are also provided and where appropriate, tables for $F(x)$ will be described. If tables are not used, then the actual form of $F(x)$ will be given when it is available.

The first four sections describe common discrete probability models; the remaining sections handle the continuous case.

You should note that this is not an exhaustive set of models, but serves only as an introduction to some of the more commonly used probability models. Other texts and references can supplement this list if a model is inadequate for the data or population that you are studying.

Where We Are Going in This Chapter

In this chapter we present a variety of the most common probability density functions along with their various properties, such as their means, variances, and moment-generating functions. For each distribution you should learn how to graph it or at least be able to identify its shape; know whether the graph is symmetric or skewed or under what conditions it may be either; be able to determine appropriate probabilities by tables or direct computation; and in general be able to describe its various properties.

We will present the pdf's in two broad groups: (1) those whose random variables are discrete, and (2) those that are continuous. You should keep in mind that later on in the text we will assume one of these distributions as a model for a population from which we will sample. We will then use the data from the sample obtained to estimate or test hypotheses about the various unknown parameters involved in the distribution we assume as our model.

The statistical-inferential model, Figure 4.1 indicates the names of the *pdf*'s that are presented in this chapter and could be used to model the parent population.

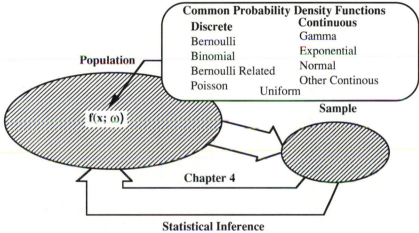

Figure 4.1 SIM—Chapter 4.

Chapter Exercise

The quality of perishable items is a constant concern for both the supplier and the retail outlet of the perishable item. Commercial Bakeries, Inc. has studied the sales and shelf life of its bread extensively. They have found that the number of loaves that are sold within 48 hours of baking out of 100 produced follows a binomial distribution with $p = 0.95$. Those not sold are returned to a day-old bread outlet and sold for 70% of retail cost. The number of loaves sold in the day-old outlet in n loaves stocked is also binomial with $p = 0.90$.

(a)   What is the probability that of 100 loaves baked, 100 will be sold at the regular retail outlet?

(b)   What is the probability that 95 will be sold at the regular retail outlet and of the remaining 5, all will be sold in the day-old bread outlet? If the retail cost of a loaf of bread is $0.90, what are the expected sales figures for 100 loaves baked?

One of the crucial characteristics of a loaf of bread is its moisture content.   If it is too high then it hastens the rate of spoilage. If it is too low then the loaf tends to dry out too quickly and thus becomes stale. It has been found that the ideal moisture content of baked bread is 45% with anything over 50% being too high and under 40% being too low.

(c)   If the moisture content has been found to be approximately normally distributed with a mean of 45% and a standard deviation of 4%, what is the likelihood of a loaf being baked with too high a moisture content? Of 100 loaves produced, how many would you expect to be found in the ideal range of 40-50% moisture content?

4.1   THE BERNOULLI DISTRIBUTION

Introduction

Many types of measurements result in only one of two possible outcomes, success or failure, male or female, defective or nondefective, and so on. Such measurement processes can be called Bernoulli trials. If the outcome of the measurement process is unpredictable, then we can denote the probability of a *success* with the letter p, and the probability of a *failure* as $1 - p$. Such a simple, single observation experiment produces a random variable that follows what is referred to as the Bernoulli distribution.

Definitions

DEFINITION 4.1: The Bernoulli distribution is a probability distribution associated with the discrete random variable X which takes on the values 0 and 1 with probability $1 - p$ and p. The probability distribution is defined as:

$$f(x) = \begin{cases} p^x(1-p)^{1-x} & x = 0,1, \ 0 < p < 1 \\ 0 & \text{elsewhere} \end{cases}$$

The mean and variance of the Bernoulli Distribution are:

$$\mu = p$$

$$\sigma^2 = V(X) = p(1 - p).$$

The distribution is skewed right (positively skewed) if p is less than 0.5, is skewed left (negatively skewed) if p is greater than 0.5 and is symmetric if $p = 0.5$. The moment-generating function for the Bernoulli random variable is

$$M_X(t) = [(1 - p) + e^t p]$$

DISCUSSION

The Bernoulli distribution is associated with any random experiment dealing with two mutually exclusive outcomes such as tossing a coin, recording the sex of an offspring on a single birth, noting whether a light bulb works or doesn't work, asking a yes-no question, answering a true-false question on an exam, and so forth.

In general, the two outcomes can be denoted as a "success" or a "failure". If we are interested in whether a success occurs in a single repetition of the random experiment, then we could let the random variable X equal a 1 for a "success" and a 0 for a "failure".

If we assume that the probability of a success is p, then the probability of a failure will be $1 - p$, and the probability density function for the random variable X is

$$f(x) = p^x (1-p)^{1-x} \qquad \text{for} \qquad x = 0,1.$$

The parameter of this distribution is the probability of success, p. Once the value of p is known, the distribution is completely specified and can be plotted. For this reason we say the Bernoulli distribution is a one-parameter distribution with p being the parameter. It is easy to show that the mean and variance of the Bernoulli distribution are : $\mu = p$ and $\sigma^2 = p(1-p)$. (See exercise 1.) Figure 4.2 shows the graph of $f(x)$ for values of p of 0.4, 0.5 and 0.7.

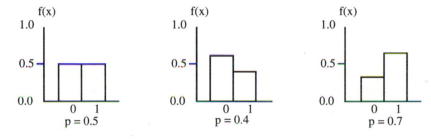

Figure 4.2 Graphs of the Bernoulli distribution for values of p of 0.4, 0.5, and 0.7 respectively.

Note that in Figure 4.2, the distribution is positively skewed when $p = 0.4$, is symmetric for $p = 0.5$, and is negatively skewed when $p = 0.7$. Notice also that $0 \le p \le 1$ and that V(X) is a maximum if $p = 0.5$. (For instance, if $p = 0.1$ or 0.9, then $p•(1-p) = (0.1)•(0.9) = 0.09$. But if $p = 0.5$, then $p•(1-p) = 0.25$. You might check other values for p and compute $p(1-p)$.)

EXAMPLES

EXAMPLE 1. A space mission to the moon is planned. The probability of success for this mission is 0.95. Find the expected value and the variance for the random variable, X, if $X = 1$ denotes a success of this mission and $X = 0$ denotes a failure.

SOLUTION: This random variable is Bernoulli with $p = 0.95$ and $1 - p = 0.05$. Therefore, E(X) = 0.95, and V(X) = (0.95)(0.05) = 0.0475. The standard deviation is $\sqrt{0.0475} = 0.2179$.

EXAMPLE 2: The seal on plastic quart bottles of oil is found to be broken in 1 out of every 500 quarts inspected. Define a Bernoulli random variable for this example and find its expected value.

SOLUTION: A given quart of oil, by inspection, either has a faulty seal or not. Let $Y = 0$ if the seal is good and 1 if the seal is broken. Therefore, $p = \dfrac{1}{500}$. Therefore, $E(Y) = \dfrac{1}{500}$ and $V(Y) = \dfrac{1}{500} \cdot \dfrac{499}{500}$.

EXAMPLE 3:   In a marketing research study of the users of a particular brand of gasoline, participants in the study were asked if they intended to buy a new car within the next six months. Define a Bernoulli random variable for this example and find its expected value.

SOLUTION: A respondent to the question either reply with a yes or no (or some answer other than yes). Let $X = 1$ if the answer is yes and 0 if the answer is not yes. If p is the true proportion of respondents who answer yes, the $E(X) = p$ and $V(X) = p \cdot (1 - p)$.

EXERCISES

1.   Derive the mean and variance of the Bernoulli distribution. Show that $\mu = p$ and $\sigma^2 = p(1 - p)$.

2.   Verify that the moment-generating function for the Bernoulli distribution is

$$M_X(t) = [(1 - p) + e^t p].$$

3.   Using the mgf of the Bernoulli distribution, show that the mean and variance of the distribution is p and $p(1 - p)$, respectively.

4.   A fair die is rolled. What is the expected value and the variance of a random variable X if: (a) A success is associated with rolling a "six"? (b) A success is associated with an even number? (c) A success is associated with rolling a number *larger* than 4?

5.   Ralph Hardluck has the opportunity to win one million dollars. All he has to do is pick an ace from a deck of cards while blindfolded. (Remember that a deck has 52 cards in it of which 4 are aces.) What is the expected value and the variance associated with the random variable associated with Ralph's winning the money? That is, a success denotes a win, and a failure denotes a loss.

6.   In the processing of regular 35mm film, it has been found that 3 exposures out of 20 are overexposed. Define a Bernoulli random variable for this situation.

7.   When asked on a student survey what their gpa is, 60% of students report a value that is higher than the gpa listed on their student records. Define a Bernoulli random variable to reflect the tendency to overstate the gpa. What is its expected value and variance.

8.   In a vegetable processing plant, corn arriving for canning is classified as either *too young, young and tender, mature, too mature* or *old*. Producers are then paid according to a scale with the classification—*young and tender*—receiving the highest return. Define a Bernoulli random variable on this multi-classification system with $Y = 1$ for the *young and tender* classification.

9.   An urn contains 4 red and 6 black balls in it. A ball is randomly selected from the urn. What is

the expected number of red balls to appear? What is the variance?

10. An urn contains 4 red and 6 black balls in it. A second urn contains 5 red and 5 black balls and a third urn contains 8 red and 2 black balls. A ball is selected from each of the three urns, randomly. What is the expected number of red balls in the three balls selected? (Hint: Let $Y_1 = 1$ if a red ball is selected from urn 1, and 0 otherwise, ..., and $Y_3 = 1$ if a red ball is selected from urn 3. Then let $Y = Y_1 + Y_2 + Y_3$. Note that Y is the sum of three independent Bernoulli random variables. Apply the principles of section 3.10.)

11. For the sampling situation described in exercise 9, what is the variance of $Y = Y_1 + Y_2 + Y_3$?

12. Prove that the maximum for the variance of the Bernoulli random variable occurs when $p = \dfrac{1}{2}$.

4.2   THE BINOMIAL DISTRIBUTION

Introduction

Consider a sequence of n independent and identical Bernoulli random variables. Suppose we are interested in their sum; i.e., the total number of successes in the n trials. For instance, instead of simply tossing a coin once to observe whether a heads or tails occurs, we toss it 10 times and observe the total number of heads (or tails). Or we may observe 20 births in a hospital and report the number of births in the 20 which are female. The random variable associated with the number of successes in the sequence of n Bernoulli random variables is called a binomial random variable.

More generally, if the random variable X_1 is Bernoulli with parameter p and we conduct the experiment that produces a Bernoulli outcome (a 0 or 1 occurs) we say we have a "Bernoulli trial." Next, if we repeat the experiment independently of the first, then we say we have two indpendent and identical Bernoulli trials. (They are identical since the parameter p is the same for both trials.) The repetition of a basic Bernoulli experiment n times, each repetition being indepenentt of any other repetition will then give us n identical and independent Bernoulli trials. If we define a new random variable as Y, the total number of successes in the n Bernoulli trials, Y will follow the distribution called a binomial distribution.

There are many real-world examples that meet these assumptions closely enough (with a coin-tossing experiment serving as a prototype) so that the *binomial distribution* becomes an acceptable modelling distribution for them. This section introduces the formal properties of the binomial distribution and the examples and exercises will provide some practical applications of this distribution.

Definitions

DEFINITION 4.2: The *binomial distribution* is the pdf of a random variable X which is the sum of n independent and identical Bernoulli random variables, (that is, $X = X_1 + ... + X_n$ where each Bernoulli random variable, X_i, has the same probability of success, p). As such the binomial random variable is discrete and takes on any value from 0 to n. The binomial random variable X represents the total number of successes that occur in the n "Bernoulli trials" and has the probability density function

$$f(x;n,p) = \begin{cases} \dfrac{n!}{x! \bullet (n-x)!}\, p^x (1-p)^{n-x} & x = 0, 1, 2, ..., n \\[2mm] 0 & \text{elsewhere} \end{cases}$$

The distribution has a mean and variance of:

$$\mu = E(X) = np$$

$$\sigma^2 = V(X) = np(1-p),$$

and its moment-generating function is:

$$M_X(t) = [(1-p) + e^t p]^n.$$

The distribution is negatively skewed if p exceeds 0.5, is positively skewed if p is less than 0.5 and is symmetric if $p = 0.5$. The parameters of the binomial distribution are n and p which, if known, completely specify the distribution. In almost all cases, the number of independent Bernoulli trials, n, will be known though one may not know the probability of success on a single trial.

DISCUSSION

Many experiments possess common, basic characteristics of multiple Bernoulli random variables. For example, observing the number of heads in 10 tosses of a coin (10 Bernoulli trials), counting the number of male births in a list of n births at a local hospital, the number of correct answers on a 20-question true-false test assuming the test preparer makes questions that are independent in content from one another and such that the probability of getting a correct answer is the same from question to question (20 Bernoulli trials) , and so on. When these above situations are encountered then we can formulate the probability density function in a straight-forward fashion if we can assume they are associated with independent and identical Bernoulli random variables.

To understand the formulation of this probability distribution, first consider a sequence of x successes followed by a sequence of n-x failures in a total of n observations. If the probability of a success on a given trial is p, and if all trials are independent, then the probability for this particular sequence of x successes followed by n-x failures would be the product:

$$\left. \begin{array}{cc} x \text{ Successes} & n-x \text{ Failures} \\ S\,S\,...\,S & F \quad F \quad ... \quad F \\ p \bullet p \bullet\, ... \bullet p & (1-p) \bullet (1-p) \bullet\, ... \bullet (1-p) \end{array} \right\} = p^x (1-p)^{n-x}.$$

However, this simply gives the probability for that specific sequence of x successes and $n-x$ failures. It is easy to confirm that any other sequence containing x p's and $n-x$ $(1-p)$'s produces the same probability and that there are

$$\frac{n!}{x!(n-x)!} = \binom{n}{x}$$

different sequences that produce x successes and $n - x$ failures. (See Chapter 2, Appendix A, Theorem 2.4 and the notation of Theorem 2.5.) Therefore, the probability of getting x successes in n independent and identical Bernoulli trials is:

$$P(X = X) = f(x;n,p) = \begin{cases} \binom{n}{x} p^x (1-p)^{n-x} & x = 0, 1, 2, ..., n \\ 0 & \text{elsewhere} \end{cases}$$

Note that the notation $f(x;n,p)$ is simply a clarifying notation to indicate that the parameters in the distribution are n and p.

To evaluate a given probability you substitute the appropriate value of x into the equation above. For example if $n = 10$ and $p = 0.2$, suppose we want to find the probability of getting 4 successes. Therefore, we would get:

$$f(4;10,0.2) = \binom{10}{4}(0.2)^4 (1 - 0.2)^{10-4} = \frac{10!}{4! \cdot (10-4)!}(0.2)^4 (1 - 0.2)^{10-4} = 0.0881.$$

The derivation of the expected value and variance can be obtained by applying the definitions: $E(X) = \Sigma x f(x;n,p)$, and $V(X) = E(X^2) - [E(X)]^2$. But the mean and variance are much more simply derived by noting that the binomial random variable X, the number of successes in n Bernoulli trials, is the sum of n independent and identical Bernoulli random variables. That is, the binomial random variable X is

$$X = X_1 + X_2 + ... + X_n,$$

where the X_i are Bernoulli random variables each with identical parameters p. Then, from Chapter 3, Section 10, we get

$$E(X) = E(X_1 + X_2 + ... + X_n) = E(X_1) + ... + E(X_n)$$

$$= p + p + ... + p = np$$

and

$$V(X) = V(X_1 + X_2 + ... + X_n) = V(X_1) + ... + V(X_n)$$

$$= p(1-p) + p(1-p) + ... + p(1-p) = np(1-p)$$

GRAPHS

You can see in Figure 4.3 below some of the graphical properties of skewness and symmetry for $p = 0.2$, 0.5 and 0.9 with $n = 5$. Though this is only three graphs out of an infinite number of graphs that could be made, the basic features you see generalize logically for other values of n and p.

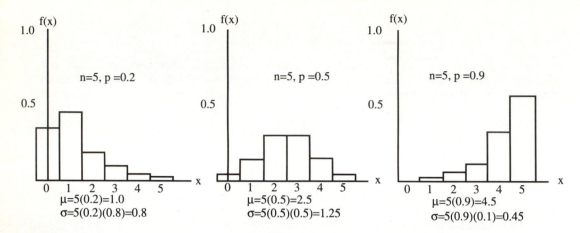

Figure 4.3 Graphs of the binomial distribution for $n = 5$ and p of 0.2, 0.5, and 0.9 respectively.

TABLES

The cumulative distribution for the binomial distribution is defined as:

$$F(x) = \sum_{k=0}^{x} \frac{n!}{k! \cdot (n-k)!} p^k (1-p)^{n-k}$$

for integer x. However, this expression can become very tedious to evaluate for all the possible values of x, n and p that may arise. Therefore, it is customary to prepare a selected set of tables for various combinations of the values of x, n and p. (With today's easy access to fairly powerful pocket calculators and desktop computers, the need for extensive binomial tables is much less important than in the past. It becomes a fairly simple function to program on a computer in order to get the desired probability.)

In the appendix at the back of this book, the distribution function $F(x)$ is tabled for values of n = 1 to n = 20, over the values of p = 0.01, 0.05, 0.10, ..., 0.45 and 0.50. This table then can be used to find various probabilities of interest. In addition, it can also be used to find quantiles for the binomial distribution. It should also be noted that for this, and every other discrete random variable X that takes on integer values x, that

$$f(x) = F(x) - F(x-1).$$

That is, the probability $P(X = x)$ is the difference of two cumulative probabilities. A portion of the binomial tables, Table A in the in appendix is presented here as Table 4.2. This shows the table values for n = 10, 11 with the values of p going from 0.05 up to 0.50 in the increments shown.

Table 4.2 Cumulative binomial probabilities for n = 10, 11, and selected values of p.

n	x	0.01	0.05	0.10	0.15	0.20	0.25	0.30	0.35	0.40	0.45	0.50
10	0	0.9044	**0.5987**	0.3487	0.1969	0.1074	0.0563	0.0282	0.0135	0.0060	0.0025	0.0010
	1	0.9957	0.9139	0.7361	0.5443	0.3758	0.2440	0.1493	0.0860	0.0464	0.0233	0.0107
	2	0.9999	0.9885	0.9298	0.8202	0.6778	0.5256	0.3828	0.2616	0.1673	0.0996	0.0547
	3	1.0000	0.9990	0.9872	0.9500	0.8791	0.7759	0.6496	0.5138	0.3823	0.2660	0.1719
	4	1.0000	0.9999	0.9984	0.9901	0.9672	0.9219	0.8497	0.7515	0.6331	0.5044	0.3770
	5	1.0000	1.0000	0.9999	0.9986	0.9936	0.9803	0.9527	0.9051	0.8338	0.7384	0.6230
	6	1.0000	1.0000	1.0000	0.9999	**0.9991**	0.9965	0.9894	0.9740	0.9452	0.8980	0.8281
	7	1.0000	1.0000	1.0000	1.0000	0.9999	0.9996	0.9984	0.9952	0.9877	0.9726	0.9453
	8	1.0000	1.0000	1.0000	1.0000	1.0000	1.0000	0.9999	0.9995	0.9983	0.9955	0.9893
	9	1.0000	1.0000	1.0000	1.0000	1.0000	1.0000	1.0000	1.0000	0.9999	0.9997	0.9990
	10	1.0000	1.0000	1.0000	1.0000	1.0000	1.0000	1.0000	1.0000	1.0000	1.0000	1.0000
11	0	0.8953	0.5688	0.3138	0.1673	0.0859	0.0422	0.0198	0.0088	0.0036	0.0014	0.0005
	1	0.9948	0.8981	0.6974	0.4922	0.3221	0.1971	0.1130	0.0606	0.0302	0.0139	0.0059
	2	0.9998	0.9848	0.9104	0.7788	0.6174	0.4552	0.3127	0.2001	0.1189	0.0652	0.0327
	3	1.0000	0.9984	0.9815	0.9306	0.8389	0.7133	0.5696	0.4256	0.2963	0.1911	0.1133
	4	1.0000	0.9999	0.9972	0.9841	0.9496	0.8854	0.7897	0.6683	0.5328	0.3971	0.2744
	5	1.0000	1.0000	0.9997	0.9973	0.9883	0.9657	0.9218	0.8513	0.7535	0.6331	0.5000
	6	1.0000	1.0000	1.0000	0.9997	0.9980	0.9924	0.9784	0.9499	0.9006	0.8262	0.7256
	7	1.0000	1.0000	1.0000	1.0000	0.9998	0.9988	0.9957	0.9878	0.9707	0.9390	0.8867
	8	1.0000	1.0000	1.0000	1.0000	1.0000	0.9999	0.9994	0.9980	0.9941	0.9852	0.9673
	9	1.0000	1.0000	1.0000	1.0000	1.0000	1.0000	1.0000	0.9998	0.9993	0.9978	0.9941
	10	1.0000	1.0000	1.0000	1.0000	1.0000	1.0000	1.0000	1.0000	1.0000	0.9998	0.9995
	11	1.0000	1.0000	1.0000	1.0000	1.0000	1.0000	1.0000	1.0000	1.0000	1.0000	1.0000
	...	...	...	...	...	...	...	...	...	...	...	...

EXAMPLES

EXAMPLE 1: Suppose a 10 question, five-part multiple choice test is given and that a person selects answers by spinning a spinner with 5 equal sections on it. (That is, the probability of getting a correct answer on any question is $\frac{1}{5}$ = 0.20.) If a passing grade is obtained only if 7 or more correct answers are obtained, what is the chance of getting a passing grade by using the spinner?

SOLUTION: The random variable X is the number of correct answers on the 10 question exam. Under the conditions given, the selection of an answer to each question constitutes a Bernoulli trial with p = 0.20. There are 10 independent Bernoulli trials, therefore X is a binomial random variable with n = 10 and p = 0.20. Therefore, we compute the following:

$$P(X \geq 7) = 1 - P(X \leq 6) = 1 - F(x = 6; n = 10, p = 0.20)$$

$$= 1 - 0.9991 = 0.0009$$

The entry used from Table 4.2 is shown in bold-faced type.

EXAMPLE 2: The probability of a particular knee injury for high school football players suffering an injury of any kind is about 0.70. A set of 20 hospital records dealing with high school football injuries is being examined. Suppose this set of 20 records represents a random sample of hospital records dealing with high school football injuries.   What is the probability that no records will be found that deal with the particular knee injury of interest?

SOLUTION: Let the random variable X denote the number of knee injuries in the 20 hospital records looked at. In this case, we want to know the $P(X = 0)$ when $n = 20$ and $p = 0.70$. However, the tables don't have probabilities for $p = 0.70$. But with a little ingenuity this problem is easily solved. Suppose we let Y denote the number of injuries that are *not* knee injuries. In addition, $X + Y = 20$. Therefore, $X = 0$ implies that $Y = 20$, where Y is binomially distributed with $n = 20$ and $p = 0.30$. Thus, we can write

$$f(0) = P(X = 0 \mid n = 20, p = 0.70) = P(Y = 20 \mid n = 20, p = 0.30)$$

$$= F_Y(20) - F_Y(19) = 1.000 - 1.000 = 0.000.$$

EXAMPLE 3: A photoelectric cell was manufactured to be sensitive to 40 lumens of light and has been shown to have a probability of 0.05 of not detecting the given amount of lumens. If 10 photoelectric cells are tested for sensitivity to a fixed source of 40 lumens of light, what is the probability that none will fail? At least 1 will fail?

SOLUTION: Let us assume that the 10 photoelectric cells under test are independent of one another all having the same probability 0.05 of failing. If X is the number of cells that fail then X is a binomial random variable with $n = 10$ and $p = 0.05$, and we are interested in $P(X = 0)$ and $P(X \geq 1)$. We can evaluate these expressions by either consulting the tables or using the formula. Using the formula we get:

$$P(X = 0) = \frac{10!}{0! \cdot (10 - 0)!}(.05)^0 (1 - 0.05)^{10-0} = (0.95)^{10} = 0.5987$$

(This entry in Table 4.2 is shown in bold-faced type.)

$$P(X \geq 1) = 1 - P(X \leq 0) = 1 - P(X = 0) = 1 - 0.5987 = 0.4013$$

EXAMPLE 4: Using the information from example 3, calculate $P(2 \leq X < 5)$.

SOLUTION: Because the random variable is descrete, we can solve the above probability expression as shown:
$$P(2 \leq X < 5) = P(2 \leq X \leq 4) = P(1 < X \leq 4) = F(4) - F(1) = 0.9999 - 0.9139 = 0.0860$$

EXERCISES

1. From previous studies it has been found that about 20% of the drivers in a local community wear seat belts when driving. In a follow-up study a student researcher observed 20 cars passing through an intersection. What is the probability that fewer than 10 drivers will be observed wearing seat belts? Fewer than 5? More than 15? Comment on the reasonableness of using the binomial distribution as a model for this study.

2. In a statistical study of word usage, the non-contextual words like "the", "and", "of", and so forth, occur at different rates for different writers/authors. A given author being investigated uses such words on the average of 40% of the time. If a given passage of 50 words is examined what is the probability of finding exactly 20 such "non-contextual" words? Exactly 30?

3. In a "large section" probability course, the instructor has 10 homework assignments due during the semester. Students submit their homework on a random basis as determined by the last digit of their social security number. On the day homework is due, the instructor generates a random number between 0 and 9 to decide who submits homework on that day. What is the probability that a particular student will never be selected? What is the probability of being selected once? Twice? More than twice? What is the probability of being selected every time?

4. Three different professors give a test to their classes on a particular day. One test is a 10 question, True-False test with 7 or more correct answers necessary to pass. The second test is a 10 question, 5-part multiple choice test with 7 or more correct answers necessary to pass. The third test is a 10 question, 10-part multiple choice test with 7 or more correct answers necessary to pass. What is the probability of a person passing each of the tests by guessing? Comment on the assumptions of the binomial distribution in each of the cases described.

5. Compute the mean and standard deviation of each of the random variables under the conditions described in exercise 4.

6. A popular game show on TV asks participants to estimate the price of a common household good without exceeding it retail price. Assume that participants have an approximate knowledge of the price of the object but ultimately it comes down to a guess with a 50:50 chance of being under or over the retail value. If 25 participants are followed and you find that only 8 of the participants exceed the retail value, is it reasonable to conclude they have no skill and are essentially guessing? [Hint: Calculate the probability of getting 8 or fewer participants getting an estimate that is too high. If this is a very small probability would you conclude chance is the sole explanation?]

7. In a taste test concerning a popular regular soft drink versus its "diet" counterpart 20 testers were given one of each in random order and asked to correctly identify the regular versus the diet drink. If there is no difference or if the testers have no skill, then the probability of a correct identification would be argued to be 0.50. Suppose 15 correct identifications were made. Would you conclude there is a detectable difference? (Note: If 15 correct identifications implies a detectable difference then so would 16 or more. Therefore, calculate $P(X \geq 15)$ in order to answer this question.)

8. In the sports world, the phrase "on any given day" is used frequently to imply that one team or player could beat another even if the other is usually a superior team. Suppose we assume that two tennis players A and B meet head to head 15 times during the course of the tennis season. If they are considered equal in ability, what is the probability that player A will beat the other 10 or more times?

9. Under the assumptions of exercise 8, how many times, at least, would you expect player A to beat player B before you would conclude that player A has more than a 50% chance of beating player B? [Hint: What is the value of k such $P(A \text{ wins} \geq k \text{ times}) \leq 0.05$ when $n = 15$ and $p = 1/2$. Here, k is a "rare event" outcome which if it occurs we might be persuaded to think the A is better than B.]

10. Suppose a computer ranking scheme gives one player a 70% probability of beating another player "on any given day." What is the probability that the superior player will win at least 8 of 10 matches played?

4.3   OTHER DISTRIBUTIONS BASED ON BERNOULLI TRIALS

Introduction

A Bernoulli trial is a very common experiment. With ingenuity, any single experimental outcome can be dichotomized and described as a two-event outcome. If a series of Bernoulli trials are performed, then with the appropriate assumptions, a binomial distribution results. There are three other distributions that are discussed in this section that are related to a sequence of Bernoulli trials. However, there are subtle, but crucial differences between them and those of the binomial distribution that lead to different probability distributions. The distributions that we examine here are called the *geometric*, the *negative binomial* and the *hypergeometric* distributions. The assumptions and properties of these three distributions are discussed in this section.

Definitions

DEFINITION 4.3: Suppose that the random variable X denotes the number of trials until the first success occurs in a sequence of independent Bernoulli trials. Then X is a *geometric random variable* and has the probability density function

$$f(x) = \begin{cases} (1-p)^{x-1}p & x = 1,2,\dots \\ 0 & \text{elsewhere} \end{cases}$$

where p is the probability of a success on a given Bernoulli trial. The distribution has a mean and variance of:

$$\mu = E(X) = \frac{1}{p}$$

$$V(X) = \frac{1-p}{p^2}$$

Is moment-generating function is:

$$M_X(t) = \frac{pe^t}{1-(1-p)e^t}$$

DEFINITION 4.4: Suppose that the random variable X denotes the number of trials until the *rth* success occurs in a sequence of independent Bernoulli trials. Then X is a *negative binomial* random variable and has the probability density function

$$f(x) = \begin{cases} \binom{x-1}{r-1}(1-p)^{x-r}p^r & x = r, r+1,\dots \\ 0 & \text{elsewhere} \end{cases}$$

The distribution has a mean and variance equal to

$$\mu = E(X) = \frac{r}{p}$$

$$V(X) = \frac{r(1-p)}{p^2}$$

Its moment-generating function is

$$M_X(t) = \left[\frac{pe^t}{1-(1-p)e^t}\right]^r$$

DEFINITION 4.5: Suppose a finite population of N units contains K units denoted as a "success" and $N-K$ units denoted as a "failure." A sample of n units is selected randomly, *without replacement,* and the random variable X denotes the number of successes in the sample. This random variable is said to have the hypergeometric distribution, defined to be:

$$f(x) = \begin{cases} \dfrac{\binom{K}{x} \cdot \binom{N-K}{n-x}}{\binom{N}{n}} & x = 0,1,\ldots,min(n,K) \\[6pt] 0 & \text{elsewhere} \end{cases}$$

The distribution has a mean and variance of:

$$\mu = n \cdot \frac{K}{N}$$

$$V(X) = \frac{n(N-n)}{N-1} \frac{K}{N}\left(1-\frac{K}{N}\right)$$

A simple expression for the moment-generating function is not available.

DISCUSSION

The Geometric Distribution

Suppose a grade school student, being introduced to beginning ideas of probability, is asked to toss a coin until the first head occurs. Anticipating what *could* happen, the student may only have to toss the coin once, or then again, it may take two tosses, or three, and so forth. It is assumed that each toss is independent of the preceding tosses, that the probability of a head remains the same from toss to toss and is equal to some probability p.

Thus, the experiment consists of a sequence of repeated, independent Bernoulli trials each having common probability p of a head. The random variable is *not* the number of heads that occurs in n tosses of the coin, but rather is the number of tosses of the coin to produce the first head. Such a random variable has a probability density function called the geometric probability density function.

The form of the density function is easy to derive from basic arguments. First, consider a simple Bernoulli random variable Y which takes on one of two outcomes: $Y = 0$ (call it a "failure" or "tails") and $Y = 1$ (call it a "success" or "heads"). Suppose that if on the first occurrence of this experiment a failure occurs, we repeat the experiment. If a failure occurs again, we repeat it a third time, and we continue to repeat the experiment until a "success" finally occurs. Assume the experiments are conducted independently, and the probability of a success p is the same from trial to trial. Then, if we let X = the number of trials needed to produce the first success, X is distributed according to the geometric probability density function.

To evaluate $P(X = x)$, the probability that x trials are needed in order to get the first success, note that this occurs only if the first $x - 1$ experiments resulted in a failure and the xth experiment resulted in a success. The probability of $x - 1$ failures is $(1 - p)^x$, and the probability of a success on the xth trial is p. Therefore, the probability that it takes x trials to produce the first success is simply the product of these two expressions because of the independence of one trial to another. This produces the pdf as follows:

$$P(X = x) = f(x) = \begin{cases} (1 - p)^{x-1}p & x = 1, 2, \ldots \\ 0 & \text{elsewhere} \end{cases}$$

The mean of the geometric random variable is obtained by applying the basic definition for the expected value in the following steps:

$$E(X) = \sum_{x=1}^{\infty} xp(1 - p)^{x-1}$$

$$= p \sum_{x=1}^{\infty} xq^{x-1}$$

where $q = 1 - p$. Then we recognize that $\dfrac{d}{dq}(q^x) = xq^{x-1}$ so that we can write:

$$E(X) = p \sum_{x=1}^{\infty} \frac{d}{dq}(q^x)$$

$$= p\frac{d}{dq}\left(\sum_{x=1}^{\infty} q^x \right) \qquad \text{(the sum of derivatives is the derivative of the sum)}$$

$$= p\frac{d}{dq}\left(\frac{q}{1-q} \right) \qquad \text{(the sum of a geometric series)}$$

$$= \frac{p}{(1-q)^2} = \frac{p}{p^2} = \frac{1}{p} = \mu$$

The variance of the geometric random variable is most easily obtained by an argument using conditioning which we have not examined sufficiently to justify developing the derivation. The outcome, however, is the very simple expression:

$$V(X) = \frac{1-p}{p^2}.$$

EXAMPLES

EXAMPLE 1: Suppose a balanced coin is tossed until a head occurs. What is the probability that it takes 5 trials before the first head occurs?

SOLUTION: For this problem, we have a geometric random variable X and we want to find $P(X = 5)$. The outcome of interest is written $\{TTTTH\}$. It is easy to obtain the probability as:

$$P(TTTTH) = P(X = 5) = \left(\frac{1}{2}\right)^4 \cdot \frac{1}{2} = \left(\frac{1}{2}\right)^5 = \frac{1}{32}$$

EXAMPLE 2: For the example of 3.1, what is the probability that at least 5 trials are needed to produce the first head?

SOLUTION: First note that the probability that at least 5 trials occur before the first head is the same as having the first 4 trials produce tails. Therefore,

$$P(X \geq 5) = P(\text{First 4 trials are tails}) = (1 - \tfrac{1}{2})^4 = \frac{1}{16}.$$

Note that this argument can be generalized to show that

$$P(X \geq x) = (1 - p)^{x-1} \text{ and } P(X \leq x - 1) = 1 - (1 - p)^{x-1}$$

The Negative Binomial Distribution

Suppose we ask the question, "How many tosses of a coin will it take in order to get 5 heads?" This question it is very much like the question, "How many tosses of a coin will it take in to get 1 head?" From the preceding discussion, the latter question is answered by using the geometric distribution with $p = \dfrac{1}{2}$ if the coin is balanced. Therefore, the former question should be seen as a generalized version of a geometric random variable. If we want to know the probability that it will take exactly 10 tosses of a coin to get 5 heads, we must recognize the truth of two statements: it takes exactly 10 tosses of a coin to get 5 heads if, (1) in the first 9 tosses of the coin exactly 4 heads are obtained, and (2) the 10th toss produces a head. If tosses are independent with identical probabilities, then the probability of statement 1 is obtained by using a binomial probability for $n = 9$, $x = 4$, and $p = 0.5$ for a balanced coin. The probability for statement 2 is $p = 0.5$. Therefore, the probability that it will take 10 tosses to produce 5 heads is the product of these two probabilities, i.e.:

$$P(X = 10) = \left[\binom{9}{4}(0.5)^4(1 - 0.5)^{9-4} \right] \cdot (0.5) = 0.12305$$

A general expression that follows this same argument is developed by first defining the random variable X as the number of trials needed to produce the rth success in a sequence of repeated, independent Bernoulli trials. Next, note that the xth trial produces the rth success only if the there are $r - 1$ successes in the first $x - 1$ trials, followed with the xth trial producing a success. The probability of $r - 1$ successes in $x - 1$ trials is the binomial probability:

$$\binom{x-1}{r-1}(1 - p)^{x-r}p^{r-1}$$

The probability of a success on the xth trial is p. Therefore, because of independence of trials, the probability that it requires x trials to produce the rth success is the product of these two probabilities and becomes:

$$P(X = x) = \binom{x-1}{r-1}(1-p)^{x-r}p^r \quad \text{for } x = r, r+1, \ldots$$

This is the expression for the negative binomial distribution.

To find the mean and variance of this distribution by using the "expected value" approach is very involved, but becomes simple by defining a new set of random variables. Let $Y_1, Y_2, \ldots, Y_r$ denote respectively the number of trails, to obtain the first success, the additional number of trials to obtain the 2nd success, the number of additional trials to obtain the 3rd success, and finally the additional number of trials to obtain the rth success. Each Y_i is a geometric random variable with common parameter p. Therefore, $X = Y_1 + Y_2 + \ldots + Y_r$; that is, the negative binomial random variable can be viewed as the sum of r independent and identically distributed geometric random variables. Therefore,

$$E(X) = E(Y_1 + Y_2 + \ldots + Y_r) = \frac{1}{p} + \frac{1}{p} + \ldots + \frac{1}{p} = \frac{r}{p}$$

and

$$V(X) = V(Y_1 + Y_2 + \ldots + Y_r) = \frac{1-p}{p^2} + \frac{1-p}{p^2} + \ldots + \frac{1-p}{p^2} = \frac{r(1-p)}{p^2}$$

For computational convenience, note again that the negative binomial probability density function is the product of a binomial pdf and the geometric parameter p, i.e.

$$P(X = x) = \left[\binom{x-1}{r-1}(1-p)^{x-r}p^{r-1}\right] \cdot p = [Bin(x = r-1 | n = x-1, p)] \cdot p$$

Therefore, the binomial formula or tables can be used to obtain the expression in brackets above, which is then multiplied by the value p.

EXAMPLES

EXAMPLE 3: From a list of automobile accident records, of which 20% are associated with a compact car, records are selected until 3 records are obtained involving compact cars. What is the probability that it will take 10 selections in order to obtain the 3rd record?

SOLUTION: Assume that the records are in random order and that the chance that the next record selected involves a compact car remains constant with probability 0.2. If the random variable X denotes the number of trials until the 3rd compact car record is found ("success"), then we want to find the probability that $X = 10$. To find this probability we have:

$$P(X = 10) = \left[\binom{10-1}{3-1}(1-0.2)^{10-3}(0.2)^{3-1}\right] \cdot (0.2) = (0.3029) \cdot (0.2) = 0.0604$$

EXAMPLE 4: For the study described in example 3 what is the expected number of trials to get the 3rd record for a compact car?

SOLUTION: The mean of the negative binomial is:

$$E(X) = \frac{r}{p} = \frac{3}{0.2} = 15 \text{ records.}$$

The Hypergeometric Distribution

Suppose a jar has N marbles in it of which K are red and $N - K$ are black. The marbles in the jar are stirred until we are convinced that the marbles are distributed randomly throughout the jar. Next, n marbles are selected from the jar, one at a time. As each marble is selected it is set aside before the next marble is selected. (This is called sampling without replacement and has the same probability structure as reaching into the jar and simultaneously taking a handful of n marbles, assuming a random distribution throughout the jar.)

After n marbles are selected there will be $N - n$ marbles remaining in the jar. The number of red marbles in the n selected is counted and is the random variable X. This random variable is called a hypergeometric random variable and is distributed according to the hypergeometric probability density function.

Each selection of a marble also constitutes a "Bernoulli trial" resulting in $X = 0$ if the marble is black or $X = 1$ if the marble is red. However, the sequence of n selections, or Bernoulli trials, is not a set of independent and identical trials. The probability that $X = 1$ changes from trial to trial because the sampling is without replacement. Therefore, there is no particular advantage to this view in deriving the distribution and its properties. Rather, basic counting principles from Chapter 2, Appendix A are used.

To find the $P(X = x)$ we compute:

$$P(X = x) = \frac{\binom{K}{x} \cdot \binom{N-K}{n-x}}{\binom{N}{n}}$$

The form of this expression is argued as follows: If n marbles are randomly selected from N, there are a total of $\binom{N}{n}$ equally likely possible combinations of n that could turn up. (See the counting rules for combinations in Appendix A of Chapter 2). If a given combination is going to have x red marbles in it, then there will also be a total of $n - x$ black marbles. The total possible ways of choosing x red marbles from the K in the jar and then choosing $n - x$ black marbles from the $N - K$ black marbles in the jar is the product:

$$\binom{K}{x} \cdot \binom{N-K}{n-x}$$

Therefore, of the $\binom{N}{n}$ equally likelihood possible combinations, $\binom{K}{x} \cdot \binom{N-K}{n-x}$ of them have the characteristic of having x red and $n - x$ black marbles. The probability is then the ratio of these two numbers which is given above.

The mean and variance of this random variable is derivable using basic principles but won't be covered in detail here. The expressions for them are:

$$\mu = n \cdot \frac{K}{N}$$

$$V(X) = \frac{n(N-n)}{N-1} \frac{K}{N}\left(1 - \frac{K}{N}\right)$$

Notice that if we let $p = \frac{K}{N}$, the proportion of red marbles, then

$$\mu = np$$

and

$$V(X) = np(1-p)\frac{(N-n)}{(N-1)},$$

both of which are expressions very close in form to the mean and variance of a binomial random variable. In fact, the hypergeometric random variable is the sum of n dependent Bernoulli random variables with the probability changing from trial to trial as the number of marbles remaining in the jar is reduced by one for each addition selection.

Instead of marbles in a jar which are red or black, the more conventional terminology for the hypergeometric distribution is to refer to a population of N units of which K are "successes" and $N - K$ are "failures." The random variable X represents the number of successes that are obtained if n units are randomly selected without replacement from the population of N units. The random variable can take on the possible values of 0, 1, 2, ... up to the smaller of n or K. That is, the number of successes in the sample cannot be larger than the sample size or the number of successes within the population, whichever is smaller.

EXAMPLES

EXAMPLE 5: A teenager takes two recharged AA batteries from the recharger and puts them in her backpack, forgetting there are 4 other AA uncharged batteries in the backpack that are identical to the 2 that have been recharged. After jogging to school, with the batteries being completely intermixed, she takes 2 batteries from the backpack and puts them in her portable AM/FM cassette tape player. What is the probability she will get both of the recharged batteries?

SOLUTION: Let the random variable X be the number of recharged batteries selected in the 2 batteries chosen. Then X will be a hypergeometric random variable with $N = 6$, $K = 2$, $N - K = 4$, and $n = 2$. We find the probability to be:

$$P(X = 2) = P(X = x) = \frac{\binom{2}{2} \cdot \binom{4}{0}}{\binom{6}{2}} = \frac{1}{15}.$$

EXAMPLE 6: For the situation described in example 5, what is the expected number of recharged batteries that will be obtained?

SOLUTION: The mean of the hypergeometric distribution with $N = 6$, $K = 2$, $N - K = 4$, and $n = 2$, computes to be:

$$\mu = n \cdot \frac{K}{N} = 2 \cdot \frac{2}{6} = \frac{4}{6} = \frac{2}{3} \text{ of a battery.}$$

EXERCISES

1. A "rare population" is a small sub-group of a larger population. A rare population would show up with low frequency when the larger population is sampled. Suppose a certain defect occurs in manufactured parts with approximate probability of 3%. How large a sample would you expect to take in order to find 1 part with the defect? Suppose we needed 5 such parts for more extensive testing. How many would you expect we have to take in order for 5 to show up?

2. For the conditions described in exercise 1, what is the probability that the first defect would be found in the first 100 observations? What is the probability that it will take a sample of 500 parts in order for 5 defects to be found?

3. Suppose that a new process for manufacturing computer floppy 3.5 inch diskettes still has some "bugs" in it. In a group of 10 diskettes, if 3 are defective, what is the probability that a customer who would purchase 5 diskettes would get exactly 1 of the defective diskettes?

4. For the situation described in exercise 3 what is the probability of getting at least 1 defective diskette?

5. Suppose that the situation described in exercise 3 is modified by sampling 5 items randomly from the production process, knowing that the probability of a defective diskette is 30%. What is the probability of getting at least 1 defective diskette in this case?

6. Suppose in a batch of 30 diskettes, 9 are defective. If a sample of 5 diskettes is selected, what is the probability of getting at least 1 defective diskette? Calculate this same probability assuming 5 diskettes are selected from a group of 50 diskettes of which 15 are defective.

7. Compare your probabilities of exercise 6 with the probability obtained in exercise 5. Is there reason to believe that the binomial distribution is a limiting or approximating distribution for the hypergeometric as $N \rightarrow \infty$ where p in the binomial is $\dfrac{K}{N}$?

8. If X is a geometric random variable, show that $P(X = n + k \mid X > n) = P(X = k)$. Interpret the implications of this relationship in terms of "memory-less" experiments.

9. For a hypergeometric random variable determine $P(X = k + 1) \div P(X = k)$. What does this expression suggest as a formula for computing $P(X = k + 1)$ if $P(X = k)$ has already been computed?

10. Suppose a population consists of N households of which K represent households that have a single head of household. A sample of n households is to be selected. What is the probability that the sample will have no "single head of household" households in the sample? If such a sample were to occur and N were reasonably large, what would you think about the method of drawing the sample?

4.4   THE POISSON DISTRIBUTION

Introduction

The Poisson distribution is used in modelling many populations. The random variable X is defined to be the number of occurrences of an event in a given unit of time or space where: (1) the number of occurrences occurring in non-overlapping intervals of length t are assumed to be independent; (2) the probability of exactly one occurrence in a particular small interval is proportional to the value λt; and (3) the probability of two or more occurrences in a sufficiently small interval is essentially zero. Under these assumptions, the probability density function given below represents the limiting probability, $P(X = x)$.

The Poisson distribution also serves as an approximating distribution to the binomial distribution for large n and small p. These results are presented in the following.

Definitions

DEFINITION 4.6: Let X be a random variable representing the number of independent occurrences of an event that occurs at a constant rate over time or space. Then X is said to have a Poisson distribution with probability function

$$f(x;\lambda) = \begin{cases} \dfrac{\lambda^x e^{-\lambda}}{x!} & x = 0,1,2,... \\ \\ 0 & \text{elsewhere} \end{cases}$$

The only parameter in the distribution is λ which, if known, completely specifies the probability distribution. The mean and variance of the distribution are

$$\mu = E(X) = \lambda$$

$$\sigma^2 = V(X) = \lambda$$

The Poisson distribution is positively skewed for all λ. However, the amount of the skewness decreases as λ gets larger.

The moment-generating function is given by

$$M_X(t) = e^{(\lambda(e^t - 1))}$$

DISCUSSION

The Poisson distribution describes the probability properties of an event that occurs independently over time and are such that the probability of more than one event occurring in some small interval of time is negligible. From experience and also from theoretical considerations, it has been found that many natural phenomena possess the probability characteristics of the Poisson distribution. Typical examples are the number of phone calls arriving at a telephone exchange in a given time span, the number of flaws in a given area of a bolt of cloth, the number of bacteria colonies in a given area in some growing medium, the number of births or arrivals in a given time span. It finds great usage in the study of failures and failure rates in reliability studies.   It also serves as an excellent approximation to binomial probabilities under the proper conditions.

The parameter λ in the distribution, being the mean, would represent the mean number of occurrences per unit time or unit of space. The mean and variance are obtained most easily by finding $E(X)$ and $E[X(X-1)] = E(X^2) - E(X)$ and then algebraically working with these results to find σ^2.

If X is Poisson with λ representing the mean number of occurrences per unit interval, then the expected number of occurrences in an interval of length t is λt. For this situation the expression for the number of occurrences, X, in the interval of length t is

$$f(x;\lambda) = \begin{cases} \dfrac{\lambda t^x\, e^{-\lambda t}}{x!} & x = 0,1,2,... \\[2ex] 0 & \text{elsewhere} \end{cases}$$

In this situation we can think of t as representing the "unit interval" with λt as the mean instead of λ. Therefore, if X is the number of telephone calls coming through a switch   each 10 minute interval. And the average is 15 calls per 10 minutes, then for a 30 minute period the expected number of calls would be $\lambda = 15 \cdot 3 = 45$, or for a 5 minute period the expected number of calls would be $\lambda = 15 \cdot \dfrac{1}{2} = 7.5$, etc.

Graphs

Several graphs of the Poisson distribution are given below in Figure 4.4. Note that the distrbution has a single mode and as λ gets larger the skewness becomes less pronounced.

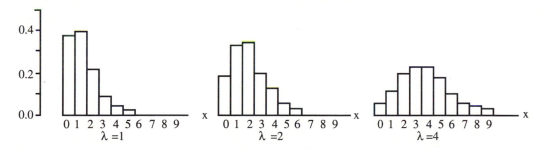

Figure 4.4 Graphs of the Poisson probability distribution.

Tables

With most pocket calculators you can compute the Poisson probabilities directly. However, the calculations could become tedious. It is also easy to program it on a personal computer. Nevertheless, a table of the cumulative distribution is given in the appendix of this book. Note again that since the

Poisson random variable is discrete and integer-valued, that $f(x;\lambda) = F(x) - F(x-1)$. In this table, λ ranges from 0.1 to 10.0 in increments of 0.1. A selected part of this table follows.

Table 4.2 Cumulative values for the Poisson distribution for values of l between 4.1 and 5.

x	4.1	4.2	4.3	4.4	4.5	4.6	4.7	4.8	4.9	5
0	0.0166	0.0150	0.0136	0.0123	0.0111	0.0101	0.0091	0.0082	0.0074	0.0067
1	0.0845	0.0780	0.0719	0.0663	0.0611	0.0563	0.0518	0.0477	0.0439	0.0404
2	0.2238	0.2102	0.1974	0.1851	0.1736	0.1626	0.1523	0.1425	0.1333	0.1247
3	0.4142	0.3954	0.3772	0.3594	**0.3423**	0.3257	0.3097	0.2942	0.2793	0.2650
4	0.6093	0.5898	0.5704	0.5512	0.5321	0.5132	0.4946	0.4763	0.4582	0.4405
5	0.7693	0.7531	0.7367	0.7199	0.7029	0.6858	0.6684	0.651	0.6335	0.6160
6	0.8786	0.8675	0.8558	0.8436	0.8311	0.8180	0.8046	0.7908	0.7767	0.7622
7	0.9427	0.9361	0.9290	0.9214	0.9134	0.9049	0.8960	0.8867	0.8769	0.8666
8	0.9755	0.9721	0.9683	0.9642	0.9597	0.9549	0.9497	0.9442	0.9382	0.9319
9	0.9905	0.9889	0.9871	0.9851	0.9829	0.9805	0.9778	0.9749	0.9717	0.9682
10	0.9966	0.9959	0.9952	0.9943	0.9933	0.9922	0.9910	0.9896	0.9880	0.9863
11	0.9989	0.9986	0.9983	0.9980	0.9976	0.9971	0.9966	0.9960	0.9953	0.9945
12	0.9997	0.9996	0.9995	0.9993	0.9992	0.9990	0.9988	0.9986	0.9983	0.9980
13	0.9999	0.9999	0.9998	0.9998	0.9997	0.9997	0.9996	0.9995	0.9994	0.9993
14	1.0000	1.0000	1.0000	0.9999	0.9999	0.9999	0.9999	0.9999	0.9998	0.9998
15	1.0000	1.0000	1.0000	1.0000	1.0000	1.0000	1.0000	1.0000	0.9999	0.9999
16	1.0000	1.0000	1.0000	1.0000	1.0000	1.0000	1.0000	1.0000	1.0000	1.0000

For instance, if $\lambda = 4.5$, then $P(X \leq 3) = F(3) = 0.3423$. (The probability is shown in bold-face in Table 4.2.)

EXAMPLES

EXAMPLE 1: The number of red blood cells for a fixed volume of blood is a random variable that occurs with a constant rate. If the average number of cells for a given volume of blood is 9 red cells in a normal person, determine the probability that the number of red cells of a normal person will be within one standard deviation of the the mean? Within two standard deviations of the mean?

SOLUTION: In this problem $\lambda = 9$, therefore $\mu = 9$ and $\sigma = 3$. Therefore we want to find

$$P(9 - 3 \leq X \leq 9 + 3) = P(6 \leq X \leq 12) = F(12) - F(5) = 0.8758 - 0.1157 = 0.7601$$

In addition we want to find

$$P(3 \leq X \leq 15) = F(15) - F(2) = 0.9780 - 0.0212 = 0.9568$$

(Probabilities such as these would help to establish what a doctor would call a "normal" range of red blood cells for a healthy adult.)

Note: The probability in this case of having an observation within 1 standard deviation of the mean is about 75% and for 2 standard deviations is about 95-96%. These results might be compared to Chebyshev's inequality and to what we will refer to as the "empirical rule" for the normal distribution.

Poisson Approximation to the Binomial

The Poisson distribution can be shown to be a limiting case of the binomial distribution under the conditions that, as n approaches $+\infty$ and p approaches zero, the product $n{\bullet}p$ remains constant. Therefore, Poisson probabilities serves as a reasonable approximation to binomial probabilities when n is large, p is small and $np = \lambda$ is of moderate size, usually less than about 5 or 6 or smaller.

EXAMPLE 2: A particular type of birth defect occurs in 8 out of every 100 births. An investigator studying the occurrence of this defect in a small remote African village, observed only one instance of the birth defect in 100 observed births. Does this result seem surprising under the circumstances?

SOLUTION: The binomial probability model would seem the most likely model to use to assess that likelihood of one or fewer observed birth defects. However, we shall use the Poisson approximation to this probability. In this case, $\lambda = np = 100{\bullet}(.08) = 8$. Therefore, from the Poisson tables we find that

$$P(X \leq 1 \mid \lambda = 8) = F(1;1) = 0.0030$$

(The exact probability can be computed and is found to be $P(X \leq 1 \mid n = 100, p = 0.08) = 0.0023$.)

EXAMPLE 3: Suppose that the frequency of a particular customer complaint about a product is 1 out of every 10 complaint calls. In the next 10 calls, what is the probability of at most 1 call of the particular type?

SOLUTION: The probability called for is obtained as follows:

$$P(\text{At most 1 call}) = P(X \leq 1) = P(X = 0 \text{ or } X = 1)$$

$$= \binom{10}{0}(0.1)^0(0.9)^{10} + \binom{10}{1}(0.1)^1(0.9)^9 = 0.7361.$$

The Poisson approximation for $\lambda = 10(0.1) = 1$ is $e^{-1} + e^{-1} = 0.7358$.

EXERCISES

1. The number of errors per page on a manuscript averages 0.7. What is the probability of a randomly selected page having no errors? 1 error? 2 errors? More than 2 errors? Assume the Poisson probability model in order to answer this question.

2. Show that the expected value of X is λ. Also find $E[X(X - 1)]$. Then use this result to determine the variance of the Poisson distribution. [Hints: To find $E(X)$ use the definition and cancel the X's in numerator and denominator, then change the variable of summation to $Y = X - 1$, recognizing that the total probability sums to 1. Use the same principle for finding $E[X(X - 1)]$ letting $Y = X - 2$. Then recall that $V(X) = E(X^2) - [E(X)]^2$.

3. An entomologist is studying the life cycle of a particular arachnid. Under controlled conditions a cluster of eggs hatches a spider on the average of 2 per hour. What is the probability of observing no eggs hatching in the next hour? more than 2 eggs hatching? Assume the Poisson probability model to answer this question.

4. The Poisson distribution serves as an approximation to the binomial under certain conditions (large n and small p). Under the following situations compute the exact binomial probabilities of getting 0, 1, and 2 successes using the binomial tables and then compute the Poisson approximation. Next, compute the percentage error— $\dfrac{|exact - approx.|}{exact} \cdot 100\%$.

    Case 1: $n = 20, p = .05$;   Case 2: $n = 50; p = .02$;   Case 3: $n = 100, p = .01$.

5. In a sampling inspection program where plastic parts are being manufactured that go into dishwasher motors, 100 items are periodically sampled and inspected. The process is expected to produce 8 defectives in every 1000 parts produced. In the sample drawn, 2 defectives are observed. Calculate the probability of such an event happening and decide whether you would recommend that the process be halted for adjustments.

6. Suppose the number of times that an individual contracts a cold in a given year is a Poisson random variable with parameter $\lambda = 5$. Suppose a new drug has been marketed that reduces the Poisson parameter to $\lambda = 3$ for 75% of the population. For the other 25% of the population the drug has no appreciable effect on colds. Suppose an individual using the drug for a year contracts 2 colds in a given year. How likely is it that the drug was beneficial to him or her?

7. For three different Poisson random variables with $\mu = 9$, $\mu = 16$ and $\mu = 20$, compute $P(\mu - \sigma < X < \mu + \sigma)$.

8. For the three cases described in exercise 7, compute: $P(\mu - 2\sigma < X < \mu + 2\sigma)$ and $P(\mu - 3\sigma < X < \mu + 3\sigma)$. Is there a trend or pattern that can be observed in the probabilities you obtain?

9. A used computer tape is tested for bad records. The computer operator found on the average 2.1 bad records per 40 feet of tape. What is the probability of finding no bad records in an 80 foot length of tape?

10. Let X equal the number of chocolate chips in a chocolate-chip cookie. Sixty-two observations of X yielded the following frequencies for the possible outcomes of X:

Outcome (x):	0	1	2	3	4	5	6	7	8	9	10
Frequency:	0	0	2	8	7	13	13	10	4	4	1

    Use these data to graph the empirical probability density function and then superimpose a Poisson pdf having a mean of $\lambda = 5.6$. Does it appear that the data follow a Poisson distribution with $\lambda = 5.6$?

4.5   THE DISCRETE UNIFORM DISTRIBUTION

Introduction

Many commonly encountered experiments produce outcomes which can be assumed to occur with equal probability. For instance, rolling a balanced die, tossing a balanced coin, spinning a spinner, choosing a ball from an urn are all examples of such experiments. Random variables with this probabilistic behavior produce what is referred to as a uniform distribution of probability. In addition, we also expect the values that the random variable takes on to be separated by the same distance, providing a symmetric distribution, though this is not absolutely essential.

A random variable of this type has a pdf that is very simple and easy to work with. The properties of this distribution are presented and discussed in this section.

Definitions

DEFINITION 4.7: A random variable X is referred to as a discrete uniform random variable if it has the following probability function

$$f(x) = \begin{cases} \dfrac{1}{k} & \text{for } x = x_1, x_2,..., x_k \\ 0 & \text{elsewhere} \end{cases}$$

Of particular interest is the case when X takes on the integer values of $1, 2, ..., k$. This is the case that we will treat more completely, though the more general case can be dealt with quite easily. The parameter of the distribution is the value of k, which will be known in most cases where the uniform distribution is examined.

The mean and variance of the distribution for the case when $x = 1, 2, 3, ..., k$ are

$$\mu = E(X) = \frac{(k+1)}{2}$$

$$\sigma^2 = V(X) = \frac{(k^2-1)}{12}$$

The moment-generating function takes the following form:

$$M_X(t) = \frac{e^t(1 - e^{kt})}{k(1 - e^t)}$$

DISCUSSION

As indicated in the introduction of this section, the uniform discrete distribution is encountered in several practical settings—the tossing of a fair coin, the rolling of a balanced die, etc. It is examined here mainly for its simplicity and the fact that it handles the probability properties of several commonly encountered real-world experiences. The particular case when X takes on integer values from $1, ..., k$ is presented in detail.

Graphs

The graph of a uniform discrete random variable shows up as a "bar graph" with bars of identical heights. For instance, if $f(x;5) = \dfrac{1}{5}$ for $x = 1, 2, 3, 4, 5$, then its pdf would appear as Figure 4.5.

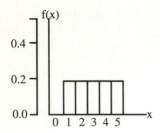

Figure 4.5 A graph of a uniform discrete probability density function for $X = 1, 2, 3, 4, 5$.

Because of symmetry, its mean is 3 which is also $\dfrac{(k+1)}{2} = \dfrac{(5+1)}{2} = 3$. Its variance would be

$$\sigma^2 = V(X) = \dfrac{(k^2-1)}{12} = \dfrac{25-1}{12} = 2.$$

Tables

Tables for this distribution are unnecessary due to its simplicity. All probabilities can be calculated in a straight forward manner.

EXAMPLES

EXAMPLE 1: Suppose the random variable X takes on the values 0 and 1 with probability $\dfrac{1}{2}$ each. What is the mean and variance for this random variable?

SOLUTION: By definition,

$$\mu = E(X) = 0(\tfrac{1}{2}) + 1(\tfrac{1}{2}) = \tfrac{1}{2}$$

$$\sigma^2 = V(X) = \left[0^2(\tfrac{1}{2}) + 1^2(\tfrac{1}{2})\right] - (\tfrac{1}{2})^2 = \tfrac{1}{4}.$$

[Note: The above random variable is also a Bernoulli random variable with $p = \dfrac{1}{2}$. Therefore, the results of Section 1 can also be used to show that the mean and variance are as determined above. Also, observe that the formulas given in definition 4.6 for a mean and variance above don't apply since this random variable, X, takes on the values of 0 and 1, not 1 and 2.]

EXAMPLE 2: Suppose a jar contains 10 chips number 0, ..., 9 respectively. Suppose the random variable X is the number that appears on a randomly selected chip. What is $E(X)$? What is $V(X)$?

SOLUTION: The random variable defined is a uniform random variable taking on values between 0 and 9 with $f(x) = \dfrac{1}{10}$. Therefore, its mean and variance are

$$E(X) = 0(\tfrac{1}{10}) + 1(\tfrac{1}{10}) + 2(\tfrac{1}{10}) + \dots + 9(\tfrac{1}{10}) = \dfrac{45}{10} = 4.5$$

$$V(X) = 0^2(\frac{1}{10}) + 1^2(\frac{1}{10}) + 2^2(\frac{1}{10}) + \ldots + 9^2(\frac{1}{10}) - \left[\frac{45}{10}\right]^2$$

$$= \frac{285}{10} - \frac{2025}{100} = \frac{825}{100} = 8.25$$

(The standard deviation of X is $S.D.(X) = \sqrt{8.25} = 2.87$.)

EXAMPLE 3: Can the formulas given in the definition section be used to compute the mean and variance for the random variable described in example 2?

SOLUTION: The formulas:

$$\mu = E(X) = \frac{(k+1)}{2} \qquad \text{and} \qquad \sigma^2 = V(X) = \frac{(k^2-1)}{12}$$

cannot be used since the formulas apply to the case where X takes on values 1, 2, 3, etc. However, we can create a transformation where $Y = X - 1$. With this transformation, if X goes from 1-10 with equal probabilities of $\frac{1}{10}$ each, then Y will take on values from 0-9 with probabilities of $\frac{1}{10}$ each. The random variable Y defined in this way has the same probability distribution as described in Example 2 and for the formulas has $k = 10$. Therefore,

$$E(Y) = E(X-1) = E(X) - 1 = \frac{(k+1)}{2} - 1 = \frac{11}{2} - 1 = 4.5.$$

And,

$$V(Y) = V(X-1) = V(X) = \frac{(k^2-1)}{12} = \frac{10^2-1}{12} = \frac{99}{12} = 8.25.$$

Therefore we get the same values as above.

EXERCISES

1. Consider the random variable given in Example 1 above where the random variable X takes on the values 0 and 1 with probability $\frac{1}{2}$ each.. Let $Y = X + 1$. Find the probability density function of the random variable Y and its mean and variance. Explain how the mean and variance of Y are the same or different from the mean and variance of X. Use the formulas in the definition to derive the mean and variance of Y.

2. Consider a simple five-part spinner used in a child's board game. What are the mean and variance of the random variable X if the 5 sections are numbered 1 through 5? What if the sections on the spinner are numbered 2, 4, 6, 8, and 10 and we use Y to denote this random variable?   Note that $Y = 2X$. Use this relationship to explain the differences in means and variances between Y and X. What is the correlation between X and Y?

3. What is the mean and variance of the random variable U if a balanced die is rolled and U denotes the number of dots on the top-most face?

4. Suppose a jar has 4 chips in it each with a number from 1 to 4 on it. If a chip is selected randomly from this jar and X is the number observed on the chip, what is $E(X)$ and $V(X)$?

5. Suppose two chips are selected randomly and with replacement from the jar described in exercise 4 above. List all the possible sample pairs that could occur in this situation and compute their mean, $\bar{X} = \dfrac{X_1 + X_2}{2}$. (There are 16 possible sample pairs, each having an equal probability.) Construct the probability density function for the means by finishing filling out the following table with the correct probabilities.

$\bar{X}$	1.	1.5	2.0	2.5	3.0	3.5	4.0
$f(\bar{x})$	$\dfrac{1}{16}$	$\dfrac{2}{16}$					

6. Using the table created in exercise 5, compute $E(\bar{X})$ and $V(\bar{X})$.

7. Using the principles of section 3.10, verify your computations of exercise 6 using the mean and variance obtained in exercise 4.

8. What is the probability of getting an observation between 2 and 3 inclusive from the pdf of exercise 4? What is the probability of getting an observation between 2 and 3 inclusive from the pdf of exercise 5? What does your intuition suggest would happen to this probability if a sample of size 4 were taken?

9. Consider the probability density function given in Example 2 above. Suppose the jar is sampled twice with replacement. Thus, the first chip might produce a 3 and the second chip might produce a 5. Compute the mean of the two outcomes for all possible samples of two, creating a new random variable $\bar{X} = \dfrac{X_1 + X_2}{2}$. What is $E(\bar{X})$ and $V(\bar{X})$ for this sampling situation?

10. For the probability density function of Example 2 and exercise 9, suppose a sample of n were selected *with replacement*. The mean of the sample $\bar{X}$ is computed. What is $E(\bar{X})$ and $V(\bar{X})$ for this sampling situation?   [Hint: By sampling with replacement, the individual X's are independent.]

11. For the probability density function of Example 2, if a chip is selected at random from this population, what is the probability that you will get an observation between 4 and 5 inclusive? What is this probability if a sample of size 2 is selected with replacement? What does your intuition suggest happens to this probability for a sample of size 3, 4, ..., n and so forth?

12. From the conditions described in Example 2 and exercise 10, what is $E(\bar{X})$ and $V(\bar{X})$ for $n = 3, 4,$ ..., n? In a short paragraph explain how these results relate to the conclusions about probabilities of exercise 11.

4.6   THE CONTINUOUS UNIFORM DISTRIBUTION

Introduction

In this and the following sections, we will examine several continuous random variables and their pdf's. The simplest of these to examine is the uniform distribution. This random variable is assumed to vary continuously over some finite domain say a to b with the probability distributed uniformly over that domain. The shape of the distribution is rectangular. If the values of a and b are 0 and 1 respectively, then we refer to the distribution as being uniformly distributed on the "unit interval".

Definitions

DEFINITION 4.8: A random variable X is said to be uniformly distributed over the interval (a,b) if the probability density function of X is given by

$$f(x) = \begin{cases} \dfrac{1}{b-a} & a \le x \le b \\[2mm] 0 & \text{elsewhere} \end{cases}$$

The parameters of the distribution are "a" and "b", one or both of which may be known or unknown depending on the situation. The mean and variance of the distribution are

$$\mu = E(X) = \frac{a+b}{2}$$

$$\sigma^2 = V(X) = \frac{(b-a)^2}{12}$$

The moment-generating function takes the following form:

$$M_X(t) = \frac{(e^{tb} - e^{ta})}{t(b-a)} \quad t \ne 0$$

DISCUSSION

The *uniform distribution* is interesting in part because of its simplicity. It is easy to derive and easy to demonstrate its properties. It also plays a central role in the generation of random numbers, whether for selecting random samples or for simulation and Monte Carlo studies. There are also various natural phenomenon which display the probability characteristics of the uniform distribution, such as round-off error, spinners in games of chance, etc. The distribution is symmetric so that the mean and median are equal. It has no mode. See Figure 4.6 to verify these conclusions.

Graphs

Figure 4.6 shows two uniform distributions. The graph on the left is for a general lower and upper value for a and b; the graph on the right results when $a = 0$ and $b = 1$, and is called the uniform distribution on the unit interval. The shaded area in the graph on the left represents the probability, $P(c \le X \le d)$, and would be found by integrating between c and d. However, because of the rectangular shape of the uniform distribution, simple properties of geometry can be used to find probabilities.

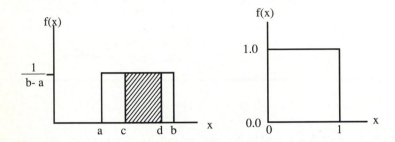

Figure 4.6 Graphs of the uniform probability density function—general case and unit uniform. Shaded area represents $P(c \le X \le d)$.

Tables

Tables of probabilities are unnecessary for the uniform distribution because of its simplicity. The distribution function of the uniform pdf is

$$F(x) = \begin{cases} 0 & x < a \\ \dfrac{x-a}{b-a} & a \le x \le b \\ 1 & x > b \end{cases}$$

It is easy to show that

$$P(c < X < d) = F(d) - F(c) = \frac{d-c}{b-a}$$

The quantiles of the distribution are also easily found through the following relationship:

$$X_q = a + (b-a)q.$$

EXAMPLES

EXAMPLE 1: A uniform distribution of particular interest is when $a = 0$ and $b = 1$ which is referred to as the uniform distribution over the unit interval. Find $f(x)$ for this particular situation and determine μ and σ^2.

SOLUTION: For this distribution, $a = 0$ and $b = 1$. Inserting these values into the general expression for $f(x)$, we get:

$$f(x) = \begin{cases} 1 & 0 < x < 1 \\ 0 & \text{elsewhere} \end{cases}$$

which has a mean and variance of

$$\mu = E(X) = \frac{a+b}{2} = \frac{0+1}{2}$$

$$\sigma^2 = V(X) = \frac{(b-a)^2}{12} = \frac{(1-0)^2}{12} = \frac{1}{12}.$$

EXAMPLE 2: For the uniform distribution on the unit interval described in example 1, find a general expression for $F(x)$ and evaluate $P(c < X < d)$ in terms of $F(x)$.

SOLUTION: The general form of the uniform distribution is given earlier. Upon inserting $a = 0$ and $b = 1$ into the expressions we get:

$$F(x) = \begin{cases} 0 & x < 0 \\ \frac{x-0}{1-0} = x & 0 \leq x \leq 1 \\ 1 & x > 1 \end{cases}$$

To find $P(c < X < d)$, we can write:

$$P(c < X < d) = F(d) - F(c) = d - c \quad 0 < c < d < 1.$$

EXAMPLE 3: Find the median, and 1st and 3rd quartiles for the uniform distribution on the unit interval.

SOLUTION: To find quantiles we use the relationship:

$$X_q = a + (b-a)q = 0 + (1-0)q = q.$$

Therefore, the median and quartiles are:

$$X_{0.5} = 0.5 \qquad X_{0.25} = 0.25 \qquad X_{0.75} = 0.75.$$

[Note that this is the distribution that is the theoretical model for generating random numbers in the interval (0,1). This process will be discussed more completely in Section 11 of this chapter.]

EXERCISES

1. Suppose X is a uniformly distributed random variable where $a = -k$ and $b = k$. Determine $f(x)$, and find its mean and variance. From these results explain what happens to the mean and variance of this distribution as k gets larger? What is the median and interquartile range for this distribution?

2. Suppose X is uniformly distributed with $a = 0$ and some value, b. Determine the general properties for this distribution, that is, find $f(x)$, $F(x)$, μ, σ^2, and X_q.

3. For the distribution defined in exercise 1, derive the mean and variance using the principles of expectation.

4. For the random variable X of exercise 2, derive the mean and variance using the principles of expectation.

5. For the pdf $f(x) = \dfrac{1}{b-a}$, derive the expressions for the mean and variance using the principles of expectation.

6. For the pdf $f(x) = \dfrac{1}{b-a}$, derive the expressions for the moment-generating function using the principles of expectation.

7. As part of an inspection program associated with metal washers used in shock absorbers, the diameter is measured by calipers. The variation in diameter is assumed to be uniformly distributed with a mean of 2 centimeters and a variance of 0.25 centimeters^2. What are the values for a and b for this uniform distribution? [Hint: can you use the results of exercise 1 to help solve this problem?]

8. A random number generator on a calculator generates numbers from a uniform distribution on the unit interval; i.e., $f(x) = 1$ for $0 \le x \le 1$. A student generates 5 such numbers and computes their mean. What is $E(\bar{X})$ and $V(\bar{X})$ for this sampling situation?

9. What is the probability of a random number generator producing a number between 0.4 and 0.6?

10. A random number generator is used to generate 36 uniform random numbers. The mean of these numbers is then computed. Use Chebyshev's inequality to assess how likely it would be to get a mean greater than 0.8 or less than 0.2.

4.7   THE GAMMA DISTRIBUTION

Introduction

The *gamma distribution* is associated with a random variable X that is continuous and positive. It has a single mode and is skewed right and can adequately model many populations that produce positive values over a continuous domain.

In this section we introduce the functional form of the distribution and many of its properties. Under certain conditions, its probabilities can be derived fairly simply, so we will not table the distribution, though tables are available in other sources. Also several special cases of the gamma distribution, namely, the exponential and the chi-square distributions are referred to.

Definitions

DEFINITION 4.9: The random variable X is distributed as the gamma distribution if its probability density function is

$$f(x) = \begin{cases} \dfrac{1}{\Gamma(\alpha)\theta^\alpha} x^{\alpha-1} e^{-\frac{x}{\theta}} & 0 \le x < \infty; \theta > 0; \alpha > 0 \\ \\ 0 & \text{elsewhere.} \end{cases}$$

where the gamma function $\Gamma(\alpha)$ has the property

$$\Gamma(\alpha) = (\alpha - 1)\Gamma(\alpha - 1)$$

and which is tabled extensively elsewhere for values of α between zero and one.

The parameters of the distribution are α and θ which are called the shape and scale parameters, respectively. That is, for a fixed α, the basic shape of the distribution stays the same, but changes as α changes. As θ is varied, the scale simply changes. See the graphs that follow in Figure 4.7 for graphs that demonstrate these properties.

The mean and variance of the distribution are

$$E(X) = \alpha\theta$$

$$V(X) = \alpha\theta^2$$

The moment-generating function takes the following form:

$$M_X(t) = \frac{1}{(1 - \theta t)^\alpha}$$

DISCUSSION

All of us spend a portion of our lives waiting in lines. We wait for service at our favorite restaurant, for service at the check out counter of the local grocery store. We wait for a busy phone line to open up. We wait for tickets for our favorite sports activity. We wait for elevators, phone calls, for our spouses, and on and on. Because of scientific as well as idle interest in "waiting times", the probability properties of waiting times, called waiting time distributions have been studied extensively. One particular distribution that is included in such studies is the gamma distribution, a very useful distribution for modeling a variety of populations. It is used in the study of waiting time distributions as well as in the study of time-to-failure models.

As can be seen from examination of the definition 4.9, it depends on two parameters, α and θ. When α is integer-valued, it is called the *Erlang* probability model; and if $\alpha = 1$, it is known as the *exponential* or *negative exponential* distribution.

In another situation if we let $\alpha = \dfrac{v}{2}$ and $\theta = 2$, then we get a one-parameter family called the *chi-square distribution* where the parameter v is called the *degrees of freedom*. The probability density function of the chi-square distribution is given by

$$f(x) = \begin{cases} \dfrac{1}{\Gamma(v/2)2^{v/2}} x^{v/2-1} e^{-\frac{x}{2}} & 0 < x < \infty; v > 0 \\ \\ 0 & \text{elsewhere.} \end{cases}$$

This distribution plays a very important role in statistical inference. The distribution will be discussed in more detail in Chapter 6 and chapters following. Usually we use the notation χ_v^2 to denote a random variable that follows a chi-square distribution with parameter v.

Graphs

The graphs of Figure 4.7 and 4.8 show the gamma distribution for a few values of α and θ. Note that we have selected a value for α, the shape parameter, then varied the value of θ, the scale parameter. (When $\alpha = 1$, we get the exponential distribution which is discussed separately in the next section.)

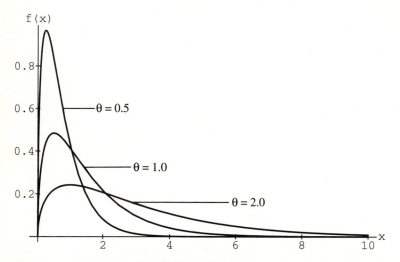

Figure 4.7 Gamma probability density function: $\alpha = 1.5$, $\theta = 0.5$, 2, and 4.

Looking at Figure 4.7 with $\alpha = 1.5$ the graphs are seen to be definitely right-skewed with a single mode. As θ goes from 0.5 to 2 to 4, the curve becomes more elongated and flatter. As you examine Figure 4.8 with $\alpha = 2$, a clearly different shape is produced but the right-skewness is still present. As before, as θ increases, the shape gets "stretched" to the right. These outcomes are predictable based on the relationship of the mean and variance of the gamma distribution to the parameters, α and θ. For the curves of Figure 4.7 the mean and variance are $\dfrac{3\theta}{2}$ and $\dfrac{3\theta^2}{2}$, respectively. Therefore, the mean and variance gets larger as θ gets larger.

For the curves of Figure 4.8, the means and variances are 2θ and $2\theta^2$, respectively.   As θ increases so does the mean and variance.

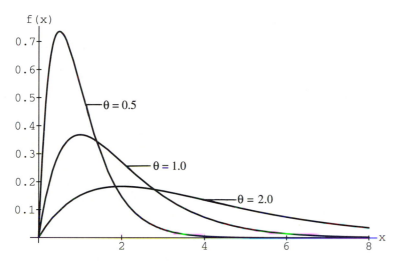

Figure 4.8 Gamma probability density function: $\alpha = 2$, $\theta = 0.5$, 2, and 4.

Tables

Tables for the gamma function are available elsewhere but are not included in this book. The distribution function does not simplify into a nice closed form except for integer α, as indicated in the next paragraph. Otherwise you must consult other sources if tables of this function are required. (As noted above, the chi-square distribution is a special case of the gamma distribution. This distribution is extensively tabled and we have tables of quantile values for the quantiles: 0.005, 0.010, 0.025, 0.050, 0.100, 0.900, 0.950, 0.975, 0.990, and 0.995 and degrees of freedom 1 to 30, 35,40 45,50,60, 70, 80, 90, 100.)

When α is integer-valued the distribution function $F(x)$ of the gamma distribution can be integrated by parts with the following result:

$$F(x;\alpha,\theta) = 1 - \left[1 + \frac{x}{\theta} + \frac{x^2}{2!\theta^2} + \cdots + \frac{x^{\alpha-1}}{(\alpha-1)!\theta^{\alpha-1}}\right]e^{-x/\theta}$$

If the term, $e^{-x/\theta}$, is brought inside the brackets, the sum can be recognized as a sum of Poisson terms with $\lambda = \frac{x}{\theta}$, with the sum starting at 0 and going to $\alpha - 1$.

Therefore, we can write

$$F_G(x;\alpha,\theta) = 1 - F_P(\alpha - 1; \lambda = \frac{x}{\theta})$$

where $F_G(x; \alpha, \theta)$ denotes the gamma distribution function and $F_P(\alpha - 1; \lambda = \frac{x}{\theta})$ is the Poisson distribution function.

EXAMPLES

EXAMPLE 1: Suppose X is a gamma random variable with $\alpha = 5$ and $\theta = 2$. Calculate the probability $P(X \le 4) = F(4)$?

SOLUTION: Since α is integer we have that

$$F_G(4; \alpha = 5, \theta = 2) = 1 - F_P(5 - 1; \lambda = \frac{x}{\theta} = \frac{4}{2}) = 1 - F_P(4; \lambda = 2)$$

$$= 1 - 0.9473 = 0.0527$$

where 0.9473 is obtained by entering the Poisson table with $\lambda = 2$ and $x = 4$. Remember: using the Poisson to get the probability is not an approximating approach. The probability is exact, within rounding error.

EXAMPLE 2: Consider a chi-square random variable where $\nu = 2$. Find the probability of getting a chi-square value within (a) one, and (b) two standard deviations of the mean.

SOLUTION: Since a chi-square random variable is a gamma with $\alpha = \frac{\nu}{2}$ and $\theta = 2$, then for the case of $\nu = 2$, we have that $\alpha = 1, \theta = 2$; and also, $\mu = \alpha\theta = 2$ and $\sigma = \sqrt{\alpha\theta^2} = 2$. Therefore, for the "one-standard-deviation" interval we get

$$P(2 - 2 \le X \le 2 + 2) = F_G(4) = 1 - F_P(0; \lambda = 2)$$

$$= 1 - 0.1353 = 0.8647.$$

For the "two-standard-deviation" interval we get

$$P(2 - 4 \le X \le 2 + 4) = F_G(6) = 1 - F_P(0; \lambda = 3)$$

$$= 1 - 0.0498 = 0.9502.$$

EXAMPLE 3: Consider a chi-square random variable where $\nu = 8$. Find the probability of getting a chi-square value within one and two standard deviations of the mean for this case.

SOLUTION: If $\nu = 8$, then $\alpha = \frac{\nu}{2} = \frac{8}{2} = 4, \theta = 2$ and $\mu = \alpha\theta = 2 \cdot 4 = 8$ and $\sigma = \sqrt{\alpha\theta^2} = 4$. Therefore, for the "one-standard-deviation" interval we get

$$P(8 - 4 \le X \le 8 + 4) = F_G(12) - F_G(4) = 1 - F_P(3; \lambda = 6) - [1 - F_P(3; \lambda = 2)]$$

$$= F_P(3; \lambda = 2) - F_P(3; \lambda = 6)$$

$$= 0.8571 - 0.1512 = 0.7059.$$

For the "two-standard-deviation" interval we get, after similar simplification,

$$P(8 - 8 \leq X \leq 8 + 8) = F_G(16) - F_G(0)$$

$$= [1 - F_P(3; \lambda = 8)] - 0 = 1 - 0.0424 = 0.9576.$$

EXAMPLE 4: Suppose the copper wire used in electrical transmission has a flaws in the prescribed diameter that will occur on the average of once every 50 feet of length. Suppose the distance between flaws is distributed as a gamma random variable with $\alpha = 5$ and $\theta = 10$. What is the probability that the next flaw in the wire will not occur within the next 70 feet of wire?

SOLUTION: Let the random variable X denote the number of feet until the next flaw occurs. Therefore, we want to determine $P(X \geq 70)$. We do this as follows:

$$P(X \geq 70) = 1 - F_G(70; \alpha = 5, \theta = 10)$$

$$= 1 - [1 - F_P(5 - 1; \lambda = \frac{70}{10} = 7)]$$

$$= F_P(4; \lambda = 7) = 0.1730$$

EXERCISES

1. Simplify the expression for the gamma distribution when $\alpha = 2$ and $\theta = 4$. Determine the mean and variance for this particular gamma random variable. Express $F(x)$ in terms of the corresponding Poisson probability and evaluate $F(3)$.

2. The time of arrival of the next car at an intersection stoplight on a lightly traveled country intersection is assumed to follow a gamma distribution with parameters 3 and 2 for α and θ, respectively. Thus, the mean time to arrival is 6 minutes. What is the probability that a traffic observer will have to wait more than 10 minutes for the next car to arrive? If the traffic study being done takes 3 minutes to complete after observing the present vehicle, what is the probability of a car arriving before the completion of the traffic survey form?

3. When an experimental animal is injected with a dose of an experimental drug, the time to the occurrence of a drug-related reaction is assumed to follow the gamma distribution with parameters $\alpha = 4$ and $\theta = 3$ with mean time to death at 12 minutes. What is the probability of a death before 6 minutes? Longer than 20 minutes?

4. Use the information given in exercise 3 to determine the median time to death?

5. Use the moment-generating function of the gamma distribution to show that the mean is $\alpha\theta$ and the variance is $\alpha\theta^2$.

6. Use the results of this section to shown that the mean and variance of the chi-square distribution is $E(X) = v$ and $V(X) = 2v$.

7. For a chi-square random variable with $v = 18$, what is the probability of finding an observation

within one standard deviation of the mean? Two standard deviations of the mean? Compare these results with example 2 above.

8. Recognizing that the chi-square random variable is a special case of the gamma distribution, what does the moment-generating function look like for this case?

9. Suppose a random variable X has the moment-generating function:

$$\frac{1}{(1 - 6t)^4}$$

What distribution does this random variable possess and what is its mean and variance?

10. Suppose X is a gamma random variable with parameters $\alpha = 2$, and $\theta = 3$. Write the expression for f(x). Use integration by parts to find $F(x)$. Show that this expression can be written as:

$$1 - \left[e^{-x/3} + \frac{x}{3} e^{-x/3} \right]$$

11. Use the moment-generating function to find the value of $E(X^3)$, $E(X^2)$ and $E(X)$ to find a measure of skewness for the gamma probability density function. [Remember that the measure of skewness defined in Chapter 3 was $\alpha_3^* = \frac{\mu_3}{(\sigma)^3}$.]

4.8   THE (NEGATIVE) EXPONENTIAL DISTRIBUTION

Introduction

The *negative exponential* or simply the *exponential distribution* is a special case of the gamma distribution. Specifically, if $\alpha = 1$, the gamma distribution reduces to the simpler form of the exponential distribution. Therefore, the general properties that hold for the gamma distribution also hold for the exponential distribution. However, because of the simple form of the expression, the problems of integration are much easier to handle. Applications of the exponential distribution and its properties are presented in the following pages.

Definitions

DEFINITION 4.10: A random variable that is exponentially distributed has a probability density function given by

$$f(x) = \begin{cases} \dfrac{1}{\theta} e^{-\frac{x}{\theta}} & 0 \le x < \infty \text{ and } \theta > 0 \\[2ex] 0 & \text{elsewhere} \end{cases}$$

The parameter of the distribution is θ which represents the average length of time between two independent Poisson events. It is also known as the "mean time between failures" in reliability theory. The exponential distribution is a gamma distribution with $\alpha = 1$; therefore, the general properties of the gamma distribution are carried across to the exponential distribution. Therefore, by letting $\alpha = 1$ in the expressions for the mean and variance of the gamma distribution we get

$$E(X) = \theta$$

$$V(X) = \theta^2$$

Thus, the mean and standard deviation of the exponential distribution equals the same value. The moment-generating function takes the following form (see exercise 5):

$$M_X(t) = \frac{1}{(1 - \theta t)}$$

DISCUSSION

As already stated, the exponential distribution is a special case of the gamma distribution; it finds particular application in the study of reliability. The random variable X in this case could represent the time-to-failure of a part under study with θ being the mean time-to-failure, or the mean length of life of the part or component under study. (It should not be concluded, however, that the length of life is always exponentially distributed. But it is of sufficient interest and has enough applicability to merit our interest.)

There is a direct connection between a Poisson random variable with parameter λ and an exponential random variable. Suppose X is Poisson with parameter λ and T is the time until the occurrence of the first Poisson event. Then, T is a continuous, nonngative random variable and we can write its distribution function as:

$$F(y) = P(T \le t) = 1 - P(T > t) = 1 - P(X = 0 \text{ in the interval } [0,t])$$

$$= 1 - \frac{(\lambda t)^0 e^{-\lambda t}}{0!} = 1 - e^{-\lambda t}$$

Therefore,

$$f(t) = F'(t) = \lambda e^{-\lambda t}$$

which can be recognized as an exponential random variable where $\lambda = 1/\theta$. As a result if the average number of Poisson events per unit of time is λ then the average time to the occurrence of the next Poisson event is $1/\lambda$.

Graphs

The graph of the exponential distribution maps as an exponential function. The function will approach the y axis from the right at the point $\frac{1}{\theta}$. The mean of the distribution is denoted by the Greek letter theta (θ) as previously shown, so as θ gets larger, the exponential gets flatter and more elongated. That is, it approaches the x-axis more slowly; thus the value of θ is more a scale or location parameter than a shape parameter. (The shape is the same for all θ.) The graphs of Figure 4.9 represent this perspective.

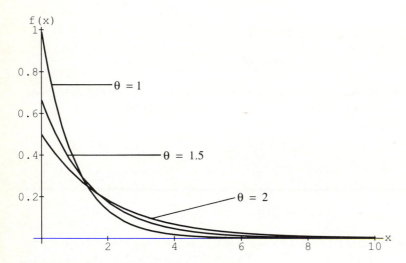

Figure 4.9 Graphs of the exponential distribution.

Tables

To prepare a typical distribution function-type table we need $F(x)$ which can be shown, by integrating over the proper region, to be of the form:

$$F(x) = 1 - e^{-x/\theta}$$

This is a very easy function to evaluate using a pocket calculator, thus a set of tables is unnecessary. Also remember that for any continuous random variable X, $P(a < X < b) = F(b) - F(a)$.

Again because of the simplicity of $F(x)$ it is also very easy to find any of the quantile values. That is, if we set $F(x_q) = q$ and solve the resulting equation for x_q in terms of q we end up with

$$X_q = \theta\left[\ln\frac{1}{(1 - q)}\right]$$

(This is also the equation for generating random numbers from the exponential distribution. See Section 11 of this chapter.) Thus, the quantile scores are easy to obtain, as are the quartiles, median, interquartile range, and so forth.

EXAMPLES

EXAMPLE 1: A manufacturer of computer components warrants them for 90 days. If the length of life is assumed to be exponential with mean length of life of 800 days, what is the probability of a particular part failing during the warranty period. (Assume that multiple failures of the same part is essentially impossible.)

SOLUTION: The random variable of interest is the time to failure X which has an exponential distribution with parameter value of 800. We find

$$P(X \le 90 \mid \theta = 800) = F(90)$$

$$= 1 - e^{-90/800}$$

$$= 1 - 0.8936 = 0.1064,$$

a fairly high probability of a failure occurring.

EXAMPLE 2: For the situation described in Example 1, what mean length of life is needed in order to reduce the probability to about 0.05, that is, so that only about 5% of items will need repair during the warranty period?

SOLUTION: In this situation, the unknown value is the parameter θ. We want $F(90;\theta) = 0.05$. Therefore,

$$0.05 = 1 - e^{-90/\theta}$$

Therefore, solving for θ we get

$$\theta = -\frac{90}{\ln(0.95)} = 1755.$$

This result tells us that we have to more than double the mean length of life of the part to reduce the percentage of failures within 90 day from 10% to 5%. This represents quite a challenge to the engineers involved.

EXAMPLE 3: What is the median for the pdf defined in Example 1?

SOLUTION: The median is $X_{.5}$, which by the equation above is:

$$X_q = \theta \left[ln\frac{1}{(1-q)} \right] \qquad \text{so that} \quad X_{0.50} = 800 \left[ln\frac{1}{(1-0.5)} \right] = 554.52 \text{ days.}$$

Note that the mean for this distribution is 800 days compared to a median of 554.52 days. This would imply a strong tendency to positive skewness which is clearly seen in the graphs of the exponential distribution.

EXERCISES

1. Suppose that the random variable X follows an exponential distribution with parameter θ. Determine the value of the median as a function of θ. Determine the mean deviation for this random variable.

2. Determine the probability of an exponentially distributed random variable falling within a standard deviation of the mean, within 2 standard deviations of the mean? Evaluate these expressions for θ of 2, 4, 8, and 16, respectively.

3. The time to failure of a new long-life light bulb is assumed to be exponentially distributed with mean-time-to-failure of 1500 hours. What is the probability of a light bulb lasting longer than 2000 hours? Less than 1000 hours?

4. Derive the mean and variance of the exponential distribution in general and confirm it using the moment-generating function.

5. Derive the moment-generating function for the exponential distribution.

6. Use the moment-generating function for the exponential distribution to obtain $E(X)$, $E(X^2)$ and $E(X^3)$, the first three moments of the exponential random variable.

7. Obtain the coefficient of skewness for the exponential distribution using the three moments obtained in exercise 6. What happens to the skewness of the exponential distribution as the parameter θ gets larger? [Remember that the coefficient of skewness is $\alpha_3^* = \dfrac{\mu_3}{(\sigma)^3} \cdot$]

8. Show that the general expression for $F(x)$ of an exponential distribution is $F(x) = 1 - e^{-x/\theta}$. Use this expression to provide a general form for calculating $P(a < X < b)$.

9. Suppose that a random sample of size n is selected from an exponential distribution. The mean of these n observations is then computed. What is $E(\bar{X})$ and $V(\bar{X})$?

10. A random variable has the moment-generating function $\dfrac{1}{1-12t} \cdot$. What is the mean, variance, and measure of skewness for this random variable?

11. For the random variable defined in exercise 10 find the median and interquartile range.

12. For the random variable defined in exercise 10 find $P(6 < X < 12)$.

4.9   THE NORMAL DISTRIBUTION

Introduction

The *normal distribution* or Gaussian distribution, by which it is also known, is one of the most commonly used probability density functions. Many natural phenomena seem to possess the familiar "bell-shape" curve, such as test scores for a large group of people, measurement error, height and weight distributions, and so on. The distribution also serves as a limiting distribution for many random variables when those random variables can be thought of as being the sum or mean of other more fundamental measurements. Thus, the normal distribution probabilities can often provide good approximations to other probability distributions under the proper conditions. (See the Central Limit Theorem in Chapter 6.)

In this section you will be introduced to the functional form of the distribution along with the graphical and probability properties it possesses.

Definitions

DEFINITION 4.11: The random variable X is said to be *normally distributed* if it has the following probability density function

$$f(x) = \frac{1}{\sigma\sqrt{2\pi}} e^{-\frac{[x-\mu]^2}{2\sigma^2}} \qquad -\infty < x < \infty;$$

$$-\infty < \mu < \infty; \qquad\qquad \text{and} \qquad\qquad 0 < \sigma < \infty$$

The parameters of the distribution are "μ" and "σ", with "μ" representing a location parameter and "σ" representing a scale parameter.

The mean and variance of the distribution are

$$E(X) = \mu \qquad\qquad \text{and} \qquad V(X) = \sigma^2$$

and the moment-generating function takes the following form:

$$M_X(t) = e^{\left(\mu t + \frac{\sigma^2 t^2}{2}\right)}$$

Notationally, to say that X is normally distributed with mean μ and variance σ^2 we write: X is $N(\mu,\sigma^2)$.

DEFINITION 4.12: The random variable Z is said to be *standard normal* if it has the following probability density function

$$f(z) = \frac{1}{\sqrt{2\pi}} e^{-\frac{z^2}{2}} \quad -\infty < z < \infty;$$

In this case it is easy to see that

$$E(X) = 0 \qquad \text{and} \qquad V(X) = 1$$

and we say that Z is $N(0,1)$.

DISCUSSION

Many data sets and experimental phenomena possess a bell-shaped distribution. Heights, weights, test scores, measurement error, all tend in this direction though there may be minor deviations from perfect symmetry and bell-shapedness. For such distributions, the normal distribution often serves as an appropriate probability model. It is a perfectly symmetric, bell-shaped distribution with a single mode. It is one of the most common distributions discussed in statistical and probability literature. When someone "grades on the curve" it is the normal distribution "curve" that is being referred to.

The mathematical and probability properties of this curve are briefly described in the definition. The curve is symmetric which can be seen by an examination of the function $f(x)$:

$$f(x) = \frac{1}{\sigma\sqrt{2\pi}} e^{-\frac{[x-\mu]^2}{2\sigma^2}}$$

The only place the variable x shows up is in the exponent and there it takes the form $(x - \mu)^2$. This is an expression symmetric around μ. Since all the rest of the terms in $f(x)$ are constants, the function itself is symmetric around μ. Also by taking the first derivative, the value of μ is found to be a maximum. Because $f(x)$ is symmetric with a single mode (maximum) at μ, its mean, median and mode are equal to μ. (The mean and variance of the random variable X can be formally obtained by using the moment-generating function. You will be asked to do this in the section exercises.)

It is also obvious that $f(x)$ is a positive function, i.e., $f(x) > 0$. To prove that its area is equal to one we can apply the following argument.

Let

$$I = \int_{-\infty}^{\infty} \frac{1}{\sigma\sqrt{2\pi}} e^{-\frac{[x-\mu]^2}{2\sigma^2}} dx$$

Next let $z = (x - \mu)/\sigma$ so that $dx = \sigma dz$, and we get

$$I = \int_{-\infty}^{\infty} \frac{1}{\sqrt{2\pi}} e^{-\frac{z^2}{2}} dz$$

Since $I > 0$, it is true that if $I^2 = 1$, then $I = 1$. So we can write

$$I^2 = \left[\int_{-\infty}^{\infty} \frac{1}{\sqrt{2\pi}} e^{-\frac{z^2}{2}} dz \right] \left[\int_{-\infty}^{\infty} \frac{1}{\sqrt{2\pi}} e^{-\frac{w^2}{2}} dw \right] = \frac{1}{2\pi} \left[\int_{-\infty}^{\infty} e^{-\frac{z^2}{2}} dz \right] \left[\int_{-\infty}^{\infty} e^{-\frac{w^2}{2}} dw \right]$$

$$= \frac{1}{2\pi} \int_{-\infty}^{\infty} \int_{-\infty}^{\infty} e^{-\frac{(z^2 + w^2)}{2}} dz dw.$$

Changing to polar coordinates by letting $z = r \cos \theta$ and $w = r \sin \theta$, we have

$$I^2 = \frac{1}{2\pi} \int_{0}^{2\pi} \int_{0}^{\infty} e^{-\frac{r^2}{2}} r \, dr \, d\theta = \frac{1}{2\pi} \int_{0}^{2\pi} d\theta = \frac{1}{2\pi} 2\pi = 1$$

The normal distribution is a two-parameter family, and because of the nature of its functional form, one parameter value is also the mean of the distribution. In addition, the other parameter is the variance of the distribution. Thus, μ and σ are the symbols used to represent the parameters in the distribution. As "μ" increases or decreases, the distribution shifts right or left; as "σ" increases or decreases the distribution peaks or flattens, the scale diminishing or expanding, respectively.

The random variable X is continuous and theoretically can take on any value between $-\infty$ and $+\infty$. Probabilities are equal to "areas" under the curve over appropriate intervals. Therefore, to determine a probability the function must be integrated over the appropriate limits. However, the function $f(x)$ has the property that it cannot be integrated in a closed form over a set of finite limits. (The integral over the entire real line *can* be obtained and has been shown to equal one above.)

If the random variable X is standardized (that is, we define $Z = (X-\mu)/\sigma$ which we have previously shown to have a mean of 0 and variance of 1, then by an appropriate substitution and change of variable as shown above, we get

$$f(z) = \frac{1}{\sqrt{2\pi}} e^{-\frac{z^2}{2}} \qquad -\infty < z < \infty$$

which is the same form as a random variable with $\mu = 0$ and $\sigma = 1$. This particular distribution is called the standard normal distribution. The standard normal distribution serves as our basic distribution of reference as will be shown later.

Graphs

The graph of the distribution is symmetric about the mean, and has large dispersion or small dispersion depending on the value of the variance. The distribution has a single mode and is bell-shaped with mean, median and mode all equal. Graphs for different values for the mean and variance are given later. The following graphs demonstrate some of the properties of the normal distribution. Figure 4.10 shows the effect of changing the means from 3 to 8 to 10 as the variance is held constant. As the figure shows, the normal curve simply shifts its location; the shape remains the same for an unchanging variance.

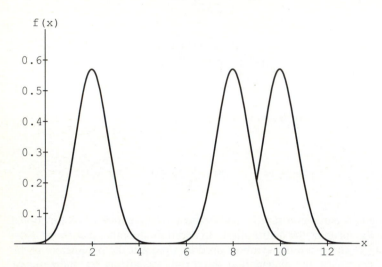

Figure 4.10 Normal distributions—differing means.

Figure 4.11 shows the effect of increasing the variance from 0.50 to 0.75 to 1.00 while holding the mean constant at $\mu = 3$. The shape of the curve gets flatter and flatter as the variance increases, which we would naturally expect as the variation of the distribution is increased, while still requiring the total area (probability) under the curve to equal one. Changing both μ and σ simultaneously would effect both a change in location and a change in shape in a way that your intuition would predict.

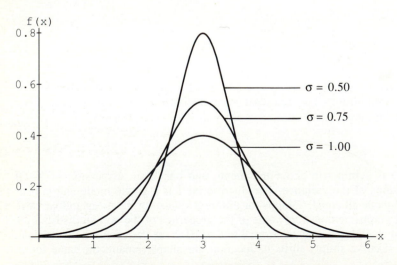

Figure 4.11 Normal distributions with different variances.

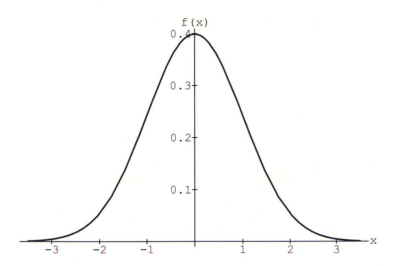

Figure 4.12 The standard normal distribution.

Figure 4.12 shows a graph with the horizontal and vertical scales appropriate for a standard normal distribution. Note that essentially all of the area is contained between −3 to +3 and the mean is equal to zero.

Tables

To be consistent with all that has preceded this section, we want to determine the form for $F(x)$ in a nice, simple expression, or develop a table of values for $F(x)$ which can be consulted. To determine $F(x)$ we would proceed in the following fashion:

$$F(x) = P(X \le x) = \int_{-\infty}^{x} \frac{1}{\sigma\sqrt{2\pi}} e^{-\frac{[t-\mu]^2}{2\sigma^2}} dt.$$

If we make the change of variable $z = \dfrac{t-\mu}{\sigma}$ so that $dt = \sigma dz$, we get the following simplification.

$$F(x) = \int_{-\infty}^{\frac{x-\mu}{\sigma}} \frac{1}{\sqrt{2\pi}} e^{-\frac{z^2}{2}} dz.$$

Note that the expression to be integrated is the standard normal distribution referred to earlier. Thus, the area under the normal curve to the left of x for a general normal distribution having arbitrary mean and variance equals the area under the standard normal distribution to the left of $z = \dfrac{x-\mu}{\sigma}$. Therefore, we merely have to develop the probability properties of the standard normal distribution in order to determine probabilities of interest for any normally distributeed random variable X. [Another way to denote this relationship symbolically is:

$$F_X(x;\mu,\sigma) = F_Z(z = \frac{x - \mu}{\sigma}; 0,1)$$

where $F_X(x;\mu,\sigma)$ denotes the distribution function of X which is normally distributed with mean μ and standard deviation σ, and $F_Z(z;0,1)$ denotes the distribution function of Z which is a standard normal random variable with a mean of 0 and a standard deviation of 1. It is also common to use the notation $N(\mu,\sigma)$ and $N(0,1)$ to refer to the two distributions of the random variables X and Z above.]

    This property—that any probability for a general normal distribution can be related to the area under a standard normal distribution—is shown in Figures 4.13 and 4.14 that follow. In Figure 4.13 a normal distribution having a mean of 50 and a standard deviation of 10 is shown. Also identified is the area (probability) under the curve between 35 and 60. Along the horizontal axis, values range from 20 to 80. In addition, the corresponding "standardized" values ranging from –3 to +3 are also shown.

    Figure 4.14 shows the same area under the standard normal distribution scaled to the same shape and size. (Note, however, that the vertical scale of Figure 4.13 is different than the vertical scale of 4.14. If both normal distributions were plotted to the same vertical scale, the shapes and sizes of the two figures would appear to be *quite* different.) The probabilities represented by the shaded areas in both diagrams are obviously the same as the theory says they should be.

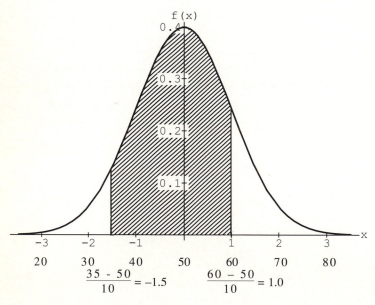

Figure 4.13 Normal distribution; $\mu = 50$, $\sigma = 10$ with standardized values for $X = 35$ and $X = 60$.

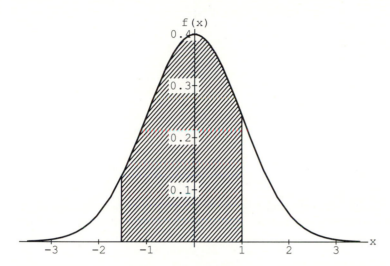

Figure 4.14 The standard normal distribution—corresponding area shaded.

However, as mentioned before, the normal distribution cannot be integrated over definite limits. For this reason, numerical methods have to be used to determine the distribution function $F(x)$, for either a general normal distribution *or* the standard normal distribution. But, for the reasons mentioned above, it is only necessary to prepare tables for the *standard normal distribution.* Table D in the appendix at the end of the book gives values for $F(x)$ for the standard normal distribution. It should also be noted that even though the random variable Z is defined over the entire real line, it is necessary only to catalog $F(x)$ between approximately -3 and $+3$. A portion of Table D is shown here as Table 4.3. This part of the table corresponds to areas for negative values for Z.

Table 4.3 A portion of Table D: cumulative probabilities for the standard normal distribution.

z	0.09	0.08	0.07	0.06	0.05	0.04	0.03	0.02	0.01	0.00
-3.0	0.0010	0.0010	0.0011	0.0011	0.0011	0.0012	0.0012	0.0013	0.0013	*0.0013*
2.9	0.0014	0.0014	0.0015	0.0015	0.0016	0.0016	0.0017	0.0018	0.0018	0.0019
-2.8	0.0019	0.0020	0.0021	0.0021	0.0022	0.0023	0.0023	0.0024	0.0025	0.0026
-2.7	0.0026	0.0027	0.0028	0.0029	0.0030	0.0031	0.0032	0.0033	0.0034	0.0035
-2.6	0.0036	0.0037	0.0038	0.0039	0.0040	0.0041	0.0043	0.0044	0.0045	0.0047
-2.5	0.0048	0.0049	0.0051	0.0052	0.0054	0.0055	0.0057	0.0059	0.0060	0.0062
-2.4	0.0064	0.0066	0.0068	0.0069	0.0071	0.0073	0.0075	0.0078	0.0080	0.0082
-2.3	0.0084	0.0087	0.0089	0.0091	0.0094	0.0096	0.0099	0.0102	0.0104	0.0107
-2.2	0.0110	0.0113	0.0116	0.0119	0.0122	0.0125	0.0129	0.0132	0.0136	0.0139
-2.1	0.0143	0.0146	0.0150	0.0154	0.0158	0.0162	0.0166	0.0170	0.0174	0.0179
-2.0	0.0183	0.0188	0.0192	0.0197	0.0202	0.0207	0.0212	0.0217	0.0222	0.0228
-1.9	0.0233	0.0239	0.0244	*0.0250*	0.0256	0.0262	0.0268	0.0274	0.0281	0.0287
...	...	...	...	...	...	...	...	...	...	...

To use Table 4.3, a z-score is obtained and the table is entered for that z-score. The entry found for that z-score represents the area under the standard normal curve to its left. That is, $P(Z \le z)$ = $F(z)$ = {Value in the body of table for z}. For instance $F(-3.00) = 0.0013$, and $F(-1.96) = 0.0250$.

See the italicized table values in Table 4.3. [Note that the combination of row and column heading give z-scores with accuracy of two places to the right of the decimal.]

For example, to use Table D to find the probability between the points a and b for a general normal distribution, we merely standardize those two values as follows:

$$\frac{a-\mu}{\sigma} \quad \text{and} \quad \frac{b-\mu}{\sigma},$$

and find the area between the corresponding standardized values from Table D, the standard normal distribution table. That is, if we want $P(a < X < b)$ we would use the relationship:

$$P(a < X < b) = F_X(b) - F_X(a) = F_Z\left(\frac{b-\mu}{\sigma}\right) - F_Z\left(\frac{a-\mu}{\sigma}\right)$$

By examination of Table D, the following simple generalization can be made that is a useful summary of the area properties of a normal distribution. This is often referred to as the "empirical rule" of the normal distribution.

Table 4.4 Empirical rule for the normal distribution.

The proportion of the area between the mean and k standard deviations from the mean	
k	**Proportion of area**
1	approx. 68%
2	approx. 95%
3	approx. 100%

EXAMPLES

EXAMPLE 1: Suppose that a set of test scores appear to be reasonably bell-shaped with a mean of 70 and a standard deviation of 10. If we assume a normal distribution for the test scores, what percentage of scores would be 90 or above? Between 80 and 90?

SOLUTION: The standardized score for 90 is $Z = \frac{(90-70)}{10} = 2$. Therefore, the percentage of scores above 90 is the same percentage as the percentage of Z-scores above 2. To find the probability, we would write:

$$P(X \geq 90) = P\left[\frac{X-70}{10} \geq \frac{90-70}{10}\right]$$

$$= P(Z \geq 2) = 1 - F_Z(2)$$

$$= 1 - .9772 = 0.0228$$

Therefore, we would expect about 2.28% of the scores to be above 90.

By a similar argument we would write

$$P(80 \leq X \leq 90) = F(90) - F(80)$$

$$= F_Z(\frac{90-70}{10} = 2) - F_Z(\frac{80-70}{10} = 1)$$

$$= 0.9772 - 0.8413 = 0.1359$$

or about 13.59%

EXAMPLE 2: Suppose that a set of test scores appeared to be reasonably bell-shaped with a mean of 70 and a standard deviation of 10. The instructor decides to grade "on the curve" giving 5% A's and E's (F's), 20% B's and D's, and the remaining middle 50% C's. What will be the break points for each grade category on the regular scale of 0 to 100 points? What is the interquartile range for this set of test scores?

SOLUTION: In this problem we are being asked for the quantile scores of $X_{0.05}$, $X_{0.25}$, $X_{0.75}$, $X_{0.95}$. (Why is that? Draw a diagram and note the cumulative areas.) Because of the symmetry of the normal distribution, if we find $X_{0.05}$ and $X_{0.25}$ we should immediately be able to determine the other two values.

Now using the basic relationship that $Z_q = \frac{(X_q - \mu)}{\sigma}$ and then solving for X_q we get:

$$X_q = \mu + Z_q\sigma$$

Therefore, we need $X_{0.05}$, $X_{0.25}$, $X_{0.75}$, and $X_{0.95}$ which are, respectively, -1.645, -0.675, 0.675, and 1.645. As a result, the various quantile values (grade division points) are

$$X_{0.05} = 70 + (-1.645)10 = 53.55$$

$$X_{0.25} = 70 + (-0.675)10 = 63.25$$

$$X_{0.75} = 70 + (0.675)10 = 76.75$$

$$X_{0.95} = 70 + (1.645)10 = 86.45$$

The interquartile range (IQR) is $X_{0.75} - X_{0.25} = 76.75 - 63.25 = 13.5$.

EXAMPLE 3: A battery is to be guaranteed for a given number of hours such that replacement without charge or full money back is promised if the battery fails before that time. If it is assumed that battery life is normally distributed with a mean length of life of 20 hours and standard deviation of 0.5 hours, what must be the guaranteed length of life if the manufacturer stipulates that they can to replace or accept returns with full money back for at most 1% of all batteries sold?

SOLUTION: Let the value $X_{0.01}$ denote the guaranteed length of life. That is, only 1% of the values are less than $X_{0.01}$. Using the same argument as given in example 2 we have that:

$$X_{0.01} = \mu + Z_{0.01}\sigma = 20 + (-2.327) \bullet 0.5 = 18.836 \text{ hours.}$$

Therefore, we might recommend to the manufacture that they guarantee the battery for 18.8 hours, knowing that slightly less than 1% will fail sooner than that.

EXERCISES

1. A clothing manufacturer specializing in men's sweaters, studied the heights of adult males and is willing to assume that they are normally distributed with $\mu = 70$ and $\sigma = 3$. If they were to establish their size categories of small, medium and large solely on height with small being 66 inches and smaller and tall being 76 inches and taller, in what proportions should they manufacture their various sweater sizes?

2. An insurance company bases their premiums for some specialized groups on the basis of risk relative to heart disease. Since excess weight is highly correlated with having heart disease, the weights of premium holders is one characteristic to be examined. Previous studies suggest that weight is normally distributed with a mean of 165 pounds and a standard deviation of 20 pounds. In considering this information, does the normal pdf seem reasonable as a model of weights? Assuming that it will suffice, and anyone weighing over 200 pounds would be considered obese, what is the proportion of obese persons in the population described? (Are there other factors that should enter the picture in an investigation such as described?)

3. If the pdf of X is

$$f(x) = \frac{1}{\sqrt{32\pi}} e^{-\frac{[x + 10]^2}{32}} \qquad -\infty \le x \le \infty;$$

what is its mean and variance?

4. For the pdf of exercise 3 write the expression for its moment-generating function.

5. Suppose the moment-generating function of the random variable X is $MX(t) = e^{(7t + 12t^2)}$. What is the mean and variance for the random variable X?

6. Write the expression for the pdf of the random variable defined in exercise 5.

7. For a standard normal distribution, what is the value of the interquartile range?

8. For a standard normal distribution, let Z_q denote the qth quantile value. By the symmetry of the normal distribution, what is the connection between Z_q and Z_{1-q}? For the following values of q, go to the table and find the value of Z_q and Z_{1-q}.

q	Z_q	Z_{1-q}
0.25		
0.20		
0.10		
0.05		
0.025		
0.01		

9. If Z is standard normal (denoted as $N(0,1)$) find:
   (a) $P(|Z| < 1)$ (b) $P(|Z| < 1.96)$
   (c) $P(|Z| < 2)$ (d) $P(|Z| < 3)$
   Compare the results of (a), (c) and (d) to the results of Table 4.4.

10. Develop an expression to find the interquartile range for a normal probability density function having a mean μ and standard deviation σ. (Take into account the symmetry of the normal distribution in simplifying this expression.)

11. Upon reading that a normal distribution often resulted when the random variable X equaled the sum or mean of other random variables, a student decided to investigate this property. She created an arbitrary population using old scrabble chips, writing numbers between 20 and 50 on them. She then took a random sample of 5 chips and computed the mean. She repeated this 50 times, creating 50 "means", each mean being computed on the basis of 5 randomly selected chips. She then carefully accounted for all 100 chips and found that the mean of the 100 was 40 with a variance of 16. What is the expected value of the sample mean, theoretically. Use the results of section 3.10, Chapter 3 to answer this question. What is the variance of the sample mean $V(\overline{X})$? If, in fact, a normal distribution occurs from such a "simulation", what is the probability of getting a sample mean that is less than 30?

4.10   THE LOGNORMAL DISTRIBUTION

Introduction

The *lognormal* distribution is another commonly encountered probability distribution.   Its presentation here is rather brief, however. The purpose in presenting it is to give you some insight into another pdf that finds application in a variety of disciplines.   A short explanation of its properties is given and its connection to the normal distribution is also provided. No tables are provided; thus you will not be expected to evaluate a probability. However, you should be able to find the mean and variance of the lognormal distribution on the basis of the mean and variance of the "normally distributed" random variable.

Definitions

DEFINITION 4.13: If the random variable X has a normal distribution with mean μ_x and variance σ_x^2 and if we let $\ln Y = X$, then $Y = e^X$ will have a lognormal distribution with probability density function

$$f(y) = \begin{cases} \dfrac{1}{y\sigma_x\sqrt{2\pi}}\,e^{-\frac{1}{2}\left(\frac{[\ln y - \mu_x]}{\sigma_x}\right)^2} & y \geq 0 \\[2em] 0 & \text{elsewhere} \end{cases}$$

The parameters of the distribution are μ_x and σ_x^2. However, they are not location and scale parameters in this distribution in the same sense as they were in the normal distribution. They are the mean and variance of the random variable X, but not of the random variable Y.

The mean and variance of the random variable Y are

$$\mu_y = E(Y) = e^{[\mu_x + \frac{\sigma_x^2}{2}]};$$

$$\sigma_y^2 = V(Y) = e^{[2\mu_x + \sigma_x^2]}(e^{\sigma_x^2} - 1).$$

DISCUSSION

It is often of interest to study the probability properties of the random variable $Y = e^X$ rather than the variable X, because the behavior of the variable under study appears to vary in an exponential pattern. If the random variable X is distributed normally, then through the transformation of X to Y, Y is distributed as the lognormal distribution. The regular normal density function can be thought of as describing the behavior of a series of additive effects; the lognormal distribution seems best suited to describe the behavior of a series of multiplicative effects. By its nature, Y is always positive and is therefore to be expected to be skewed positively. One such lognormal distribution is shown in Figure 4.15 below.

The lognormal distribution has found application in engineering, the physical sciences, the life sciences, etc. and is felt by some to be as commonly encountered as the regular normal distribution.

Graphs (tables are not provided)

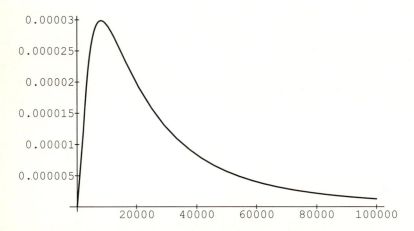

Figure 4.15 The lognormal distribution.

As can be noted from the graph and by the definition of the random variable, the lognormal distribution would be associated with a positive random variable. It is skewed right and has a single mode.

EXERCISES

1. An environmentalist measured, in yards, lengths of beach segments on some coasts in California bounded by cliffs. She knew that this type of data followed a lognormal distribution so she transformed all of her data to a log scale. She found that the mean and variance for her transformed data were 3.6 and 0.09, respectively. Years later, she decided that she wanted to know the average beach length and the variance in yards. Unfortunately, though, the original data had been deleted from her computer. Find the mean, in yards, without recourse to the original data. Also find the variance.

2. A geologist measures the length of streams in drainage basins. He wishes to determine the average stream length and to find the variance associated with the stream lengths. The following lengths are recorded:

3.41	3.14	3.48	3.97	3.83
2.95	2.81	2.81	2.52	2.79
2.58	2.98	2.95	2.70	2.88
2.78	2.73	2.56	2.88	2.52
3.69	3.98	3.92	4.20	4.88
2.63	2.53	2.89	3.30	3.30
3.15	3.07	3.34	3.32	3.49
2.30	2.31	2.07	2.29	2.25
2.40	2.02	2.72	2.77	2.74
3.20	3.46	3.36	3.03	3.41

   a) Construct a bar graph of the stream data using class widths of 0.50, i.e., 2.0-2.5, 2.5-3.0, ..., 4.5-5.0. Comment on the bar chart.
   b) For the grouped data, find the average stream length and the variance.

3. After examining the stream data, the geologist of problem 2 decides that the data seem to follow a lognormal distribution.

   a) Transform the above data to $L = ln(x)$.
   b) Create a bar chart for the transformed data and comment on its appearance. Use class widths of 0.1, i.e., 0.7-0.8, ..., 1.5-1.6.
   c) Find the mean and variance for the grouped lognormal data.
   d) Using the mean and variance from part (c), calculate the mean and variance for the raw data. Do you arrive at the same answer you calculated in part 2b?

4. The particle size, by weight, of some stream sediments is distributed as a lognormal distribution. For the following data, find μ_x, $V(x)$, μ_y, and $V(y)$ if $X = ln(y)$.

Class limits	Class midpoint(x_i)	Frequency
0.0-1.0	0.5	4
1.0-2.0	1.5	14
2.0-3.0	2.5	38
3.0-4.0	3.5	42
4.0-5.0	4.5	23
5.0-6.0	5.5	6

5. Permeability of sedimentary rocks (sandstone, limestone) follows a lognormal distribution. For the following data, find μ_x, $V(x)$, μ_y, and $V(y)$ if $X = ln(y)$.

Class limits	Class midpoint(x_i)	Frequency
2.0-2.4	2.2	5
2.4-2.8	2.6	17
2.8-3.2	3.0	26
3.2-3.6	3.4	18
3.6-4.0	3.8	4

4.11   THE BIVARIATE NORMAL DISTRIBUTION

Introduction

The *bivariate normal distribution* is a joint probability density function commonly utilized when dealing with 2 random variables. In this presentation, only a brief look at this joint distribution will be taken. Particular characteristics of interest will be the to examine the marginal and conditional distributions that are associated with this joint distribution. No problem assignments other than recognizing the relationship of parameters of the joint distribution to parameters of the marginal and condition will be required.

Definitions

DEFINITION 4.14: Let the random variables X and Y be jointly distributed having the following joint probability function . If so, they will be said to be bivariate normal random variables.

$$f(x,y) = \frac{1}{2\pi\sigma_x\sigma_y\sqrt{1-\rho^2}} \, e^{-\frac{1}{2(1-\rho^2)}\left\{ \left(\frac{x-\mu_x}{\sigma_x}\right)^2 - 2\rho\left(\frac{x-\mu_x}{\sigma_x}\right)\left(\frac{y-\mu_y}{\sigma_y}\right) + \left(\frac{y-\mu_y}{\sigma_y}\right)^2 \right\}}$$

$$-\infty < x, \ y < +\infty ; \quad \sigma_x, \sigma_y > 0; \quad -1 \le \rho \le 1, \quad -\infty < \mu_x, \mu_y < +\infty$$

The parameters of this distribution are marginal distribution means, variances and the correlation coefficient, ρ. Completely specifying all of these parameters will completely identify the probability density function.

The marginal distributions from this bivariate distribution are both normal probability density functions. That is, the marginal density of the random variable X is a regular normal probability density function with mean μ_x and variance σ_x^2. The marginal distribution of Y is a normal probability density function with mean and variance of μ_y and σ_y^2 . That is:

$$E(X) = \mu_x \qquad\qquad\qquad\qquad V(X) = \sigma_x^2.$$

$$E(Y) = \mu_y \qquad\qquad\qquad\qquad V(Y) = \sigma_y^2.$$

$$Cov(X,Y) = \rho\sigma_x\sigma_y$$

If the covariance in the joint probability density function is zero then X and Y will be said to be independent. (The bivariate normal distribution is the only distribution in common usage where the covariance/independence relationship works both ways. That is, if X and Y are independent then the covariance is zero. If the covariance is zero then we can conclude the random variables X and Y are independent.)

The conditional distribution of Y given $X = x$ is a normal distribution with a mean and variance of:

$$\mu_{Y|X} = \mu_Y + \rho\,\frac{\sigma_Y}{\sigma_X}(x - \mu_x)\;; \qquad \sigma_{Y|X}^2 = \sigma_y^2\,(1 - \rho^2)$$

(The conditional distribution of X given Y is also normal with mean and variance obtained simply by switching the subscripts of the expressions given above.)

DISCUSSION

The bivariate normal distribution is a logical extension of the univariate normal distribution and plays a vital role in any study of multivariate statistics. It is as commonly encountered in the bivariate setting as the regular normal distribution is encountered when studying a single random variable X. Our purpose here is not to study it in great detail but to define it, recognize it as a logical extension from the univariate case, and discuss its main properties. We will not calculate probabilities unless the variables involved are independent. In that case the probabilities are simply products of probabilities obtained using the regular standard normal distribution tables.

Graph

A graph of a bivariate normal distribution is given below as Figure 4.16. Note that this graph is a symmetric, three-dimensional bell-shaped figure (in this graph $\rho = 0$ and X and Y are independent). As you follow the lines of the plot you will recognize the familiar bell-shaped curve of the normal distribution of section 4.9.

If X and Y are not independent, the figure will be "stretched" in the $(+x, +y)$, $(-x, -y)$ direction if the covariance is positive (see Figure 4.17, Graph b), and in the $(+x, -y)$, $(-x, +y)$ quadrants if the covariance is negative (see Figure 4.17, Graph a). Note that any vertical slice perpendicular to the x,y plane will display a two-dimensional figure that is bell-shaped. This gives some insight as to why the marginal and conditional distributions are also normal univariate distributions.

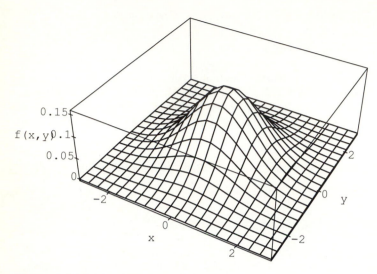

Figure 4.16 A bivariate normal distribution.

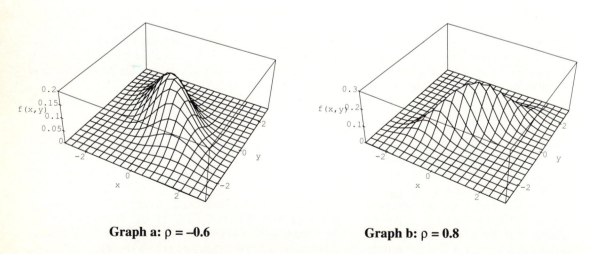

Graph a: $\rho = -0.6$ **Graph b:** $\rho = 0.8$

Figure 4.17 Bivariate normal distributions with $\rho = -0.6$ and $\rho = 0.8$.

EXAMPLES

EXAMPLE 1: Suppose the random variables X and Y denote the height and weight of adult males. The means and variances of the two random variables are, respectively: $\mu_x = 71$ in. , $\sigma_x^2 = 9$ in^2, $\mu_y = 175$ lbs., $\sigma_y^2 = 625$ lbs^2 and the correlation, $\rho = 0.70$.

(a)   What is the probability that a randomly selected person will be taller than 72 inches in height?

(b)   What is the probability that a randomly chosen person will weigh less than 145 lbs.

(c)   Given that a person is 66 inches, what is the probability that they will weigh less than 145 lbs.?

SOLUTION:

(a)   What is the probability that a randomly selected person will exceed 72 inches in height? The marginal distribution for heights is normally distributed with mean 71 and variance of 9, i.e., $X \approx N(71,9)$ Therefore,

$$P(X \geq 72) = 1 - F_X(72)$$

$$= 1 - F_Z[\frac{72-71}{3}] = 1 - F_Z(0.33)$$

$$= 1 - 0.6293 = 0.3707$$

(b)   What is the probability that a randomly chosen person will weigh less than 145 lbs. By an argument similar to part (a) we conclude the $Y \approx N(175,625)$. Therefore,

$$P(Y \leq 145) = F_Y(145)$$

$$= F_Z[\frac{145-175}{25}]$$

$$= F_Z(-0.2)$$

$$= 0.1151.$$

(c)   Given that a person is 66 inches, what is the probability that they will weigh less than 145 lbs.?

By the results above, the conditional distribution of weight given height is normal with mean equal to

$$\mu_{y|x} = \mu_y + \rho \frac{\sigma_y}{\sigma_x}(x - \mu_x) ; \qquad \sigma_{y|x}^2 = \sigma_y^2 (1 - \rho^2)$$

If X = 66 we have that

$$\mu_{y|x} = 175 + .70(\frac{25}{3})(66 - 71) = 145.83 \text{ lbs.} \qquad \sigma_{y|x}^2 = 625(1 - 0.49) = 318.75$$

Therefore

$$P(Y \leq 145 | X = 66) = F_{Y|66}(145)$$

$$= F_Z[\frac{145 - 145.83}{17.854}] = F_Z(-0.05) = 0.4801.$$

(Notice the difference in probabilities from part (b) to part (c). Why is that? Also note that the mean depends upon x but the variance doesn't.)

EXERCISES

1. Suppose we are willing to assume that the grade distribution in mathematics and statistics courses is jointly distributed as the bivariate normal distribution with means of 70 and 68, variances of 9 and 10, respectively, and that the correlation is .65. What proportion of math grades exceed 75? Suppose a person has a grade of 80 in statistics, what is the probability that their math grade will exceed 80 as well?

2. The measurements of the high and low temperatures over the years at a particular weather station on June 3 is assumed to be bivariate normal with parameter values as follows:

	Low		High
Means	88		53
Variances	25		36
Correlation		0.8	

    What is the probability that the high temperature will exceed 90 degrees Fahrenheit? What is the probability the low temperature will drop into the 30's?

3. Using the data from exercise 2, what the probability that if the low is 40 degrees, the high will exceed 90?

4. Suppose the correlation is zero for the data of exercise 2. What is the joint probability that the high will exceed 90 degrees and the low will drop into the 30's? [Hint: Think through what a correlation of zero would imply.]

5. In a college health fitness program, let X denote the weight in kilograms of a male freshman at the beginning of the program and let Y denote the weight change during a semester. Suppose that X and Y have a bivariate normal distribution with $\mu_x = 72.30$, $\sigma_x = \sqrt{110.25}$, $\mu_y = 2.80$, $\sigma_y = \sqrt{2.6}$ and $\rho = -0.55$.

4.12 GENERATING RANDOM VALUES FROM VARIOUS DISTRIBUTIONS

Introduction

In this section we present the theory and methodology of generating random observations (a random sample) from any particular distribution of interest. This serves two purposes. One, it is instructive in that we get a better understanding of how a random sample from a particular distribution will behave. We can change the parameter values in order to see how they will effect the distribution of the sample. We can vary the sample size to better understand the influence of sample size on the sampling process.

Secondly, many real world experiences too complex to analyze mathematically can be simulated by generating random "values" from the appropriate distribution(s). Complex integrals can even be approximated using this same principle. In studying many complex processes where the processes involve outcomes which seem subject to chance, it has been possible, with the use of high-speed computers, to study these complex processes, arriving at satisfactory solutions, by simulating them on the computer, where a "nice" mathematical solution hadn't been found. For instance, suppose you were faced with trying to predict the total service time involved in making a deposit or withdrawal from your credit union, facing various stages of waiting. First, you arrive at a parking spot which may or may not have a parking space immediately available. If not, you wait for a vacancy. Then you park and walk to the bank which requires going across the street controlled by a traffic light which you may have to wait for. Then you stand in line until a teller is free to process your request which may vary depending upon the size of your transaction. Upon completion you have to go across the street to your car before you leave and have completed your banking transaction.

It may be possible to describe the waiting time for each of these series of tasks by some appropriate probability density function, but the inter-relationship of them may be more complicated than you are willing to attempt to deal with.

As an alternative, the sequence of tasks might be simulated by generating random values from the various distributions involved and by repeating this a large number of times, so that the average waiting time could be determined by experimentation. However, to implement this approach it is necessary to be able to generate numbers from the various distributions of interest. This is the subject of this section.

Definitions

Conceptual Approach: Associated with every probability density function $f(x)$ is its distribution function $F(x)$.   Note that $F(x)$ is such that

$$0 \leq F(x) \leq 1$$

If we generate a random value of X from $f(x)$, say x_o , there is an associated value $u_o = F(x_o)$ where u_o is a number between 0 and 1. It is possible to prove that u_o defined in this manner (see Example 4 in section 5.2) possesses the probability characteristics of the uniform distribution on the unit interval 0 to 1. Thus for every x from $f(x)$, there is one corresponding value of u (namely $u_o = F(x_o)$) from the uniform distribution. This being the case, if we generate a random value u_o from the uniform, then we can work the opposite direction, producing a corresponding random value x_o by tracking it backwards through its distribution function $F(x)$. (This path is demonstrated in Figure 4.18.)

Therefore, to generate a random sample of values from some pdf, $f(x)$, do the following:

1.   Find the distribution function, $F(x)$ associated with $f(x)$.

2.   Generate a random number from the uniform pdf—most computers provide such a random number generator. Then set up the following equation:

$$u = F(x)$$

where u denotes a uniform random value and x denotes the corresponding "random value" from $f(x)$. If $F(x)$ has a manageable inverse function $F^{-1}(x)$ then we can solve the above equation for x in terms of u; i.e.,

$$x = F^{-1}(u)$$

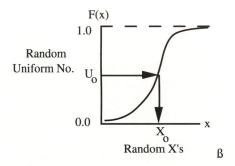

Figure 4.18 Generating random numbers from $f(x)$.

We then insert the uniform random number, u, into the function above. The x produced represents an observation randomly selected population defined by the distribution $f(x)$. (Note: not all functions have nice, tractable $F(x)$'s that allow solving for the inverse. In those cases, the use of cumulative tables will work, though not as easily.)

DISCUSSION

Computer simulations of sampling a particular probability distribution are very useful in a variety of settings. Consider playing a game of blackjack using a computer game program. After a few games you decide on a particular strategy that you think will provide you a winning edge. To test your theory you simply play a sufficient number of times, using your strategy, to convince you one way or another that your strategy works or does not work. (The question of how many times to play the game is a question that we will resolve to some extent in Chapter 7 and 8.) In this particular example, the computer is generating "cards" with the same probabilities that would be expected if a real deck of cards were being used. We thus generate a specified probability distribution using the computer's random number generator.

In this section, we extend this idea beyond a fairly simple case to the more complicated cases of generating values from the specific distributions presented in this chapter as if we were drawing a random sample from those distributions. To illustrate how this process works consider two very simple cases.

CASE 1: Simulating the toss of a balanced coin.

When asked to simulate the toss of a coin (a single Bernoulli trial) using a random number generator that produces numbers from the uniform distribution, most individuals will suggest that you let numbers less than 0.5 correspond with a tail, numbers greater than 0.5 producing a head. This suggestion is a procedure that will, in fact, produce "heads" and "tails" with the correct probability behavior. Figure 4.19 below demonstrates more completely the operation of such a system along with a connection of the procedure to the cumulative distribution of the Bernoulli random variable where X(Head) = 1, X(Tail) = 0.

In Figure 4.19, a "tail" is associated with the lower half of the area for the uniform distribution, and a head is associated with the upper half of the area.

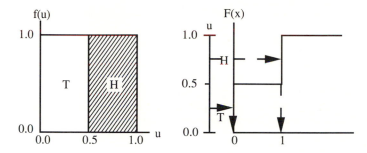

Uniform Distribution   **Cumulative Bernoulli Distribution**

Figure 4.19 Generating a random Bernoulli outcome.

Note that this same idea is shown by using a vertical axis divided into equal halves which is then connected to the cumulative distribution function of X. This same system is shown in Table 4.5 where a table of the cumulative distribution is used rather than a graph.

Table 4.5 Cumulative distribution for Bernoulli random variable, $p = 0.5$.

x	$f(x)$	$F(x)$	Rule	Value of Random Variable
0	0.5	0.5	If $U \le 0.5$ then	$X = 0$
1	0.5	1.0	If $0.5 < U \le 1$ then	$X = 1$

CASE 2: Simulating tossing a balanced coin twice.

We can simulate tossing a coin twice by implementing Case 1 procedures two separate times. However, another procedure, as an extension of Case 1 is given here to provide a more general process. Figure 4.20 shows how we set up the uniform distribution to generate the values of X with the correct probabilities. The uniform distribution in the left of Figure 4.20 is partitioned in three parts so that each area of each part corresponds to the correct probability associated with $f(x)$. That is, $f(0) = 0.25$, $f(1) = 0.50$ and $f(1) = 0.25$.   The figure on the right of Figure 4.20 shows the connection between the uniform distribution and how the random numbers coming from it to produce the binomial random digits. The cumulative distribution of the binomial distribution for $n = 2$ and $p = 0.5$ is the crucial link in this process.

We simply locate the value of u on the vertical axis and trace it horizontally until it intersects with the $F(x)$ function. The corresponding value of x on the horizontal axis is the simulated value for the random variable, X.

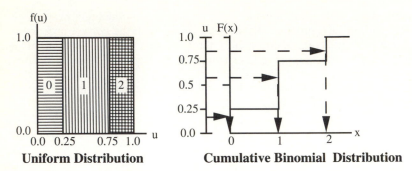

Uniform Distribution   Cumulative Binomial Distribution

Figure 4.20 Generating a random Bernoulli outcome.

Table 4.6 shows how to accomplish the same thing by using a table of the cumulative distribution function $F(x)$. In this table you see both $f(x)$, $F(x)$ and the the rule to apply to simulate the binomial random values. You can verify, by examining the table, that the values of X will be produced with the appropriate probabilities.

Table 4.6 Cumulative distribution for binomial random variable, $n = 2$, $p = \dfrac{1}{2}$.

x	$f(x)$	$F(x)$	Rule	Value of Random Variable
0	$\dfrac{1}{4}$	$\dfrac{1}{4} = 0.25$	$U \leq 0.25$	$X = 0$
1	$\dfrac{2}{4}$	$\dfrac{3}{4} = 0.75$	$0.25 < U \leq 0.75$	$X = 1$
2	$\dfrac{1}{4}$	$\dfrac{4}{4} = 1.00$	$0.75 < U \leq 1.00$	$X = 2$

The procedure developed in Case 1 and Case 2 is completely general and can be used for any distribution that has its cumulative distribution $F(x)$ graphed or tabled. In the case that a distribution is tabled, the table can be used to simulate the random variables by finding the value in the table equal to or greater than the uniform random variable u. The x value in the table meeting this condition is the simulated value for the random variable, X.

The table and graphical procedures just illustrated have an equivalent algebraic solution. This is accomplished by setting up the equation $u = F(x)$ where u denotes a value from the uniform distribution and $F(x)$ is the cumulative distribution function for the random variable X whose values we want to randomly generate. Solving this equation for X in terms of u, which solution is denoted by $x = F^{-1}(u)$, provides the algebraic method for transforming from a uniform value u to another value x.

SOME SPECIAL CASES OF INTEREST

The following explains how to generate random variables from some of the distributions discussed in this chapter. In most cases, we try to use the algebraic approach, but for the discrete distributions no convenient expression exists for the cumulative distribution $F(x)$ and for the normal distribution, $F(x)$ has to be obtained by numerical integration. In these cases, other arguments have to be provided to accomplish our task. We assume in all cases that we are capable of generating a random observation U from the unit-uniform distribution.

THE UNIFORM ON THE INTERVAL (a,b):

By definition, $f(x)$ and $F(x)$ are defined as follows:

$$f(x) = \frac{1}{(b-a)} \qquad\qquad a \le x \le b$$

$$F(x) = \frac{(x-a)}{(b-a)} \qquad\qquad a \le x \le b$$

Setting $u = F(x)$ and solving for x gives the result: $x = a + u(b-a) \quad a \le x \le b$

THE EXPONENTIAL DISTRIBUTION:

By definition, $f(x)$ and $F(x)$ are defined to be:

$$f(x) = \frac{1}{\theta} e^{-x/\theta}$$

$$F(x) = 1 - e^{-x/\theta}$$

Therefore, letting $u = 1 - e^{-x/\theta}$ and solving for x yields

$$x = -\theta[\ln(1-u)]$$

THE NORMAL DISTRIBUTION:

The normal distribution has no closed form for $F(x)$ and therefore an inverse function $F^{-1}(u)$ cannot be obtained. However, the table of the standard normal distribution can be used to convert a uniform variate u to a standard normal variate z. Enter the *body* of the standard normal table, Table C in the appendix of the book, with the value of u assuming that u can be bounded by adjacent values a and b in the so that $a < u \le b$. The values of a and b are associated with unique z-values by looking at the margins of the table. Let them be denoted by z_a and z_b where $z_a < z_b$. Applying the principles of this section, then any value of u satisfying the condition $a < u \le b$ generates the standard normal value z_b. See Example 3 below.

THE BINOMIAL DISTRIBUTION:

Since the binomial distribution is a sum of Bernoulli trials with probability of success equal to p, then one way to generate a binomial random variate from the binomial distribution with n and p as the parameters is to generate n Bernoulli random variates and add them. To do this, generate n uniform values and score them as follows:

If: $0 \le u \le p$ then let $X = 1$.
If: $u > p$, the let $X = 0$.

Let $Y = \sum_{i=1}^{n} X$, which will then be a binomial random value. (Note: It takes $n{\bullet}k$ uniform values to generate k binomial random values.)

THE POISSON DISTRIBUTION:

We have argued that the time between Poisson occurrences is exponentially distributed. If λ is the average number of occurrences in time t, then the average time between occurrences in the time interval t is $1/\lambda$. Therefore, if we want to simulate a Poisson distribution having parameter λ, generate successive exponential random values with $\theta = \frac{1}{\lambda}$. Form the cumulative sum of the exponential values until the sum of the values exceeds the unit of time t. If it takes $x + 1$ terms in the sum to finally exceed t, then the Poisson variable is the maximum number of values x in the cumulated sum that didn't exceed the time t.

EXAMPLES

EXAMPLE 1: Generate a set of values that behave as if they came from an exponential distribution with $\theta = 4$.

SOLUTION: The simplest way to do this is to set up the equation $u = 1 - e^{-x/4}$ which we then solve for x. Doing this we get:

$$e^{-x/4} = 1 - u,$$

$$\frac{x}{4} = \ln(1 - u),$$

Therefore,

$$x = 4\ln(1 - u).$$

So if the random number we generate is 0.834, then the exponential random number is

$$x = 4\ln(1 - 0.834) = 7.183$$

EXAMPLE 2: Suppose the uniform random number generated on a computer is 0.252. Use this number to generate a corresponding random number from the following distributions: Exponential with $\theta = 5$. Uniform with $a = 2$ and $b = 7$. Poisson with $\lambda = 5$ (use the Poisson table). Binomial with $n = 5, p = 0.2$ (use the binomial table).

SOLUTION: Using the results of this section we would get:

Exponential—
$$x = -5 \bullet \ln(1 - 0.252)) = 1.452$$

Uniform—
$$x = 2 + (0.252)(7 - 2) = 3.260$$

Poisson—
Entering the Poisson table with $\lambda = 5$ we find the that 0.252 is between the table entries of 0.1247 and 0.2650 which correspond to x values of 2 and 3 respectively. Therefore, the Poisson value is the

larger of these and is 3. (We could generate a series of exponential random values with $\theta = \frac{1}{5}$. We would then sum them until their sum exceeds 1. The Poisson random value would be one less than the *number* of exponentials in the sum.

Binomial—
Entering the binomial table with $n = 5$ and $p = 0.2$ we find the that 0.252 is less than the table entry of 0.3277 which corresponds to an x value of 0. Therefore, the binomial value is the value 0. (We could also generate 5 Bernoulli random variables with $p = 0.2$. The binomial random value would be the sum of the Bernoulli random numbers.)

EXAMPLE 3: Suppose a uniform random variate of $u = 0.3317$ is to be used to generate a standard normal value. What is the standard normal value it would generate?

SOLUTION: From the body of Table C of the standard normal distribution we find that $u = 0.3317$ is bounded by the values 0.3300 and 0.3336. The z-scores of these two values are $z = -0.44$ and -0.43. Therefore, any value of u where $0.3300 < u \le 0.3336$ would generate the larger of the two z-scores, or -0.43. Thus, the standard normal variate associated with $u = 0.3317$ is $z = -0.43$.

EXAMPLE 4: Suppose $Z = -0.43$ is a standard normal random number. With what value would this be associated for the random variable X that is $N(60,36)$?

SOLUTION: The inverse expression for a z-score transformation is $X = \mu + Z\sigma$. Since the random variable X has a mean of 60 and a standard deviation of $\sqrt{36} = 6$, then

$$X = \mu + Z\sigma = 60 + (-.43)*6 = 57.42$$

Therefore, a random $Z = -0.43$ translates into a random X of 57.42 under the conditions given.

Statistical Software Programs

Most statistical software has the ability to generate uniform random numbers, and many of these incorporate subprograms to generate random numbers from the common distributions—usually those we have studied in this chapter such as the normal, exponential, binomial, and Poisson. If you have access to any computer software programs you would find it instructive to generate a sample from the various distributions, create a histogram and compare your results to the expected shape based on what you have learned in this chapter. Figures 4.21 through 4.23 show the results for samples of three different sizes from a uniform distribution on the interval from 0 to 10, a binomial distribution with $n = 10$ and $p = 0.5$, a Poisson distribution with $\lambda = 5$ and a normal distribution with $\mu = 5$ and $\sigma = 2$. Notice that the theoretical mean of all four distributions equals 5 though the standard deviations vary from distribution to distribution.

Figure 4.21 shows the histograms that resulted when $n = 10$ observations were sampled from each of the four distributions. Figure 4.22 shows the histograms when $n = 30$ and Figure 4.23 shows the histograms for $n = 400$. Notice that with $n = 10$ and even for $n = 30$, it is difficult to recognize the parent distributions or distinguish one from another based on the histograms of the samples. It is not until the sample was very large that the distributions were recognizable and distinguishable from one another by looking at the histogram. Therefore, for small sample sizes, you would wonder how sensitive our techniques might be that are based on one distribution versus another?

One question remains after looking at these three figures. What happens between $n = 30$ and

$n = 400$?   Is there a point somewhere between them that would provide visual recognition for the distribution under consideration simply by looking at the histogram of the sample? Is that point the same for all distributions? Are there more objective methods to help us answer these questions than a subjective visual inspection?

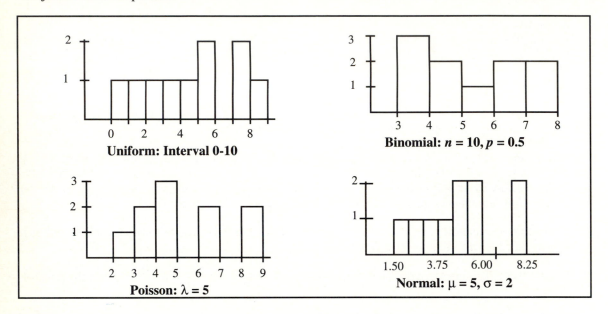

Figure 4.21 Histograms for randomly generated data from 4 distributions: Sample size is $n = 10$.

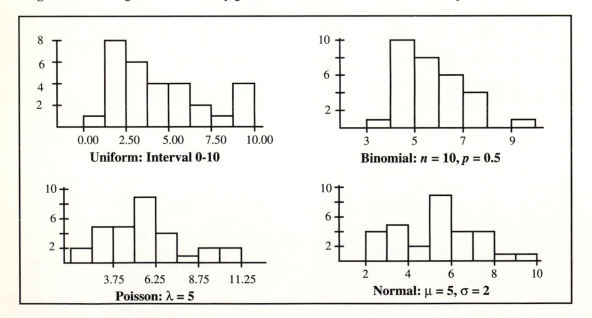

Figure 4.22 Histograms for randomly generated data from 4 distributions: Sample size is $n = 30$.

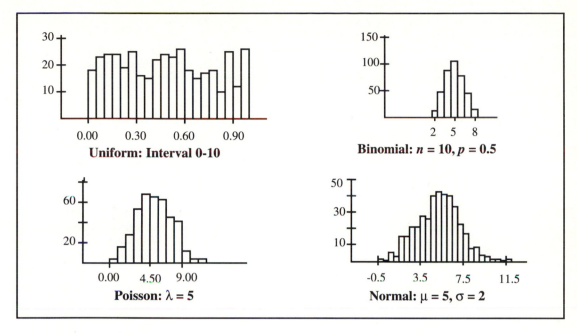

Figure 4.23 Histograms for randomly generated measurements from 4 distributions: Sample size is $n = 400$.

EXERCISES

1. A random number generator on a calculator produced the value 0.3490. Use this number to generate random numbers from a binomial distribution with $n = 5$ and $p = 0.1, 0.2, 0.3, 0.4$, and 0.5. Use the binomial tables in the appendix.

2. For the uniform random number of 0.5668 generate a Poisson random number for values of λ of 1, 1.5, 2, 2.5 and 3. Use the Poisson tables in the appendix.

3. Suppose X is a gamma random variable with $\alpha = 2$ and $\theta = 4$. Suppose a uniform random number generated on a calculator was 0.9343. Go as far as the tables will allow in getting a random number for the gamma distribution.

4. Use the standard normal table to generate a random number from a normal distribution having a mean of 50 and standard deviation of 12 if a uniform random number is 0.8882.

5. Generate a set 10 of exponential random variables using the value $\theta = 0.2$. Then use these random numbers to generate an observation from a Poisson distribution with $\lambda = 5$. Recall that if X is Poisson with parameter λ, then Y, the time between Poisson events, is exponential with parameter $\theta = \dfrac{1}{\lambda}$.

6. "Invent" a probability density function of the form, $f(x) = kx^a$ over some appropriate region. Determine $F(x)$ and plot both $f(x)$ and $F(x)$ on a graph. Generate a random sample of 30

observations from your distribution and make a histogram of the values you obtained. Compare this histogram to the function and note any obvious discrepancies.

7. Suppose the following 5 random numbers are uniformly distributed. Convert each of the 5 numbers into a Bernoulli random variable with $p = 0.6$. Random numbers:

    0.1247   0.5067   0.6236   0.4033   0.1564

8. Use the five random numbers in the exercise 7 and the 5 Bernoulli random variables to generate one binomial random variable with $n = 5$ and $p = 0.6$.

9. Generate 5 binomial observations from a binomial with $n = 3$ and $p = 0.4$ using 15 uniform random numbers implementing the concept of 15 Bernoulli trials.

10. It can be shown that the sum of n independent Exponential random variables with parameter θ, is gamma with parameters θ and $\alpha = n$. Generate 20 exponential random variables and make a histogram. Then create 10 pairs of these and let $Y = X_1 + X_2$, the sum of the two numbers in each pair. This should behave as a random sample of 10 observations from a Gamma distribution with the same θ and $\alpha = 2$.

11. Computer problem: If you have access to a computer with a random number generator and the ability to sort a column of data, generate a large number of random numbers, say 100, sort them and make a histogram having 10 equally spaced class intervals. Does the shape of the histogram look uniform?

4.13   ANALYSIS OF THE CHAPTER EXERCISE

The quality of perishable items is a constant concern for both the supplier and the retail outlet of the perishable item. Commercial Bakeries, Inc. has studied the sales and shelf life of its bread extensively. They have found that the number of loaves that are sold within 48 hours of baking out of 100 produced is modelled by a binomial distribution with $p = 0.95$. Those not sold are returned to a day-old bread outlet and sold for 70% of retail cost. The number of loaves sold in the day-old outlet in n loaves stocked is also binomial with $p = 0.90$.

(a)   What is the probability that of 100 loaves baked, 100 will be sold at the regular retail outlet?

(b)   What is the probability that 95 will be sold at the regular retail outlet and of the remaining 5, all will be sold in the day-old bread outlet? Under the preceding situation, what are the expected sales revenue figures for 100 loaves baked if the retail cost of a loaf of bread is $0.90?

     One of the crucial characteristics of a loaf of bread is its moisture content. If it is too high then it hastens the rate of spoilage. If it is too low then the loaf tends to dry out too quickly and thus becomes stale. It has been found that the ideal moisture content of baked bread is 45% with anything over 50% being too high and under 40% being too low.

(c)   If the moisture content has been found to be normally distributed with a mean of 45% and a standard deviation of 2%, what is the likelihood of a loaf being baked with too high a moisture content? Of 100 loaves produced, how many would you expect to be found in the ideal range of 40-50% moisture content?

SOLUTION:

(a)   What is the probability that of 100 loaves baked, 100 will be sold at the regular retail outlet?

Since the distribution of loaves sold is assumed to be binomial with $p = 0.95$ and $n = 100$, then the probability of selling all 100 loaves will be:

$$P(X = 100) = \binom{100}{100}(.95)^{100}(.05)^0 = 0.00592,$$

not a very large probability.

(b)   What is the probability that 95 will be sold at the regular retail outlet and of the remaining 5, all will be sold in the day-old bread outlet?   Under the preceding situation, what are the expected sales revenue figures for 100 loaves baked if the retail cost of a loaf of bread is set at $0.90 per loaf?

Let X denote the number of loaves sold fresh, and Y the number of loaves sold in the day old bread store. Note that $X + Y \le 100$. We want to calculate:

$$P(X = 95, Y = 5) = P(X = 95)P(Y = 5 \mid X = 95) = [\binom{100}{95}(0.95)^{95}(0.05)^5][\binom{5}{5}(0.9)^5(0.1)^0]$$

Let us approximate the probability in the first set of brackets by the Poisson approximation (the term in the second set of brackets can be obtained directly by a calculator). First note that 95 loaves sold means 5 loaves unsold, therefore, $P(X = 95) = P(\text{Unsold} = 5) = P(Z = 5 \mid \lambda = 5)$ where we are letting the random variable Z denote the number of unsold loaves. Thus, from the Poisson table we get the probability, 0.1755. Thus we have,

$$P(X = 95, \& \ Y = 5) = P(X = 95)P(Y = 5 \mid X = 95) = (0.1755)(.59049) = 0.1036.$$

Relative to the expected sales figures on 100 loaves of bread, again let X denote the number of loaves sold fresh, and Y the number of loaves sold in the day old bread store. Note that $X + Y \le 100$. Also, let Z denote the number of fresh loaves unsold. Note that $X + Z = 100$. Now, sales revenue, S, will be: $S = \$.90X + (.7)(\$.90)Y$

Therefore, $E(S) = E[\$.90X + (.7)(\$.90)Y] = 0.9(100)(.95) + (.7)(.9)E(Y)$

Now to get $E(Y)$ a principle of conditional expectation has to be used to be rigorous, but intuitively, we would argue that Y would behave as a binomial random variable with $n = 5$ and $p = .9$ in the long-run average sense. Therefore, $E(Y) = 5(.9) = 4.5$. Thus $E(S) = \$88.335$.

(c)   If the moisture content has been found to be normally distributed with a mean of 45% and a standard deviation of 2%, what is the likelihood of a loaf being baked with too high a moisture content? Of 100 loaves produced, how many would you expect to be found in the ideal range of 40-50% moisture content?

Let the random variable M stand for the moisture content. Then we want

$$P(M \geq 50) = 1 - F(50) = 1 - F_Z[(\frac{50 - 45}{2})] = 1 - F_Z(2.5) = 0.0062.$$

Thus, very few loaves will be outside the upper moisture limit—only 62 out of every 10,000 loaves.

The second question asks for the probability that the moisture content, M will be within acceptable moisture limits of 40-50%. That is we want

$$P(40 < M < 50) = F(50) - F(40) = F_Z(2.5) - F_Z(-2.5) = 0.9938 - 0.0062 = 0.9876$$

Thus, out of every 100 loaves baked, we would expect about 98 to 99 of them to be acceptable relative to moisture content.

CHAPTER SUMMARY

The purpose of this chapter was to introduce a few of the most common discrete and continuous probability functions that are encountered in practical applications. The Bernoulli random variable and its generalized expression in the binomial distribution is one of the most versatile of the distributions that we discussed. The negative binomial, the geometric and the hypergeometric are distributions based on repeated Bernoulli trials as well. However, it is important to distinguish one random variable from the other. The binomial the random variable is associated with a fixed number of trials and represents the number of successes that occur in the n trials that occur. The geometric and negative binomial, on the other hand, deal with the number of trials to obtain a certain number of successes.

The hypergeometric random variable is more like a binomial random in that it represents the number of successes in n trials. However, the Bernoulli trials are not independent and identical.

The Poisson is the other prominent discrete random variable that we covered in this chapter and has a wide variety of applications one of which is to approximate binomial probabilities. The approximation improves as the value of n is large and the value of p is small in the binomial distribution.

We examined several continuous-type probability density functions, the first being the gamma probability density function. This distribution is associated with a random variable X that is always positive and is a function of two parameters, α and θ. If $\alpha = 1$, the gamma distribution is called the negative exponential or simply the exponential distribution. If X is an exponential random variable it depends only on one parameter, θ.

Next we examined the most frequently discussed distribution called the normal distribution. (It is also referred to the Gaussian distribution in honor of Carl Friedrich Gauss who was one of the early developers to give an explicit expression for the normal probability density function.) The normal distribution has extensive applications and has the interesting property that the parameters turn out to be the mean and standard deviation of the distribution.

Other distributions briefly covered were the lognormal distribution and the bivariate normal distribution. Both of these distributions have important applications in modeling real-world data.

All of the distributions discussed in this chapter have parameter values that have bearing on the shape and location of the distribution. Each distribution has a unique moment-generating function in the sense that if we know the form of the pdf we know the distribution or if we know the form of the moment-generating function we know the distribution. The moments of the distribution can be obtained by appropriate integration of the pdf or by taking derivatives of the moment-generating function.

As each distribution was discussed, the form of the pdf, its mean and variance and its moment-generating function were given and for the duration of your study of these materials it would be to your advantage to either memorize some or all of those properties and expressions or to have them for ready access as we move into later chapters. They are summarized below in Table 4.7 for easier reference.

Table 4.7 $f(x)$, μ, s^2 and $M_X(t)$ for a few common probability density functions.

Name: Parameters	Probability Density Function	Mean	Variance	Moment-Generating Function
Bernoulli: $0 < p < 1$	$p^x(1-p)^{1-x}$ $x = 0,1$	p	$p(1-p)$	$[(1-p) + e^t p]$
Binomial: n,p $0 < p < 1$	$\binom{n}{x} p^x (1-p)^{n-x}$ $x = 0,1,2,...,n$	np	$np(1-p)$	$[(1-p) + e^t p]^n$
Geometric: $0 < p < 1$	$(1-p)^{x-1}p$ $x = 1,2,...$	$\dfrac{1}{p}$	$\dfrac{1-p}{p^2}$	$\dfrac{pe^t}{1-(1-p)e^t}$
Negative Binomial: r,p $0 < p < 1$	$\binom{x-1}{r-1}(1-p)^{x-r}p^r)$ $x = r, r+1,...$	$\dfrac{r}{p}$	$\dfrac{r(1-p)}{p^2}$	$\left[\dfrac{pe^t}{1-(1-p)e^t}\right]^r$
Hypergeometric: N,K,n	$\dfrac{\binom{K}{x}\cdot\binom{N-K}{n-x}}{\binom{N}{n}}$ $x = 0,1,...,\min(n,K)$	$n\cdot\dfrac{K}{N}$	$\dfrac{n(N-n)}{N-1}\dfrac{K}{N}\left(1-\dfrac{K}{N}\right)$	
Poisson: $\lambda > 0$	$\dfrac{\lambda^x e^{-\lambda}}{x!}$ $x = 0,1,2,...$	λ	λ	$e^{(\lambda(e^t-1))}$

Table 4.7   (continued)

Name: Parameters	Probability Density Function	Mean	Variance	Moment-Generating Function
Uniform: $a < b$	$\dfrac{1}{b - a}$ $a \le x \le b$	$\dfrac{a + b}{2}$	$\dfrac{(b - a)^2}{12}$	$\dfrac{(e^{tb} - e^{ta})}{t(b - a)}$ $t \neq 0$
Gamma: $\alpha, \theta \quad \theta > 0$	$\dfrac{1}{\Gamma(\alpha)\theta^{\alpha}} x^{\alpha - 1} e^{-\frac{x}{\theta}}$ $0 \le x < \infty$	$\alpha\theta$	$\alpha\theta^2$	$\dfrac{1}{(1 - \theta t)^{\alpha}}$
Exponential: $\theta > 0$	$\dfrac{1}{\theta} e^{-\frac{x}{\theta}}$ $0 < x < \infty$	θ	θ^2	$\dfrac{1}{(1 - \theta t)}$
Normal: $\mu, \sigma \quad \sigma > 0$	$\dfrac{1}{\sigma\sqrt{2\pi}} e^{-\frac{[x - \mu]^2}{2\sigma^2}}$ $-\infty < x < \infty;$	μ	σ^2	$e^{\{\mu t + (\frac{\sigma^2 t^2}{2})\}}$

We concluded this chapter by showing you how to generate random numbers from any distribution by using the cumulative distribution function $F(x)$. This procedure can be used with discrete random variables for which we produce a cumulative table, or for continuous random variables where $F(x)$ can be manipulated to produce an inverse function $F^{-1}(x)$. However, many statistical software programs provide the capability of generating values for the most common pdf's.

CHAPTER EXERCISES

1. In order to insure product safety, a large pharmaceutical firm uses a sampling procedure to detect faulty seals on its popular over-the-counter pain reliever. On a regular basis, 200 bottles are randomly selected from the production line and inspected for defective seals and/or packaging. It is known from historical data that about .5% of packages will show up defective in some way. Suppose in a given sample 4 defective seals were found? Would you conclude that something is wrong with the packaging machines? Or would you simply attribute this to chance. (Note that if 4 seals were an indication of something wrong, 5 defective seals would also lead to the same conclusion, and so on. Therefore, you want to calculate the probability of getting 4 or more defective seals in a sample of 200.)

2. It was stated that the Poisson distribution provides an approximation to binomial probabilities whenever n is large and p is small. A friend noticed that for reasonably large n and $p = 0.5$, that the binomial distribution is symmetric and very much "normal" looking. So she suggested that

the normal distribution might provide an adequate approximation to binomial probabilities for large n and p close to 0.5. To check this property out you decided to examine a few cases, and to determine what "large n" means, and "p close to 0.5" means. You decided to look at both an "interior" type of probability and a "tail" area probability. Do this by filling out the following table.

Comparing the normal and binomial probabilities

	Interior Region $P[np - 1.5\sqrt{np(1-p)} \le X \le np + 1.5\sqrt{np(1\ p)}]$ Binomial Normal	Tail Area $P[X \le np - 2\sqrt{np(1-p)}]$ Binomial Normal
$n = 5$ $p = 0.2$		
$p = 0.4$		
$p = 0.5$		
$n = 10$ $p = 0.2$		
$p = 0.4$		
$p = 0.5$		
$n = 15$ $p = 0.2$		
$p = 0.4$		
$p = 0.5$		
$n = 20$ $p = 0.2$		
$p = 0.4$		
$p = 0.5$		

On the basis of the above results what recommendations would make, or what further studies would you suggest?

3. As you enter the checkout stand in a grocery store, you find that there are 4 people ahead of you with the first person in line just starting to be served. If the time it takes to serve a customer is exponential with $\theta = 4$ minutes, what is the probability that it will take longer than 20 minutes to serve the 4 people in front of you? What is the probability it will take less than 10 minutes to serve the 4 people in front of you?

4. The specifications for a capacitor are that its life must be greater than 1500 hours. The life is known to be normally distributed with mean of 2500 hours. The invoice price realized from each capacitor is $10.00; however, a failed unit must be replaced at a cost of $4.00 to the company. Two manufacturing processes can produce capacitors having satisfactory means lives, but the standard deviation of process A is 900 hours and that for process B is 500 hours. The cost of producing a part by process A is $6.00, while the cost of producing a part by process B is $8.00.

(a)   What proportion of parts will not meet specifications from process A? From process B?

(b)   Letting "Profit = Invoice price – Cost of manufacturing – Replacement cost",what is the "expected profit" for a part manufactured by process A? Process B?

5.   A college professor has noticed that grades on each of two quizzes have a bivariate normal distribution with the following parameters:

$$\mu_x = 75,\ \mu_y = 83,\ \sigma_x^2 = 25,\ \sigma_x^2 = 16,\ \rho = 0.8$$

If a student receives a grade of $X = 80$ on the first quiz, what is the probability that he will do better, $Y \geq 80$, than that on the second one?

6.   The timing device used to control the "intermittent" phase of the windshield wipers on an automobile uses a "clock" that activates the wipers on the average of every 3 seconds. However, there is variation in the clock and the standard deviation is .3 seconds. A timing device or "clock" is considered defective if it deviates from the mean by more than .5 seconds. (a) What proportion of "clocks" will be found to be defective? (b) If a sample of 20 clocks is to be drawn every half hour, what is the probability that no defective clocks will be found?

7.   Using the information from exercise 6, a quality control engineer wants to find 4 clocks that are classified to be defective and subject them to further testing. What is the expected number of clocks that will have to be tested in order to find 4 defective clocks? What is the probability that 100 trials will be required to obtain the 4 defective? What is the probability that 100 trials will be required to obtain the 1 defective?

8.   The following set of data was generated using principles of random number generation. One of the probability density functions from this chapter was used as the population. First, make as logical a guess as you can about the distribution that was used to generate this data. Estimate or guess the value of the parameter(s) associated with your distribution of choice. Use any defensible logic you can to make your decision.

49.0662	15.2768	23.3803	2.2905	13.7978
37.7604	33.9633	22.8145	16.0873	43.4627
40.6222	10.2427	36.0867	26.9065	20.2895
16.0575	5.6385	14.9561	36.7405	13.6624
7.3373	13.9306	30.7167	34.9067	27.5957
20.0064	40.2106	64.0425	6.9909	50.5328
20.7794	3.8205	7.1160	2.7182	31.7776
7.6585	9.3867	6.0724	5.3690	22.0052

$n = 40$ MEAN = 22.552   ST.DEV. = 15.1

9.   Suppose a characteristic shows up in a given population in 10% of the population units. What is the value of n to give a probability of at least 90% of getting at least 1 element in the population with the characteristic in a random sample of the population. [Hint: use a binomial model for this problem and set-up an equation and solve for n.]

10.   Solve the problem addressed in problem 9 assuming the characteristic occurs 50% of the time; 1% of the time.

5

THE DISTRIBUTION OF A FUNCTION OF RANDOM VARIABLES

OBJECTIVES

When you complete this chapter you should understand how to derive the probability density function for a transformation of random variables associated with a given pdf. In addition, you should understand why this is an important task to be able to perform. Once you have obtained such a pdf, you should be able to find means, variances, and probabilities of interest from the resulting distributions as long as those distributions are mathematically manageable.

INTRODUCTION

Suppose we have a random sample from some pdf, defined by $f(x)$. The observations in the sample can be represented by the random variables, $X_1, ..., X_n$. The problem facing the researcher with this data is how to analyze the sample data and make some useful conclusions based on them.

From what we have done so far in this text, to analyze a set of data you would first organize the data in some way and then compute a statistic such as a mean and standard deviation, or some other measure of location and variability. We would then compare what we see in the sample with what our experience would suggest we would see if the mean and variance were computed on the random variables from the pdf $f(x)$. We might ask how likely it is to get the particular sample mean and standard deviation from the population defined by $f(x)$. This requires a probability computation. However, you must recognize that having summarized the data by computing the mean and standard deviation, we have altered the probability structure so that we are no longer working with $f(x)$ as it stands. Rather we would have to use some other probability density function to obtain a correct probability.

For instance, suppose X is the time to failure of a battery installed in a portable stereo tape player. If the player requires 6 size D batteries, the player will fail as soon as any one of the batteries fail. If $X_1, X_2, ..., X_6$ denotes the time to failure of each individual battery and Y denotes the time to failure of the player, then $Y = \text{Minimum}(X_1, X_2, ..., X_6)$. That is, the stereo fails when the first battery fails.

We may be willing to assume that we know $f(x)$ for the batteries, but *we want to calculate probabilities about the random variable Y*—which requires us to know $f_y(y)$, the probability density function of the *minimum* of six random variables. The connection between $f(x)$ and $f_y(y)$ is not immediately obvious and intuitive though they will be connected. In this chapter, we will present several methods to determine the distribution of Y when Y is some function of the random variables $X_1, \ldots, X_n$.

Where We Are Going in This Chapter

In the course thus far we have learned how to calculate probabilities concerning a random variable X ior about two random variables X and Y when given their respective probability density functions. However, most of the time we deal not with just one random variable, but a "sample" of "n" random variables. In addition we do not restrict our questions about the values of the random variables individually. Rather we ask probability questions about some function of the random variables. For example, we want to know if the sum of the random variables exceed a given value, or if the mean of the random variables has changed either up or down. In order to answer these questions, we need to know the probability density function of the "new" random variable which is the "sum" or "mean" of a set of other random variables.

In this chapter we present three methods for determining the distribution of a function of a random variable or set of random variables. The presentation will be at a fairly low level, the intent being to introduce you to the concept but not to make you an expert. The purpose of this chapter is to show the methods that are available to us to determine the distribution of a transformed random variable or some function of a set of random variables. We will go only so far as to indicate the direction of derivation for rather simple cases with the more complex processes to be treated in a higher level text.

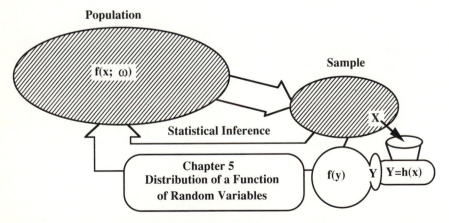

Figure 5.1 Statistical Inferential Model (SIM) for Chapter 5.

The general idea of this chapter is demonstrated in Figure 5.1 above which suggests that we start out with a population modelled by the pdf $f_x(x)$. A sample is taken, producing the random variable X which is then transformed by the function $Y = h(X)$. We then want to determine the probability density function, denoted by $f_Y(y)$, of the transformed random variable Y. [Recognize that X could be a vector of values and the transformation $h(X)$ is an operation on a set of random variables.

For instance the X may denote a sample of observations and the transformation amounts to computing the sample mean, or the sample variance, the maximum, or minimum, and so forth.]

Chapter Exercise

Suppose a sample of electronic parts are being bench tested for reliability. Previous studies show that the "time-to-failure" of the parts is exponential with parameter 5 (hundreds of hours). If 5 parts are being tested what is the probability that the mean "time to failure" of the 5 parts will exceed 5 (500 hundred hours)? What is the probability that the mean "time to failure" of the 5 parts will be between 4 and 6 hundred hours?

5.1   THE DISTRIBUTION FUNCTION TECHNIQUE

Introduction

In this section we present a method for determining the distribution of transformed variables that using the cumulative distribution function $F(y)$ For instance, if X is a continuous random variable with pdf $f(x)$, and $Y = h(X)$ is the transformed random variable, we define the cumulative distribution function $f_y(y)$ of the transformed random variable, Y, and re-express it in terms of a probability involving the original, untransformed random variable X. If this probability statement can be evaluated in terms of an integral involving the random variable X and its pdf, then the pdf of Y can be obtained by taking the derivative of $F_y(y)$. This approach is described and demonstrated in the following section.

Definitions

The Distribution Function Technique

> **DEFINITION 5.1:** Suppose the random variables $X_1, ..., X_n$. are continuous random variables with a given joint probability density function $f(x_1, ..., x_n)$. If Y is some function of the random variables; i.e. ,
>
> $$Y = h(X_1, ..., X_n).$$
>
> then the probability density function of the random variable Y will be obtained by first finding $F(y) = F(h(X_1, ..., X_n) \leq y)$ and then differentiating to get
>
> $$f(y) = \frac{dF(y)}{dy}.$$

DISCUSSION

The distribution function technique presented in this section is implemented by trying to find the cumulative distribution function of a random variable $Y = h(X)$ defined as some function of another random variable X whose probability properties we know. If we can succeed in finding $F(y)$, then we find $f(y)$ by simply taking the derivative. That is: $f(y) = \dfrac{dF(y)}{dy}$.

For instance we may know that the random variable X is an exponentially distributed random variable but we are interested in the transformation $Y = \sqrt{X}$. We know the pdf of X, but we want the pdf of Y, $f(y)$. If we can find $F(y)$, taking its derivative will produce $f(y)$. The distribution function technique will be demonstrated with several examples to make it clearer for you.

EXAMPLES

EXAMPLE 1: Suppose the manufacturer of cubes used in making a Rubic cube has controlled the process such that the length of an edge, X, varies randomly with the probability density of X given by

$$f(x) = \begin{cases} 6x(1-x) & 0 < x < 1 \\ 0 & \text{otherwise} \end{cases}$$

find the probability density function of $Y = X^3$, the volume of the cube. That is, letting $Y = h(X) = X^3$, what is $f_Y(y)$?

SOLUTION: Using the distribution function technique we want to find the cumulative distribution function of the random variable Y. We take the following steps.

$$F_Y(y) = P(Y \le y) = P(X^3 \le y)$$

$$= P(X \le y^{1/3})$$

$$= \int_0^{y^{1/3}} 6x(1-x)dx = 6\left[\frac{x^2}{2} - \frac{x^3}{3}\right]\Big|_0^{y^{1/3}}$$

This yields the expression: $F_y(y) = 3y^{2/3} - 2y.$

Taking the derivative, we get for the pdf

$$f_y(y) = \frac{2}{y^{1/3}} - 2$$

Therefore to address probability questions concerning the volume of the cubes, we can use the pdf for the random variable $Y = X^3$.

The graphs of both $f(x)$ and $f(y)$ are given below as Figure 5.2 and Figure 5.3. Notice that they are quite different in shape but both have the basic properties of any pdf. They are positive functions and the area under the curve equals 1.

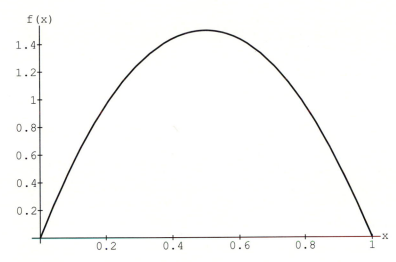

Figure 5.2 Graph of the pdf: $f(x) = 6x(1 - x)$, $0 < x < 1$.

EXAMPLE 2: Suppose the random variable X has the pdf

$$f(x) = 3x^2 \text{ for } 0 < x < 1$$

and let $Y = X^3$. What is the pdf of Y?

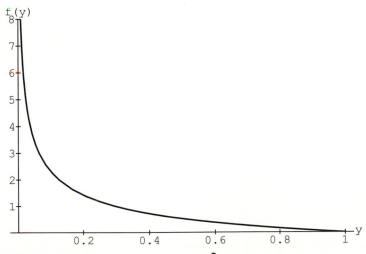

Figure 5.3 Graph of the pdf: $f_y(y) = \dfrac{2}{y^{1/3}} - 2$.

SOLUTION: Using the distribution function technique we get

$$F(y) = P(Y < y) = P(X^3 < y) = P(X < (y)^{\frac{1}{3}}) = \int_0^{y^{\frac{1}{3}}} 3x^2 \, dx$$

$$= \frac{3x^3}{3} \Bigg|_0^{y^{\frac{1}{3}}} = y.$$

Therefore, $f(y) = \dfrac{dF(y)}{dy} = 1$ where $0 < y < 1$. This pdf is the uniform distribution on the unit interval.

A graph of $f(x)$ is given in Figure 5.4 below. A graph of the unit uniform distribution is given in Chapter 4. Again, an obvious difference is present between the two graphs of the pdf's involved in this example

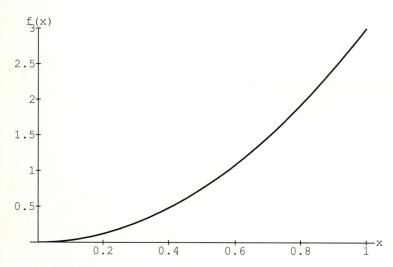

Figure 5.4 Graph of $f(x) = 3x^2$ for $0 < x < 1$.

EXAMPLE 3: For the distribution function given in Example 3 what is the mean of X and the mean of Y? Therefore, find $E(X)$ and $E(Y)$.

SOLUTION: We find $E(X)$ by solving the following integral.

$$E(X) = \int_0^1 x(3x^2)) dx = 3\left(\frac{x^4}{4}\right) \Bigg|_0^1 = \frac{3}{4}$$

To find $E(Y)$ we note that Y is uniform on the interval 0 to 1 and therefore, $E(Y) = \dfrac{1}{2}$. However, we could also find $E(Y)$ by integrating as shown:

$$E(Y) = E(X^3) = \int_0^1 x^3(3x^2))dx = 3\left(\frac{x^6}{6}\right)\Big|_0^1 = \frac{3}{6} = \frac{1}{2}$$

EXAMPLE 4: Suppose that X_1 and X_2 are independent exponential random variables with $\theta = 4$. Therefore, the joint distribution is given by

$$f(x_1, x_2) = \frac{1}{4^2} e^{-\frac{(x_1 + x_2)}{4}} \quad x_1 > 0, \ x_2 > 0$$

Find the density function of the sum, $Y = X_1 + X_2$. The graph of the joint density is given below in Figure 5.5.

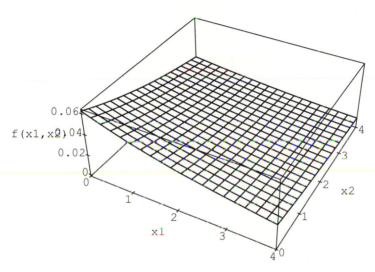

Figure 5.5 Graph of the joint distribution: $f(x_1, x_2) = \dfrac{1}{4^2} e^{-\frac{(x_1 + x_2)}{4}} \quad x_1 > 0, \ x_2 >. \, 0$

SOLUTION: We denote the cumulative distribution function of Y, $F(y)$ as:

$$F_y(y) = P(X_1 + X_2 \leq y)$$

$$= \int_0^y \int_0^{y-x_2} \frac{1}{4^2} e^{-\frac{(x_1 + x_2)}{4}} \, dx_1 dx_2.$$

See Figure 5.6 to more clearly identify this region. The region of integration in Figure 5.6 is the triangular area defined on the x_1, x_2 plane, and the integral will give the volume under the surface for that triangular region.

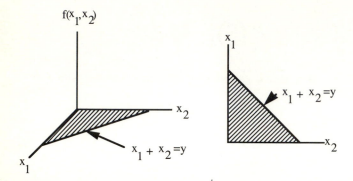

Figure 5.6 Region of integration for example 4: $X_1 + X_2 \leq Y$.

Carrying out the integration we finally get

$$F_Y(y) = 1 - e^{-y/4} - \frac{ye^{-y/4}}{4}.$$

To get the pdf we take the derivative of F with respect to y, getting

$$f_Y(y) = \frac{1}{4^2}y\, e^{-y/4}$$

which can be recognized as a gamma distribution with $\alpha = 2$ and $\theta = 4$. The graph of this pdf is shown in Figure 5.7 below.

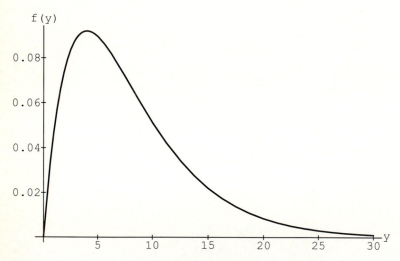

Figure 5.7 Graph of gamma pdf with $\alpha = 2$ and $\theta = 4$.

Thus the sum of these two independent exponential random variables is distributed as a gamma random variable with the specific parameter values as given. (Note that if we had simply used θ above instead of 4, the results would still apply and would be more general. That is, the sum of 2 independent exponential random variables with parameter θ becomes gamma with $\alpha = 2$ and parameter θ. Further, we could then implement an inductive proof to show that the sum of n independent exponential random variables becomes gamma with $\alpha = n$ and θ. However, a simpler method to prove this result is given in section 5.3 which follows.)

EXERCISES

1. Find the probability density of $Y = X^2$ if the probability density of X is given by

    $$f(x) = 2xe^{-x^2}, \quad x > 0$$

2. Use the results of exercise 1 to find $E(Y) = E(X^2)$.

3. If X has an exponential distribution with the parameter θ, find the probability density of $Y = \ln X$.

4. Find $E(Y) = E(\ln X)$ for the exponential distribution given in exercise 3.

5. If X is distributed uniformly for $0 < x < 1$, find the probability density of $Y = (X)^{1/2}$.

6. Find the probability density of $Y = X^3$ if the probability density of X is given by

    $$f(x) = \frac{x}{2}, \quad 0 < x < 2$$

7. Find the $E(Y)$ by using the pdf of Y found in exercise 6.

8. Find the probability density of $Y = 8X^3$ if X has the probability density

    $$f(x) = 2x, \quad 0 < x < 1$$

9. Suppose that X_1 and X_2 are independent exponential random variables with $\theta = 4$. Therefore, the joint distribution is given by

    $$f(x_1, x_2) = \frac{1}{4^2} e^{-\frac{(x_1 + x_2)}{4}} \quad x_1 > 0, \ x_2 > 0$$

    Find the density function of the sum, $Y = \dfrac{X_1 + X_2}{2}$. [Hint: Follow the procedure shown in example 2, step-by-step.]

10. For the random variable Y defined in exercise 9, find its mean and variance by recognizing the family of pdf's with which it is associated. Also, show that the procedure of section 3.10 provides the same results.

5.2   THE CHANGE OF VARIABLE TECHNIQUE

Introduction

This section presents a slightly less involved method of obtaining the distribution of a transformed random variable. It is an outgrowth of the distribution function method of section 5.1, but we present it here as a separate method. By working the problems at the end of this section and comparing results with the last section, you will appreciate the more direct approach that this method represents.

THEOREM

THEOREM 5.1:   Let $f(x)$ be the probability density function of a continuous random variable X. Let $Y = h(X)$ be a differentiable function of X and either an increasing or decreasing function of x (but not both) for all values within the range of X for which $f(x) \neq 0$. That is, Y is a monotonic function of x.

$$f_y(y) = f(h^{-1}(y)).|(h^{-1}(y))'|$$

where $x = h^{-1}(y)$ is the inverse of the function $g(x)$ and $|h'^{-1}(y)|$ is the absolute value of the first derivative of the inverse function taken with respect to, y, that is, dx/dy. (This absolute value of dx/dy is called the *Jacobian* of the inverse function.)

DISCUSSION

The procedure of this section can be shown to be a derivation of the distribution function method discussed in section 5.1. But it represents a less involved method under the conditions appropriate to its use, so we present it as a separate method. The best way to understand it is to look at a few examples as follows.

EXAMPLES

EXAMPLE 1:   If X has the exponential distribution given by

$$f(x) = \begin{cases} e^{-x} & x > 0 \\ 0 & \text{elsewhere} \end{cases}$$

find the probability density function of the random variable

$$Y = (X)^{1/2}$$

A graph of the exponential function as defined is shown in Figure 5.8 below.

SOLUTION: In this case $Y = (X)^{1/2}$ is monotonically increasing for $x > 0$, and the inverse function (solving for X) is $X = Y^2$. Taking the derivative with respect to y and then taking its absolute value gives:

$$\left| \frac{dx}{dy} \right| = |\, 2y \,|$$

Since this derivative is positive for all values of y when $x > 0$ then the Jacobian is $2y$. Therefore by the theorem above,

$$f_y(y) = e^{-y^2} \cdot 2y \; = \; 2y \, e^{-y^2} \quad \text{for } y > 0$$

and is 0 elsewhere.

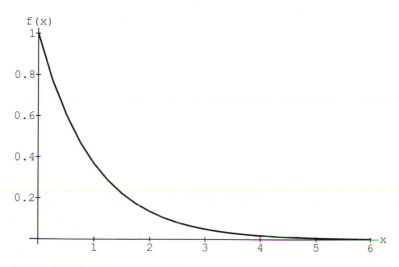

Figure 5.8 Graph of $f(x) = e^{-x}$.

This particular probability distribution is referred to as the Weibull Distribution and its graph is given in Figure 5.9 below.

EMPIRICAL VERIFICATION

By generating random values from a particular distribution we can demonstrate, empirically, the effect of transformations of random variables. We do this by generating a large set of observations from the distribution $f(x)$ which we then transform by the specified transformation. We examine the resulting distribution of transformed random variables by constructing a histogram or by computing moments of various kinds to assess the impact of the given transformation.

EXAMPLE 2: If X has the exponential distribution given by

$$f(x) = \begin{cases} e^{-x} & x > 0 \\ 0 & \text{elsewhere} \end{cases}$$

generate a set of exponential random numbers, and transform them by the function $y = (x)^{1/2}$ and make a histogram of the transformed variables, y.

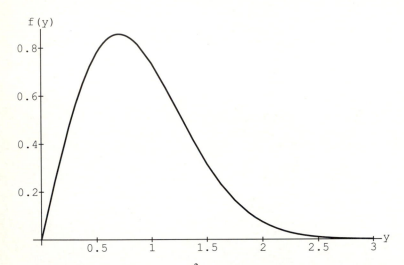

Figure 5.9 Graph of $f(y) = 2ye^{-y^2}$: "Weibull" distribution with parameter 2.

SOLUTION: We first generate 1000 uniform random numbers which are then converted to exponential random numbers. From section 4.12 this is done by letting $x = -\ln(1 - u)$. (Histograms of these 1000 uniform and "converted" exponential random numbers are shown in Figure 5.10 and 5.11 that follows.) The 1000 exponential random numbers are then subjected to the transformation given above, i.e., we take the square root of each one. A histogram of these numbers is then shown as Figure 5.12. These histograms should be compared to the theoretical results presented in example 1 above and shown as graphs in Figures 5.8 and 5.9.

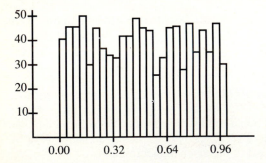

Figure 5.10. Histogram of 1000 uniform random numbers.

In comparing Figures 5.10–5.12 with one another, it is important to note the differences in both the horizontal and vertical scales if you are to make a valid comparison. of their relationships. Note that the uniform numbers are scaled between 0 and 1, the exponential numbers are scaled between 0 and about 7, and the Weibull between 0 and about$\sqrt{7}$ which would be expected through the square root transformation.

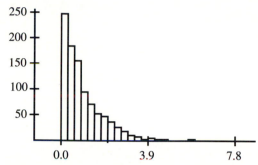

Figure 5.11. Histogram of 1000 exponential random numbers with $f(x) = e^{-x}$.

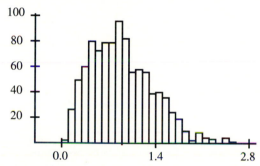

Figure 5.12. Histogram of 1000 Weibull random numbers: $Y = \sqrt{X}$.

EXAMPLE 3: Suppose the manufacturer of cubes used in making a Rubic cube has controlled the process such that the length of an edge, X, varies randomly with the probability density of X given by

$$f(x) = \begin{cases} 6x(1-x) & 0 < x < 1 \\ 0 & \text{otherwise} \end{cases}$$

Find the probability density function of $Y = X^3$, the volume of the cube. (See example 1 from section 5.1 of this chapter.)

SOLUTION: In the case, with $Y = X^3$, solving for X yields $X = Y^{1/3}$. The Jacobian, involving the derivative $\dfrac{dx}{dy}$, turns out to be

$$\left| \frac{dx}{dy} \right| = \frac{1}{3} y^{-2/3} \text{ for } 0 < x < 1$$

Therefore, we get as the pdf of Y for y between 0 and 1.

$$f_y(y) = 6(y^{1/3})(1 - y^{1/3})\frac{y^{-2/3}}{3} = 2(y^{-1/3} - 1) = 2(y^{-1/3}) - 2$$

This is the same pdf as obtained in section 5.1 using the distribution function technique.   Graphs of $f(x)$ and $f_y(y)$ are given in section 5.1, Figures 5.2 and 5.3.

EXAMPLE 4: If $F(x)$ is the distribution function of the continuous random variable X, find the probability density function of the random variable Y defined to be

$$Y = F(x).$$

SOLUTION: Using the principle that $\dfrac{dx}{dy} = \dfrac{1}{\dfrac{dy}{dx}}$ we can proceed as follows:

Since $y = F(x)$, then $\dfrac{dy}{dx} = f(x)$ so that $\dfrac{dx}{dy} = \dfrac{1}{f(x)}$. Now since $x = h^{-1}(y)$ and $\dfrac{dx}{dy} = \dfrac{1}{\dfrac{dy}{dx}}$, we write

$$f_y(y) = f(h^{-1}(y))|(h^{-1}(y)')|$$

$$= f(x)|\frac{dx}{dy}|$$

as long as the right side is expressed in terms of y. But for this particular case it can be seen that

$$f_y(y) = f(x)|\frac{dx}{dy}| = f(x)(\frac{1}{f(x)}) = 1 \text{ whenever } 0 < y < 1$$

However, we recognize $f_y(y) = 1$ as the uniform distribution. (This result is very important and serves as the underlying principle involved in using the uniform distribution in generating random numbers from any distribution. This result confirms that the cumulative distribution function for *any* random variable is distributed as a uniform random variable on the unit interval.)

EXERCISES

Use the change of variable technique of this section to solve the following problems. Compare your results with the answers you obtained in working these same problems in section 5.1.

1.   Find the probability density of $Y = X^2$ if the probability density of X is given by

$$f(x) = 2xe^{-x^2}, \quad x > 0$$

2.   If X has an exponential distribution with the parameter θ, find the probability density of $Y = \ln X$.

3.   Generate a sample of 100 observations from the distribution of the random variable X for exercise 2, make a histogram of the 100 observations and compare with the parent pdf. Then transform each of the 100 values by the transformation indicated and make a histogram of these "transformed" values and compare to the theoretical results obtained. [Hint: Use the technique of

section 4.12. Generate 100 uniform random numbers, and convert them by using the inverse function of $F(x)$.]

4. If X is distributed uniformly for $0 < x < 1$, find the probability density of $Y = (X)^{1/2}$.

5. Generate a sample of 100 observations from the distribution of the random variable X for exercise 4, make a histogram of the 100 observations and compare with the parent pdf. Then transform each of the 100 values by the transformation indicated and make a histogram of these "transformed" values and compare to the theoretical results obtained.

6. Find the probability density of $Y = X^3$ if the probability density of X is given by

$$f(x) = \frac{x}{2}, \quad 0 < x < 2$$

7. Generate a sample of 100 observations from the distribution of the random variable X for exercise 6, make a histogram of the 100 observations and compare with the parent pdf. Then transform each of the 100 values by the transformation indicated and make a histogram of these "transformed" values and compare to the theoretical results obtained.

8. Find the probability density of $Y = 8X^3$ if X has the probability density

$$f(x) = 2x, \quad 0 < x < 1$$

9. Suppose X has the pdf given by
$$f(x) = \frac{3}{16}x^2 \qquad -2 < x < 2$$
If $Y = \sqrt{X}$, why can't you use the method of this section to find the pdf of Y?

10. For the pdf of the random variable X, define a random variable $Y = h(X)$ that will meet the conditions of the theorem for this section. Find the pdf of Y for the function of your choice.

5.3   THE MOMENT-GENERATING FUNCTION TECHNIQUE

Introduction

In this section we present a technique for identifying the probability density function for the sum of independent random variables. This method depends on the fact that there is a one-to-one relationship between a pdf and its moment-generating function, a result that is proved in more advanced texts. The theorem that follows shows how to obtain the moment-generating function of a sum of independent random variables. If the resulting mgf is recognizable as that of a particular distribution, then the pdf of the sum is immediately known. The need to know what the pdf is for a sum of independent random variables is common in statistics.

THEOREM

THEOREM 5.2: If $X_1, ..., X_n$ are independent random variables and $Y = a_1 X_1 + ... + a_n X_n$, then

$$M_Y(t) = \prod_{i=1}^{n} M_{Xi}(a_i t)$$

where $M_{Xi}(a_i t)$ is the moment-generating function of X_i at the point a_i. (The symbol $\prod_{i=1}^{n}$

signifies the product of the respective mgf's.)

PROOF: The moment-generating function of Y is by definition

$$E(e^{tY}) = E[e^{t(a_1 X_1 + ... + a_n X_n)}]$$

$$= E(e^{ta_1 X_1} \bullet e^{ta_2 X_2} \bullet ... \bullet e^{ta_n X_n}) \qquad \text{(law of exponents)}$$

$$= E(e^{ta_1 X_1}) \bullet E(e^{ta_2 X_2}) \bullet ... E(e^{ta_n X_n}) \qquad \text{(independence)}$$

$$= \prod_{i=1}^{n} M_{Xi}(a_i t)$$

DISCUSSION

Many of the computations that we perform on a set of random variables is expressed as a sum or mean. In addition, the probability questions that we raise concern themselves with the sum or the mean. Therefore, if a sample is drawn from a population defined by $f(x)$ with this sample being denoted by $X_1, ..., X_n$, we will often be interested in knowing whether the mean or sum of these random variables will likely exceed or be smaller than some value. If we ask about the mean, then

$$Y = \frac{X_1 + ... + X_n}{n}$$

in which case, in Theorem 3.1, the a's are $\frac{1}{n}$. If we ask about the sum, then the a's are equal to 1 and the above theorem can be applied accordingly. To show more directly the steps and arguments taken, consider the following examples.

EXAMPLES

EXAMPLE 1: Find the probability distribution of the sum of n independent random variables $X_1, ..., X_n$ each having a Poisson distribution with common parameter, λ.

SOLUTION: In this example $Y = a_1X_1 + \dots + a_nX_n$ where the a_i's are 1's. In addition, the moment-generating function of the random variable X_i which is Poisson is

$$M_{X_i}(t) = e^{\lambda(e^t - 1)}$$

Therefore, the moment-generating function of the sum is the product of the separate moment-generating functions:

$$M_Y(t) = \prod_{i=1}^n e^{\lambda(e^t - 1)} = e^{n\lambda(e^t - 1)}$$

But we recognize this is as a Poisson mgf with parameter $n\lambda$. Therefore, the distribution of the sum of n independent and identical Poisson random variables is itself Poisson with the parameter value equal to $n\lambda$.

EXAMPLE 2: Find the probability distribution of the mean of n independent and identically distributed exponential random variables where $\theta = 5$.

SOLUTION: The mgf of the exponential is $M_{X_i}(t) = \dfrac{1}{(1 - 5t)}$. Also, when expressing a mean as a linear combination, $a_1X_1 + \dots + a_nX_n$, the value of a_i must be $\dfrac{1}{n}$ for all i. That is,

$$\bar{X} = \frac{X_1 + \dots + X_n}{n} = a_1X_1 + \dots + a_nX_n$$

only if $a_i = \dfrac{1}{n}$. Therefore the mgf of a mean of exponential random variables is

$$M_Y(a_it) = M_Y(\tfrac{1}{n}t) = \prod_{i=1}^n \frac{1}{(1 - 5(\frac{1}{n}t))} = \frac{1}{(1 - \frac{5}{n}t)^n}$$

But this is the mgf of a gamma distribution with parameters $\alpha = n$ and $\theta = \dfrac{5}{n}$.

EXAMPLE 3: Suppose each one of a set of independent random variables $X_1, \dots, X_n$ follows a Bernoulli distribution with common parameter $p = 0.3$. What is the probability density function of the sum of these Bernoulli random variables?

SOLUTION: The sum is $Y = X_1 + X_2 + \dots + X_n$ where the coefficient for each X is a 1. Thus, in the general expression the a_i's are all equal to 1. The moment-generating function of Y is the product of the moment-generating functions of each X. Therefore, we get

$$M_Y(t) = \prod_{i=1}^n [(1 - 0.3) + 0.3pe^t] = [(1 - 0.3) + 0.3e^t]^n$$

Upon examination of the expression on the right, we recognize it as the moment-generating function of the binomial distribution with n trials and parameter $p = 0.3$.

EXERCISES

1. What is the probability density function (pdf) of the random variable X which has the moment-generating function:

$$M_X(t) = \frac{1}{\left(1 - \frac{1}{8}t\right)}$$

2. What is the pdf of the random variable X which has the following moment-generating function:

$$M_X(t) = e^{69t + \frac{36}{100}t^2}$$

3. What is the pdf of the random variable $Y = \sum X_i$ if $X_1, ..., X_n$ are independent and identically distributed random variables and if each X_i is Poisson with common parameter λ?

4. What is the pdf of the random variable $Y = \sum X_i$ if $X_1, ..., X_n$ are independent and identically distributed random variables and if each X_i is exponential with common parameter θ?

5. What is the pdf of the random variable $Y = \sum X_i$ if $X_1, ..., X_n$ are independent and identically distributed random variables and if each X_i is gamma with common parameter α and θ?

6. What is the pdf of the random variable $Y = \sum X_i$ if $X_1, ..., X_n$ are independent and identically distributed random variables and if each X_i is normal with common parameter μ and σ?

7. What is the pdf of the random variable $Y = \frac{\sum X_i}{n}$ if $X_1, ..., X_n$ are independent and identically distributed random variables and if each X_i is Poisson with common parameter λ?

8. What is the pdf of the random variable $Y = \frac{\sum X_i}{n}$ if $X_1, ..., X_n$ are independent and identically distributed random variables and if each X_i is exponential with common parameter θ?

9. What is the pdf of the random variable $Y = \frac{\sum X_i}{n}$ if $X_1, ..., X_n$ are independent and identically distributed random variables and if each X_i is gamma with common parameters α and θ?

10. What is the pdf of the random variable $Y = \frac{\sum X_i}{n}$ if $X_1, ..., X_n$ are independent and identically distributed random variables and if each X_i is normal with common parameters μ and σ?

5.4   ANALYSIS OF THE CHAPTER EXERCISE

Suppose a sample of electronic parts are being bench tested for reliability. Previous studies show that the "time to failure" of the parts is exponential with parameter 5 (hundreds of hours). If 5 parts are being tested what is the probability that the mean "time to failure" of the 5 parts will exceed 5 (500 hundred hours)? What is the probability that the mean "time to failure" of the 5 parts will be between 4 and 6 hundred hours?

SOLUTION: Let the random variables X_1, X_2, ..., X_5 denote the time to failure of each of the 5 parts. The distribution of each X_i is exponential with parameter 5. We are interested in the random variable $Y = \overline{X}$, the mean of the 5 random variables. The random variable Y is of the form $\sum a_i X_i$ where $a_i = \dfrac{1}{5}$ and is the sum of independent and identically distributed random variables.

$$\text{(Note that: } \overline{X} = \frac{X_1 + X_2 + \dots + X_5}{5} = \frac{1}{5}X_1 + \frac{1}{5}X_2 + \dots + \frac{1}{5}X_5$$

To find the pdf of the random variable Y we can use the moment-generating functions technique, noting that

$$M_Y(t) = \frac{1}{(1 - \frac{5t}{5})^5} = \frac{1}{(1 - \frac{5}{5}t)^5}$$

However, by examination, it can be observed that this is the mgf of a gamma pdf with parameters $\alpha = 5$ and $\theta = 1$. Therefore, to answer the probability questions, we would use the gamma distribution with the parameters $\alpha = 5$ and $\theta = 1$, as defined above. Thus, the probability that the "mean time to failure" will exceed 5 is

$$P(Y > 5) = 1 - F_G(5;5,1) = 1 - [1 - F_P(4, \lambda = \frac{5}{1}) = 0.4405$$

The probability that the "mean time to failure" will be between 4 and 6 hundred hours is:

$$P(4 < Y < 6) = F_G(6;5,1) - F_G(4;5,1)$$

$$= [1 - F_P(4; \lambda = \frac{6}{1})] - [1 - F_P(4; \lambda = \frac{4}{1}) = F_P(4; \lambda = \frac{4}{1}) - F_P(4; \lambda = \frac{6}{1})$$

$$= 0.6288 - 0.2851 = 0.3437$$

CHAPTER SUMMARY

In this chapter we have examined, without extensive depth, the techniques that are commonly used to find the probability density function of a transformed random variable such as $Y = \sqrt{X}$, $Y = 4X^2$, $Y = e^X$ and so forth. In addition, the moment-generating function was used to find the probability density function of linear combinations of random variables. The most common examples of the application of this technique was to the sum $Y = X_1 + X_2 + X_3$ or mean $Y = X_1 + X_2 + X_3$ of a set of independent random variables. These techniques are used to prove the results of Chapter 6 that follows. The proofs in Chapter 6 will generally not be given explicitly, but a reference to the type of technique used to complete the proof will be indicated.

In the Appendix to this chapter is another technique based on the principle of a Taylor's series expansion that provides an approximation to the mean and variance of functions that are too complicated to derive analytically, or when only the mean and variance of the original pdf is known, the exact form of the pdf being unknown.

CHAPTER EXERCISES

1. Pieces of pipe are joined together for a fire-extinguisher system in a large high school under construction. The lengths of pipe are assumed to be normally distributed with a mean of 20 feet and a standard deviation of .25 inches. A particular length of pipe is needed to be 59 feet, 11 and .75 inches. If three lengths of pipe are randomly selected, what is the probability that their combined lengths will be as least as long or longer than the section needed without any patching or joining of a 4th piece of pipe? Let $Y = X_1 + X_2 + X_3$, where the X's are normally distributed as given above. {See exercise 1d of section 5.3.}

2. The diameter of wire to be used in a particular electrical application is variable due to the manufacturing process. It is assumed to be uniformly distributed with a mean of 0.1 cm ± 0.02 cm. However, the crucial electrical property of the wire is associated with the cross-sectional area of the wire. What is the pdf for the associated cross-sectional area? What is the mean and variance of the cross-sectional area?

3. If the number of minutes that a dentist takes to take care of a patient is exponential with $\theta = 45$ minutes, what is the probability that it will be longer than an hour and 20 minutes before you exit if the patient immediately before you is just being seated in the dentist's chair?

4. The number of persons buying tickets for an afternoon movie is assumed to be Poisson with $\lambda = 3$ customers per minute. What is the probability that in a 5 minute period they will sell more than 20 tickets? Less than 10?

5. Suppose a random variable X denotes an index of success developed by an educational psychologist and has the distribution

$$f(x) = 6x/(1 + x)^4 \quad 0 < x < \infty$$

An index of failure could be the reciprocal $Y = 1/X$. What is the probability distribution function of Y? What interesting relationship do you note between $f(x)$ and $f(y)$?

6. Using principles of random number generation presented in Chapter 4, generate 100 samples of 3 observations each from an exponential distribution with $\theta = 2$. Form the 100 sums: $Y = X_1 + X_2 + X_3$ and plot a histogram of these 100 sums. Compare it with the theoretical results that should apply in this case.

7. Using principles of random number generation presented in Chapter 4, generate 100 samples of 4 observations each from a normal distribution with $\mu = 5$ and $\sigma = 2$. Form the 100 means: $Y = \dfrac{[X_1 + X_2 + X_3 + X_4]}{4}$ and plot a histogram of these 100 means. Compare it with the theoretical results that should apply.

8. If $X_1, \ldots, X_n$ are independent Poisson random variables with parameters $\lambda_1, \lambda_2 \ldots \lambda_n$, what is the distribution of $Y = \bar{X}$?

9. If $X_1, \ldots, X_n$ are independent normal random variables with parameters $\mu_1, \mu_2, \ldots \mu_n$, and $\sigma_1^2, \sigma_2^2, \ldots \sigma_n^2$, what is the distribution of $Y = \bar{X}$?

10. If $X_1, \ldots, X_n$ are independent normal random variables with parameters $\mu_1, \mu_2, \ldots \mu_n$, and $\sigma_1^2, \sigma_2^2, \ldots \sigma_n^2$, what is the distribution of $Y = X_1 + X_2 + \ldots + X_n$?

APPENDIX:

APPROXIMATING THE MEAN AND VARIANCE OF FUNCTIONS OF RANDOM VARIABLES—THE "DELTA METHOD"

Introduction

In this chapter we have discussed three ways to determine the probability density function for some transformation of a random variable. Suppose, however, as we apply these methods in a particular case, the mathematics becomes too difficult or even impossible to solve in terms of the new probability density function $f_Y(y)$.

   If $E(Y)$ and $V(Y)$ would provide useful information, it is possible to avoid the mathematical problems and obtain an approximation that might serve our purposes. That is the purpose of this section.

   We only have to know the expected value $E(X)$ and the variance $V(X)$ for a random variable X. Our objective is to transform X using the transformation $Y = h(X)$ and then find $E(Y)$ and $V(Y)$. If we find that a straight forward evaluation of $E(Y = h(X))$ and $V(Y = h(X))$ might be non-integrable or, at the least, very difficult to integrate and if we can be satisfied with an approximation, then the following method, referred to as the Delta method, can be used. This method uses a Taylor's series expansion of the function $h(X)$ to provide the approximation.

Definitions

DEFINITION 5.2: An approximation to the mean and variance of a function of a random variable, denoted by $h(X)$, is obtained by applying the first few terms of a Taylors' series expansion to $h(X)$. The resulting approximations are given as:

$$E(h(X)) \approx h(\mu) + \frac{h''(\mu)\sigma^2}{2}$$

$$V(h(X)) \approx [h'(\mu)]^2\sigma^2.$$

The expression for $E(h(X))$ will tend to be a little more accurate than that for $V(h(X))$ since the former uses more of the Taylors' series terms in it than the latter.

DISCUSSION

Let $h(X)$ represent some function of X associated with a continuous pdf. We are interested in finding the expected value and variance for $h(X)$. (Typically, the resulting integral is to difficult to solve or is even non-integrable, making an approximation an acceptable alternative.) Let μ and σ^2, which values we know, denote the mean and variance, respectively, for the random variable X. We expand the function $h(X)$ about the mean μ using a Taylor's series expansion. This produces the following expression:

$$h(X) = h(\mu) + h'(\mu)(X - \mu) + \frac{h''(\mu)(X - \mu)^2}{2!} + R$$

where R is the remainder term, consisting of the rest of the terms in the series expansion, and the terms $h'(\mu)$ and $h''(\mu)$ denote the first and second derivatives of $h(X)$, each term evaluated at μ.

Now the expected value of $h(X)$ could be written:

$$E(h(X)) = h(\mu) + h'(\mu)E(X - \mu) + \frac{h''(\mu)E[(X - \mu)^2]}{2!} + E(R)$$

$$\approx h(\mu) + 0 + \frac{h''(\mu)\sigma^2}{2}.$$

We assume that $E(R)$ is approximately zero. Next we write $h(X)$ as $h(X) = h(\mu) + h'(\mu)(X - \mu) + R_1$ The variance can now be written as:

$$V(h(X)) = V(h(\mu)) + V(h'(\mu)(X - \mu)) + V(R_1)$$

$$\approx 0 + [h'(\mu)]^2\sigma^2$$

Therefore we have the following two expressions that will give approximate means and variances for the function $h(X)$, the goodness of the approximation depending on the size of the remainder terms in the Taylors expansion. They are

$$E(h(X)) \approx h(\mu) + \frac{h''(\mu)\sigma^2}{2},$$

$$V(h(X)) \approx [h'(\mu)]^2\sigma^2$$

EXAMPLES

EXAMPLE 1: Suppose a random variable X has a mean and variance of $\mu = 1.5$ and $\sigma^2 = 0.15$. What is the mean and variance of $Y = \frac{1}{X}$?

SOLUTION: Since we are interested in $Y = \frac{1}{X}$, let $h(x) = \frac{1}{x}$. By the results of this section we have for $h(\mu)$ and its 1st and 2nd derivative:

$$h(\mu) = \frac{1}{\mu} = \frac{1}{1.5} = 0.6667$$

$$h'(\mu) = -\frac{1}{(\mu^2)} = -\frac{1}{1.5^2} = -0.4444$$

and

$$h''(\mu) = \frac{2}{(\mu^3)} = \frac{2}{(1.5^3)} = 0.5926.$$

Therefore,

$$E(\frac{1}{X}) \approx h(\mu) + \frac{h''(\mu)\sigma^2}{2} = \frac{1}{1.5} + \frac{(\frac{2}{1.5^3})(0.15)}{2!} = 0.7111$$

For the variance we have

$$V(\frac{1}{X}) \approx [h'(\mu)]^2\sigma^2 = (\frac{-1}{1.5^2})^2(0.15) = 0.0296.$$

EXAMPLE 2: Suppose the pdf of the random variable X in Example 1 is:

$$f(x) = \begin{cases} (\frac{3}{8})x^2 & 0 < x < 2 \\ \\ 0 & \text{elsewhere} \end{cases}$$

Find the exact value of $E(\frac{1}{X})$ and $V(\frac{1}{X})$ to compare against the approximations.

SOLUTION: For this particular function it is easy to find the mean and variance for the random variable $\frac{1}{X}$ directly, though for other pdf's it would not necessarily be so. To do so, we get

$$E(\frac{1}{X}) = \int_0^2 \frac{1}{x}(\frac{3x^2}{8}))dx = 0.75$$

compared to the approximation of 0.71111. The variance is:

$$V(\frac{1}{X}) = \int_0^2 \frac{1}{x^2}(\frac{3x^2}{8})dx - (0.75)^2 = 0.19$$

compared to 0.0296.

It can be seen that for this example that the approximation for the mean of $h(x)$ is not too bad but the approximation for the variance is much less satisfactory. In general it will be found that if the variance of the distribution is large and the mean is relatively small this approximation may be less than satisfactory. In addition, the variance uses fewer terms of the Taylor's expansion to obtain the approximation, and will be a poorer approximation.

<hr>

EXERCISES

1. The surface tension of a liquid is represented by T (dyne/cm) and under certain conditions, $T = 2(1 - 0.005X)^{1.2}$, where X is the liquid temperature in degrees centigrade. Suppose X has the pdf

$$f(x) = 3000x^{-4} \quad x > 10$$

   and is 0 elsewhere. Calculate approximations to E(T) and V(T).

2. For the random variable T described in exercise 1, find the exact mean and variance and compare to the approximations obtained in that exercise. You can solve this analytically or by using numerical integration on a calculator or by other means available.

3. The acidity of a certain product, measured on an arbitrary scale, is given by the relationship

$$A = (3 + 0.05G)^2$$

   where G is the amount of one of the constituents, having the probability distribution

$$f(g) = \frac{1}{32}(5g - 2) \quad 0 < g < 4$$

   and is 0 elsewhere. Find approximations to E(A) and V(A).

4. For the random variable A described in exercise 3, find the exact mean and variance and compare to the approximations obtained in that exercise. You can solve this analytically or by using numerical integration on a calculator or by other means available.

5. The concentration of reactant in a chemical process is a random variable having the pdf

$$f(r) = 6r(1 - r) \quad 0 < r < 1$$

   and is 0 elsewhere. The profit associated with the final product is $P = \$1.00 + \$3.00R$. Find approximations to E(P) and Var(P).

6. For the random variable R described in exercise 5, find the exact mean and variance and compare to the approximations obtained in that exercise. You can solve this analytically or by using numerical integration on a calculator or by other means available.

7. The demand for anti-freeze in a season is considered to be a random variable X, with density

$$f(x) = 10^{-6} \quad 10^6 < X < 2 \times 10^6$$

   and is 0 otherwise, where X is measured in liters. The evaporation rate is given by $R = e^{-x}$. Find approximations to E(R) and V(R).

8. For the random variable R described in exercise 7, find the exact mean and variance and compare to the approximations obtained in that exercise. You can solve this analytically or by using numerical integration on a calculator or by other means available.

9. Body weight is assumed to be related to the cube of height by the relationship $Y = kX3$, where Y is the body weight and X is the height and k is an appropriate scaling constant. If $\mu_X = 70"$ and $\sigma_X = 3"$, what is the "delta method" approximation to the mean and variance of body weight?

10. An estimate of wildlife abundance using a principle of capture-tag-recapture is of the form $Y = \dfrac{k}{X}$ where Y is the number of wildlife to be estimated and X is the number of tagged wildlife in a sample of n wildlife recaptured. Write an expression for the mean μ_y and variance σ_y^2 of Y in terms of the mean and variance of X, μ_x and σ_x^2.

6

SAMPLING DISTRIBUTIONS

OBJECTIVES

When you complete this chapter, you should understand the concept of the sampling distribution of a statistic and relate it to the process of "repeated sampling". You should be able to create an empirical sampling distribution of, say, the sample mean, median, variance, etc. and be able to find the expected value and variance of this distribution of sample statistics.

If the parent population is a normal distribution, then you should know what the sampling distribution is for a z-score, squared z-score, sum of squared z-scores, the sample variance (appropriately transformed), the ratio $\frac{(\bar{x} - \mu)}{s/\sqrt{n}}$, and the ratio of two variances.

You should be able to explain the connection between the concepts of a sampling distribution from the normal distribution and the generalization that the Central Limit Theorem provides for the distribution of a linear function of random variables and its sampling distribution.

INTRODUCTION

One of the most common statistical procedures with which most people are familiar is in the arena of pre-election polls and public opinion polls. Because of its familiarity, the sample survey, a general term, for which the election poll is a specific case, can be used as a good introduction to some of the ideas of this chapter. In it, we have a "population" of voters whose opinions and voting behavior is sought. We draw a "sample" from this "population" of voters. We then compute a sample statistic using the data collected. For instance, we may want to know the proportion of voters who favor the incumbent president, "if the election were held today."

However, it is well recognized that the results of the election poll will be in error due to the particular sample chance has produced. Therefore, along with the estimate of the proportion favoring the candidate that the sample produces, a statement about the maximum error to be expected in the estimate is necessary. Included with a statement about the error is a probability statement that reflects our confidence that the the sample selected is well behaved and not unusual or misleading in any particular way.

For instance, after collecting and analyzing the data, the data might show that the incumbent has 42% of the vote with a margin of error of 3%, with 95% confidence that the margin of error will not exceed 3%.

In this illustration we assume the response of each voter interviewed follows a Bernoulli distribution with unknown parameter p, where p stands for the probability of a vote in favor of our candidate. We select a random sample of n observations from this population (each observation representing one Bernoulli trial). This produces n random variables, $X_1, X_2, ..., X_n$. Using these n random variables, we might define the function $Y = h(X_1, X_2, ..., X_n) = \dfrac{(X_1 + ... + X_n)}{n}$ whose probability distribution we need to know in order to predict the probability that Y will fall within 3% of the true parameter value. Thus we need to know how to determine the probability distribution of a random variable Y where that random variable Y is a function of sample values. (That is we need to know the sampling distribution, $f_Y(y)$, of the sample proportion Y.)

The answer to this and similar questions to it are to be answered in this chapter.

Where We Are Going in This Chapter

The statistical-inferential model (SIM) in Figure 6.1 below indicates the perspective that will be taken in this chapter. We either assume that a sample is drawn from a population that is normally distributed, or that a large sample size is selected from some other distribution. (Typically, a sample is considered large if more than 30 observations are taken, though this is only a rule-of-thumb.) With the sample in hand, we might compute the sample mean $\bar{x}$ as a reflection of what the mean of the population, μ, might be. Being realistic, however, we know the sample mean is not likely to equal the true mean, and thus we can only say how probable it is that the sample mean is close to the true mean. To be able to compute such a probability, we have to know the probability laws that predict the behavior of means obtained from sampling from a normal distribution.

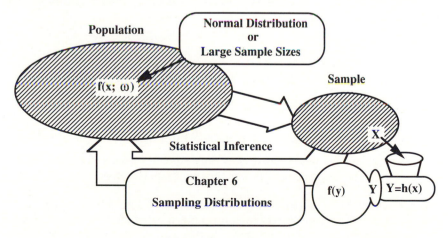

Figure 6.1 SIM—Sampling distribution for a mean of random variables.

One way of discovering these probability characteristics is to examine the distribution of the sample mean by repeatedly sampling from the population, using the same sample size n each time from which the mean $\bar{x}$ is computed, until it becomes clear what the distribution of the sample means is. However, it is easier to determine the probability distribution $f_{\bar{x}}(\bar{x})$ by using the methods presented in Chapter 5. (We will demonstrate the idea of "all repeated samples" in the following sections to confirm the theoretical conclusions, but we would resort to it only if the mathematics of the Chapter 5 techniques become too difficult to manage.)

In this chapter, we state the distribution, $f_Y(y)$, for a function, $Y = h(x_1, x_2, ..., x_n)$ computed from a set of random variables where the random variables are produced by sampling a *normal* distribution as the "parent population". However, we will not show the derivation of $f_Y(y)$. Rather we will simply identify the problem and with a theorem, state a conclusion. (In the homework exercises of Chapter 5, you have already proved some of the findings that we will restate here.) In those cases where a normally distributed parent population (pdf) is not assumed, we must rely on "large sample theory" and what is known as the Central Limit Theorem, to provide us with the direction to go.

Chapter Exercise

The assembly time of a certain part consists of two steps each of which is exponentially distributed with $\theta = 5$ minutes. What is the distribution of total assembly time? What is the probability that the total assembly time will exceed 15 minutes? If the assembly time of a certain part consists of 20 steps, each of which is exponentially distributed with $\theta = 5$, what is the approximate probability that the total assembly time will exceed 120 minutes? What is the probability that the total assembly time would be less than 90 minutes?

6.1   THE CONCEPT OF A SAMPLING DISTRIBUTION

Introduction

This section presents the concept of a *sampling distribution*. Crucial to this understanding is what is meant by a random sample. This is formally defined in this section. Also, we will refer to many of the "sample statistics" that were defined in Chapter 1, since their probability behavior is of primary interest in the context of "sampling" a particular population.

You can put these ideas into better perspective if you think about how to assess the probability behavior of sampling a given population. For instance, if a random sample is selected from a population and the sample mean $\bar{X}$ is computed, how do we predict what value or range of values it is likely to take on. Will it be larger or smaller than the true mean μ? How close is it likely to be to μ?

One way to resolve this problem is to conduct an experiment. To do this: (1) "create" a known population, (2) choose a sample of a given size from this known population, and (3) compute the sample statistic whose characteristics you want to study. Replace all the items you have drawn from the population (your sample) in order to restore the population to its original state. Now repeat steps 2 and 3 a large number of times, keeping track of the various values your sample statistic takes on. Examining the distribution of these sample values by means of a histogram, numerical measures such as means and variances, etc., will allow you to understand how the sampling properties tend to behave relative to the characteristics known about the population you constructed. We will illustrate this process in this chapter.

However, we would have to wonder if the properties observed in this experimental approach were unique to the population you created, or whether they are more generally applicable. Thus, a more mathematical treatment of this problem would be helpful.

The distribution function technique, the change of variable technique and the moment-generating function technique presented in the last chapter provide the mathematical link to help understand the principles of a "sampling distribution."

Definitions

DEFINITION 6.1: A *random sample* is a set of independent and identically distributed random variables associated with a common distribution. We denote these random variables by $X_1, X_2, ..., X_n$.

DEFINITION 6.2: If $X_1, X_2, ..., X_n$ constitute a sample of observations and $Y = h(X_1, X_2, ..., X_n)$, where $h(X_1, X_2, ..., X_n)$ defines a function on $X_1, X_2, ..., X_n$, then Y is a *sample statistic*. A sample statistic is itself a random variable since it is a function of random variables.

DISCUSSION

Suppose we have a population modelled by a probability density function $f(x)$. If we select a set of n measurements $X_1, X_2, ..., X_n$, from this distribution such that they are all mutually independent of one another, then we will claim that we have a random sample from the pdf, $f(x)$. The measurements $X_1, X_2, ..., X_n$, having come from the same pdf, are said to be identically distributed. Therefore, the marginal distributions for each random variable would be denoted by $f(x_1), f(x_2),...,f(x_n)$ and because of their independence, their joint distribution is the product of their marginal distributions; i.e.,

$$f(x_1, x_2, ..., x_n) = f(x_1)f(x_2)\bullet \, ... \, \bullet f(x_n)$$

If we define some function Y on these random variables, say, $Y = h(X_1, X_2, ..., X_n)$, the resulting value is called a sample statistic. Lest this sound too obscure, note that we have done this many times without specifically describing things that way. For instance, sample statistics with which you are already familiar are:

DESCRIPTION	$h(X_1, X_2, ..., X_n)$
Sample mean:	$\overline{X} = \dfrac{(X_1 + X_2 +... +X_n)}{n}$
Sample variance-st.dev:	$s^2, \; s_x^2$
Sample median:	$X_{0.5}$
Maximum of sample:	$Max(X_1, ..., X_n)$

DESCRIPTION	$h(X_1, X_2, ..., X_n)$
Minimum of sample:	$\text{Min}(X_1, ..., X_n)$
Sample quantiles:	X_q
Sample z-score:	$\dfrac{(X - \mu)}{\sigma}$
Squared z-scores:	$\left[\dfrac{(X - \mu)}{\sigma}\right]^2$
Sample t-score:	$\dfrac{(\bar{X} - \mu)}{(s/\sqrt{n})}$

(The last three items in the list—the two z-scores and the t-score–are useful sample statistics only if the parameter values in the expression are known)

Since all of the above sample statistics are functions of random variables, they will have a probability distribution that describes their probabilities over their possible range of values. We can denote the respective probability distributions by a subscript such as to clarify things when necessary. For instance we might use the following notation:

Sampling Distribution	Notation
For the sample mean:	$f_{\bar{X}}(x)$
For the sample median:	$f_{X_{0.5}}(x)$

It is the purpose of this chapter to identify the probability distributions of the most common sample statistics under reasonable assumptions. These probability distributions are called the *sampling distributions* of the sample statistics. We can derive these sampling distributions theoretically by the techniques of chapter 5, and/or demonstrate them empirically by the principle of repeated sampling.

In the next sections we will first demonstrate the sampling distribution of several common sample statistics by computer simulations. This is a *repeated sampling* approach. We will then state, by theorem, the theoretical distribution of various sample statistics assuming the sample came from a normally distributed population (pdf).

6.2 THE SAMPLING DISTRIBUTION OF THE MEAN AND THE CENTRAL LIMIT THEOREM

Introduction

In this section we consider the sampling distribution of the sample mean under two conditions: (1) that the population being sampled is normally distributed with some mean and variance; and (2) that the population being sampled is any population with a finite mean and variance. In the first case, we will see that the sampling distribution of the mean is also normally distributed.

In the second case we find that the distribution of the sample mean is approximately normal for large samples. This latter case is summarized in the Central Limit Theorem and represents one of the most important theorems in all probability theory.

Theorems

THEOREM 6.1: Let the random variables $X_1, ..., X_n$ represent a random sample of size n from a normal distribution with mean, μ, and variance, σ^2. The sampling distribution of the mean $f_{\overline{X}}(x)$ of these random variables is normally distributed with mean and variance of

$$\mu_{\overline{x}} = E(\overline{X}) = \mu$$
$$\sigma_{\overline{x}}^2 = V(\overline{X}) = \frac{\sigma^2}{n}$$

METHOD OF PROOF: The moment-generating function technique is used to prove this theorem which was proved in Chapter 5, section 5.3, exercise 10.

THEOREM 6.2 (THE CENTRAL LIMIT THEOREM): If $X_1, ..., X_n$ are independent and identically distributed random variables with common mean μ and variance σ^2, then the distribution of

$$Z = \frac{(\overline{X} - \mu)}{\sigma/\sqrt{n}} = \frac{(\sum_1^n X_i - n\mu)}{\sqrt{n\sigma^2}}$$

is the standard normal distribution in the limit as $n \rightarrow \infty$. (Note that $Y = X_1 + ... + X_n = \sum_1^n X_i$ has a mean $n\mu$ and variance $n\sigma^2$.)

Method of Proof: The moment-generating function of Z is written and expanded in a power series which can be shown, in the limit, to approach the moment-generating function of the standard normal distribution.   The proof is given at the end of this section for the mathematically curious.

DISCUSSION

Suppose that a population of numbers is normally distributed with a mean of 50 and a standard deviation of 5. A random sample of 4 observations is to be taken from this population and a mean of 54 is obtained. Is such an outcome or one larger than that likely to occur under the assumed conditions?   To answer that question we want to calculate the probability:

$$P(\overline{X} \geq 54 \mid \mu_x = 50 \text{ and } \sigma_{\overline{x}}^2 = 25)$$

One way to estimate this probability would be to randomly sample 4 observations from a normal distribution, repeatedly, say for a 1000 times and then count how many times out of the 1000 a mean greater than 54 was obtained. However, this would be a very tiresome project unless you had a computer available to do the job.

The other alternative is to use the moment-generating function technique to determine the sampling distribution of X so that we could compute the probability exactly. In the following, we illustrate both methods.

Empirical Demonstration

Setting:

A random sample of four ($n = 4$) observations is taken from a normally distributed population having a mean of 50 and a standard deviation of 5.

Sampling procedure for empirical demonstration:
(1)   A random sample of size four will be selected from the above defined "parent population";
(2)   The mean will be computed for this sample;
(3)   Steps 1 and 2 will be repeated sufficient times (1000) to give a good representation of the distribution of the sample means.

Figure 6.2 that follows shows what the theoretical normal distribution, $N(50,25)$, looks like along along with a histogram of 1000 randomly selected *individual* numbers from the same normal distribution. Note that the sample mean and the sample standard deviation for these 1000 numbers are very close to the theoretical mean and standard deviation of 50 and 5, respectively. (We refer to this distribution as the "parent population" which is be sampled *repeatedly* to produce sample means as "offspring.")

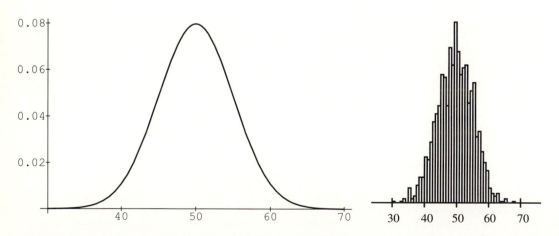

Figure 6.2 Theoretical normal distribution [$N(50,5)$] and histogram of a random sample of 1000 taken from it: Parent population. Histogram: Mean = 49.808, St.Dev. = 4.7851, no. of means = 1000.

After applying the three steps of the sampling procedure described above, Table 6.1 is produced which gives a partial list of the 1000 groups of 4 observations each and the means that were produced. Next, Figure 6.3 shows the histogram of these 1000 means. This histogram of means—the offspring population—is the empirical "sampling distribution of the sample mean." Theoretically, the sampling distribution of means would be normally distributed with a mean of 50 and a standard deviation of $\dfrac{\sigma}{\sqrt{n}} = \dfrac{5}{\sqrt{4}} = 2.5$.

Table 6.1 1000 Random samples with $n = 4$ from parent population with their means.

ROW	X_1	X_2	X_3	X_4	MEAN
1	49.0633	62.1284	55.1217	57.3368	**55.9126**
2	54.1985	57.2848	42.9818	48.6289	**50.7735**
3	51.1674	57.3354	49.8496	46.6351	**51.2469**
4	55.9193	56.4848	53.1603	44.2274	**52.4480**
...	...	...	...	...	**...**
1000	47.9271	50.1947	50.0475	45.7800	**48.4873**

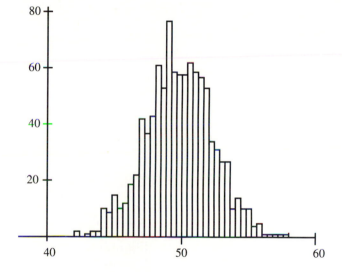

1000 means
Mean of means = 50.190
Standard Deviation of means = 2.4797

Figure 6.3 Distribution of 1000 means, each with $n = 4$ from a $N(50,5)$.

Note from Figure 6.3 that the "offspring" population is also bell-shaped but with a smaller standard deviation than the parent population and verifies Theorem 2.1 quite clearly. Specifically, from the theorem, since the parent population is normally distributed with mean and standard deviation of 50 and 5 respectively, the distribution of means should be normally distributed with a mean and standard deviation of 50 and $\dfrac{\sigma}{\sqrt{n}} = \dfrac{5}{\sqrt{4}} = 2.5$. Our empirical distribution of means shows a mean of 50.190 and a standard deviation of 2.4797—fairly good agreement.

To answer the question we asked previously, "What is the probability that a mean from the parent population would exceed 54?" we take the two approaches suggested. The first approach is to look at the distribution of means and find the proportion that are 54 or greater. This was done by examining the MEAN column of Table 6.1. The number of means at least as large as 54 was found to be 63 out of the 1000 means. Therefore, $\dfrac{63}{1000} = 0.063$ is an approximate probability.

Using Theorem 2.1 we can calculate the probability exactly as:

$$P(\overline{X} \geq 54 \mid \mu_x = 50 \text{ and } \sigma_{\overline{x}}^2 = 25) = 1 - P(\overline{X} \leq 54 \mid \mu_x = 50 \text{ and } \sigma_{\overline{x}}^2 = 25)$$

$$= 1 - F_Z\left(\frac{54 - 50}{5/\sqrt{4}} = 1.60\right)$$

$$= 1 - 0.9452 = 0.0548.$$

This probability and the empirical probability are reasonably close to be convincing.

[Note: A question that often arises in student's minds is what would happen if instead of taking 1000 samples of size $n = 4$ and computing their means, 5000 or 10,000 samples of size 4 were selected? What would be different? Figure (b) on the right of Figure 6.4 below shows what happened when 10,000 samples of size $n = 4$ were selected, and a histogram of the 10,000 means was produced. Figure (a) on the left is from Figure 6.3 and is a histogram of the means of 1000 samples of size $n = 4$.

Note that theoretically the "sampling distribution of the means" is still $N(50, 5/\sqrt{4})$. As you compare the histograms of (a) with (b), you observe histograms that are essentially the same in shape and location, but that (b) is smoother and less ragged on the edges than (a).]

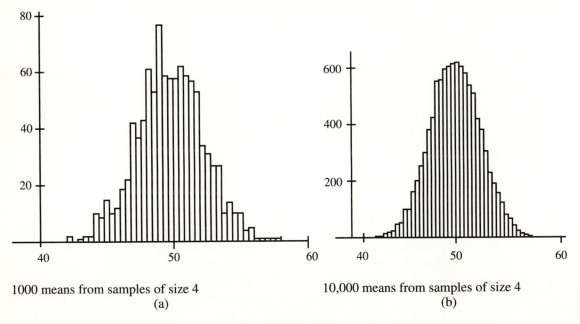

1000 means from samples of size 4   10,000 means from samples of size 4
              (a)   (b)

Figure 6.4 Two histograms of means: (a) 1000 replications versus (b) 10,000 replications.

The Central Limit Theorem

Suppose we now ask the question: If a random sample of 4 observations is drawn from *any* distribution having a mean of 50, what is the probability that the sample mean will exceed 54? To answer this question, the Central Limit Theorem suggests that we can use the normal distribution to approximate this probability regardless of the distribution from which we are sampling, and the approximation improves as the sample size increases.

For example, if a sample is drawn from a uniform distribution, gamma distribution, Poisson distribution, or any of a number of other useful distributions used in modeling real world populations, the probability properties of the sample mean, appropriately standardized, can be approximated by probabilities from the standard normal distribution. The approximation improves as the sample size gets larger. (Usually, $n = 30$ is sufficiently large for the most common distributions we have examined; and for symmetric distributions, adequate approximations can be obtained for samples as small as 6 to 10.)

To demonstrate this property suppose we select a sample of 4 observations from an exponential distribution with $\theta = 50$; that is, from a distribution with $\mu_x = 50$ and in this case, $\sigma_x = 50$.

We want to determine the probability that $\bar{X} \geq 54$. To obtain this probability we take three different approaches:

(1)   We can calculate the probability exactly knowing that the distribution of the mean of four exponential random variables is distributed as a gamma distribution with $\alpha = 4$ and $\theta = \dfrac{50}{4}$. (See Chapter 5, section 5.3, Example 2 and/or exercise 8.)

(2)   We can create an empirical distribution by repeated sampling and get an estimate of the probability.

(3)   We can use a normal approximation to the probability by applying the Central Limit Theorem.

We look at each case in turn to compare them.

Case (1): The gamma distribution calculation

If $Y = \bar{X}$ where $X_1, ..., X_4$ are exponential with $\theta = 50$, then the exact distribution of Y, using a moment-generating function proof, is a gamma distribution with $\alpha = 4$ and $\theta = \dfrac{50}{4}$. Therefore, we make the following computation:

$$P(\bar{X} \geq 54) = 1 - F_G(54; \alpha = 4, \theta = \frac{50}{4})$$

$$= 1 - [1 - F_P(3; \lambda = 54/(\frac{50}{4}))]$$

$$= F_P(3; \lambda = 4.3) = 0.3772.$$

This is the exact probability, within rounding error for the value of λ.

Case (2): The empirical distribution

Setting: Suppose we have an exponential distribution with $\theta = 50$, a fairly skewed and definitely non-normally distributed population. The sampling distribution of 1000 mean of samples of size 4 will be created. Though this is not a "large n" case ($n = 4$), the tendency towards normality will be obvious upon examination of the histogram. The probability that $\overline{X}$ exceeds 54 will be estimated by counting the number of means out of the 1000 computed that are at least as large as 54.

Method of Sampling: One thousand samples of size 4 each are first generated and the mean is computed for each of the 1000 samples. The sampling distribution is then created. Each observation in each sample of 4 is exponentially distributed with $\theta = 50$.

The population from which the samples are drawn is shown in Figure 6.5 below. A histogram of the 1000 means is then shown in Figure 6.6 The estimate of the probability that $\overline{X} \geq 54$ is obtained by counting the number of means greater than 54. Of the 1000 means, 367 were equal to or greater than 54. Thus, we estimate the probability empirically to be $\dfrac{367}{1000} = 0.367$.

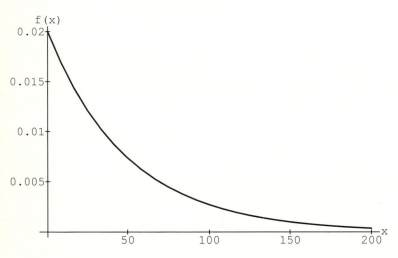

Figure 6.5 Distribution for exponential probability function with $\theta = 50$.

Case 3: Normal approximation by using the central limit theorem.

Since the population sampled has $\mu_x = 50$ and $\sigma_x = 50$, then

$$P(\overline{X} \geq 54) = 1 - F(54) \doteq 1 - F_Z\left(\frac{54 - 50}{50/\sqrt{4}}\right)$$

$$= 1 - F_Z(0.16) = 1 - 0.5636 = 0.4364.$$

This probability is an approximate probability to be compared with the exact value of case 1 of 0.3772, and the estimate by "repeated sampling" of case 2 which is 0.3670. The normal approximation leaves something to be desired, but remember that the sample size is only 4 in this case and thus the probability would be expected to be only approximate. Figure 6.7 shows a comparison of the gamma distribution with $\alpha = 4$ and $\theta = 12.5$ which produces the exact probability under our assumptions and the $N(50,25)$ curve that is used in the central limit theorem approximation.

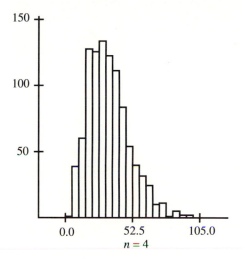

Figure 6.6 Distribution of 1000 sample means, each mean is computed from $n = 4$ observations from an exponential probability density function with $\theta = 50$.

You can see why the normal curve overestimates the probability that we seek. You can see from the figure that the normal curve would underestimate probabilities in the range from 10 to 50.

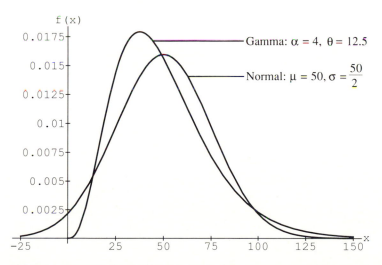

Figure 6.7 Comparison of the normal curve used to approximate the gamma curve.

A Further Demonstration of the Central Limit Theorem

In Figure 6.8 and 2.8 that follows a visual demonstration is provided to illustrate the Central Limit Theorem. In frame (a) of Figure 6.8 a histogram is given of a sample of 1000 random values chosen from an exponential distribution with $\theta = 50$. In frame (b) a histogram of 1000 means is shown; each mean is computed from a $n = 2$ random values chosen from the same exponential distribution. Frames (c) and (d) show histograms of 1000 means each, with each mean being computed from samples of $n = 4$ and $n = 6$ observations; each observation is taken from the same exponential

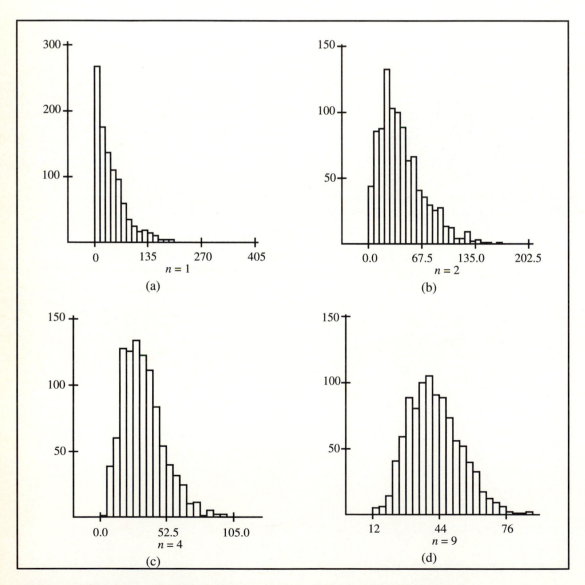

Figure 6.8 Histograms of 1000 means for varying sizes of samples: $n = 1, 2, 4, \& 9$.

distribution. In Figure 6.9 the demonstration continues for 1000 means based on individual samples of 16 and 25 each.

Notice that as you progress from frame to frame the distribution becomes less skewed, more symmetric and bell-shaped. For $n = 25$ you would be hard pressed to distinguish this histogram from a histogram of 1000 observations from a true normal distribution. Such a normal distribution is given in frame (g) of Figure 6.9 and there is little difference to be noted between frame (f) and (g) other than that which would be attributable to "sampling variability"—the irregularities due to the different samples that chance would produce.

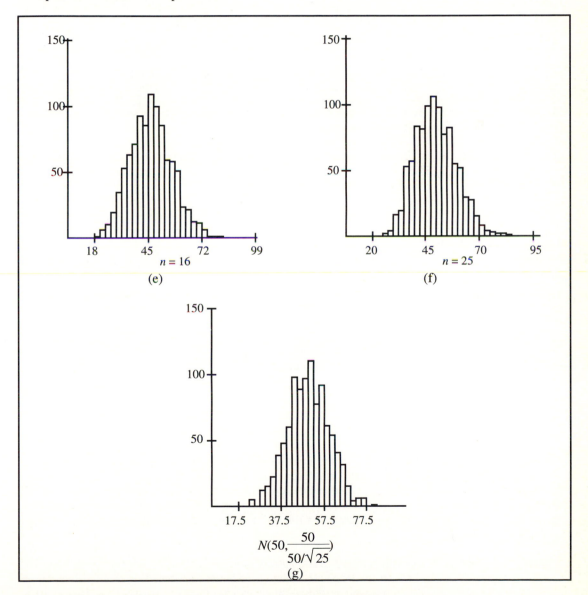

Figure 6.9 Histograms of 1000 means for varying sizes of samples.

EXAMPLES

EXAMPLE 1:   It has been reasonable to approximate the length of pregnancies by a normal distribution with a mean of 280 days and a standard deviation of 12 days. A new "morning sickness" pill is being tested on a group of 16 volunteers. Their average date to delivery was 275 days. How likely is this, assuming the pills have no tendency to cause pre-mature births?

SOLUTION: In this case the true mean and standard deviation are assumed to be 280 and 12 days respectively. The mean of a sample of sixteen observations from this population should behave as if it came from a normal distribution with a mean of 280 and a standard deviation of $12/\sqrt{16} = 3$ days. Therefore the probability of getting a mean as small as 275 or smaller is:

$$P(\overline{X} \leq 275|\, \mu = 280, \sigma = 3) = F_Z[\frac{(275-280)}{3}] = F_Z(-1.67) = 0.0475$$

This probability is small enough to raise serious questions regarding the assumption of no link between the medication and premature delivery.

EXAMPLE 2:   The city administrator in a small city conducted a survey of home and apartment dwellers relative to their satisfaction with city services. From a recent government census it was known that the ratio of apartment-dwellers to home-owners was 1:3. The survey was included with the monthly utility bill and only 1500 questionnaires were returned out of 5000 sent out. The number of apartment-dwellers in the 1500 was 330. Aside from other obvious problems with this study, does the ratio seem inconsistent with what chance might produce? That is, does it seem that apartment dwellers are under-represented in the returns?

SOLUTION: In this problem, each returned questionnaire can be thought of as a Bernoulli trial producing either an apartment-dweller with $p = \frac{1}{(1 + 3)} = 0.25$, or a home-owner with probability $(1 - p) = 0.75$. Recall that for the Bernoulli distribution that $\mu = p = 0.25$ and $\sigma^2 = p(1 - p) = (0.25)(0.75)$ and X is 0 or 1. Therefore, the mean of 1500 returns is $\frac{330}{1500} = 0.22$ or 22%. The probability of getting this many apartment dwellers or fewer by chance is

$$P(\overline{X} \leq 0.22 \,|\, \mu = 0.25, \sigma = \sqrt{\frac{(.25)(.75)}{1500}}) = F[\frac{(.22 - .25)}{.011}]$$

$$\cong F_Z(-2.68) = 0.0037$$

This probability is obtained by using the central limit theorem and is only an approximation. But with 1500 observations it should be a reasonable approximation. It suggests that there is reason to rule out chance as the explanation of the under-representation of apartment dwellers in the 1500 returned questionnaires. Therefore, one might look for other logical explanations for the under-representation such as lack of interest in the subject matter of the questionnaire, the "temporary" nature of their living arrangements, etc.

EXAMPLE 3:   Engineers, studying a rechargeable battery have found that the average time between charges is 8 hours. They have been challenged to improve the battery. After some changes have been implemented, 16 of the new batteries were tested and the average time to recharge of the 16 was 10.5 hours. Can we conclude that an improvement has been made? Or could we attribute this to chance?

SOLUTION: Suppose we assume the time to recharge is exponentially distributed with $\theta = 8$. Then the mean of 16 exponentially distributed random variables would be distributed as a gamma distribution with $\alpha = 16$ and $\theta = \dfrac{8}{16} = \dfrac{1}{2}$.

Therefore,

$$P(\bar{X} \geq 10.5) = 1 - F_G(10.5; \alpha = 16, \theta = \tfrac{1}{2})$$

$$= F_P\left(15; \lambda = \frac{10.5}{1/2} = 21\right)$$

Our table doesn't include this case, so let's approximate it using the Central Limit Theorem. We get,

$$P(\bar{X} \geq 10.5) = 1 - F(10.5)$$

$$= 1 - F_Z(\frac{10.5 - 8}{2} = 2) = 1 - 0.8944 = 0.1056$$

This probability is large enough that we could attribute the mean of 10.5 hours to chance rather than to an improvement of product design. (The actual gamma distribution probability is 0.111. The normal approximation is very close and to two places is right on. The comparison between the exact gamma distribution curve and the approximating normal curve are shown in Figure 6.10 that follows.)

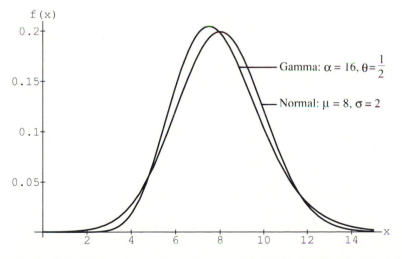

Figure 6.10 Comparison of actual gamma distribution with approximating normal distribution.

EXERCISES

1. A random variable X is gamma with parameters $\alpha = 3$ and $\theta = 4$.
    (a) What proportion of the distribution is within one standard deviation of the mean; i.e., $P(\mu - \sigma < X < \mu + \sigma)$? (What is the mean and variance for this random variable?)
    (b) A random sample of size 6 is selected from the gamma distribution above. What is the probability (approximately) that the mean of the sample will fall within the same interval ias part
    (a)? That is, what is $P(\mu - \sigma < \overline{X} < \mu + \sigma)$? Approximate this probability with the standard normal distribution.
    (c) Compare your answers to (a) and (b). What does this tell us about the tendency of the sample mean to center itself around the mean of the population?

2. Using the information presented in exercise 1 above, what is the probability that the sum will fall below 200? Use the standard normal distribution to approximate this probability.

3. A sample of 15 observations is taken from a chi-square distribution with a parameter value of 15. What is an approximate probability that the mean of the sample will be less than 9 or more than 21?

4. Suppose 50 Bernoulli trials are conducted with each Bernoulli trial having probability $p = 0.4$ for a Success. If $Y = X_1 + \ldots + X_{50}$, what is $P(Y \geq 25)$? Note that Y will be binomial with $n = 50$ and $p = 0.4$, $\mu = 50(0.4)$ and $\sigma^2 = .50(0.4)(0.6)$. Use a normal approximation to this probability assuming the normal distribution has a mean a variance equal to that of the binomial random variable Y.

5. Suppose a random sample of 400 registered voters is selected and the proportion favoring the incumbent president for re-election is calculated. If 55% of the electorate favor the re-election of the incumbent, what is the chance that the sample proportion will be less than 50%? (This is the same as asking for the probability that Y is less 200 when Y is binomial with $n = 400$ and $p = 0.55$. See exercise 4.)

6. The time between a call for a particular type of computation on a large, main-frame computer system is averages 3 minutes and is assumed to follow an exponential distribution. Suppose the average time for a random sample of 9 such calls on a Monday following a three-day weekend was 5 minutes. Is this an unusual occurrence based on your information? Approximate the probability using the standard normal distribution, applying the central limit theorem.

7. Using the information given in exercise 6, compute the exact probability under the assumptions given, knowing that the mean of 9 exponential random variables is a gamma random variable

8. A genealogist has been studying the male–female mix of families of size 5 in the census records of 1910. A sample of 20 such families produced a mean of 2 males per family. Is this an unusual result considering that the binomial distribution with $n = 5$ and $p = 0.5$ models the number of boys per family? Use a central limit theorem approximation for this probability.

9. In 50 blood samples from patients having a blood disorder the average number of particles per unit of blood was found to be 13. If the usual number is assumed to be modelled by a Poisson distribution with $\lambda = 10$, would you consider the sample results to be unusual?

10. Create a probability density function of your own choosing, with whatever properies you desire relative to symmetry or skewness.   Try to beat the central limit theorem by showing that for samples of size thirty, the proportion of means found within 1 standard deviation of the mean is not 68%, or the proportion of means within 2 standard deviations of the mean is not 95%. "Not 68%" means less than 65% or more than 71%. "Not 95%" means less than 93% or greater than 97%. You must have 200 means, each mean based on 30 observations each. (You must generate 30•200 random values.)

Proof of the Central Limit Theorem

We intend to show that the moment-generating function of the random variable

$$Z = \frac{(\bar{X} - \mu)}{\sigma/\sqrt{n}} = \frac{(\sum_1^n X_i - n\mu)}{\sigma\sqrt{n}}$$

tends to the moment-generating function of the standard normal distribution as n tends to infinity. First, note that Z can be written as

$$Z = \frac{(\sum_1^n X_i - n\mu)}{\sigma\sqrt{n}} = \sum_1^n \frac{1}{\sqrt{n}} \frac{(X_i - \mu)}{\sigma} = \sum_1^n \frac{1}{\sqrt{n}} Z_i$$

where $Z_i = \frac{(X_i - \mu)}{\sigma}$. Since by assumption, the random variables X_i are independent, then Z is a sum of independent, identically distributed random variables. From the results of section 5.3 it follows that the moment-generating function of Z is the product of the moment-generating function of $\frac{1}{\sqrt{n}} Z_i$. That is,

$$M_Z(t) = \prod_{i=1}^n M_{Z_i}(\frac{t}{\sqrt{n}}) = \left[M_{Z_i}(\frac{t}{\sqrt{n}}) \right]^n = \left\{ E\left[\exp(\frac{t Z_i}{\sqrt{n}}) \right] \right\}^n$$

where we use the notation: $\exp(x) = e^x$.

Expanding $\exp(\frac{t Z_i}{\sqrt{n}})$ in a Taylor series expansion we get (see your calculus text)

$$\exp(\frac{t Z_i}{\sqrt{n}}) = 1 + \frac{1}{\sqrt{n}} Z_i + \frac{t^2}{2n} Z_i^2 + \frac{t^3}{3!n^{3/2}} Z_i^3 + \dots$$

Next we take the expected value, recalling that $E(Z_i) = 0$, and $Var(Z_i) = E(Z_i^2) = 1$ for $i = 1, 2, \dots,$

$$E[\exp(\frac{t\,Z_i}{\sqrt{n}})] = E\left[\, 1 + \frac{1}{\sqrt{n}}\,Z_i + \frac{t^2}{2n}\,Z_i^2 + \frac{t^3}{3!n^{3/2}}\,Z_i^3 + \ldots \,\right]$$

$$= 1 + 0 + \frac{t^2}{2n} + \frac{t^3}{3!n^{3/2}}\,E[Z_i^3] + \ldots$$

Therefore,

$$M_Z(t) = \left[\, 1 + \frac{t^2}{2n} + \frac{t^3}{3!n^{3/2}}\,E[Z_i^3] + \ldots \,\right]^n = \left\{ 1 + \frac{1}{n}\left[\frac{t^2}{2} + \frac{t^3}{3!\sqrt{n}}\,E[Z_i^3] + \ldots \right]\right\}^n$$

$$= \left(1 + \frac{v}{n} \right)^n$$

where

$$v = \frac{t^2}{2} + \frac{t^3}{3!\sqrt{n}}\,E[Z_i^3] + \ldots$$

Now, in the limit as n goes to infinity we have

$$\lim_{n\to\infty} M_Z(t) = \lim_{n\to\infty} \left(1 + \frac{v}{n} \right)^n$$

However, by definition

$$\lim_{n\to\infty} \left(1 + \frac{v}{n} \right)^n = e^v$$

In addition, all of the term in v go to zero in the limit except the first term because they contain positive powers of n in their denominators. Therefore,

$$\lim_{n\to\infty} M_Z(t) = e^{t^2/2}$$

which is the moment-generating function for a standard normal distribution. This means that the limiting distribution of Z is standard normal, completing the proof of the Central Limit Theorem.

6.3   THE CHI-SQUARE (χ^2) DISTRIBUTION

Introduction

In this section we discuss in more detail, the chi-square distribution that was introduced in Chapter 4. In Chapter 4 it was noted that the chi-square distribution was a special case of the gamma distribution. In this section we will emphasize the connection it has with the normal distribution which will justify its use in the chapters that follow. Symbolically, we will use the Greek symbol chi with a square on it to denote this distribution and random variable so that $\chi 2$ denotes the chi-square random variable.

Definitions/Theorems

DEFINITION 6.3: A chi-square random variable denoted by $\chi 2$ is a special case of a gamma random variable with $\alpha = \dfrac{v}{2}$, and $\theta = 2$ (see section 4.7). As such it has the following form for its probability density function:

$$f(x) = \begin{cases} \dfrac{1}{\Gamma(v/2)2^{v/2}} x^{v/2-1} e^{-\frac{x}{2}} & 0 \le x < \infty; v > 0 \\ \\ 0 & \text{elsewhere.} \end{cases}$$

This distribution has a mean and variance of:

$$\mu = v$$

and

$$\sigma^2 = 2v$$

THEOREM 6.3: Let the random variable Z be distributed as a standard normal random variable, i.e., Z is normally distributed with $E(Z) = 0$ and $V(Z) = 1$. Then the distribution of Z^2 will be chi-square with $v = 1$, called the degrees of freedom (df).

METHOD OF PROOF: This result can be proved by the distribution function technique of section 5.1. The proof is given at the end of this section.

THEOREM 6.4: Let $X_1, ..., X_n$ be independent chi-square random variables with degrees of freedom of $K_1, K_2, ..., K_n$ respectively. Then

$$Y = X_1 + ... + X_n$$

will be chi-square with $v = K_1 + ... + K_n$ degrees of freedom.

METHOD OF PROOF: The moment-generating function technique provides an easy proof of this theorem. Therefore, you are asked to prove it in exercise 1 at the end of this section.

THEOREM 6.5: Let X_1 and X_2 be independent chi-square random variables; suppose that X_1 has n_1 degrees of freedom and $X_1 + X_2$ is chi-square with n degrees of freedom where $n > n_1$. Then X_2 will be chi-square with $n - n_1$ degrees of freedom.

METHOD OF PROOF: The moment-generating function technique provides the proof of this theorem. Therefore, you are asked to prove it in exercise 2 at the end of this section.

THEOREM 6.6: Suppose $\bar{x}$ and s^2 are the sample mean and variance respectively of a random sample of size n selected from a normally distributed population with mean μ and variance σ^2. Then

(a) $\bar{x}$ and s^2 are independent;

(b)   The random variable $\dfrac{(n-1)s^2}{\sigma^2}$ has a chi-square distribution with $v = n - 1$ degrees of freedom.

METHOD OF PROOF: The proof of part one is beyond the scope of this text. The proof of part 2 relies upon the application of Theorem 3.2 and 3.3 and will be laid out, step by step, in the exercises of this section.

DISCUSSION

In the chapters that follow this we will assume, in most cases, that we draw a sample from a normal distribution. If the random variable X is such a random variable; that is, it is normally distributed with mean, μ, and standard deviation, σ, then we can easily standardize this random variable letting $Z = \dfrac{(X - \mu)}{\sigma}$ which will be distributed as a standard normal. Theorem 6.3 above states that Z^2 will be distributed as a chi-square random variable with 1 degree of freedom; that is, $v = 1$. This distribution is displayed in Figure 6.11 below. The pdf for this particular chi-square distribution is given by

$$f(x) = \begin{cases} \dfrac{1}{2^{1/2}\Gamma(1/2)}x^{1/2 - 1}\,e^{-x/2} & x > 0 \\[2mm] 0 & \text{elsewhere} \end{cases}$$

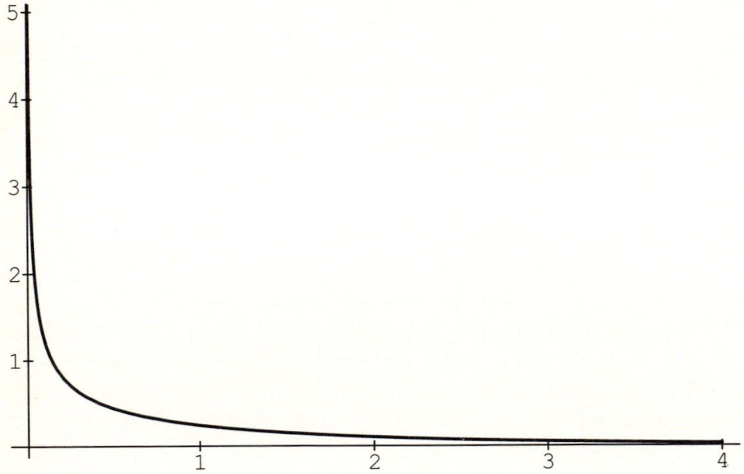

Figure 6.11 χ^2 distribution, 1 degree of freedom.

An empirical demonstration of this property is shown in the following histograms using a computer to generate the data. Figure 6.12 shows a histogram of 1000 standard normal Z-scores along with a histogram of the squares of those 1000 z-scores. Notice the agreement between the theoretical distribution of Figure 6.11 and empirical distribution on the right of Figure 6.12.

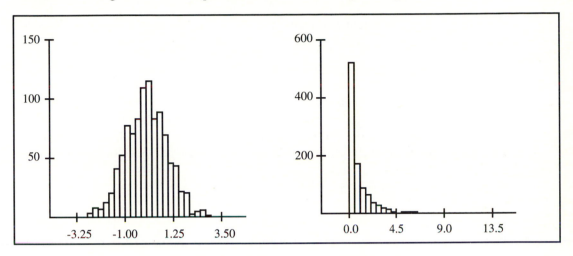

Figure 6.12 Histograms of 1000 standard normal scores and their squares.

In Figure 6.13 below the shape of three chi-square distributions are shown. The parameter value ν takes on the values: 2, 4, and 8. Since the mean for the chi-square distribution equals the parameter ν, you can see the curve shift to the right and become a little flatter for increasing values of

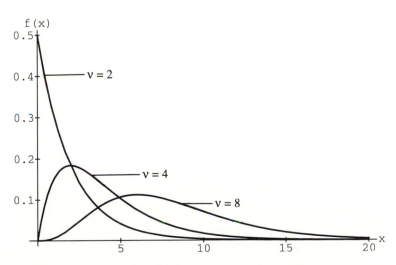

Figure 6.13 Chi-square distributions for ν = 2, 4 and 8.

v . Also, the standard deviation equals $\sqrt{2v}$ and so as the parameter v gets larger, the curve also spreads out more from tail to tail.   (Recall the parameter is also called the "degrees of freedom.")

Tables

Because for each value of v a different chi-square distribution results, it becomes impractical to provide a table of probabilities as complete as that of the standard normal distribution.   In the appendix at the end of the book, given as Table D, an abbreviated table is given. A portion of that table is given below as Table 6.2 and 6.3.   Row headings are the various values of v, the degrees of freedom; column headings denote the various quantile values to be reported and the body of the table records the quantile scores. Thus, the table entry under $\chi_{.050}$ and $v = 8$, shown in bold face in Table 6.2 below, is 2.733 and indicates $P(\chi^2 \leq 2.733) = 0.05$; that is, the area to the left of a chi-square value of 2.733 when the degrees of freedom are 8 is 5% of the area.   We often use the notation $\chi^2{}_{0.050,8}$ to denote this value and therefore we would say, $\chi^2{}_{0.050,8} = 2.733$. (More generally, the notation $\chi^2{}_{q,v}$ denotes the qth quantile of the chi-square distribution with v degrees of freedom.)

Since were typically are interested in the tails of the distribution, that is the only part of the distribution that is reported extensively.

Table 6.2 The left-tail portion of Table D of chi-square values.   (See appendix tables.)

v	$\chi_{.001}$	$\chi_{.005}$	$\chi_{.010}$	$\chi_{.025}$	$\chi_{.050}$	$\chi_{.100}$
1	0.000	0.000	0.000	0.001	0.004	0.016
2	0.002	0.010	0.020	0.051	0.103	0.211
3	0.024	0.072	0.115	0.216	0.352	0.584
4	0.091	0.207	0.297	0.484	0.711	1.064
5	0.210	0.412	0.554	0.831	1.145	1.610
6	0.381	0.676	0.872	1.237	1.635	2.204
7	0.598	0.989	1.239	1.690	2.167	2.833
8	0.857	1.344	1.646	2.180	**2.733**	3.490
9	1.152	1.735	2.088	2.700	3.325	4.168
10	1.479	2.156	2.558	3.247	3.940	4.865
11	1.834	2.603	3.053	3.816	4.575	5.578
12	2.214	3.074	3.571	4.404	5.226	6.304
13	2.617	3.565	4.107	5.009	5.892	7.042
14	3.041	4.075	4.660	5.629	6.571	7.790
15	3.483	4.601	5.229	6.262	7.261	8.547
16	3.942	5.142	5.812	6.908	7.962	9.312
17	4.416	5.697	6.408	7.564	8.672	10.085
18	4.905	6.265	7.015	8.231	9.390	10.865
19	5.407	6.844	7.633	8.907	10.117	11.651
20	5.921	7.434	8.260	9.591	10.851	12.443
...	...	...	...	...	...	...

Table 6.3 The right-tail portion of Table D of chi-square values.   (See appendix tables.)

v	$\chi_{.900}$	$\chi_{.950}$	$\chi_{.975}$	$\chi_{.990}$	$\chi_{.995}$	$\chi_{.999}$
1	2.707	3.843	5.026	6.637	7.881	10.829
2	4.605	5.991	7.378	9.210	10.597	13.816
3	6.251	7.815	9.348	11.345	12.838	16.266
4	7.779	9.488	11.143	13.277	14.860	18.467
5	9.236	11.071	12.833	15.086	16.750	20.515
6	10.645	12.592	14.449	16.812	18.548	22.458
7	12.017	14.067	16.013	18.475	20.278	24.322
8	13.362	15.507	17.535	20.090	21.955	26.124
9	14.684	16.919	19.023	21.666	23.589	27.877
10	15.987	18.307	20.486	23.213	25.188	29.588
11	17.275	19.678	21.923	24.728	26.757	31.264
12	18.549	21.029	23.340	26.221	28.300	32.910
13	19.813	22.365	24.739	27.692	29.819	34.528
14	21.065	23.687	26.122	29.145	31.319	36.123
15	22.308	24.999	27.492	30.582	32.801	37.697
16	23.543	26.299	28.849	32.004	34.267	39.252
17	24.770	27.590	30.195	33.413	35.718	40.790
18	25.991	28.872	31.530	34.809	37.156	42.313
19	27.205	30.147	32.856	36.195	38.582	43.820
20	28.413	31.413	34.173	37.570	39.997	45.315
...	...	...	...	...	...	...

EXAMPLES

EXAMPLE 1:   Assume that the random variable X is normally distributed with mean 50 and variance 25. What is the distribution of

$$\chi^2 = \frac{(X-50)^2}{25}?$$

What is the probability that $\chi^2 \geq 4$?

SOLUTION: Since X is normally distributed, then χ^2 is a squared standard normal score which will be distributed as a chi-square random variable. Therefore to calculate $P(\chi^2 \geq 4)$ we must consult a table of chi-square values. That is:

$$P(\chi^2 \geq 4) = 1 - P(\chi^2 \leq 4) = 1 - F_{\chi^2}(4)$$

where F_{χ^2} denotes the distribution function for the chi-square distribution. And as near as our tables will allow us to determine it, this probability is between 0.025 and 0.05. We could report this outcome as $0.025 < p < 0.05$, where p denotes the probability we were trying to compute. (The exact probability using a statistical calculator gives $p = 0.0455$.)

EXAMPLE 2: Test scores are believed to be approximately normally distributed with a mean of 70 and a standard deviation of 12.   A sample of 25 students take the test and the sample standard deviation is $s = 18$.   What is the probability of such a result or one more extreme under the assumptions given?

SOLUTION: Under the conditions described, we would expect the sample standard deviation to be close to 12. If a sample standard deviation of 18 is extreme then larger numbers would be more extreme. Therefore, we want to compute the probability of getting a sample standard deviation as large or larger than 18. We compute:

$$P(s \geq 18 \mid \sigma = 12) = P(s^2 \geq 18^2)$$

$$= P\left(\frac{(n-1)s^2}{\sigma^2} \geq \frac{24 \cdot 18^2}{12^2}\right) = 54)$$

$$= P(\chi^2 \geq 54) < 0.001 \text{ (as close as tables will allow.)}$$

(An exact result is $P(\chi^2 \geq 54) = 0.00043$)

EXAMPLE 3: Find the probability that $|Z| \geq 1.96$ if Z is a standard normal random variable.

SOLUTION: $P(|Z| \geq 1.96) = P(Z \geq 1.96 \text{ or } Z \leq -1.96) = 0.025 + 0.025$. Note that when $|Z| \geq 1.96$, then $Z^2 \geq 1.96^2 = 3.84$. But Z^2 is distributed as a chi-square random variable with $v = 1$ degree of freedom. Therefore,

$$P(|Z| \geq 1.96) = P(\chi^2 \geq 3.84) = 0.05 \text{ (from the table.)}$$

This relationship can be generalized because a squared z-score is a chi-square with 1 degree of freedom. Therefore,
$$P(|Z| \geq k) = P(\chi^2 \geq k^2) \text{ and } P(|Z| \leq k) = P(\chi^2 \leq k^2)$$

EXERCISES

1.   Prove Theorem 6.4 given at the beginning of this section.

2.   Prove Theorem 6.5 given at the beginning of this section.

3.   For a chi-square distribution find $\chi^2_{.025}$, $\chi^2_{.05}$, $\chi^2_{.95}$ and $\chi^2_{.975}$ for degrees of freedom of 5, 20, and 40. Make a rough sketch of each of the three curves and place the table values on your curve. Locate the mean on each sketch.

4.   Suppose that $X_1, X_2, ..., X_n$ constitute a random sample of size 20 from a normal distribution with mean, μ and variance, 25. What is the probability that the sample variance will exceed 40? (Hint: First write the probability statement as $P(s^2 > 40)$ then transform the left-hand side of the inequality to a chi-square random variable.)

5.   Find the probability that a random sample of 25 observations, from a normal population with variance $\sigma^2 = 6$, will have a sample variance s^2 greater than 9.1? Between 3.462 and 10.745?

6.   Suppose an inoculation from a "gun-type" injection system is supposed to inject 2 cc's of serum. The manufacturer claims the variance is .15 $(cc's)^2$. Twenty subjects are hypothetically inoculated using the injection system. (The gun is used on a device that captures the inoculant and measures it.) The sample variance turns out to be .24 $(cc's)^2$. What is the probability that the sample variance could be that large under the manufacturer's claim? Would you say there is reason to doubt the manufacturer? What assumptions have you made?

7. An instructor, historically, has had a standard deviation of about 10 on exams that she has given. The first exam using a large section format produced a standard deviation of 19.1. This section had 75 students in it. Would you conclude that is ordinary variation, explainable by chance? What assumptions did you make?

8. Suppose that $X_1, X_2, ..., X_n$ constitute a random sample of size n from a normal distribution with mean, μ and variance, σ^2. Prove that the random variable

    $$\sum_{i=1}^{n} \frac{(X_i - \mu)^2}{\sigma^2}$$

    is chi-square with n degrees of freedom.

9. Suppose that $X_1, X_2, ..., X_n$ constitute a random sample of size n from a normal distribution with mean, μ and variance, σ^2. Prove that $\dfrac{n(\bar{X}-\mu)^2}{\sigma^2}$ is chi-square with 1 degree of freedom.

10. Verify the following identity

    $$\sum_{i=1}^{n} (X_i - \mu)^2 = \sum_{i=1}^{n} (X_i - \bar{X})^2 + n(\bar{X} - \mu)^2$$

    (Hint: First write $\displaystyle\sum_{i=1}^{n} (X_i - \mu)^2 = \sum_{i=1}^{n} (X_i - \bar{X} + \bar{X} - \mu)^2$. Then simplify the right side of this expression by squaring and summing.)

11. Suppose that $X_1, X_2, ..., X_n$ constitute a random sample of size n from a normal distribution with mean, μ and variance, σ^2. Show that the random variable

    $$\frac{(n-1)s^2}{\sigma^2}$$

    is chi-square with $df = n-1$. (Hint: Use the results of exercises 6-8 to show that

    $$\sum_{i=1}^{n} \frac{(X_i - \mu)^2}{\sigma^2} = \frac{(n-1)s^2}{\sigma^2} + \frac{n(\bar{X} - \mu)^2}{\sigma^2}$$

    whose parts can be argued to be independent chi-square random variables.)

12. Recognizing that the chi-square distribution is a gamma distribution with $\alpha = \nu/2$ and $\theta = 2$, what is the mean and variance of the chi-square distribution?

Proof of Theorem 6.3

Let $Y = Z^2$ where Z is $N(0,1)$. We apply the distribution function approach to this problem as follows:

$$F(u) = P(Y \le u) = P(Z^2 \le u) = P(-\sqrt{u} \le Z \le \sqrt{u}).$$

With $u \ge 0$ this is

$$F(u) = \int_{-\sqrt{u}}^{\sqrt{u}} \frac{1}{\sqrt{2\pi}} e^{-z^2/2} \, dz = 2 \int_{0}^{\sqrt{u}} \frac{1}{\sqrt{2\pi}} e^{-z^2/2} \, dz.$$

Changing the variable of integration by letting $z = \sqrt{y}$ so that $dz = \dfrac{1}{2\sqrt{y}}$, we get

$$F(u) = \int_{0}^{u} \frac{1}{\sqrt{2\pi y}} e^{-y/2} \, dy \qquad\qquad 0 \le u$$

Therefore, $f(u) = \dfrac{dF(u)}{du}$ which gives, from one form of the fundamental theorem of calculus,

$$f(u) = \frac{1}{\sqrt{\pi}\sqrt{2}} u^{1/2 - 1} e^{-u/2}, \qquad\qquad 0 < u < \infty$$

However, by examination, remembering that $\Gamma(1/2) = \sqrt{\pi}$, we can recognize that this is a chi-square distribution with $\nu = 1$ degrees of freedom.

6.4   STUDENT'S t DISTRIBUTION

Introduction

Another prominent distribution associated with normal distribution theory is the t distribution. This is used whenever you sample from a normal distribution with unknown variance and the sample variance is used in place of the true variance when the sample mean is standardized. The t distribution depends upon a single parameter ν called the degrees of freedom. Cumulative tables are available for the most commonly needed quantiles.

The distribution obtained its name from a paper written by W. S. Gosset. He was employed in an Irish brewery that wouldn't allow publication by its staff. Thus, Gosset published his work "secretly" under the name "Student" to circumvent this company policy. The distribution has been referred to as Student's t distribution as a result.

Theorems

THEOREM 6.7: If Y and Z are independent random variables, Y has a chi-square distribution with v degrees of freedom, and Z has a standard normal distribution, then the distribution of

$$t = \frac{Z}{\sqrt{(Y/v)}}$$

has a distribution called the t distribution whose pdf is as follows:

$$f(t) = \frac{\Gamma\left(\frac{v+1}{2}\right)}{\Gamma\left(\frac{v}{2}\right)\sqrt{\pi v}} \left(1 + \frac{t^2}{v}\right)^{-\frac{(v+1)}{2}}, \quad -\infty < t < \infty, \ v > 0$$

This is a one-parameter family where v is called the degrees of freedom.

Method of Proof: The distribution function technique provides the for proof of this theorem.

THEOREM 6.8: Suppose $X_1, ..., X_n$ represent a random sample from a normally distributed population with mean μ and standard deviation σ. The distribution of the random variable

$$\frac{\overline{X} - \mu}{s/\sqrt{n}}$$

will be the t distribution with $v = n - 1$ degrees of freedom.

METHOD OF PROOF: The proof of this result is a direct application of Theorem 6.7 and is left for the exercises at the end of the section.

THEOREM 6.9: Suppose $X_1, ..., X_n$ and $Y_1, ..., Y_n$ represent two independent samples from two normally distributed populations with means μ_x and μ_y but with a common variance σ^2. The distribution of the random variable

$$\frac{(\overline{X} - \overline{Y}) - (\mu_x - \mu_y)}{\sqrt{s_p^2 \left(\frac{1}{n} + \frac{1}{m}\right)}}$$

will be the t distribution with $v = (n - 1) + (m - 1)$ degrees of freedom, and where

$$s_p^2 = \frac{(n-1)s_x^2 + (m-1)s_y^2}{(n-1) + (m-1)}.$$

METHOD OF PROOF: The proof is a direct application of Theorem 6.7 and is left to the exercises.

DISCUSSION

The t distribution is a bell-shaped distribution, symmetric around zero. It has more area in the tails and less in the central region than does the standard normal distribution. But as ν approaches infinity, the t distribution approaches the normal distribution. Figure 6.14 shows three t distribution curves for the cases when $\nu = 1$, 3, and 7. Included is a standard normal curve for comparison with the three t distribution curves.

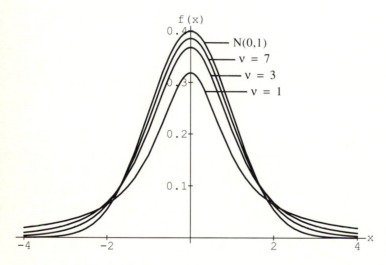

Figure 6.14 Plot of three t distribution curves compared to the standard normal curve.

Tables

A table of probabilities for various quantile values of the t distribution is given as Table E in the appendix. Because the distribution is symmetric around 0, the quantile score for the 0.5 quantile is 0; i.e. 50% of the area is to the left of a t-value of 0. Also because of symmetry only the right tail of the distribution is recorded. A portion of table E for the t distribution is given below as Table 6.4. To illustrate the use of this table, suppose the random variable T has a t distribution with 10 degrees of freedom and we want to find the value t such that $P(T \le t) = 0.975$; i.e., t is the 0.975 quantile for the t distribution with 10 degrees of freedom. Looking in the table with $\nu = 10$ and finding the column headed by $t_{0.975}$ we see the value 2.229. (This value is shown in bold-face in Table 6.4.) We often use the notation $t_{0.975,10}$ to denote this value and therefore we would say, $t_{0.975,10} = 2.229$. (More generally, the notation $t_{q,\nu}$ denotes the qth quantile of the t distribution with ν degrees of freedom.)

Table 6.4 Portion of Table E: table of the *t* distribution.

ν	$t_{0.500}$	$t_{0.800}$	$t_{0.900}$	$t_{0.950}$	$t_{0.975}$	$t_{0.990}$	$t_{0.995}$	$t_{0.999}$
1	0	1.376	3.078	6.314	12.706	31.821	63.657	318.313
2	0	1.061	1.886	2.920	4.303	6.965	9.925	22.327
3	0	0.978	1.638	2.353	3.182	4.541	5.841	10.215
4	0	0.941	1.533	2.133	2.777	3.747	4.604	7.173
5	0	0.919	1.476	2.016	2.571	3.366	4.032	5.893
6	0	0.906	1.440	1.944	2.448	3.144	3.708	5.208
7	0	0.896	1.415	1.895	2.365	2.999	3.500	4.786
8	0	0.889	1.397	1.860	2.307	2.897	3.356	4.502
9	0	0.883	1.383	1.834	**2.263**	**2.822**	3.251	4.297
10	0	0.879	1.372	1.813	**2.229**	2.764	3.170	4.144
11	0	0.875	1.364	1.796	2.202	2.719	3.106	4.025
12	0	0.872	1.356	1.783	2.179	2.682	3.055	3.930
13	0	0.870	1.350	1.771	2.161	2.651	3.013	3.853
14	0	0.868	1.345	1.762	2.145	2.625	2.977	3.788
15	0	0.866	1.341	1.753	2.132	2.603	2.947	3.733
16	0	0.864	1.337	1.746	2.120	2.584	2.921	3.687
17	0	0.863	1.334	1.740	2.110	2.568	2.899	3.646
18	0	0.862	1.331	1.734	2.101	2.553	2.879	3.611
19	0	0.861	1.328	1.730	2.094	2.540	2.861	3.580
20	0	0.860	1.326	1.725	**2.086**	2.529	2.846	3.552
21	0	0.859	1.323	1.721	2.080	2.518	2.832	3.528
22	0	0.858	1.321	1.718	2.074	2.509	2.819	3.505
23	0	0.857	1.320	1.714	2.069	2.500	2.808	3.485
24	0	0.857	1.318	1.711	2.064	2.493	2.797	3.467
25	0	0.856	1.317	1.709	2.060	2.486	2.788	3.451
26	0	0.855	1.315	1.706	2.056	2.479	2.779	3.435
27	0	0.855	1.314	1.704	2.052	2.473	2.771	3.421
28	0	0.854	1.313	1.702	2.049	2.468	2.764	3.409
29	0	0.854	1.312	1.700	2.046	2.463	2.757	3.397
30	0	0.854	1.311	1.698	2.043	2.458	2.750	3.386
...	...	...	...	...	...	...	...	...
∞	0	0.841	1.282	1.645	1.960	2.327	2.576	3.091

By an examination of Table 6.4 you can confirm that the *t* distribution approaches the normal distribution as the degrees of freedom increases. Simply choose a column and scan down it as the degrees of freedom get larger. You can confirm that the numbers in a given column seem to settle into a particular value. That limiting value is given by the bottom row of the table with degrees of freedom equal to ∞. However, note that these are the same quantile values we get for the standard normal distribution. Thus, by inspection you can conclude that the limiting distribution as ν approaches ∞ for the *t* distribution is the standard normal distribution. (Thus, the bottom row of the *t* table can be used to get various quantiles of interest for the normal distribution as well.)

Application

A common statistical situation under which we will need to consult the *t* distribution is as follows: Suppose we have a random sample of *n* observations from a normal distribution having an unknown variance. We want to compare the sample mean to the theoretical mean and thus we standardize the mean in the following way.

$$\frac{\overline{X} - \mu}{s/\sqrt{n}}$$

Note that we have used the sample standard deviation in place of the unknown theoretical standard deviation σ.

It can be shown, using Theorem 6.1 of this section, that this expression follows the t distribution with $v = n - 1$ degrees of freedom. (See exercise 3 in this section.)

There is also a two sample variation on this same idea. For instance, suppose we have two independently selected random samples, each having come from a normal distribution with means μ_x and μ_y respectively, and with unknown but equal variance σ^2. If we want to examine the probability about the difference of the means then we would consider a z-score such as the following:

$$\frac{(\overline{X} - \overline{Y}) - (\mu_x - \mu_y)}{\sqrt{\sigma^2(\frac{1}{n} + \frac{1}{m})}}.$$

However, the computation can't be finished because of the unknown variance. Therefore, in order to "estimate" the unknown variance, use the two separate sample variances, "pooling" them together in a "weighted average" in the following fashion,

$$s_p^2 = \frac{(n-1)s_x^2 + (m-1)s_y^2}{(n-1) + (m-1)}$$

This produces a "standardized score,"

$$\frac{(\overline{X} - \overline{Y}) - (\mu_x - \mu_y)}{\sqrt{s_p^2 (\frac{1}{n} + \frac{1}{m})}}$$

that follows the t distribution with $v = (n-1) + (m-1)$ degrees of freedom. The details of this proof are developed in the section exercises.

EXAMPLES

EXAMPLE 1: Using the table of the t distribution, find the 0.025 and 0.975 quantile values of the t distribution when there are 10 degrees of freedom and for 20 degrees of freedom; that is, we want to find: $t_{0.025,10}$, $t_{0.975,10}$, $t_{0.025,20}$, and $t_{0.975,20}$.

SOLUTION: This is a simple exercise using Table E in the appendix. See, also, Table 6.4 of this section. Entering the table for $v = 10$ degrees of freedom we find that $t_{0.975} = 2.229$ and so by symmetry, $t_{0.025} = -2.229$. (These numbers are shown in bold-face type in Table 6.4.) For 20 degrees of freedom the values are -2.086 and $+2.086$. This demonstrates the symmetry property of the t distribution. Also remember that $Z_{.025} = -1.96$ and $Z_{.0975} = 1.96$ from the standard normal distribution. Thus, you can note that the area in the tails of the t distribution beyond -1.96 and 1.96 will be larger than 0.025 for each tail, verifying the "longer" tails of the t distribution compared to the standard normal distribution. As the degrees of freedom get larger, you can also note by examining the table, that the t-values approach the corresponding normal distribution values.

EXAMPLE 2: An air quality inspector takes readings on a particular pollutant several times a day. Acceptable levels for that particular station have been 7 parts per million (ppm).   During a temperature inversion, a series of 10 readings produced a mean of 7.7 ppm with a standard deviation of .8 ppm. Would you conclude there was evidence that chance can be ruled out as the explanation of the higher levels of the pollutant? (Assume that readings are reasonably approximated by a normal distribution.)

SOLUTION: To answer this question we want to calculate the probability of getting a mean as large or larger than 7.7 ppm, assuming that true mean is 7 ppm. We do this as follows:

$$P(\overline{X} \geq 7.7 \mid \mu = 7) = P\left[\frac{\overline{X}-7}{0.8/\sqrt{10}} \geq \frac{(7.7-7)}{0.8/\sqrt{10}}\right] = P(t \geq 2.77)$$

From Table E or Table 6.4, using 9 degrees of freedom, all we can say is that this probability is somewhere between 0.025 and 0.01 since 2.77 lies between $t_{0.975}$ and $t_{0.990}$. Note the bold-faced values in Table 6.4. This probability is small enough to conclude that it is very likely there has been a real increase in the pollutant beyond that which chance could account for.

EXAMPLE 3: Two different stations keeping records on air pollution report the following results. Is there reason to suspect that there is a real difference in the parts per million of the particular pollutant being measured? (One of the stations is near a busy intersection, the other is near an industrial area.)

	Station	
	1	2
Number of readings taken	10	12
Mean (ppm)	8.3	9.2
Standard deviation (ppm)	1.2	0.9

SOLUTION: If there is no real difference in the station means, then μ_x and μ_y would be the same and thus, $\mu_x - \mu_y = 0$. If we let the random variable Y be associated with station 1 and X with station 2, then $(\overline{x} - \overline{y}) = 9.2 - 8.3 = 0.9$, and we want to compute $P(\overline{x} - \overline{y} \geq 0.9)$ assuming that $\mu_x - \mu_y = 0$. If this probability is very small—indication a rare event has occurred—then we would question whether the true means were the same. Since we don't know the true standard deviation in this problem we can't obtain a z-score. If we are willing to assume that the readings at each station are normally distributed with a constant and equal variance, then we can apply a transformation that will result in a t-score to compare against values from the t distribution table. The transformation is:

$$t = \frac{(\overline{X}-\overline{Y}) - (\mu_x - \mu_y)}{\sqrt{s_p^2 \left(\frac{1}{n} + \frac{1}{m}\right)}}$$

where

$$s_p^2 = \frac{(n-1)s_x^2 + (m-1)s_y^2}{(n-1) + (m-1)}$$

Inserting values into the formula we get:

$$s_p^2 = \frac{(n-1)s_x^2 + (m-1)s_y^2}{(n-1)+(m-1)} = \frac{(12-1)0.9^2 + (10-1)1.2^2}{(12-1)+(10-1)} = \frac{21.87}{20} = 1.0935$$

and

$$t = \frac{(\bar{X}-\bar{Y})-(\mu_x-\mu_y)}{\sqrt{s_p^2\left(\frac{1}{n}+\frac{1}{m}\right)}} = \frac{(9.2-8.3)-0}{\sqrt{1.0935\left(\frac{1}{12}+\frac{1}{10}\right)}} = \frac{0.9}{0.4477} = 2.0103$$

Thus,

$$p = P(\bar{x}-\bar{y} \geq 0.9) = P(t \geq 2.0103)$$

where t has $(12-1)+(10-1)$ degrees of freedom. From Table 6.4 the probability p is:

$$0.025 < p < 0.05.$$

Therefore we would conclude that it is very unlikely that the means of the two stations are equal and something is different in the readings. This difference could be due to errors in measurement, the assumption of normality may be wrong, the assumption of equal variances may not be appropriate, or there may be a real difference in the amount of pollutants at the two locations. (We are not so naive as to believe that $\mu_x = \mu_y$ exactly. Rather we are asking whether they are different enough to show up in an analysis of the data with samples of relatively small sizes, $n = 12$ and $m = 10$. It would seem reasonable that if samples of this size would show statistical justification of concluding the true means are different, then there must be a real difference that should be investigated further.)

EXERCISES

1. Determine the 0.95, 0.975 and 0.99 quantiles of the t distribution for 10, 20, and 30 degrees of freedom.

2. Suppose the random variable T is t distributed with 14 degrees of freedom. What is the probability of an observation being between -1.345 and 2.625?

3. Let T have a t distribution with $v = 25$ degrees of freedom. Find a value of k such that
    (a) $P(|T| \geq k) = 0.10$ (b) $P(|T| \geq k) = 0.05$
    (c) $P(|T| \geq k) = 0.02$ (d) $P(|T| \geq k) = 0.01$

4. Using Theorem 6.7, show that the expression

$$t = \frac{\bar{X}-\mu}{s/\sqrt{n}}$$

has the t distribution with $n-1$ degrees of freedom. Assume that $X_1, \ldots, X_n$ are a random sample from a normally distributed population with mean μ and variance σ^2.   (Hint: Let

$Y = \dfrac{(n-1)s^2}{\sigma^2}$ and $Z = \dfrac{\bar{X}-\mu}{\sigma/\sqrt{n}}$. Simplify the algebra that results after applying the theorem to these random variables.)

5. In 16 one-hour test runs, the gasoline consumption of an engine averaged 16.4 gallons with a standard deviation of 2.1 gallons. Is the claim that the average gasoline consumption of this engine is 12.0 gallons per hour a reasonable claim based on the data.

6. A cigarette manufacturer claims that their cigarettes have an average nicotine content of 1.70 milligrams. A random sample of 6 cigarettes was tested for nicotine content in the laboratory and the following values were obtained: 2.0, 1.6, 2.1, 1.9, 2.2, 2.0. Would you agree with the manufacturer's claim? (What assumptions do you make?)

7. Suppose a random sample of size n, denoted by $X_1, ..., X_n$, is selected from a normally distributed population having a mean μ_x and variance σ^2. Then an independent random sample of size m, denoted by $Y_1, ..., Y_m$, is selected from a separate normally distributed population having mean μ_y and the same variance σ^2. Show that $E(\bar{X}-\bar{Y}) = \mu_x - \mu_y$ and that $V(\bar{X}-\bar{Y}) = \sigma^2(\dfrac{1}{n}+\dfrac{1}{m})$ so that a standardized score would be

$$\dfrac{(\bar{X}-\bar{Y})-(\mu_x-\mu_y)}{\sqrt{\sigma^2(\dfrac{1}{n}+\dfrac{1}{m})}}$$

8. Using the same assumptions of exercise 6, if s_x^2 and s_y^2 are the respective sample variances of the two independent samples $X_1, ..., X_n$ and $Y_1, ..., Y_m$, argue that

$$\dfrac{(n-1)s_x^2+(m-1)s_y^2}{\sigma^2}$$

is chi-square distributed with $v = (n-1)+(m-1)$. (See the theorems in section 6.3.)

9. Using the results of exercises 6 and 7, simplify the algebra to show that

$$\dfrac{(\bar{X}-\bar{Y})-(\mu_x-\mu_y)}{\sqrt{s_p^2\left(\dfrac{1}{n}+\dfrac{1}{m}\right)}}$$

is t distributed with $v = (n-1)+(m-1)$ where

$$s_p^2 = \dfrac{(n-1)s_x^2+(m-1)s_y^2}{(n-1)+(m-1)}$$

10. The processing speed of two different computer chips is being compared. A random sample of 10 chips each is selected and a given job is processed by each chip. The sample results are as follows:

	Chip	
	A	B
Sample Mean (milli-secs)	40.0	36.0
Sample Standard Deviation (milli-secs)	2.3	2.9
Sample Size	10.0	10.0

Assuming normal distributions and a common variance, what is the probability of having means differ by 4 units or more, that is what is the probability that

$$|\bar{X} - \bar{Y}| \geq 4?$$

11. Let Z_1, Z_2, and Z_3 be independent random variables from a standard normal distribution, that is they are a random sample from a $N(0,1)$.

   (a) What is the distribution of $Y = Z_1^2 + Z_2^2$?

   (b) What is the distribution of $U = \dfrac{Z_3}{\sqrt{(Z_1^2 + Z_2^2)/2}}$?

6.5   FISHER'S F DISTRIBUTION

Introduction

The F distribution is encountered typically when investigating the ratio of two sample variances. Under the appropriate normality conditions as defined below, the ratio will be distributed according to the F distribution. This distribution has two parameters, v_1 and v_2, called the degrees of freedom of the numerator and degrees of freedom of the denominator, respectively. The distribution is positive and skewed right and has a single mode. A limited table is presented in the appendix as Table F.

Theorems

THEOREM 6.10: If U and V are independent random variables having chi-square distributions with v_1 and v_2 degrees of freedom, then the distribution of

$$F = \frac{(U/v_1)}{(V/v_2)}$$

is the F distribution given by

$$f(F) = \frac{\Gamma[(v_1 + v_2)/2](v_1/v_2)^{v_1/2}}{\Gamma(\frac{v_1}{2})\Gamma(\frac{v_2}{2})} \frac{F^{v_1/2 - 1}}{(1 + v_1 F/v_2)^{(v_1 + v_2)/2}}$$

$$0 < F < \infty$$

which has two parameters, v_1 and v_2, called the degrees of freedom of the numerator and denominator, respectively.

METHOD OF PROOF: The change of variable technique would be used to prove this result.

THEOREM 6.11: Let $X_1, ..., X_n$ be independent, normally distributed random variables having mean μ_x and variance σ_x^2. Let $Y_1, ..., Y_m$ be independent normally distributed random variables having mean μ_y and variance σ_y^2. The ratio, $\frac{s_x^2}{\sigma_x^2} / \frac{s_y^2}{\sigma_y^2}$ follows the F distribution with $v_1 = n - 1$ and $v_2 = m - 1$, respectively.

PROOF: The proof is developed in the exercises of this section.

THEOREM 6.12: If $F_\alpha(v_1, v_2)$ denotes the α quantile for the F distribution with v_1 and v_2 degrees of freedom, then the following relationship holds

$$F_\alpha(v_1, v_2) = \frac{1}{F_{1-\alpha}(v_2, v_1)}$$

METHOD OF PROOF: First note that from Theorem 6.10 if F is $F(v_1, v_2)$ then $1/F$ will obviously be $F(v_2, v_1)$. Next, $P(F \leq F_\alpha(v_1, v_2)) = \alpha$, and $P(\frac{1}{F} > \frac{1}{F_\alpha(v_1, v_2)}) = \alpha$. From the complement, $P(\frac{1}{F} < \frac{1}{F_\alpha(v_1, v_2)}) = 1 - \alpha = P(\frac{1}{F} < F_{1-\alpha}(v_2, v_1))$ by definition. Therefore Theorem 6.12 holds.

DISCUSSION

The F random variable as the ratio of two positive random variables (independent chi-square random variables each divided by its degrees of freedom) will itself be positive. As such the probability density function given in Theorem 6.10 will produce a right-skewed distribution, having a single mode. The skewness becomes less pronounced as both degrees approach infinity.

The distribution is a function of two parameters, the degrees of freedom for the numerator denoted by v_1, and the degrees of freedom of the denominator denoted by v_2.

Figures 6.15 and 1.16 show the shapes of the F distribution for various combinations of degrees of freedom. In Figure 6.15 the degrees of freedom for the numerator and denominator are kept the same and are increased from 1 to 20 in the increments shown. In this figure you can see as the degrees of freedom get larger the shape of the distribution becomes less elongated and skewed.

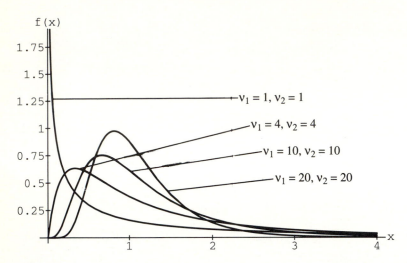

Figure 6.15 F distribution for equal numerator and denominator degrees of freedom: 1, 4, 10, and 20.

Figure 6.16 show what happens as the numerator degrees of freedom and denominator degrees of freedom are interchanged for several cases. The shapes can be seen to be different, but again the presence of a single mode is clear along with a lessening of the skewness as the degrees of freedom get larger.

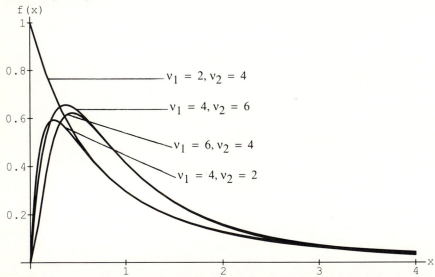

Figure 6.16 F distribution for alternating numerator and denominator degrees of freedom: $v_1 = 2$, $v_2 = 4$; $v_1 = 4$, $v_2 = 2$; $v_1 = 4$, $v_2 = 6$ and $v_1 = 6$, $v_2 = 4$.

Since the F distribution is the ratio of two independent chi-square distributions divided by their degrees of freedom and since the mean of a chi-square random variable equals the degrees of freedom, we would expect this ratio to settle in around the value 1. This tendency can be seen in both Figures 6.15 and 6.16.

Tables

Extensive tables for the *F* distribution would occupy many pages because of the many possible combinations of degrees of freedom, each pair producing a unique curve of probabilities. Therefore, we have included a limited set of tables, shown as Table F in the appendix, a portion of which is included here as Table 6.5. The entries in the body of the table represent the quantile scores for the three quantiles: 0.95, 0.975 and 0.99. The various rows in the table are associated with the denominator degrees of freedom. In Table 6.5 below, these numbers vary from 1 to 10. The column headings correspond to the numerator degrees of freedom, going from 1 to 11 in Table 6.5.

Table 6.5 A portion of Table F, the *F* distribution table for various combinations of numerator and denominator degrees of freedom.

F(f)	v_2	Numerator Degrees of Freedom, v_1										
		1	2	3	4	5	6	7	8	9	10	11
0.95	1	161.4	199.5	215.7	224.6	230.2	233.6	235.8	237.4	238.6	239.5	240.2
0.975		647.8	796.7	854.6	885.2	903.5	915.7	924.2	930.3	934.8	938.2	941.4
0.99		3980.	4843.	5192.	5377.	5491.	5570.	5613.	5653.	5683.	5706.	5712.
0.95	2	18.51	19.00	19.16	19.25	19.30	19.33	19.35	19.37	19.38	19.4	19.40
0.975		38.51	39.00	39.17	39.25	39.30	39.33	39.35	39.37	39.39	39.4	39.41
0.99		98.50	99.00	99.17	99.25	99.30	99.33	99.35	99.37	99.38	99.4	99.41
0.95	3	10.13	9.55	9.28	9.12	9.01	8.94	8.89	8.85	8.81	8.79	8.76
0.975		17.44	16.04	15.44	15.10	14.89	14.73	14.62	14.54	14.47	14.42	14.37
0.99		34.12	30.82	29.46	28.71	28.24	27.91	27.67	27.49	27.35	27.23	27.13
0.95	4	7.71	6.94	6.59	6.39	6.26	6.16	6.09	6.04	6.00	5.96	5.94
0.975		12.22	10.65	9.98	9.60	9.36	9.20	9.07	8.98	8.90	8.84	8.79
0.99		21.20	18.00	16.69	15.98	15.52	15.21	14.98	14.80	14.66	14.55	14.45
0.95	5	6.61	5.79	5.41	5.19	5.05	4.95	4.88	4.82	4.77	4.74	4.70
0.975		10.01	8.43	7.76	7.39	7.15	6.98	6.85	6.76	6.68	6.62	6.57
0.99		16.26	13.27	12.06	11.39	10.97	10.67	10.46	10.29	10.16	10.05	9.96
0.95	6	5.99	5.14	4.76	4.53	4.39	4.28	4.21	4.15	4.10	4.06	4.03
0.975		8.81	7.26	6.60	6.23	5.99	5.82	5.70	5.60	5.52	5.46	5.41
0.99		13.75	10.92	9.78	9.15	8.75	8.47	8.26	8.10	7.98	7.87	7.79
0.95	7	5.59	4.74	4.35	4.12	3.97	3.87	3.79	3.73	3.68	3.64	3.60
0.975		8.07	6.54	5.89	5.52	5.29	5.12	4.99	4.90	4.82	4.76	4.71
0.99		12.25	9.55	8.45	7.85	7.46	7.19	6.99	6.84	6.72	6.62	6.54
0.95	8	5.32	4.46	4.07	3.84	3.69	3.58	3.50	3.44	3.39	3.35	3.31
0.975		7.57	6.06	5.42	5.05	4.82	4.65	4.53	4.43	4.36	4.30	**4.24**
0.99		11.26	8.65	7.59	7.01	6.63	6.37	6.18	6.03	5.91	5.81	5.73
0.95	9	5.12	4.26	3.86	3.63	3.48	**3.37**	3.29	3.23	**3.18**	3.14	3.10
0.975		7.21	5.71	5.08	4.72	4.48	4.32	4.20	4.10	4.03	3.96	3.91
0.99		10.56	8.02	6.99	6.42	6.06	5.80	5.61	5.47	5.35	5.26	5.18
0.95	10	4.96	4.10	3.71	3.48	3.33	**3.22**	3.14	3.07	3.02	2.98	2.94
0.975		6.94	5.46	4.83	4.47	4.24	**4.07**	3.95	3.85	3.78	3.72	3.66
0.99		10.04	7.56	6.55	5.99	5.64	**5.39**	5.20	5.06	4.94	4.85	4.77

For instance, if an F distribution has numerator degrees of freedom of 6 and a denominator degrees of freedom of 10, then from the table the area to the left of 3.22 is 95%, to the left of 4.07 is 97.5% and to the left of 5.39 is 99%. (See the bold-faced values in Table 6.5.)

Application

Theorem 6.11 describes how we will apply the F distribution in this text. (The F distribution also plays a dominant role in the statistical analysis technique called the *analysis of variance*. See Chapter 10 for a brief introduction to this technique.) That is, we have two independent samples from normally distributed populations. For one reason or another we want to compare the variances of the two samples. To compare these two variances, we compute their ratio, which according to the theorem would have a probability that would behave as a random variable from the F distribution. We would consult a table of the F distribution, a selection of which we have in the Appendix as Table F. Because we tabulate only the quantiles of 0.95, 0.975 and 0.99 of the F distribution, Theorem 6.12 become very important to give us the other tail of the distribution. Consider several examples of the use of the tables and applications of the F distribution.

EXAMPLES

EXAMPLE 1: Suppose an F random variable has numerator and denominator degrees of freedom of 6 and 9 respectively. What is the probability of getting an F value greater than 3.37?

SOLUTION: To calculate $P(F \geq 3.37)$ we enter Table 6.5 under the column heading of 6 and the row degrees of freedom of 9. We find the number 3.37 associated with the 0.95 quantile. Therefore,

$$P(F \geq 3.37) = 1 - 0.95 = 0.05 \text{ or } 5\%.$$

EXAMPLE 2: If a sample of size 10 from a normally distributed population is selected, after which a second independent sample of size 10 is selected from the very same population, what is the chance that the second sample will produce a sample variance more than twice as large as the first sample's variance?

SOLUTION: We want to know the probability that s_2^2 will be more than 2 times as large as s_1^2. We can write this equivalently as $P(s_2^2/s_1^2 \geq 2)$. Applying the Theorem 6.11 under the condition of sampling the same normally distributed population we would note that $\dfrac{[s_2^2/\sigma^2]}{[s_1^2/\sigma^2]} = s_2^2/s_1^2$ is a random variable that follows the F distribution with 9 and 9 degrees of freedom. Therefore, consulting the F table we can only say that this probability exceeds 0.05. $[P(F \geq 2) \geq P(F > 3.18) = 0.05$ for 9 and 9 degrees of freedom.]

EXAMPLE 3: What is the 0.025 quantile score for an F distribution with numerator degrees of freedom of 8 and denominator degrees of freedom of 11?

SOLUTION: Note that the table does not include the left tail of the distribution, but using Theorem 6.12 we can solve this problem. By the theorem, we have the following relationship:

$$F_{0.025(8,11)} = \frac{1}{F_{1-0.025(11,8)}} = \frac{1}{F_{0.975(11,8)}} = \frac{1}{4.24} = 0.24$$

EXERCISES

1. Find the 0.01, 0.05, 0.95 and 0.99 quantile values for $F(5,5)$, $F(10,15)$, and $F(20,20)$. (The numbers in a the parentheses denote the numerator and denominator degrees of freedom, respectively.)   Note what happens to the F values as the quantile values get larger; as the degrees of freedom get larger. You might sketch a rough graph for each F distribution associated with a given pair of degrees of freedom to demonstrate these properties.

2. Suppose the random variable F has an F distribution with numerator and denominator degrees of freedom of 12 and 12 respectively. Find numbers a and b such that:
    (a) $P(a \le F \le b) = 0.90$ (b) $P(a \le F \le b) = 0.95$

3. Suppose the random variable F has an F distribution with numerator and denominator degrees of freedom of 9 and 11 respectively. Find numbers a and b such that:
    (a) $P(a \le F \le b) = 0.90$ (b) $P(a \le F \le b) = 0.95$

4. Let X_1, ..., X_n be independent, normally distributed random variables having mean μ_x and variance σ_x^2. Let Y_1, ..., Y_m be independent normally distributed random variables having mean μ_y and variance σ_y^2

    (a) What is the distribution of $U = \dfrac{(n-1)s_x^2}{\sigma_x^2}$?   What is the distribution of $V = \dfrac{(m-1)s_y^2}{\sigma_y^2}$?

    (b) Provide an argument that the ratio, $\dfrac{s_x^2/s_y^2}{\sigma_x^2/\sigma_y^2}$ follows the F distribution with $v_1 = n-1$ and $v_2 = m-1$, respectively.

5. Under the assumptions of exercise 4, how does the expression simplify if the random variables come from the same normal population?

6. A machine is supposed to manufacture a metal rod for use in a small gasoline engine with a variance no greater than 0.16 cm2. A quality control inspector samples 8 rods from the process regularly and inspects the diameters and computes the variances of the measurements taken. The last two samples produced variances of 0.27 and 0.36 cm2. What is the likelihood of these sample variances occurring under the assumption that the true variance is 0.16 cm2? Is it reasonable to conclude that the two sample variances came from a population with the same variance?

7. The distribution of scores on one test in a mathematical statistics course had a standard deviation of 10.1. The scores on the second test had a standard deviation of 20.3. The means of the two test where about the same. If we assume the scores for both tests are normally distributed, how likely is it that the standard deviations would be as different as they are assuming the variability of test scores is the same for both tests? (Hint: Don't look at the difference of test scores, but rather look at the ratio of the scores to compare them.)

8. Provide an argument that justifies Theorem 6.12 of this section. Do this in the context of the description of exercise 4.

9. Suppose we have a random sample of 4 observations from a standard normal distribution. Compute $U = \dfrac{Z_1^2 + Z_2^2}{2}$ and $V = \dfrac{Z_3^2 + Z_4^2}{2}$. What is the distribution of the numerators of U and V? Are U and V independent.

10. Using the information from exercise 9, what is the distribution of $W = \dfrac{U}{V}$?

6.6   THE NORMAL APPROXIMATION TO BINOMIAL PROBABILITIES

Introduction

As we studied the binomial distribution in section 4.2 of Chapter 4, we were constrained to a fairly limited table of binomial probabilities. However, in section 4.3 we found that for large n and small p, the Poisson distribution could be used to approximate binomial probabilities, expanding our ability to obtain binomial probabilities. But many practical binomial problems still remain out of reach without extensive computations.

In this section we extend the coverage of the binomial distribution by using a normal distribution approximation to the binomial probabilities. This approximation is justified under the assumption of a large sample size and p close to 0.5.

THEOREM

THEOREM 6.13: If the random variable X is a binomial random variable with values n and p then

$$P(a \leq X \leq b) \doteq P(\frac{a - np}{\sqrt{np(1 - p)}} \leq Z \leq \frac{b - np}{\sqrt{np(1 - p)}})$$

where Z is a standard normal random variable. The approximation is best when n is large and p is close to 0.5.

PROOF: Regard the random variable X as the sum of n independent and identically distributed Bernoulli random variables with parameter p; that is, $X = X_1 + X_2 + \ldots + X_n$ and $E(X_i) = p$, and $V(X_i) = p(1 - p)$. By the central limit theorem, X will be approximately normally distributed with mean $\mu = np$ and variance $\sigma^2 = np(1 - p)$. Therefore, $\dfrac{X - np}{\sqrt{np(1 - p)}}$ will be approximately standard normal.

DISCUSSION

Suppose an opinion poll is conducted in a small city with a random sample of 400 registered voters being selected. The gender of the respondent is recorded for further information and analysis. Suppose we want to calculate the probability that the sample results between 190 and 210 males in the sample. This problem can be viewed as a binomial distribution problem with $n = 400$ and $p = 0.5$, assuming there is a 50:50 split in the sexes in the community. Each respondent is either male or female with probability of 0.5, and random sampling guarantees the independence and identical distribution of the 400 random variables X_i where

$$X_i = \begin{cases} 0 & \text{respondent is female} \\ 1 & \text{respondent is male} \end{cases}$$

The random variable $X = X_1 + X_2 + \ldots + X_{400}$ is therefore, the sum of 400 independent and identically distributed Bernoulli random variables with $p = 0.5$. As such, X is binomially distributed with

$$\mu = 400(0.5) = 200$$

and

$$\sigma = \sqrt{400 \bullet 0.5(1 - 0.5)} = 10$$

We want to calculate $P(190 \leq X \leq 210)$ but Table A doesn't accommodate a sample of size 400. Therefore, we propose to approximate this probability by applying the central limit theorem as follows:

$$P(190 \leq X \leq 210) \doteq P\left(\frac{190 - np}{\sqrt{np(1-p)}} \leq Z \leq \frac{210 - np}{\sqrt{np(1-\ )}} \right)$$

$$= P\left(\frac{190 - 200}{\sqrt{400 \bullet (0.5)(1 - 0.5)}} \leq Z \leq \frac{210 - 200}{\sqrt{400 \bullet (0.5)(1 - 0.5)}} \right)$$

$$= P(-1 \leq Z \leq 1) = 0.6826.$$

So there is about a two-thirds chance of getting between 190 and 210 males in the sample.

This solution to the problem is obtained by appealing to the central limit theorem, and would provide a fairly good approximation since n is large ($n = 400$), and $p = 0.5$ which produces a symmetric distribution for the binomial distribution. There are techniques to improve the approximation using what is called a "continuity correction" in other books. But generally I have found that applying it is more bother than it is worth in most cases where I have considered using. Therefore, it is not presented here, though it can be found in other texts.

As a comparison of the accuracy of this approximation consider a couple of special cases shown in Table 6.6.

Table 6.6 Comparison of actual binomial probabilities for various values of n and p with the normal approximation to the binomial..

n	p	$\mu = np$	$\sigma = \sqrt{np(1-p)}$	Probability Expressions $P(\mu - \sigma \leq X \leq \mu + \sigma)$	$P(\mu - 2\sigma \leq X \leq \mu + 2\sigma)$
10	0.3	3	$\sqrt{2.1}$	0.7004	0.9244
	0.4	4	$\sqrt{2.4}$	0.6665	0.9817
	0.5	5	$\sqrt{2.5}$	0.6563	0.9785
20	0.3	6	$\sqrt{4.2}$	0.7796	0.8916
	0.4	8	$\sqrt{4.8}$	0.7469	0.9630
	0.5	10	$\sqrt{5.0}$	0.7368	0.9586
50	0.3	15	$\sqrt{10.5}$	0.7204	0.9567
	0.4	20	$\sqrt{12.0}$	0.6877	0.9406
	0.5	25	$\sqrt{12.5}$	0.6778	0.9672
100	0.3	30	$\sqrt{21}$	0.6740	0.9626
	0.4	40	$\sqrt{24}$	0.6416	0.9481
	0.5	50	$\sqrt{25}$	0.7288	0.9648
Normal	Approximation			0.6826	0.9544

As you look at Table 6.6 the normal probabilities for the expressions given are fairly close in all cases. You would expect the probabilities to be closer in general for the more symmetric cases and also for larger value of n. The accuracy will obviously be better or worse for different parts of the distribution depending on the section of the curve considered and also the amount of skewness over that section.

This characteristic can be understood by looking at Figure 6.17 which shows a symmetric, bell-shaped normal distribution superimposed over a slightly left-skewed binomial distribution with $n = 20$ and $p = 0.2$ so that $\mu = np = 4$, and $\sigma = \sqrt{np(1-p)} = \sqrt{3.2}$. By examining this curve we can see the normal distribution would slightly over-estimate probabilities (areas) to the left of $X = 2$ and would under-estimate probabilities (areas) over the right side of the distribution. The amount it would over-estimate or underestimate depends upon the amount of skewness the binomial distribution has and how it would relate to the symmetry of the normal distribution super-imposed on it.

EXAMPLES

EXAMPLE 1: Suppose X is binomial with $n = 70$ and $p = 0.2$. What is the mean and variance of the normal distribution that would approximate this distribution?

SOLUTION: The mean of the and variance to be used respectively are $\mu = 70(0.2) = 14.0$ and $\sigma^2 = 70(0.2)(1 - 0.2) = 11.20$.

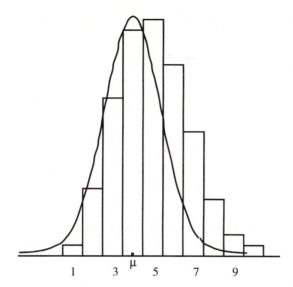

Figure 6.17 A normal distribution $N(4,3.2)$ super-imposed on a binomial distribution ($n = 20$ and $p = 0.4$) as an approximating distribution to the binomial distribution.

EXAMPLE 2: Given the binomial distribution of Example 6.1, what is the approximate probability of getting an observation between 10 and 15 inclusive?

SOLUTION: We approximate the probability by using a normal distribution with $\mu = 14$ and $\sigma = \sqrt{11.2} = 3.35$, and find the area under the normal curve between 10 and 15. Therefore, we get

$$P(10 \leq X \leq 15) = P\left(\frac{10 - 14}{3.35} \leq Z \leq \frac{15 - 14}{3.35}\right) = P(-1.19 \leq Z \leq 0.30) = 0.6179 - 0.1170$$
$$= 0.5009$$

The exact binomial probability is $P(10 \leq X \leq 15) = 0.4863$. The approximation is not very far off and would be acceptable under many applications. The approximation is especially reassuring when we recognize that the binomial distribution would be fairly skewed for $p = 0.2$.

EXERCISES

1. Find the mean and variance for the binomial random variable X with value of n and p of:.
    (a) $n = 50, p = 0.15$ (b) $n = 50, p = 0.85$
    (c) $n = 500, p = 0.15$ (c) $n = 500, p = 0.85$

2. Use the binomial formula to calculate $P(X = 4)$ for $n = 8$, $p = 0.5$; $n = 20$, $p = 0.2$ and $n = 40$ and $p = 0.1$.

3. As you look at Figure 6.17 above, notice that the rectangle for $X = 4$ starts at 3.5 and ends at 4.5. Therefore to approximate this area using a normal curve we might calculate the area between 3.5 and 4.5. Notice in exercise 2 that $\mu = np = 4$ for all cases. Approximate the same three

probabilities in exercise 2 by calculating $P(3.5 \leq X \leq 4.5)$ under the corresponding normal distribution. Remember that σ will be different for each case since n and p changes.

4. Examine your results for exercises 2 and 3 and make a general conclusion as to what it suggests.

5. For $n = 100$, $p = 0.5$ find the probability that X will take on the values, $49 \leq X \leq 51$, using the binomial formula and then approximate it with the normal approximation to the binomial.

6. Repeat the normal approximation calculation in exercise 5 by computing the probability over the range: 48.5 to 51.5. Has the approximation improved?

7. For $n = 400$, $p = 0.5$ find the probability that X will take on the values, $199 \leq X \leq 201$, using the binomial formula and then approximate it with the normal approximation to the binomial.

8. Repeat the normal approximation calculation in exercise 7 by computing the probability over the range: 198.5 to 201.5. Has the approximation improved? Has it improved as much as it did for exercise 6.

9. Give a graphical explanation as to why the adjustments in exercises 6 and 8 improved the approximation of the binomial probabilities.

10. Considering the results of exercises 6, 8 and 9, over what range would the normal approximation probability be computed if the binomial probability of interest was:

(a) $P(48 \leq X < 55)$?   (b) $P(48 < X < 55)$?
(c) $P(48 \leq X \leq 55)$?   (d) $P(a \leq X < b)$?

6.7   A BRIEF LOOK AT ORDER STATISTICS

Introduction

One class of sample statistics that are often used are the *order statistics*. Order statistics come up naturally whenever we work with data that are ordinal measurements. Methods of analyzing "ranked" data have been developed using the theory of order statistics. These procedures represent a valuable addition in statistical literature and methodology under the heading of *non-parametric* or *distribution-free* methods.

However, the theoretical development is more difficult than we want to present in a text at this level, so our presentation is admittedly brief. We introduce order statistics here only with enough discussion that you might recognize what the phrase means if you encounter it in a different setting. Chapter 10 contains two sections 10.2 and 10.3 that use this principle in testing hypotheses. (Read Chapter 8 on testing hypotheses before reading Chapter 10.) These techniques represent a very powerful set of statistical tools.

Definitions

DEFINITION 6.4: Let $X_1, ..., X_n$ be a random sample from some pdf $f(x)$. If the random variables are arranged according to their size from small to large, the resulting ordered set of random variables are called the *order statistics*. We use the notation

$$X_{(1)} \leq X_{(2)} \leq ... \leq X_{(n)}$$

i.e., $X_{(1)}$ is the smallest, $X_{(2)}$, is the next smallest, ..., and $X_{(n)}$ is the largest of the random variables.

THEOREM 6.14: For large m, the sampling distribution for the median for random samples of size $2m + 1$ is approximately normal with the mean equal to the median $X_{0.5}$ and variance $\dfrac{1}{8[f(X_{0.5})]^2 m}$. Here, $X_{0.5}$ denotes the population median which has the property

$$0.5 = \int_{-\infty}^{X_{0.5}} f(x)dx.$$

DISCUSSION

One of the first steps in analyzing a set of data is to order the data from smallest to largest. This makes it easy to find the largest and smallest observations in the set. In addition, we may want to find the quantiles and quantile related values such as the median, interquartile range, etc. Or maybe we are looking at a set of test scores where the grades are assigned by referencing the the highest score in the class. Maybe the range of the data is one of the most important characteristics that we want to examine. In all these cases, we are referring to statistics that are related to the *order statistics*. That is, we must order the data from small to large in order to pick off the values of interest.

The theory of order statistics has been developed and the pdf's of any of the order statistics can be found under reasonable assumptions as long as the pdf of the parent population is known. Much of the methodology of distribution-free/nonparametric statistics depends upon the properties of the order statistics.

To distinguish the notation of order statistics from the random variables themselves, we use the following:

Random Sample	*Ordered Sample*
$X_1, ..., X_n$	$X_{(1)} \leq X_{(2)} \leq ... \leq X_{(n)}$

The order statistics of greatest interest, and therefore those whose pdf's would be sought are:

Sample median	$X_{(m)}$ when $n = 2m + 1$
Sample minimum	$X_{(1)}$
Sample maximum	$X_{(n)}$
Range	$X_{(n)} - X_{(1)}$.

Since the median of a sample might often be used as an alternative to the mean as a measure of location, the following large sample property of the median is given. Suppose we have some probability density function $f(x)$ from which a random sample of n observations is selected. The sample can be represented by the random variables $X_1, ..., X_n$. Suppose that the sample size $n = 2m + 1$ and we find the median of the sample which is then denoted by $X_{(m)}$. If we repeat this sampling process again and again, finding the median $X_{(m)}$ each time, we can visualize the creation of a distribution of sample medians.

Theorem 6.14 above asserts that the mean of the sample medians $E(X_{(m)})$ equals the median of $f(x)$, which we denote by $X_{.5}$, and the variance of the sample median is:

$$V(X_{(m)}) = \frac{1}{8[f(X_{0.5})]^2 m}$$

In addition, the distribution will be approximately normally distributed, with the approximation improving as the size of the sample n gets large.

EXAMPLES

EXAMPLE 1: Suppose a random sample of $n = 2m + 1$ is selected from a normal distribution with mean μ and variance σ^2, i.e., X_i is $N(\mu,\sigma^2)$. What is the large sample distribution of the median of the sample?

SOLUTION: By the application of Theorem 6.14, knowing the distribution is normal, the median of $f(x)$ is μ because of symmetry of the normal distribution. Thus,

$$E(X_{(m)}) = X_{0.5} = \mu.$$

The variance of the mean is

$$V(X_{(m)}) = \frac{1}{8[f(X_{0.5})]^2 m} = \frac{1}{8[f(\mu)]^2 m}.$$

Now $f(\mu)$ for the normal distribution is:

$$f(u) = \frac{1}{\sigma\sqrt{2\pi}} e^{-\frac{[\mu - \mu]^2}{2\sigma^2}} = \frac{1}{\sigma\sqrt{2\pi}}.$$

so that

$$(f(\mu))^2 = \frac{1}{2\pi\sigma^2}.$$

Therefore,

$$V(X_{(m)}) = \frac{1}{8[f(\mu)]^2 m} = \frac{1}{8m(\frac{1}{2\pi\sigma^2})} = \frac{\pi\sigma^2}{4m}.$$

EXAMPLE 2: Suppose a random sample of $n = 2m + 1$ is selected from a normal distribution as described in Example 1. What is the sampling distribution of the mean of the n observations?

SOLUTION: By the results of section 6.2 of this chapter, the distribution of the mean of

$n = 2m + 1$ normally distributed random variables is normally distributed with a mean μ and variance $\dfrac{\sigma^2}{n} = \dfrac{\sigma^2}{2m + 1}$. Comparing the variance of the mean $\dfrac{\sigma^2}{2m + 1}$ with the variance of the median $\dfrac{\pi\sigma^2}{4m}$ for large samples we can see the variance of the mean is smaller than the variance of the median for all values of $m > 0$.

EXAMPLE 3: Suppose a large sample of $2m + 1$ is selected from a normally distributed population. What is the probability that the sample mean with fall within $k\sigma$ of the population mean? What is the approximate probability that the sample median will fall within $k\sigma$ of the population mean?

SOLUTION: Using the results of Example 2 for the sample mean we get:

$$P(\mu - k\sigma \leq \bar{X} \leq \mu + k\sigma) = P\left(\frac{(\mu - k\sigma) - \mu}{\sigma/\sqrt{2m + 1}} \leq \frac{\bar{X} - \mu}{\sigma/\sqrt{2m + 1}} \leq \frac{(\mu + k\sigma) - \mu}{\sigma/\sqrt{2m + 1}} \right)$$

$$= P(-k\sqrt{2m + 1} \leq Z \leq k\sqrt{2m + 1})$$

Using the results of Example 1 for the sample median we get:

$$P(\mu - k\sigma \leq X_{0.5} \leq \mu + k\sigma) = P\left(\frac{(\mu - k\sigma) - \mu}{\sigma\sqrt{\pi}/\sqrt{4m}} \leq \frac{X_{0.5} - \mu}{\sigma\sqrt{\pi}/\sqrt{4m}} \leq \frac{(\mu + k\sigma) - \mu}{\sigma\sqrt{\pi}/\sqrt{4m}} \right)$$

$$= P\left(\frac{-2k\sqrt{m}}{\sqrt{\pi}} \leq Z \leq \frac{2k\sqrt{m}}{\sqrt{\pi}} \right)$$

This latter probability will be smaller than the probability above for the mean. Thus, there is a higher probability that the mean will be within a certain specified distance of the true mean than there is for the median. The sample median is less reliable than the sample mean.

EXERCISES

1. Using the results of Examples 1 and 2 suppose a random sample of 17 observations is selected from a normal distribution with a mean of 50 and a standard deviation of 10. What is the size of the variance of the sample mean compared to the size the variance of the sample median?

2. Using the results of Example 3, what is the probability that the sample mean will be within 1/2 of a standard deviation of the true mean μ if the sample came from a normally distributed population with a mean of 50 and a standard deviation of 10?

3. Use the problem description of exercise 2 to calculate the probability that the sample median with fall within 1/2 of a standard deviation of 50. Compare this probability with that obtained in exercise 2.

4. Work exercises 2 and 3 if the sample size is 25. If the sample size is 37.

5. Suppose a random sample of $n = 2m + 1$ is selected from a uniform distribution with $a = 0$ and $b = 10$. What is the large sample distribution of the median of the sample? [Hint: Follow the procedure of Example 1 substituting the uniform distribution in place of the normal distribution where appropriate.]

6. Suppose a random sample of $n = 2m + 1$ is selected from a uniform distribution as described in exercise 5. What is a large sample sampling distribution of the mean of the n observations? [Hint: Apply the results of the central limit theorem to this situation.]

7. Develop an expression similar to that obtained in Example 3 for the uniform distribution described in exercises 5 and 6.

8. Evaluate the expression developed in exercise 7 for $k = 1$ and $m = 9$.

6.8   ANALYSIS OF THE CHAPTER EXERCISE

The assembly time of a certain part consists of two steps each of which is exponentially distributed with $\theta = 5$ minutes. What is the distribution of total assembly time? What is the probability that the total assembly time will exceed 15 minutes?

If the assembly time of a certain part consists of 20 steps, each of which is exponentially distributed with $\theta = 5$, what is the approximate probability that the total assembly time will exceed 120 minutes? What is the probability that the total assembly time would be less than 90 minutes?

SOLUTION: Let the random variable X_1 denote the assembly time of the first step and X_2 denote the assembly time for the second step. Therefore, the total assembly time is $Y = X_1 + X_2$. From results obtained earlier, we found that the sum of two exponential random variables with parameter θ would be gamma with parameters $\alpha = 2$ and θ. Thus the distribution of total assembly time would be distributed as a gamma random variable with $\alpha = 2$ and $\theta = 5$. We can apply previous results to obtain

$$P(Y > 15) = 1 - F_G(15) = 1 - (1 - F_P(1; \lambda = \frac{15}{5}) = F_P(1; \lambda = 3)$$

By consulting the Poisson table we get the probability: $P(Y > 15) = 0.1991$.

If the assembly consists of 20 steps, by the same reasoning as above the distribution of the total assembly time, $Y = X_1 + \ldots + X_{20}$, would be Gamma distributed with $\alpha = 20$ and $\theta = 5$. To evaluate the probability $P(Y > 120)$ or $P(Y < 90)$ would involve using a Poisson with $\lambda = 24$ and $\lambda = 18$. However, Y is a sum of independent random variables with mean of 5 and variance of 25. The central limit theorem can be used to provide an approximate probability. In this case $E(Y) = n\mu = 20 \cdot 5 = 100$ and $V(Y) = 20 \cdot 25 = 500$. Therefore, $\dfrac{Y - 100}{\sqrt{20(25)}}$ will be approximately normally distributed with mean 0 and variance 1—the standard normal distribution. Therefore,

$$P(Y > 120) \approx P\left(Z > \frac{120 - 100}{5\sqrt{20}}\right) = 1 - F(0.89) = 0.1867$$

Similarly,

$$P(Y < 90) \approx P(Z < \frac{90 - 100}{5\sqrt{20}}) = P(Z < -0.45) = 0.3264.$$

CHAPTER SUMMARY

In this chapter our purpose has been to introduce the concept of a sampling distribution of a sample statistic–the distribution that results from theoretically sampling a population (pdf) repeatedly using the same sample size, computing the same sample statistic each time and identifying the distribution that the sample statistic follows. We started out by assuming a single population that was normally distributed with mean μ and standard deviation σ from which a random sample of size n was selected. Under this assumption we have the results shown in Table 6.7.

Table 6.7 Summary of sample statistics and their distribution: one sample case.

Sample Statistic	Distribution and Degrees of Freedom
$\dfrac{(\bar{X} - \mu)}{\sigma/\sqrt{n}}$	Standard Normal
$\dfrac{(\bar{X} - \mu)}{s/\sqrt{n}}$	t $v = n - 1$ degrees of freedom (df)
$\dfrac{(n-1)s^2}{\sigma^2}$	Chi-square $v = n - 1$ degrees of freedom

Next, we had two normally distributed populations with means and variances of μ_x and μ_y, and σ_x^2 and σ_y^2, respectively. A random sample of size n was selected from the population denoted by the X and an independent random sample of size m was selected from the Y population. Under this sampling assumption we have the results shown in Table 6.8.

Of major importance to this chapter, and to the discipline of statistics and probability in general, is the central limit theorem. The importance of this theorem cannot be over-emphasized. This theorem assumes that a population has a finite mean and variance from which a random sample of size n is selected. Then if n is large, a sample statistic based on the sum of the sample observations will have a sampling distribution that is approximately normally distributed. This sample statistic can then be standardized so that approximate probabilities can be obtained from the standard normal tables.

For instance if $Y = a_1X_1 + a_2X_2 + \ldots + a_nX_n$, then

$$Z = \frac{Y - E(Y)}{\sqrt{V(Y)}}$$

will be approximately standard normal in distribution, and values from the standard normal table with provide approximate probabilities for probability questions about Y. Specifically we looked at $Y = \bar{X}$ and $Y = \Sigma\, X_i$ so that $Z = \dfrac{(\bar{X}-\mu)}{\sigma/\sqrt{n}}$ and $Z = \dfrac{(\Sigma X_i - n\mu)}{\sigma\sqrt{n}}$ would both be approximately standard normal by the central limit theorem.

Table 6.8 Summary of sample statistics and their distribution: two independent samples.

Sample Statistic	Distribution
$\dfrac{(\bar{X}-\bar{Y})-(\mu_x-\mu_y)}{\sqrt{\left(\dfrac{\sigma_x^2}{n}+\dfrac{\sigma_y^2}{m}\right)}}$	Standard Normal
$\dfrac{(\bar{X}-\bar{Y})-(\mu_x-\mu_y)}{\sqrt{s_p^2(\dfrac{1}{n}+\dfrac{1}{m})}}$ where $s_p^2 = \dfrac{(n-1)s_x^2 + (m-1)s_y^2}{(n-1)+(m-1)}$	t $\nu = n+m-2$ if $\sigma_x^2 = \sigma_y^2 = \sigma^2$
$\dfrac{s_x^2}{\sigma_x^2}\Big/\dfrac{s_y^2}{\sigma_y^2}$	F with $\nu_1 = n-1$ and $\nu_2 = m-1$

In this chapter we also looked at this theorem specifically applied to a set of X's from a Bernoulli population and used the normal distribution as an approximation to the resulting binomial probabilities.

These concepts now become essential to the results of Chapters 7-9 that follow. It would be well for you to review this section sufficiently to understand their implications before going on to the next chapter.

CHAPTER EXERCISES

1. Two independent samples of size n and m respectively are to be selected from the same normally distributed population. If μ denotes the theoretical mean of the populations and σ^2 denotes the theoretical variance, show that $E(\overline{X} - \overline{Y}) = 0$, and $V(\overline{X} - \overline{Y}) = \sigma^2(\frac{1}{n} + \frac{1}{m})$.

2. Using the results of exercise 1, construct a standard score that would follow the standard normal distribution. Appeal to various theorems, as appropriate, to justify your result.

3. Two different groups of students taking the same mathematical statistics course from the same instructor giving the same tests show the following test results on the first exam.

	Group 1	Group 2
Mean	73.7	77.6
Standard Deviation	19.1	13.4
Sample Size	95	28

    Is the difference in variances large enough to conclude it is not attributable to chance? What assumptions do you make to justify your approach?

4. Is the difference in means in exercise 3 large enough to conclude it is not due to chance? In working this problem, explicitly state the assumptions you make to complete the problem.

5. Find the expected value of $s_p^2 = \dfrac{(n-1)s_x^2 + (m-1)s_y^2}{(n-1) + (m-1)}$ assuming the sample variances were computed using two independent samples from the same, normally distributed population.

6. Suppose a population consists of only 6 elements: {3, 4, 5, 5, 6, 7}. For this small population, what is the its mean and variance?

7. Using the population described in exercise 6, list all possible samples of size 2 that could be selected. (There should be 36 of them.) Compute the mean for each of these 36 possible samples and make the "sampling distribution of the mean" by filling out the probabilities in the following table. (Fill in the numerators of the fractions in the second row of the following table. Use the denominator of 36 for all probability calculations.)

$\overline{x}$	3	3.5	4	4.5	5	5.5	6	6.5	7
$f(\overline{x})$	$\dfrac{}{36}$	$\dfrac{}{36}$	$\dfrac{}{36}$	$\dfrac{}{36}$	$\dfrac{}{36}$	$\dfrac{}{36}$	$\dfrac{}{36}$	$\dfrac{}{36}$	$\dfrac{}{36}$

8. Using the results from exercises 6 and 7, find $E(\overline{X})$ and $V(\overline{X})$ and therefore, verify that $E(\overline{X}) = \mu$, and $V(\overline{X}) = \dfrac{\sigma^2}{n}$ where μ and σ^2 where computed in exercise 6.

9. Using a calculator, or preferably a computer, generate a sample of 5 random numbers from the

unit uniform. Repeat this operation 100 times and construct a histogram of the 100 means. Does this histogram look bell-shaped?

10. Using the data generated in exercise 9, compute the mean of the 100 sample means and then compute the variance of the 100 sample means. Do these numbers agree with what you would expect when sampling a unit uniform population? (What is $E(\overline{X})$ and $V(\overline{X})$ when the unit uniform is sampled?)

6.9   APPLICATION: STATISTICAL QUALITY CONTROL

Objectives

Statistics can be thought of as an application of mathematical and probability concepts. You may have felt this was a strongly theoretical course. But the development of statistics as a discipline was an outgrowth of the need to account for and explain the variation present in observational data.

In this text we have tried to point you towards applications of the concepts as they have been presented to you. But we have never devoted any extensive space to some of the broader applications of statistics. That was in part due to having insufficient background to present some of the more elegant and important applications. However, that constraint no longer applies and in this and the next section we introduce two major areas of application, though that presentation is only to tantalize and not to satisfy your appetite. The topics presented here and in the application sections of the next few chapters generally justify a complete textbook devoted to them in order to adequately give them their due.

In this section we present some information on statistical quality control—an area of application that business is finding of vital importance to their general quality and productivity programs.

The second section deals with survey sampling and some of its concepts. This is an area of application with which most of us are familiar as newspapers and magazines report on the outcomes of election polls, opinion polls, etc.

Introduction

In any manufacturing or production process variation will be present. The producer, to remain competitive, has to be concerned about the quality of the product being produced. To insure a high quality of product an inspection program of some type is usually implemented. If it is impractical to perform a 100% inspection of items then a "sampling inspection" program should be introduced.

If the samples are drawn using scientific methods of sampling–random sampling procedures–then our knowledge of chance variation can be used to monitor the production process. We can then determine if changes in the production process are taking place that can't be attributed to chance. A simple, but very useful method used to monitor production is the quality control chart or Shewhart chart.

Interest in the procedures of quality control were strongly motivated by the need for quality supplies in support of the war effort of World War II. The federal government set up standards which manufacturers were expected to meet in order to keep quality high. These standards were developed by statisticians and mathematicians called together to develop them, recognizing the presence of variation in any manufacturing process. These standards were probability based and often assumed normality of measurements associated with the characteristics being examined. Walter Shewhart, the developer of the control chart was among those aiding this effort.

We introduce the concept of a control chart for a mean or standard deviation (range) here as one illustration of the important application of statistics and probability.

Control Charts

Suppose we are monitoring the manufacturing process of some product. Each item that is produced has certain characteristics that are crucial to its correct use, i.e., the right diameter, length, electrical resistance, etc. Statistically speaking we would be concerned whether the correct mean is being produced as well as whether the process has a small variance.

One way to monitor the process mean and process variability is to construct what are called *x*-bar charts and *s*-charts, charts that monitor the product mean and the product standard deviation. To create such charts we assume that we take regular random samples from the production process throughout the day, day after day, week after week. We then plot the sample means and standard deviations on separate charts to enable us to observe the presence of any non-random changes or shifts in process average or variation that might have occurred.

Creating Control Charts and Looking for Non-Random Influences

Assume that the characteristic of the process being monitored is normally distributed with a known mean and standard deviation. (The mean and standard deviation, μ and σ, would be obtained by collecting enough data over time to feel that the characteristics of the process are known. It may also be that the characteristic of the process has to have a given value which can be set by calibrating the process that manufacturers it.) At regular intervals during the manufacturing day, a sample of *n* items are selected (*n* remains constant throughout the day for this presentation) and the mean of the sample measurements is computed.

1. To create a control chart as in Figure 6.18 below, we place time on the horizontal axis with parallel lines to the horizontal being established at the process mean μ, and also at $\mu - 3\dfrac{\sigma}{\sqrt{n}}$ and $\mu + 3\dfrac{\sigma}{\sqrt{n}}$. We call these two lines the "lower control limit" and the "upper control limit." Under normality assumptions, the probability of a mean falling above or below these lines is very small. The combined probability is 0.0026. (See Table C for the standard normal distribution.)

2. Plot the sample mean obtained at each time period connecting them by straight lines as in Figure 6.19.

3. Evidences of non-random influences would be associated with points that (a) fall outside the 3 standard error limits, (b) produce "runs" of excessive length or "runs" that are very short. (A run is defined as a sequence of points on the chart that have a common characteristic; such as a sequence of points that are all below μ, or a sequence of points where the next point is above the last point, and the next is above it. The length of a run is the number of points until the pattern is broken.) Runs above the center line or below, or runs up or runs down may be examined. (Charts and tables for checking on runs are available to determine what would be unreasonable under an assumption of chance influences. However, we will not go into the study of runs extensively.)

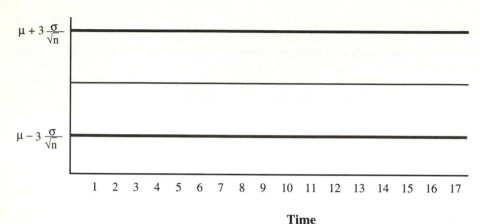

Figure 6.18 Layout for an x-bar chart.

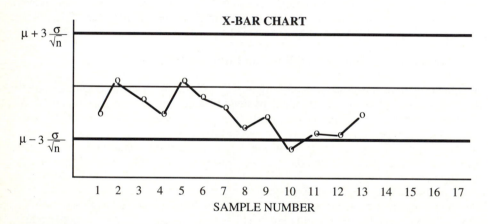

Figure 6.19 A typical x-bar chart.

As we examine Figure 6.19 several characteristics can be observed that should alert the inspector that something unusual may be happening. The first indication is that the mean of sample number 9 is below the lower control limit indicating that a very "unlikely" event has occurred. In addition, starting with sample number 6 and continuing through sample number 12, seven means can be noted that are below the mean line. This constitutes a *run* of length seven. Very simply, the probability of a sample mean being below μ is 0.5, and by independence of sampling throughout the day, the probability of a string of 7 means being below the mean is $(0.5)^7 = 0.00781$. Other "runs" that could be examined are a run "up" of length 1 from sample number 1 to 2, a run "down" of length 2 from sample number 2 to 4, a run up of length 1 from 3 to 4, a run down of length 2 from 5 to 7, up 7 to 8, down 8 to 9 , up, down, and up. Nothing here raises a question.

There are many more sophisticated procedures that have been developed over the years since the development of the Shewhart chart. But a more extensive presentation will not be given here. Rather, see one of the many books on the market such as Duncan, A. J. (1974), *Quality Control and Industrial Statistics*, 4th edition, Richard D. Irwin, Homewood, Ill.

EXAMPLE

EXAMPLE 1: In manufacturing children's tricycles a metal sleeve is manufactured to receive an axle connecting two of the wheels. See Figure 6.20. This sleeve has a diameter of 0.75 inches and the manufacturing process has a standard deviation of 0.01 inches. Every 20 minutes a sample of 5 sleeves is taken and the diameters are measured. Set up a control chart and plot it for 30 time periods. Identify any characteristics that would indicate a process change. (A portion of the data, for 8 of the 30 periods is given below in Table 6.9)

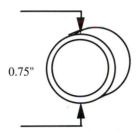

0.75"

Figure 6.20 Diagram of an axle sleeve with $\mu = 0.75$ inches and $\sigma = 0.01$ inches.

Table 6.9 Diameter data and the means of 5 randomly selected axle sleeves for 8 of 30 time periods.

Time Period	\multicolumn{6}{c}{Observation Number}
	1	2	3	4	5	Mean
1	0.7526	0.7519	0.7485	0.7513	0.7471	0.7503
2	0.7442	0.7400	0.7634	0.7426	0.7364	0.7453
3	0.7613	0.7484	0.7539	0.7548	0.7569	0.7550
4	0.7632	0.7608	0.7542	0.7429	0.7393	0.7520
5	0.7467	0.7641	0.7460	0.7364	0.7483	0.7483
6	0.7361	0.7511	0.7486	0.7432	0.7488	0.7455
7	0.7335	0.7457	0.7402	0.7667	0.7509	0.7474
8	0.7632	0.7578	0.7432	0.7632	0.7294	0.7513
...	...	...	...	...	...	...

SOLUTION: A 3-standard deviation control chart is plotted as Figure 6.21 below with the lower control limit (LCL) and upper control limit (UCL) computed as follows:

$$LCL = 0.75 - 3 \cdot \frac{(0.01)}{\sqrt{5}} = 0.737$$

$$UCL = 0.75 + 3 \cdot \frac{(0.01)}{\sqrt{5}} = 0.763$$

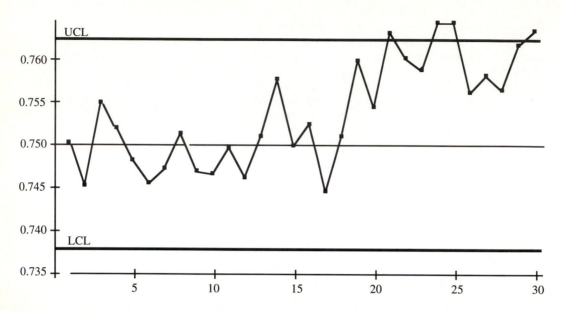

Figure 6.21 Control chart of the means for data of Example 1.

Notice first that the mean for the 21st time period is above the upper control limit, with 3 other means following it that are "out of control." This clearly signals a need to check the process to see if some identifiable cause can be found. In addition all of the observations from time period 20 on are above the mean, another indication of a rare event, requiring that the process be examined. (Is this where a change in shift was made? Did a bolt come loose on the machinery? What other potential causes for the apparent change in the process are there?)

EXAMPLE 9.2 Examining the control chart of Figure 6.21, what is the longest run length "down"? what is the longest run "up"?

SOLUTION: Scanning across the chart, from periods 3 to 6 a run down is observed of length 3. The longest run "up" is of length 2 which occurs several places, the first goes from period 6 to 8. Neither of these would seem to concern us though this is a subjective judgment at this point. (Probability tables for runs are available in other texts; see Duncan previously cited.)

EXERCISES

1. Suppose a ball bearing sleeve is to be mass produced such that the inside diameter is to be 70 millimeters. Shipments are to be sampled to assess whether the shipment meets the required average. Concern centers on whether the process average may really have shifted upward. The contract with the buyer stipulates that an acceptance sampling plan that will accept shipments that are 74 millimeters only 5% of the time will be agreeable to them (the consumer's risk). Management also has determined that they want to reject acceptable sleeves only 2.5% of the time as well (the producer's risk). What is the size of sample needed to meet these criteria if the process standard deviation is 1.5 millimeters? (See Chapter 9, Tests of Hypothesis. Hint: Determine the rejection region boundary value which will be between 70 and 74 having the appropriate areas of 0.05 and 0.025. The boundary value when $\mu = 70$ will be in terms of an

unknown n; the boundary value when $\mu = 74$ will also be in terms of an unknown n. Set these two boundary value equations equal to each other and solve for n.)

2. Based on data from the past 5 years, the manufacturers of bearings know that the mean and standard deviation of the bearing diameters should be 33.3 and 5.65 respectively. The average bearing diameter of 20 samples, each of size 5, is as follows. Construct an X-bar chart and comment on whether the process appears to be in control.

Sample	$\bar{X}$	Sample	$\bar{X}$
1	31.6	11	29.8
2	33.0	12	34.0
3	35.0	13	33.0
4	32.2	14	34.8
5	33.8	15	35.6
6	38.4	16	30.8
7	31.6	17	33.0
8	36.8	18	31.6
9	35.0	19	28.2
10	34.0	20	33.8

3. A new company has entered the market to compete with ACME Ruler Company. The initial production of 2500 rulers suggest that $\mu = 6.00$ inches and $\sigma = 0.04$ inches. A regular inspection takes place through time to see if standards are being met. You are presented with the following sample data from the most recent 10 consecutive batches ($n = 3$). Construct an $\bar{X}$ chart. Does the process seem to be in control?

BATCH

1	2	3	4	5	6	7	8	9	10
5.92	6.00	5.98	5.99	5.99	6.00	6.00	5.96	5.96	5.94
5.98	6.02	5.96	6.03	6.00	5.99	5.97	5.98	5.99	5.95
5.97	5.98	5.98	5.98	6.01	5.99	6.01	5.97	5.98	5.94

4. Using your knowledge of the binomial distribution, construct a control chart for the fraction defective for a transistor production line. Twenty samples, each of size 100, were taken. Assume that the samples are numbered in the sequence of production and that the true proportion is 35%.

Sample	No. of defects	Sample	No. of defects
1	44	11	36
2	48	12	52
3	32	13	35
4	50	14	41
5	29	15	42
6	31	16	30
7	46	17	46
8	52	18	38
9	44	19	26
10	48	20	30

5. The following data pertains to the overall lengths of a fragmentation bomb base manufactured during the war by the American Stove Company. Construct an X-bar chart using the mean and standard deviation of 0.8340 and 0.007, respectively.

Sample	Mean	Sample	Mean
1	0.8372	11	0.8380
2	0.8324	12	0.8322
3	0.8318	13	0.8356
4	0.8344	14	0.8322
5	0.8346	15	0.8304
6	0.8332	16	0.8372
7	0.8340	17	0.8282
8	0.8344	18	0.8346
9	0.8308	19	0.8360
10	0.8350	20	0.8374

6. Using the properties of the chi-square distribution such that its mean is v and its variance is $2v$, find the $E\left(\frac{(n-1)s^2}{\sigma^2}\right)$ as well as $V\left(\frac{(n-1)s^2}{\sigma^2}\right)$ and then use those results to find $E(s^2)$ and $V(s^2)$. Remember that $\frac{(n-1)s^2}{\sigma^2}$ is chi-square with $v = n - 1$.

7. Suppose the sample size is large so that s^2 would be approximately normally distributed with the mean and variance given in exercise 6. Write the expressions for such a large-sample upper and lower 3-standard deviation control limit.

8. Suppose the variation of a process is of concern and so the variance of a sample of 15 is taken from a process whose theoretical standard deviation is 16 units. Use the results of exercise 7 to compute a 3-standard deviation upper and lower control limit. Make a control chart to accommodate 10 periods using the above information.

7

ESTIMATING POPULATION PARAMETERS

OBJECTIVES

When you complete this chapter you should understand the motivation and objective for deriving estimates of parameters, specifically by the *method of moments* and *method of maximum likelihood*. For the easier cases mathematically, you should be able to obtain estimates using either or both principles. You should also be able to explain the meaning of certain good properties of estimates such as *unbiasedness*, *efficiency*, *minimum variance*, and *consistency*. (The property of *sufficiency* will be treated very lightly, and you would not be expected to understand it beyond the level of its presentation.)

You should be able to show in certain easily proved cases whether a certain estimate is unbiased or consistent. You should be able to show the general derivation of an *interval estimate* for a mean and the principle involved, and through homework assignments be able to derive other interval estimates as presented in this chapter. Given a set of data, you should be able to compute any interval estimate requested.

INTRODUCTION

The Bureau of the Census regularly provides estimates of the percent of the work force who are unemployed. The State Tax Commission has to present to the state budgetary committee an estimate of revenue that will come in from income, sales and gasoline taxes. A manufacturer is concerned with estimating the reliability of its products. A manufacturer of computer chips estimates the speed of its various processing units.

In all these examples, an estimate is made on the basis of incomplete data. In very few instances is a complete "census" of the population of interest conducted. Rather, a sample from the population is obtained and the estimate is made by analyzing the sample data.

However, such an estimate is a very inadequate number without some information concerning its reliability. That is, we need to have some idea as to how close we think our estimate is to the true population parameter, what we may call a *margin of error*. And since both the estimate and statement about its error are obtained from data from a random sample, it is also important to report how confident we are in the particular sample we've drawn and the inferences made from it.

The purpose of this chapter is to address the rationale for developing such ideas in some of the most common situations and using the most common assumptions.

Where We Are Going in This Chapter

This chapter represents the first formal discussion and development of the inferential process. The setting of the problem is this: We have a set of sample observations from some population having a known pdf $f(x;\omega)$. However, the value of the population parameter, ω, will be unknown to us (ω could represent either a single parameter or a vector of parameters). Our problem is to use the information in the sample to estimate the parameter value(s) in the population, thus completely specifying the population pdf that we have assumed. This process is indicated in Figure 7.1 below by the darker shaded region and constitutes a "statistical inference."

Our objective is to develop a defensible estimator of the population parameter(s) which is the "best" one when compared with others that might be under consideration. The estimators obtained will be called *point estimators* because they produce a single point on the real line as the estimator.

However, a point estimator provides no measure of closeness to the parameter value that we are estimating. Therefore, we then take an additional step to develop an interval which places an upper and lower value around the point estimate.   This specifies the region within which we feel, with reasonable confidence, the true parameter value will be found. This also provides a measure of precision of the point estimate being used. The confidence or probability that we have in the interval thus obtained is called the *confidence level*.

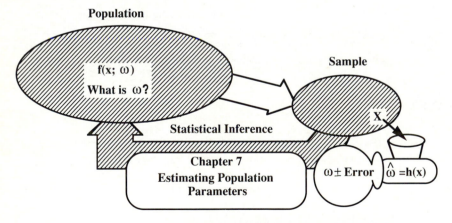

Figure 7.1 SIM—Using sample data to estimate population parameters.

Consider a more specific illustration. Suppose that we have the following sample values from a normally distributed population having a mean of zero but an unknown standard deviation:

$$X_1 = 0.253, \ X_2 = -1.64, \ X_3 = 0.055, \ X_4 = 0.899, \ X_5 = -0.137$$

What should we do with these numbers to estimate the unknown variance in the population? Should we use s^2? $\hat{\sigma}_0^2$? The $\left(\dfrac{\text{Range}}{4}\right)^2$? Which of these is the best estimator? What makes an estimator good? How close is the estimate we come up with to the true variance? Can we predict how close? And so forth.

In this chapter we will address three main concepts: (1) How do you choose an estimator of a parameter? (2) How do you decide which estimator is a good one, and is there ever a "best" one? and, (3) How do you construct an interval estimate with a specific level of confidence using a "good" estimator as determined by your answers to (1) and (2) above?

Chapter Exercise

At a particular university a study was done among the students attending the university regarding non-educational costs: housing, food, utilities, entertainment and so forth.   A random sample of 400 students was selected from the student files who producing the following summarized information. (Figures are monthly figures.)

Item	Mean	Std. Dev.
Rent	$104	$52
Utilities	$ 12	$ 4
Food	$ 68	$25

Estimate the mean amount spent on rent, utilities, and food per month per student along with a "margin of error" and a "level of confidence."   Also estimate the total money entering the community assuming all students at the university (20,000) are living in university or community housing, paying utilities, and making food purchases there as well. (If you are not willing to assume these categories are independent of one another, what additional information is needed?)

7.1   PARAMETER ESTIMATION BY THE METHOD OF MOMENTS

Introduction

In this section we present one of the simplest and intuitive methods for obtaining an estimator of a population parameter. We simply equate sample and theoretical moments to one another and solve for the respective parameters. It is a method that is intuitively appealing and persons often will take such an approach as a natural course when faced with an estimation problem.

In this section, we present the method and its logic and then demonstrate the method for several different common problems.

Definitions

The Method of Moments

DEFINITION 7.1: Let X_1, ..., X_n denote a random sample from some pdf $f(x)$ with k unknown parameters. For this distribution we derive the 1st, 2nd, ..., and kth theoretical moments about the origin denoted by $\mu_1', ..., \mu_k'$. (See Chapter 3, section 3.3.)

We then define the 1st, 2nd, ..., and kth sample moment about the origin in the following way. (See section 1.4.)

$$M_1' = \sum_{i=1}^{n} \frac{x}{n}$$

$$M_2' = \sum_{i=1}^{n} \frac{x^2}{n}$$

$$\dots$$

$$M_K' = \sum_{i=1}^{n} \frac{x^k}{n}$$

We argue that the 1st sample (empirical moment) ought to estimate the first theoretical moment, the 2nd sample moment ought to estimate the 2nd theoretical moment, ..., and the kth sample moment ought to estimate the kth theoretical moment. Since the theoretical moments will typically be functions of the parameters in the distribution, we set up the k equations:

$$M_1' = \mu_1'$$
$$M_2' = \mu_2'$$
$$\dots$$
$$M_K' = \mu_K'$$

which will produce a set of k simultaneous equations in k unknowns. Solving for these k unknowns provides a set of estimators for the k unknown parameters of the distribution.

DISCUSSION

Many times I've presented students with a set of data, telling them the data came from a gamma distribution, say, but without specifying the values of α and θ. I then ask them to use the data to estimate the values of these two parameters. After some thought, someone will always argue that we compute the sample mean and variance,$\bar{x}$ and s^2, and equate their values to the theoretical mean and variance for the gamma, $\alpha\theta$ and $\alpha\theta^2$; that is, we write

$$\bar{x} = \alpha\theta$$
$$s^2 = \alpha\theta^2$$

and solve for the values of α and θ. Doing this gives

$$\hat{\theta} = \frac{s^2}{\bar{x}} \qquad\qquad \hat{\alpha} = \frac{\bar{x}^2}{s^2}$$

as solutions. (Note that we use the notation, $\hat{\theta}$ and $\hat{\alpha}$ where the caret (^) reads as *estimator of*. Therefore, $\hat{\theta}$ means "estimator of θ"; $\hat{\alpha}$ means "estimator of α."

This intuitive argument is based on the same principle of logic that is called the method of moments. The only difference is that in the method of moments we set up the equations involving the theoretical and sample moments about the origin. Therefore, in the above example, since $E(X) = \alpha\theta$, and $E(X^2) = \alpha\theta^2 + (\alpha\theta)^2$ and the first two sample moments are $M_1 = \sum_{i=1}^{n} \frac{x}{n}$ and $M_2 = \sum_{i=1}^{n} \frac{x^2}{n}$, we have the two equations

$$\sum_{i=1}^{n} \frac{x}{n} = \alpha\theta,$$

$$\sum_{i=1}^{n} \frac{x^2}{n} = \alpha\theta^2 + (\alpha\theta)^2$$

to solve. The solution of these two equations in the two unknowns gives:

$$\hat{\theta} = \frac{s_o^2}{\bar{x}} \qquad\qquad \hat{\alpha} = \frac{\bar{x}^2}{s_o^2},$$

only a slight difference between this and the preceding results.

(Note: There are two relationships that provide some simplification in the steps above. The first is to recognize that

$$E(X^2) = \sigma^2 + [E(X)]^2$$

which for the gamma gives

$$\alpha\theta^2 + (\alpha\theta)^2.$$

The second is that

$$s_o^2 = \sum_{i=1}^{n} \frac{x^2}{n} - \bar{x}^2.$$

To further illustrate the method of moments, consider the following examples.

EXAMPLES

EXAMPLE 1: Suppose we have a normal distribution with unknown mean and variance. We would like to find estimators for the mean and variance using the method of moments and the data provided by a random sample.

SOLUTION: In this case we have 2 unknown parameters, so $k = 2$. Next, we set up 2 equations in 2 unknowns. Therefore, the two equations in two unknowns are

$$\mu_1' = M_1 \qquad\qquad \text{and} \qquad\qquad \mu_2' = M_2$$

where

$$M_1 = \sum_{i=1}^{n} \frac{x}{n} \qquad\qquad \text{and} \qquad\qquad M_2 = \sum_{i=1}^{n} \frac{x^2}{n}.$$

Solving these two equations simultaneously for μ and σ^2, we get

$$\hat{\mu} = \bar{x}$$

$$\hat{\sigma}^2 = \frac{\sum_{i=1}^{n} (x - \bar{x})^2}{n} = s_o^2$$

Thus, we see that the sample mean is our "method of moments" estimator of the mean, and s_o^2 is our "method of moments" estimator of the variance.

EXAMPLE 2: Find the method of moments estimator of the parameter p in the Bernoulli distribution.

SOLUTION: Suppose our sample consists of n Bernoulli trials, $X_1, X_2, ..., X_n$. Then

$$\mu = E(X) = p$$

and

$$M_1' = \frac{1}{n}\sum_{i=1}^{n} x_i = p_o$$

the sample proportion. The equation is $\mu = p_o$, whose solution is trivial. Therefore p_o, the sample proportion, is the method of moments estimator of p.

EXAMPLE 3: Suppose the following 16 observations were selected randomly from a gamma distribution with $\alpha = 5$ and $\theta = 3$. What are the method of moments estimators of the parameter values already known.

13.7	21.7	10.8	11.7
18.8	38.7	23.6	12.1
20.1	15.4	30.3	23.9
11.0	19.2	18.2	15.9

SOLUTION: From the results section discussion obtained above, we know that the estimators of θ and α respectively are

$$\hat{\theta} = \frac{s_o^2}{\bar{x}} \qquad \hat{\alpha} = \frac{\bar{x}^2}{s_o^2}$$

which in this case turn out to be:

$$\hat{\theta} = \frac{s_o^2}{\bar{x}} = \frac{53.35}{19.06} = 2.80$$

$$\hat{\alpha} = \frac{\bar{x}^2}{s_o^2} = \frac{363.2836}{53.35} = 6.8$$

(These data were actually generated by using a gamma distribution with $\theta = 3$ and $\alpha = 5$.)

EXERCISES

1. Fill in all all the steps of the algebra to find the method of moments estimators of the parameter values of the gamma distribution, α, and θ.

2. Suppose a random variable X comes from a gamma distribution with $\alpha = 5$ and unknown value of θ. What is the form of the estimator for θ? (Note that there is only one unknown parameter and so only one equation is needed to find an estimator of it.)

3. Suppose a random variable X comes from a gamma distribution with unknown α and $\theta = 3$. What is the form of the estimator for α? (Note that there is only one unknown parameter and so only one equation is needed to find an estimator of it.)

4. Use the following data to estimate the unknown parameter for the random variable described in exercise 2.

13.7	21.7	10.8	11.7
18.8	38.7	23.6	12.1
20.1	15.4	30.3	23.9
11.0	19.2	18.2	15.9

5. Find the estimator of the exponential parameter θ. (How many equations would you have to set up for this exercise?)

6. Use the data of exercise 4 to estimate the parameter θ of exercise 5.

7. Suppose a random variable X were normally distributed with a with a mean of 300 and an unknown standard deviation. If a random sample of size n is selected from this population what is the "method of moments" estimator of the unknown variance σ^2?

8. Use the following data to estimate the unknown variance for the normal distribution described in exercise 7.

297.74550	259.22347	337.13862	337.95599	340.33565
297.41501	304.51809	246.99511	257.12093	272.98846

The true parameter value is $\sigma^2 = 30$. What is the error in your estimate?

9. Suppose a random variable Y were Poisson distributed. If a random sample of size n is selected from this population what is the "method of moments" estimator of the unknown parameter λ?

10. The following data are a random sample from a Poisson distribution with $\lambda = 6$. What is the error in your estimate of λ using this data?

5	3	8	8	5
5	5	7	8	6
2	2	3	2	3

11. Suppose a random variable Y were uniformly distributed.   If a random sample of size n is selected from this population what is the "method of moments" estimator of the unknown parameters a and b? Remember that $\mu = \dfrac{a + b}{2}$ and $\sigma^2 = \dfrac{(b - a)^2}{12}$.

12. Suppose a random variable Y were uniform on the interval 0 to b.   If a random sample of size n is selected from this population what is the "method of moments" estimator of the unknown parameter b? Recognize that there is only one parameter in this distribution.

13. The following data represent a random sample from a uniform distribution on 0 to b. Use the results of exercise 12 to estimate b.

6.0338176	0.37314686	1.4791995	0.90570770	2.2293553
8.7742915	9.5173167	7.5421168	0.35704379	0.8350545

   (The value of b used to generate this data was 10. What is your "error of estimate?")

14. Suppose a random variable X were a Bernoulli random variable.   If a random sample of size n is selected from this population what is the "method of moments" estimator of the unknown parameter p.

15. The following data represent a random sample from a Bernoulli population. Use this data and the results from exercise 14 to estimate the unknown parameter p.

0	0	0	1	1	1	0	0	1	0
0	0	0	0	0	0	0	1	0	1

   The value of p used to generate this data is 0.3. What is the error in your estimate?

7.2   PARAMETER ESTIMATION BY THE METHOD OF MAXIMUM LIKELIHOOD

Introduction

The method of maximum likelihood represents a different philosophy about obtaining an estimator of a parameter. In many cases, it provides the same form of an estimator as the method of moments. In other cases, it provides a different estimator. (In the latter situation, we would then have to have some criteria for choosing one estimator over the other. Sections 3 and 4 provide these criteria for us.)

The argument of the method is to choose the value of the parameter that would "most likely" produce the particular sample that was obtained.

Definitions

Method of Maximum Likelihood

DEFINITION 7.2: The *likelihood function*, denoted by $L = L(x_1, ..., x_n; \omega)$ is defined as follows. Let $f(x;\omega)$ denote the probability density function of the random variable X associated with modelling a particular population. Let $X_1, ..., X_n$ represent the observations obtained from a random sample of the population. Then

$$L = L(x_1, ..., x_n; \omega) = f(x_1;\omega)f(x_2;\omega) \bullet \, ... \, \bullet f(x_n;\omega)$$

DEFINITION 7.3: Let $X_1, ..., X_n$ be a random sample from a distribution $f(x;\omega)$, and let $L(x_1, ..., x_n; \omega)$ be the likelihood function of the sample values. If $\hat{\omega} = g(x_1, ..., x_n)$ is the value of ω that maximizes the likelihood function L, then $\hat{\omega} = g(x_1, ..., x_n)$ is the maximum likelihood estimator of ω, and $\hat{\omega}$ is the maximum likelihood estimate.

NOTATION: The use of the caret "^" above is read "estimator of." Thus, $\hat{\omega}$ represents the phrase: "the estimator of ω."

DISCUSSION

Suppose a sample is drawn from a Poisson distribution resulting in the values:

$$X_1 = 2, \qquad X_2 = 1, \quad X_3 = 2, \quad X_4 = 3, \quad X_5 = 1$$

We want to know what value of λ would produce this particular sample with highest probability, or that would have "maximum likelihood" of yielding this particular sample.

Since this is a random sample, (the observations are therefore independent) the likelihood of this sample in particular will be represented by the joint probability of these outcomes which is simply the product of the separate outcomes; that is,

$$f(2,1,2,3,1) = \left(\frac{e^{-\lambda} \lambda^2}{2!} \right)\left(\frac{e^{-\lambda} \lambda^1}{1!} \right)\left(\frac{e^{-\lambda} \lambda^2}{2!} \right)\left(\frac{e^{-\lambda} \lambda^3}{3!} \right)\left(\frac{e^{-\lambda} \lambda^1}{1!} \right)$$

(This is called the likelihood function, L, for this random sample from a Poisson pdf.) By trial and error we can insert various values of λ into the above expression to see which would produce the maximum probability. If we were to restrict our set of choices to the values of 1, 1.5, 2, 2.5 and 3, then the corresponding values of the likelihood function are shown in Table 7.1. Therefore, if we restrict our consideration to only those values of λ listed, the "maximum likelihood" estimate for the parameter λ is 2.

Table 7.1 Values for the likelihood function for various values of λ.

λ	$L(x_1, ..., x_n) = f(2,1,2,3,1)$	
1.0	0.00028	
1.5	0.00089	
2.0	0.00097	*(relative maximum)*
2.5	0.00059	
3.0	0.00025	

However, since the likelihood function is a continuous function of the parameter λ, we can use the principles of maximization from the calculus to find the maximum likelihood estimator. In this case, after simplification, the likelihood function L is

$$L = \frac{e^{-5\lambda}\lambda^{(2+1+2+3+1)}}{(2!1!2!3!1!)}$$

We want to take the derivative of L and set it to zero. However, the maximum of L occurs at the same value as the maximum of its natural log, $\ln L$. Since this will often simplify the mathematics, we first take the natural logarithm of the L, and *then* take its derivative to maximize the function. This produces the following results:

$$\ln L = -5\lambda + (2 + 1 + 2 + 3 + 1)\ln\lambda - \ln(2!1!2!3!1!)$$

$$\frac{d \ln L}{d\lambda} = -5 + \frac{(2 + 1 + 2 + 3 + 1)}{\lambda} - 0 = 0$$

Solving for λ gives us.

$$\hat{\lambda} = \frac{(2 + 1 + 2 + 3 + 1)}{5} = \frac{9}{5} = 1.8$$

Notice that this is equal in the general case to

$$\hat{\lambda} = \frac{\Sigma x}{n} = \bar{x}.$$

This leads to the following methodology:

- Write the likelihood function L which we want to maximize by an appropriate choice of ω.

- If L is a function of ω and the X's, consider maximizing L directly, or take its natural log and maximize it. (The maximum can be found for most cases in this text by setting the derivative of L or its natural log equal to zero; i.e.,

$$\frac{dL}{d\omega} = 0 \ \text{ or } \ \frac{d \ln (L)}{d\omega} = 0.$$

The resulting equation is solved for ω.)

EXAMPLES

EXAMPLE 1: If $x_1, ..., x_n$ are values of a random sample from an exponential distribution, what is the maximum likelihood estimate of the parameter, θ?

SOLUTION: The likelihood function is

$$L(x_1, ..., x_n; \theta) = (\frac{1}{\theta})^n\, e^{-\Sigma x/\theta}$$

Differentiating the natural log of this function and setting to zero, yields

$$\ln(L) = -n \ln \theta - \frac{\Sigma x}{\theta}$$

$$\frac{d[\ln(L)]}{d\theta} = -\frac{n}{\theta} + \frac{\Sigma x}{\theta^2} = 0$$

Therefore,

$$\hat{\theta} = \bar{x}$$

EXAMPLE 2: Let $x_1, ..., x_n$ be values of a random sample from a continuous uniform distribution with $a = 0$ and b. What is the maximum likelihood estimator of the unknown parameter b?

SOLUTION: The process of taking derivatives and setting to zero will not help in this problem since $f(x)$ is not a function of the x's. Therefore, we have to go back to the fundamental concept of maximum likelihood. In this case, the likelihood function is

$$L = (\frac{1}{b})^n.$$

This function increases as b decreases in size. Therefore we want b to be as small as possible to maximize the value of L. However, b cannot be smaller than the largest sample value. (Otherwise, that   value couldn't have occurred in the sampling process.) Therefore, the maximum likelihood estimate of b is

$$\hat{b} = \text{Max}(X_1, ..., X_n) = X_{(n)}$$

To illustrate, consider the following figure, Figure 7.2. Suppose a random sample of 5 observations is selected from $f(x) = \frac{1}{b}$ and are positioned on the x axis accordingly. The value of b can't get any smaller than x_5, the largest of the x's in this case. Thus, the maximum likelihood estimator of b would be the observation, x_5.

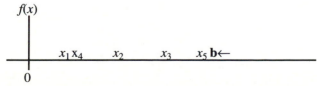

Figure 7.2 Placement of a random sample of $n = 5$ from uniform: $f(x) = \frac{1}{b}$.

EXAMPLE 3: Suppose we have the uniform distribution $f(x) = \dfrac{1}{b-a}$ with two unknown parameters. What are the maximum likelihood estimates of a and b using a random sample of n observations?

SOLUTION: We want to maximize the likelihood function L as before. The likelihood function for this pdf is:

$$L = \left(\frac{1}{b-a}\right)^n$$

Using the argument of example 2, this function will be maximized by making the expression $b - a$ as small as possible. However, $b - a$ can't be any smaller than $\max(x)-\min(x)$.

Therefore, we estimate a and b as follows:

$$\hat{a} = \text{Min}(X_1, ..., X_n) = X_{(1)}.$$

$$\hat{b} = \text{Max}(X_1, ..., X_n) = X_{(n)}.$$

EXAMPLE 4: Suppose X is $N(\mu,\sigma^2)$ with the mean and variance unknown. Suppose a random sample of size n is selected. What are the maximum likelihood estimates of the μ and σ^2?

SOLUTION: The likelihood function for the normal distribution for outcomes $x_1, x_2, ..., x_n$ is

$$L(\mu,\sigma^2) = \prod_{i=1}^{n} \frac{1}{\sqrt{2\pi\sigma^2}} \exp\left[-\frac{(x_i-\mu)^2}{2\sigma^2}\right] = \left(\frac{1}{\sqrt{2\pi\sigma^2}}\right)^n \exp\left[-\frac{\displaystyle\sum_{i=1}^{n}(x_i-\mu)^2}{2\sigma^2}\right]$$

Taking the natural log of L gives

$$\ln L(\mu,\sigma2) = -\frac{n}{2}\ln(2\pi\sigma^2) - \frac{\displaystyle\sum_{i=1}^{n}(x_i-\mu)^2}{2\sigma^2}$$

The partial derivatives with respect to μ and σ^2, when set to zero, are

$$\frac{\partial(\ln L)}{\partial\mu} = \frac{1}{\sigma^2}\sum_{i=1}^{n}(x_i-\mu) = 0$$

and

$$\frac{\partial(\ln L)}{\partial\sigma^2} = \frac{-n}{2\sigma^2} + \frac{1}{2\sigma^2}\sum_{i=1}^{n}(x_i-\mu)^2 = 0$$

The solution to the first equation gives $\hat{\mu} = \bar{x}$ which is inserted into the second equation where μ occurs. Solving the second equation for σ^2 we get

$$\hat{\sigma}^2 = \frac{1}{n}\sum_{i=1}^{n}(x_i - \bar{x})^2$$

which is the same form of the variance that we denoted by s_o^2 in Chapter 1. Therefore, in summary, the maximum likelihood estimates of the mean and variance of a random sample from a normally distributed population are:

$$\hat{\mu} = \bar{x} \quad \text{and} \quad \hat{\sigma}^2 = \frac{1}{n}\sum_{i=1}^{n}(x_i - \bar{x})^2$$

EXERCISES

1. Suppose a random variable X comes from a gamma distribution with $\alpha = 5$ and unknown value of θ. What is the form of the maximum likelihood estimator (MLE) for θ?

2. Use the following data to estimate the unknown parameter for the random variable described in exercise 1.

13.7	21.7	10.8	11.7
18.8	38.7	23.6	12.1
20.1	15.4	30.3	23.9
11.0	19.2	18.2	15.9

3. Find the MLE estimator of the exponential parameter θ.

4. Use the data of exercise 2 to estimate the parameter θ of exercise 3.

5. Suppose a random variable X were normally distributed with a with a mean of 300 and an unknown standard deviation. If a random sample of size n is selected from this population what is the MLE estimator of the unknown variance σ^2?

6. Use the following data to estimate the unknown variance for the normal distribution described in exercise 7.

297.74550	259.22347	337.13862	337.95599	340.33565
297.41501	304.51809	246.99511	257.12093	272.98846

    The true parameter value is $\sigma^2 = 30^2$. What is the error in your estimate?

7. Suppose a random variable X were normally distributed with an unknown mean μ and a standard deviation of 30. What is the MLE estimator of μ?

8. Use the data of exercise 6 to provide an estimate of μ using the results of exercise 7.

9. Suppose a random variable Y were Poisson distributed. If a random sample of size n is selected from this population what is the MLE estimator of the unknown parameter λ?

10. The following data are a random sample from a Poisson distribution with $\lambda = 6$. What is the error in your estimate of λ using this data?

5	3	8	8	5
5	5	7	8	6
2	2	3	2	3

11. The following data represent a random sample from a uniform on 0 to b. Use the results of example 2 to estimate b.

6.0338176	0.37314686	1.4791995	0.90570770	2.2293553
8.7742915	9.5173167	7.5421168	0.35704379	0.8350545

(The value of b used to generate this data was 10. What is your "error of estimate?")

12. Suppose a random variable X were a Bernoulli random variable. If a random sample of size n is selected from this population what is the MLE estimator of the unknown parameter p.

13. The following data represent a random sample from a Bernoulli population. Use this data and the results from exercise 14 to estimate the unknown parameter p.

0	0	0	1	1	1	0	0	1	0
0	0	0	0	0	0	0	1	0	1

The value of p used to generate this data is 0.3. What is the error in your estimate?

7.3   WHAT MAKES A "GOOD" ESTIMATOR

Introduction

From the last section, it can be noted that the maximum likelihood estimator of the parameter b in the uniform distribution is the maximum of the sample values. That is,

$$\hat{b} = \text{Max}(X_1, ..., X_n) = X_{(n)}$$

In addition, it is easy to show that the method of moments estimator is

$$\hat{b} = 2\bar{X}$$

Which of these two estimators should one use? How do you decide which one is better?

Some criteria need to be developed to assess which estimator is best. However, the assessment of which is best cannot be determined on the basis of one experience with it. It is necessary to see how an estimator "performs" in a "repeated sampling" setting.

In this section and section 7.4 that follows, we define several criteria that we would look for in "good" estimators—unbiasedness, efficiency, consistency and sufficiency. Good estimators will typically have one or more of these characteristics. We will define each concept and give examples of their use.

Definitions/Theorems

DEFINITION 7.4: Let $\hat{\omega}$ be any estimator of the unknown parameter ω. The mean square error of $\hat{\omega}$ is the expected value of the squared difference between $\hat{\omega}$ and ω. That is,

$$MSE(\hat{\omega}) = E(\hat{\omega} - \omega)^2 = V(\hat{\omega}) + (E(\hat{\omega}) - \omega)^2.$$

The term $(E(\hat{\omega}) - \omega)$ is called the *bias* of the estimator.

DEFINITION 7.5: An estimator $\hat{\omega}$ is said to unbiased for ω if $E(\hat{\omega}) = \omega$; that is , if the long run average value—in repeated sampling—of the estimator equals the parameter it is to estimate. The bias of an unbiased estimator is zero; that is, $E(\hat{\omega}) - \omega = 0$, obviously.

DEFINITION 7.6: An estimator is said to be a minimum variance unbiased estimators of ω if $E(\hat{\omega}) = \omega$ and its variance is smaller than the variance of any other unbiased estimator of ω.

THEOREM 7.1 (CRAMER-RAO INEQUALITY): Let $X_1, ..., X_n$ be a random sample from the distribution having a pdf $f(x;\omega)$. If $\hat{\omega}$ is an unbiased estimator of ω, then the variance of $\hat{\omega}$ satisfies the following inequality.

$$V(\hat{\omega}) \geq \frac{1}{n[E\{\{\frac{d}{d\omega}\ln f(x;\omega)\}^2\}]}$$

Any estimator that satisfies the equality condition in the above expression is called the minimum variance unbiased (MVU) estimator.

DEFINITION 7.7: Let $\hat{\omega}_1$ and $\hat{\omega}_2$ be two different estimators of the parameter ω. The estimator $\hat{\omega}_1$ will be said to be more efficient than $\hat{\omega}_2$ if $MSE(\hat{\omega}_1) < MSE(\hat{\omega}_2)$. If $\hat{\omega}_1$ and $\hat{\omega}_2$ are both unbiased, this amounts to saying $V(\hat{\omega}_1) < V(\hat{\omega}_2)$. The efficiency of $\hat{\omega}_1$ to $\hat{\omega}_2$ is measured by computing the ratio

$$\frac{MSE(\hat{\omega}_2)}{MSE(\hat{\omega}_1)}$$

DEFINITION 7.8: Let $\hat{\omega}$ be an estimator of a parameter ω and let $\hat{\omega}_1, ..., \hat{\omega}_n$ be a sequence of estimators representing $\hat{\omega}$ based on samples of sizes $1, 2, ..., n$ respectively. Then $\hat{\omega}$ is said to be a consistent estimator for ω if

$$\lim_{n \to \infty} P(| \hat{\omega}_n - \omega | \leq \varepsilon) = 1$$

for all values of ε where $\varepsilon > 0$.

THEOREM 7.2: If an estimator is unbiased for ω, then it will be consistent if its variance approaches zero as n approaches infinity. That is, if

$$\lim_{n \to \infty} V(\hat{\omega}) \to 0$$

DEFINITION 7.9: An estimator $\hat{\omega}$ is a *sufficient* estimator of the population parameter ω if it uses all the information the sample contains about ω. This condition results when the following theorem holds.

THEOREM 7.3: Let $X_1, ..., X_n$ be a random sample from a distribution with a pdf of $f(x;\omega)$. The estimator

$$\hat{\omega} = g(X_1, ..., X_n)$$

is called a sufficient estimator if and only if the likelihood function can be factored as

$$L(x_1, ..., x_n;\omega) = h_1(\hat{\omega};\omega) \cdot h_2(x_1, ..., x_n)$$

for every value of the estimator. The expression $h_2(x_1, ..., x_n)$ does not involve the parameter ω; it is a function of the sample values only and other known constants.

DEFINITION 7.10: The standard error of an estimator $\hat{\omega}$ is the square root of the variance of that estimator. In other words:

$$\text{S.E.}(\hat{\omega}) = \sqrt{V(\hat{\omega})}$$

DISCUSSION

The Mean Square Error

Suppose a population is modelled by the probability density function $f(x;\omega)$, where ω is a single parameter in the population. Next, a sample is drawn from the population of interest and, using the method of moments or maximum likelihood, an estimator $\hat{\omega}$ and an estimate is obtained. This estimate is unlikely to equal the true value of ω since the sample is not likely to represent the population exactly. (However, we would also be disappointed if the estimate differed drastically from the true value of ω.)

Accepting that our estimate is in error to some degree, we would be denote the error, symbolically, by the expression, $|\hat{\omega} - \omega|$. (Note that we take the absolute value of the difference as our error since the estimate could err by either being larger than ω or smaller than ω.) However, as we said before, the size of the error from one trial is insufficient to assess the quality of the estimator. To be fair and objective in our evaluation of our estimator, we want to examine the estimator $\hat{\omega}$ in repeated samples and consider its properties in a long-run-average sense. To do this we square the error and take the *expectation*.   (Taking the expectation implies a "long-run average" concept.) This produces what is called the mean square error (MSE), written as:

$$MSE(\hat{\omega}) = E[(\hat{\omega} - \omega)^2]$$

which after some algebraic manipulation can also be written as:

$$MSE(\hat{\omega}) = V(\hat{\omega}) + |E(\hat{\omega}) - \omega|^2$$

$$= Variance + Bias^2$$

For clarification of these ideas, consider an analogy. Suppose an unknown marksman is to be evaluated on the basis of a shooting exhibition. Observing a single shot, however close it may be to the bulls-eye, will not provide convincing evidence of the skill of the marksman. He might simply make a "lucky shot." Rather, a large number of rounds, shot under identical conditions, should be observed. One characteristic to be considered would be whether the pattern of shots seems to be centered around the center of the target ("unbiased" shooting) or whether the pattern is off-center one way or another. Another characteristic would be to examine how compact the pattern of shots is around the center of the pattern of shots (small variance) versus a large spread or variability in the pattern. (See the Figure 7.3 below.)

Shot Patterns of Marksmen

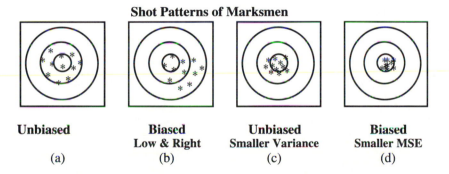

Unbiased	**Biased** Low & Right	**Unbiased** Smaller Variance	**Biased** Smaller MSE
(a)	(b)	(c)	(d)

Figure 7.3 Shot patterns to illustrate the concept of unbiasedness and variability.

Another characteristic of the marksman to consider, would be to see how the marksman shoots under different situations, kneeling, prone, and so forth. A marksman that does well under one set of conditions, may do poorly under another set of conditions, and typically we won't find one "marksman" doing better than all others under *all* conditions. In Figure 7.3 we would classify the marksmen (a) and (c) as "unbiased" marksmen but marksman (c) has less variability in the shot pattern than marksman (a).   Marksman (b) tends to shoot low and to the right and would be called "biased" as would marksman (d). However, marksman (d) has a more compact shot pattern than all the rest and it may be that, all things considered—bias and variability—marksman (d) might be superior compared to the rest.

To relate this analogy to parameter estimation, our "marksman" represents a particular estimator, based on a chosen sample size. We observe how closely the estimator/marksman comes to "hitting" the parameter value of a population. But again, one sample and one estimate would not provide convincing evidence as to how good that estimator is; rather, a large series of samples, using the same sample size, selected under identical conditions, would have to be carried out to assess what kind of properties the estimator has. Does it show unbiased characteristics? Does it have a small variance? How well does it perform when estimating the mean of the normal distribution? Does it do as well when estimating the mean of the exponential?

To answer these questions we need to know the sampling distribution of the estimator under the conditions of sampling from a given population pdf. The functional form of this distribution will allow us to answer the questions above about unbiasedness, variance, and so forth.

Unbiasedness and Efficiency

The properties of unbiasedness and efficiency defined in the Definitions are associated with the concepts of the mean and variance of the sampling distribution of an estimator. We say an estimator is unbiased if its long-run average value—its expected value—equals the parameter value. What this implies is that the value of the estimator in repeated samples averages out to be the value of the parameter; i.e., $E(\hat{\omega}) = \omega$. This is depicted in Figure 7.4. below where the sampling distribution of three unbiased estimators of ω are shown: $\hat{\omega}_1$, $\hat{\omega}_2$, and $\hat{\omega}_3$. (Note that the sampling distribution of the estimators is "centered" around ω. In addition, $MSE(\hat{\omega}) = V(\hat{\omega}) + 0^2 = V(\hat{\omega})$.)

If the three estimators are compared, then ideally we would like to choose the one which tends to be closer to the true parameter value in repeated samples. This is the property of efficiency and uses the mean square error of the estimator to measure it. For example, for the unbiased estimators $\hat{\omega}_1$, $\hat{\omega}_2$, and $\hat{\omega}_3$ in Figure 7.4 we can see that $\hat{\omega}_1$ is the more efficient than $\hat{\omega}_2$ which again is more efficient than $\hat{\omega}_3$ since $MSE(\hat{\omega}_1) < MSE(\hat{\omega}_2) < MSE(\hat{\omega}\quad)$.

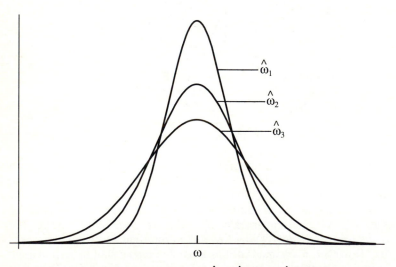

Figure 7.4 Sampling distribution of $\hat{\omega}_1$, $\hat{\omega}_2$ and $\hat{\omega}_3$: unbiased-different variances.

Once we understand that one estimator may be more "efficient" than another—having a smaller mean square error—then we must wonder if there exists an estimator that is more efficient in the given situation, the those already considered. Among the class of unbiased estimators, the Cramer-Rao inequality provides a way to answer that question.

Minimum Variance Unbiased Estimators

The Cramer-Rao inequality states that the variance of any estimator will be as least as large as the expression

$$\frac{1}{n[E\{\{\frac{d}{d\omega}\ln f(x;\omega)\}^2\}]}$$

And therefore, if we have an estimator that equals this "lower bound," then it must have the smallest variance of any unbiased estimators. Such an estimator is called the *minimum variance, unbiased* estimator (MVU). (**NOTE:** The mean square error of an estimator can be re-expressed as the sum of the variance and the square of the bias. See exercise 1. That is,

$$MSE(\hat{\omega}) = E[(\hat{\omega}-\omega)^2] = V(\hat{\omega}) + [(E(\hat{\omega}) -\omega)]^2.$$

Therefore, for an unbiased estimator, the mean square error equals the variance since the bias is zero.) Let's now consider a few examples to demonstrate these ideas more specifically.

EXAMPLES

EXAMPLE 1: Show that the sample mean from any population is an unbiased estimator of the population mean, μ. Compute the MSE for the sample mean.

SOLUTION: We have previously shown that $E(\bar{X}) = \mu$. Therefore, we can conclude that the sample mean is unbiased. Since the sample mean is unbiased the MSE is equal to the variance. Thus, it is true that

$$MSE(\bar{X}) = V(\bar{X}) = \frac{\sigma^2}{n}.$$

[Note: The standard error of the sample mean is: $S.E.(\bar{X}) = \sqrt{\frac{\sigma^2}{n}} = \frac{\sigma}{\sqrt{n}}.]$

EXAMPLE 2: A random sample of size 3 is selected from a normal distribution with mean of μ and standard deviation σ. Consider the two estimators $\hat{\omega}_1$ and $\hat{\omega}_2$, where $\hat{\omega}_1 = \frac{(X_1 + X_2 + X_3)}{3}$ and $\hat{\omega}_2 = \frac{(X_1 + 2X_2 + X_3)}{4}$. Determine whether these estimators are unbiased and also compare their efficiency.

SOLUTION: Estimator $\hat{\omega}_1$ is simply the sample mean which we have shown to be unbiased and has variance $\frac{\sigma^2}{3}$. Estimator $\hat{\omega}_2$ has expectation:

$$E(\hat{\omega}_2) = E[\frac{(X_1 + 2X_2 + X_3)}{4}] = \frac{[\mu + 2\mu + \mu]}{4} = \mu$$

and is also unbiased. Its variance is

$$V(\hat{\omega}_2) = V[\frac{(X_1 + 2X_2 + X_3)}{4}] = \frac{[\sigma^2 + 2^2\sigma^2 + \sigma^2]}{4^2} = \frac{3}{8}\sigma^2$$

The efficiency of $\hat{\omega}_1$ to $\hat{\omega}_2$ is the ratio of the variance of $\hat{\omega}_2$ to $\hat{\omega}_1$. That is,

$$\frac{[\frac{3\sigma^2}{8}]}{[\frac{\sigma^2}{3}]} = \frac{9}{8}.$$

Thus, $\hat{\omega}_1$ is slightly more efficient than $\hat{\omega}_2$.

EXAMPLE 3: Determine whether the method of moments/maximum likelihood estimator of the Bernoulli parameter p, $p_0 = \frac{\Sigma x}{n}$ is a minimum variance unbiased (MVU) estimator of p.

SOLUTION: From Section 2, the sample proportion was found to be both the method of moments and maximum likelihood estimator. It is easy to show that this estimator is unbiased as well. (See example 1.) The variance of this estimator is $\frac{p(1-p)}{n}$. To determine whether it is MVU, we apply the Cramer-Rao expression in the following steps. First, the pdf is $f(x;p) = p^x(1-p)^{1-x}$. Taking the natural log, we get

$$\ln f(x;p) = x\ln p + (1-x)\ln(1-p).$$

Next, we take the derivative of this expression, which yields

$$\frac{d}{dp}\ln f(x;p) = \frac{x}{p} - \frac{(1-x)}{(1-p)} = \frac{(x-p)}{p(1-p)}$$

We now square this result, take its expectation and multiply the expectation by n. Doing so yields

$$n E\{\frac{d}{dp}\ln f(x;p)\}^2 = n E\{\frac{(x-p)}{p(1-p)}\}^2 = \frac{n[p(1-p)]}{\{p(1-p)\}^2} = \frac{n}{p(1-p)}.$$

[Note that $E(x-p)^2 = p(1-p)$ for a Bernoulli random variable by the definition of the variance.]

Taking the reciprocal yields $\frac{p(1-p)}{n}$ as the lower bound for the variance of p_0.

However, as has been stated, $V(p_0) = \frac{p(1-p)}{n}$. Therefore, p_0 is the MVU estimator of the Bernoulli parameter p since the variance of the estimator equals the value of the lower bound.

Consistency

The principle of consistency has an intuitively logical base. An estimator is consistent if it gets closer and closer to the parameter value with probability approaching 1 as the sample size gets larger. This is a property we would want any estimator to have. Even biased estimators can have this property if the bias gets smaller as the sample size gets larger. We can express this idea with the following probability expression. First let ε represent an arbitrarily small constant and we create the interval

$\omega - \varepsilon$ to $\omega + \varepsilon$. Next, let $\hat{\omega}_n$ represent an estimator of ω based on a sample of size *n*. We form the probability expression: $P(\omega - \varepsilon \leq \hat{\omega}_n \leq \omega + \varepsilon)$. If this probability approaches 1 for increasing n then we say the estimator is consistent.

For most of the distributions discussed in this text and the estimators that we have developed from the method of moments or method of maximum likelihood, this property would hold.

The concept of consistency is displayed in Figure 7.5 below. Diagram (a) depicts the sampling distribution of a biased estimator, where $E(\hat{\omega}) \neq \omega$. Diagram (b) depicts the principle of a consistent, unbiased estimator. Notice that the proportion of area between $\omega - \varepsilon$ and $\omega + \varepsilon$ is larger for the sampling distribution of $\hat{\omega}_n$ compared to any other sampling distribution .

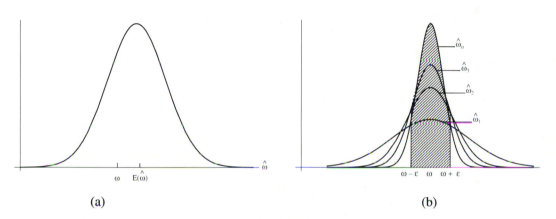

(a)	(b)

Figure 7.5 Sampling distributions of an estimator of ω: (a) Sampling distribution of biased estimator; (b) Sampling distribution of an unbiased and consistent estimator.

For unbiased estimators, it is fairly easy to prove consistency. All we have to do is show that in the limit as n approaches infinity, the variance of the estimator approaches zero. We write this symbolically by

$$\lim_{n \to \infty} V(\hat{\omega}) \to 0 .$$

Sufficiency

The principle of sufficiency, though less intuitively obvious from its definition, insures that the estimator we choose to estimate a particular parameter will not throw away any useful information that the sample provides about the parameter. For example suppose we have a normal distribution with an unknown mean and variance of 25. If we randomly select an observation from this normal distribution we may get the value: $x_1 = 68$. This provides us a little insight, more that we had before, as to the possible value of the mean. Suppose we select two more observations producing the numbers $x_2 = 65$ and $x_3 = 71$. We now have a better feeling for the possible value of the mean μ. Each additional sample value we obtain provides more and more "information" about the value of μ.

In fact, the complete set of x's, $x_1, x_2, ..., x_n$ can be called a "sufficient statistic" since they possess all the information the sample provides about the value of μ.

If we order these x's from small to large, denoted as $x_{(1)}, ..., x_{(n)}$, we still have, in the ordered set, all the information the x's provide about μ. Thus, the "order statistics" are also sufficient.

However, suppose we throw away all the x's except for the maximum value $x_{(n)}$. Does that one number contain the same information about μ as the entire set of x's did? Most of us would agree that it does not. However, suppose we form the mean $\bar{x}$ and then throw away all the x's. Have we lost any information about μ? In this case, it is possible to show that the one number $\bar{x}$ possesses as much information about the population mean μ from a normal distribution as the individual x's provide. Thus, the sample mean $\bar{x}$ is a "sufficient" statistic for the normal parameter μ.

In general then, we can regard each element in the sample as carrying a certain amount of information about the parameters of the population from which it came. As we form an estimator, it will most likely take the set of sample observations and summarize those n numbers into one number. This summary number will be a sufficient statistic if no information about the population parameter to be estimated is lost in the compression of the n numbers into one. [The definition of sufficiency utilizes the concepts of conditional distributions, but finding sufficient estimators using this definition is quite involved and we do not present it here. Suffice it to say that Theorem 7.3 given above can be proved to provide us a valid method of determining sufficiency.]

Sufficient statistics have some useful relationships with maximum likelihood estimators which are stated as follows:

• The maximum likelihood estimates will be functions of sufficient statistics.

• If an unbiased estimator for ω is a sufficient estimator, it will have smaller variance than any other unbiased estimator that is not based on the sufficient estimator for ω.

Now consider a few examples to apply these concepts.

EXAMPLES

EXAMPLE 4: Show that the sample mean is a consistent estimator of the population mean.

SOLUTION: We have shown previously that the sample mean is unbiased for μ since $E(\bar{x}) = \mu$. In addition, we know that $V(\bar{x}) = \dfrac{\sigma^2}{n}$ which approaches zero as n approaches infinity. Therefore, the sample mean is consistent when estimating the population mean.

EXAMPLE 5: Find a sufficient estimator of the Poisson parameter λ in the Poisson distribution.

SOLUTION: Assume we have a simple random sample from a Poisson distribution with observations denoted by $X_1, ..., X_n$. The likelihood function is factored as shown below.

$$L = \frac{e^{-n\lambda}\lambda^{\Sigma x}}{x_1!...x_n!} = \frac{e^{-n\lambda}\lambda^{\frac{n\Sigma x}{n}}}{x_1!...x_n!} = \frac{e^{-n\lambda}\lambda^{n\bar{x}}}{x_1!...x_n!}$$

$$\left(e^{-n\lambda}\lambda^{n\bar{x}}\right)\left[\frac{1}{x_1!...x_n!}\right]$$

Thus, it can be seen that the sample mean is a sufficient estimator of the parameter λ. It is also unbiased and consistent since

$$E(\bar{x}) = \mu = \lambda$$

and

$$V(\bar{x}) = \frac{\sigma^2}{n} = \frac{\lambda}{n},$$

which goes to zero as n goes to infinity. (It is possible to show that it is minimum variance-unbiased using the Cramér-Rao inequality, but the above property of sufficient estimators also establishes this fact.)

EXAMPLE 6: Show that the sample mean is a sufficient estimator of the Bernoulli parameter p.

SOLUTION: The Bernoulli pdf is $f(x;p) = px(1 - p)1 - x$, so the likelihood function, L is:

$$L = p^{\Sigma x}(1 - p)^{n - \Sigma x} = \left(p^{n\bar{x}}(1 - p)^{n(1 - \bar{x})}\right) \cdot (1)$$

Therefore, by the factorization theorem, the sample mean is sufficient.

EXERCISES

1. Suppose that $\hat{\omega}$ is an estimator of some parameter ω. Show that

    $$MSE(\hat{\omega}) = V(\hat{\omega}) + (E(\hat{\omega}) - \omega)^2$$

    where the second term in the right-hand side of this equation is the square of the "bias." (Hint: Start with the expression $MSE(\hat{\omega}) = E[(\hat{\omega} - E(\hat{\omega})) + (E(\hat{\omega}) - \omega)]^2$. Square the terms in the brackets and take the expectation.)

2. Show that the sample mean is always an unbiased estimator of the population mean, regardless of the population pdf.

3. Show that the sample variance s^2, is an unbiased estimator of the population variance σ^2, regardless of the population pdf for an infinitely large population. (And therefore, prove that s_o^2 is a biased estimator.) [Hint: Use the result that $\frac{(n-1)}{n} s^2 = \sum_{i=1}^{n} \frac{X^2}{n} - \bar{X}^2$. Then use the fact that $E(X^2) = \sigma^2 + \mu^2$ and $E(\bar{X}^2) = \frac{\sigma^2}{n} + \mu^2$.]

4. Prove that the sample mean $\bar{X}$ is an unbiased estimator for p in the Bernoulli distribution and find its mean square error. (Why can we call the sample mean the *sample proportion* when working with Bernoulli random variables?)

5. Show that the sample mean $\bar{X}$ is the MVU estimator of the parameter p in the Bernoulli distribution.

6. Show that the sample mean $\bar{X}$ is a consistent and sufficient estimator for the Bernoulli parameter, p.

7. Prove that the sample mean $\bar{X}$ is an unbiased estimator for θ in the exponential distribution and find its mean square error.

8. Show that the sample mean $\bar{X}$ is the MVU estimator of the parameter θ in the exponential distribution.

9. Show that the sample mean $\bar{X}$ is a consistent and sufficient estimator of the exponential parameter, θ.

10. Show that the sample mean $\bar{X}$ is unbiased and the MVU estimator of the parameter λ in the Poisson distribution.

11. Find a sufficient statistic for the normal distribution parameter, μ, assuming a constant variance. (To simplify the likelihood function expression, let the variance equal 1.)

12. Suppose we have a pdf $f(x;\mu)$ and we want to estimate the mean μ. A sample of size 3 is selected and the estimator $\hat{\mu} = \dfrac{X_1 + 2X_2 + X_3}{3}$ is proposed as the estimator of the parameter μ. Is this estimator unbiased? What is its mean square error?

13. Suppose a random sample of size n is selected from a binomial distribution with parameters n_0 and p. Show the sample mean is unbiased and sufficient for the parameter p.

14. What is the standard error for the estimator of the Bernoulli parameter p?

15. What is the standard error for the estimator of the Poisson parameter λ?

16. What is the standard error for the estimator of the exponential parameter θ?

7.4   INTRODUCTION TO INTERVAL ESTIMATORS

Introduction

The federal government regularly collects and publishes production totals for various farm products in the United States in order to administer the federal farm programs. The sources of these estimates are various agricultural surveys conducted under the direction of the United States Department of Agriculture. These estimates are, logically, subject to error. Some of this error can be estimated, surprisingly enough, using the data collected in the sample. This error is referred to as "sampling error."   Thus, instead of simply reporting that the total production of wheat was 20 million tons, it might be reported as being 20 million tons plus/minus $\frac{1}{2}$ million tons.

To see how this example relates to the preceding material of this chapter, we take the following perspective. In sections 7.1–7.3 we presented the methods for developing and evaluating estimators of parameters of a particular pdf. To accomplish this we assumed that a given population was modelled by some pdf of our choice. (For instance, maybe we assume total farm production per acre is normally distributed.) We then want to estimate some meaningful characteristic of the population such as the mean or variance. However, up to this point, we have developed only what would be called point estimates of the parameters. A point estimator provides no information about how closely our estimate comes to the true value. To have a truly meaningful estimating procedure, we need to have some measure of the error in our estimation. Such a measure we will call a the "margin of error."

One value, already discussed, that would give us an idea of the closeness of our estimator to the parameter value is the Mean square errorof our estimator. (Refer to Section 3 of this chapter.) Recall that the mean square error is related to the variance of our estimator in the following way:

$$MSE(\hat{\omega}) \;=\; E(\hat{\omega} - \omega)^2 \;=\; V(\hat{\omega}) \;+\; (E(\hat{\omega}) - \omega)^2$$

where $\hat{\omega}$ is a point estimate of the parameter ω; $V(\hat{\omega})$ is the variance of the estimator; and "$E(\hat{\omega}) - \omega$" is the bias of the estimator. Obviously, if the estimator is unbiased, the MSE equals the variance which then provides one measure of accuracy that we sought.

The objective of this section is to provide the methodology for measuring the "error of estimation" or of obtaining interval estimates. We will do this for most of the common parameters of interest assuming a normally distributed population. The principles can be extended to other distributions with appropriate modifications.

Methodology

The Methodology of Interval Estimation

1.   Select a logical estimator; typically one that is sufficient, unbiased, consistent, efficient, and so forth.

2.   Determine the sampling distribution of the above estimator such that we can find the distribution of a "pivotal quantity" $PQ(\hat{\omega} ; \omega)$ which is a random variable as an expression that has the following properties:

    (1)   It is a function of the estimator $\hat{\omega}$, and the unknown parameter ω plus other known constants, and

    (2)   The random variable PQ follows a known distribution.

3.   Using the distribution of the pivotal quantity found in the previous step, we set up a probability statement having the following form

$$P(W_{\alpha/2} < PQ(\hat{\omega} ; \omega) < W_{1-\alpha/2}) = 1 - \alpha.$$

where W_α is the α quantile of the sampling distribution associated with the statistic $PQ(\hat{\omega} ;\omega)$.

4. The expression inside the parentheses is manipulated algebraically until the following form is obtained which provides the basic interval desired.

$$P(\mathbf{ll}(\hat{\omega})\ <\ \omega\ <\ \mathbf{ul}(\hat{\omega})) = 1 - \alpha$$

where $\mathbf{ll}(\hat{\omega})$ is the lower limit of our confidence interval as a function of the statistic $\hat{\omega}$, and $\mathbf{ul}(\hat{\omega})$ is the upper limit as a function of the statistic $\hat{\omega}$.

DISCUSSION

Suppose we have a population modelled by some probability density function $f(x;\omega)$ where ω is a parameter or vector of parameters in the population whose value(s) we would like to know. To find an estimator of the parameter we use the method of moments or method of maximum likelihood to determine a point estimate of the parameter; we then draw a sample of size n from the population and compute an estimate $\hat{\omega}$ of the parameter ω. However, we anticipate that there will be some error in our estimate due to the failure of the particular sample we obtained to "mirror" the population exactly. We can denote the error as before by writing

$$\text{Error} = |\,\hat{\omega} - \omega\,|$$

It would be very desirable to be able to place a "bound" on the maximum "error of estimation" that we would get from following the preceding procedure. That is, we would hope to be able to say that Max (Error) $\leq \varepsilon$ with some probability close to one. If we can succeed in finding such a probability and value ε then we could call ε a "bound on the error of estimation." (It is also called the "margin of error.")

Our intuition would suggest that the size of the margin of error would be related to the variability of the population σ^2 and the size of the sample n we draw from the population. (We might also argue that the error would be a function of $V(\hat{\omega})$, the variance the estimator shows in repeated sampling for the size of the sample we have selected.)

Consider a more specific case. Suppose we are interested in a population where the parameter of interest is the mean μ of the population. A "method of moments" estimator of the population mean is the sample mean $\overline{X}$. We know that $E(\overline{X}) = \mu$ and $V(\overline{X}) = \dfrac{\sigma^2}{n}$. We also know that the sample mean will be approximately normally distributed by the central limit theorem. Therefore, we can write the following probability expression:

$$P\left(\mu - 3\sqrt{V(\overline{X})} \leq \overline{X} \leq \mu + 3\sqrt{V(\overline{X})}\right) = P\left(\mu - 3\sqrt{\frac{\sigma^2}{n}} \leq \overline{X} \leq \mu + 3\sqrt{\frac{\sigma^2}{n}}\right) \doteq 1$$

In other words, with practical certainty, we would expect the sample mean to be within 3 standard deviations of the true mean, a conclusion which is justified by the central limit theorem if n is large. If the sample mean is within 3 standard deviations of the true mean, then obviously the true mean will be within 3 standard deviations of the sample mean. Therefore, this bound on the error or "margin of error" for this example can be written as

$$\varepsilon = 3\sqrt{\frac{\sigma^2}{n}}.$$

This error term has been obtained through a central limit theorem argument and is approximately true for large *n*. We would like to obtain an exact expression if theoretical considerations can justify it. The methodology given in the five steps above allow us to develop, as a general approach, such exact expressions when the necessary assumptions are met.

When we have finished the steps described above in the *methodology* we will have what is commonly called an interval estimate. This interval estimate has three identifiable components: (1) a *point estimate* of the parameter; (2) a *margin of error*; and (3) a *confidence level* or *confidence coefficient* which often is expressed as a percentage. The point estimate is $\hat{\omega}$. The "margin of error" is the absolute value of the distance from the point estimate to the upper or lower limit of the confidence interval. For estimates involving means this distance is the same regardless of whether it is measured from the lower limit or the upper limit due to the symmetry usually present in the distributions used. For estimates of the variance we will find that the distances are different which poses a problem of what value to call the margin of error. In this case, rather than wrestle with this problem, we will simply publish the upper and lower limit without explicitly naming the margin of error. The confidence level/confidence coefficient is $1 - \alpha$. If expressed as a percentage, it is $100(1 - \alpha)\%$.

For example, suppose we have applied these principles to a problem where we want to estimate the average deduction on Federal tax returns. If we find that to be $2500 \pm $200, in which we express a 95% confidence, then the point estimate is $2500, the margin of error is $200 and the level of confidence is 95%.

The sections that follow this introduction examine specific cases where the methodology described here is applied.

7.5   AN INTERVAL ESTIMATOR OF THE MEAN OF A NORMAL DISTRIBUTION

Introduction

 In this section we assume that we are sampling a normally distributed population with the intent of estimating the population mean. The simplest case, which we consider first, is to assume that the population variance is known and therefore, the only unknown parameter is the mean.   Once the methodology for this procedure is understood, we will discuss how to estimate the mean when the population variance is unknown.

We repeat that our purpose is not only the estimation of the parameter but to provide a measure of confidence in our estimation by providing a *confidence interval*, a *margin of error* along with a statement of our confidence, called the *confidence coefficient*. Later sections will treat the cases involving variances, the comparison of two populations means or variances, and so forth.

Methodology

Case 1: Normal Distribution, Known Variance

Background Assumptions: Let $X_1, ..., X_n$ be a random sample from a normal distribution having an unknown mean, μ, but known variance, σ^2. We want to provide a confidence interval for the unknown mean μ.

Steps (See section 7.4):

1. Under a normal distribution assumption, a point estimator of the unknown mean μ is the sample mean $\overline{X}$. We know the sample mean is both the "method of moments" and "maximum likelihood" estimator of the population mean μ; it is minimum-variance unbiased, consistent, and sufficient. By the assumption of normality, the sample mean has a known sampling distribution which is normally distributed with mean, μ, and variance, $\dfrac{\sigma^2}{n}$.

2. Therefore, we can construct the *pivotal quantity*

$$\frac{(\overline{X} - \mu)}{\sigma/\sqrt{n}}$$

    which is distributed as the standard normal distribution.

3. The probability statement is

$$P(Z_{\alpha/2} < \frac{(\overline{X} - \mu)}{\sigma/\sqrt{n}} < Z_{1-\alpha/2}) = 1 - \alpha$$

or

$$P(-Z_{1-\alpha/2} < \frac{(\overline{X} - \mu)}{\sigma/\sqrt{n}} < Z_{1-\alpha/2}) = 1 - \alpha.$$

4. Working with the expression in parentheses, we want to isolate the parameter μ in the middle. First we multiply through by the denominator $\dfrac{\sigma}{\sqrt{n}}$. This yields

$$P(-Z_{1-\alpha/2}\frac{\sigma}{\sqrt{n}} < (\overline{X} - \mu) < Z_{1-\alpha/2}\frac{\sigma}{\sqrt{n}}) = 1 - \alpha$$

    where the term, $Z_{1-\alpha/2}\dfrac{\sigma}{\sqrt{n}}$ is the "margin of error." Next we subtract the sample mean and multiply through by a minus 1 to produce the expression.

$$P(\bar{X} + Z_{1-\alpha/2}\frac{\sigma}{\sqrt{n}} > \mu > \bar{X} - Z_{1-\alpha/2}\frac{\sigma}{\sqrt{n}}) = 1 - \alpha$$

By examination of this expression it is easy to see that the upper and lower limits of the confidence interval can be written as:

$$\bar{x} \pm Z_{1-\alpha/2}\frac{\sigma}{\sqrt{n}}$$

The point estimate is the sample mean $\bar{x}$, the margin of error is $Z_{1-\alpha/2}\dfrac{\sigma}{\sqrt{n}}$, and the confidence coefficient is $1 - \alpha$.

Case 2: Normal Distribution, Unknown Variance

Background Assumptions: Let $X_1, ..., X_n$ be a random sample from a normal distribution having an unknown mean, μ, and an unknown variance, σ^2. We desire to estimate and provide a confidence interval for the unknown mean, μ.

Steps:

1.   Use the sample mean to estimate the unknown mean, μ. Under the assumptions, the sample mean has a known sampling distribution which is normally distributed   with mean, μ, and variance, $\dfrac{\sigma^2}{n}$. However the variance is unknown which prevents us from using a standardized quantity as our pivotal quantity.

2.   The pivotal quantity we use is

$$\frac{(\bar{X} - \mu)}{s/\sqrt{n}}$$

which, by the assumptions, follows a t-distribution with $n - 1$ degrees of freedom.

3.   The probability statement is

$$P(t_{\alpha/2} < \frac{(\bar{X} - \mu)}{s/\sqrt{n}} < t_{1-\alpha/2}) = 1 - \alpha$$

or

$$P(-t_{1-\alpha/2} < \frac{(\bar{X} - \mu)}{s/\sqrt{n}} < t_{1-\alpha/2}) = 1 - \alpha$$

Working with the expression in parentheses, we isolate the parameter μ in the middle. We use the same algebraic approach as in *case 1* to finally obtain the expression

$$P(\ \bar{X} + t_{1-\alpha/2}\frac{s}{\sqrt{n}}\ >\ \mu\ >\ \bar{X} - t_{1-\alpha/2}\frac{s}{\sqrt{n}}\ ) = 1 - \alpha$$

4. The confidence interval expression obtained is:

$$\bar{x}\ \pm t_{1-\alpha/2}\frac{s}{\sqrt{n}}$$

The point estimate is the sample mean $\bar{x}$, the margin of error is $t_{1-\alpha/2}\dfrac{s}{\sqrt{n}}$, and the confidence coefficient is $1 - \alpha$.

DISCUSSION

Suppose SAT scores are assumed to be normally distributed. A high school principal wants to estimate the true high school mean SAT score using a sample of scores randomly selected from the group of students taking the test. The formulas in this section can be used, depending upon whether σ is known or unknown. From the above, the expressions to use are:

$$\bar{x}\ \pm Z_{1-\alpha/2}\frac{\sigma}{\sqrt{n}}\ \text{(if σ is known)} \qquad \text{and} \qquad \bar{x}\ \pm\ t_{1-\alpha/2}\frac{s}{\sqrt{n}}\ \text{(if σ is unknown)}.$$

Looking at these expressions, the confidence interval on the mean can be seen to be made up of two parts: (1) the point estimate of the mean $\bar{x}$, and (2) the margin of error. The choice of the point estimate is intuitive. The margin of error is made of components whose arrangement is also logical. The size of the margin of error can be seen to depend directly on the size of the standard deviation of the normal population and the level of confidence. The size of the margin of error is inversely proportional to the square root of sample size. Thus, if the standard deviation were to double, the margin of error would double; but it would require a four-fold increase in the sample size to cut the margin of error in half. The effect of increasing the level of confidence increases the size of the margin of error but it is not linear in α. The following table, Table 7.2, displays the connection between the multiplier $Z_{1-\alpha/2}$ and the level of confidence $1-\alpha$. Looking at the table we can see that to increase the level of confidence from 90% to 95% we increase the margin of error by the factor of multiplying by 1.645 and 1.96.

Table 7.2 Relationship of the level of confidence to the multiplier Z in the margin of error.

Confidence level $1 - \alpha$	α	$\dfrac{\alpha}{2}$	$1 - \dfrac{\alpha}{2}$	$Z_{1-\alpha/2}$
0.80 (80%)	0.20	0.10	0.90	1.282
0.90 (90%)	0.10	0.05	0.95	1.645
0.95 (95%)	0.05	0.025	0.975	1.960
0.99 (99%)	0.01	0.005	0.995	2.576

The concept behind an interval estimate can be demonstrated and clarified by a "repeated sampling" illustration. Suppose we want to estimate the mean of a normal distribution using a sample of size 5. If we repeatedly draw samples of 5 observations each, applying the formulas given above, we will get a set of confidence intervals.

Figure 7.6 demonstrates a set of confidence intervals obtained by repeatedly sampling the same normal distribution. Specifically, twenty samples, each of size 5, were generated from a normal distribution having a mean of 60 and standard deviation of 5. Then a 95% confidence interval was computed around each sample mean using $Z_{0.975} = 1.96$ and the formula:

$$\bar{x} \pm Z_{1-\alpha/2}\frac{\sigma}{\sqrt{n}} = \bar{x} \pm 1.96\frac{5}{\sqrt{5}}$$

These 20 confidence intervals are plotted in Figure 7.6. In the figure, the lower and upper confidence limit is denoted by o, and the location of the mean by a +.

Theoretically, 19 of the 20 confidence intervals (95% of them) should enclose the true mean of 60; only one would not. Upon close examination you can observe that in this particular computer simulation, 19 intervals enclosed the true mean of 60 and one interval (the 12th from the bottom) failed. Of course this kind of outcome would not happen every time. The 1 in 20 expectation is only a long run expectation. Any given simulation could vary from that.

If we were now to randomly select one of these confidence intervals, theoretically we would have 19 chances in 20 or a 95% chance of getting an interval that encloses the true mean. There is only a 1 in 20 or 5% chance of getting an interval that does not include the true mean, 60. The other significant thing to recognize is that the true mean is a constant value ($\mu = 60$); the variation that you see in the placement of the confidence intervals is due to the variation in the sample mean that varies from sample to sample. Therefore, the interval "jumps back and forth" relative to the true mean.

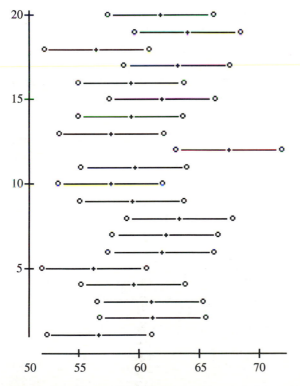

Figure 7.6 Plots of 20 means, $n = 5$, and their corresponding 95% confidence intervals.

EXAMPLES

EXAMPLE 1: The National Aeronautics and Space Administration (NASA) is continually concerned about the quality of the components it uses in its spacecrafts. They desire to estimate the mean lifetime of a particular mechanical component under weightless conditions which has had a mean lifetime of 1100 hours with a standard deviation of 40 hours when tested under normal gravitational forces. They propose to test only 10 components in the weightless environment because of the extreme cost of carrying out such tests. The following data resulted.

$$1158.3713 \quad 1219.8604 \quad 1165.1397 \quad 1264.5041 \quad 1220.0123$$
$$1217.5003 \quad 1194.6938 \quad 1201.1050 \quad 1224.2419 \quad 1263.7680$$

Construct a 90% confidence interval the mean for this data, assuming the standard deviation doesn't change in the weightless setting and that the data are normally distributed.

SOLUTION: To construct the 90% confidence interval we need the Z-score (1.645), the sample size (10), the standard deviation (40), along with the sample mean of this data (1212.9).This information is inserted into the formula which produces the following results:

$$\bar{x} \pm Z_{1-\alpha/2} \frac{\sigma}{\sqrt{n}} = 1212.9 \pm 1.645 \frac{40}{\sqrt{10}}$$

or, $1192.1 \leq \mu \leq 1233.7$. Thus, we are 90% confident that the mean lifetime is between 1192.1 and 1233.7 hours. (We could also say: "The probability that the interval obtained will enclose the true mean is 90%." However, it is *not* correct to say: "The probability that the mean is in the above interval is 90%." The random variable, the entity that shows variation is the *interval*, not the mean μ.)

EXAMPLE 2: Consider the data of example 1, but suppose we aren't willing to assume the standard deviation is the same under weightless conditions. Compute a *t*-interval for the data and compare it with the Z-interval obtained in the previous example.

SOLUTION: The mean and standard deviation of the data is $\bar{x}$ = 1212.9 and s = 35.315. Using this information the resulting 90% *t*-interval is

$$1192.4 \leq \mu \leq 1233.4$$

It can be noted that this interval is slightly wider than the Z-interval computed in example 1. (How do the *t* and Z values compare? How does the "true" standard deviation compare to the sample estimate?)

EXAMPLE 3: Report the results of example 1 in terms of the point estimate and the margin of error. Show how the margin of error changes by (1) changing the confidence level from a 90% to 95%, (2) a doubling of the standard deviation, and (3) a doubling in the sample size.

SOLUTION: From Example 1 the point estimate of the mean is the sample mean, 1212.9. The margin of error is

$$\varepsilon = 1.645 \frac{40}{\sqrt{10}} = 20.8 \text{ hours.}$$

If the confidence level were changed from 90% to 95% then the margin of error would be:

$$\varepsilon = 1.96 \frac{40}{\sqrt{10}} = 24.8 \text{ hours},$$

a slightly larger margin of error. If the standard deviation were twice as large then the margin of error would be (90% confidence)

$$\varepsilon = 1.645 \frac{2 \cdot 40}{\sqrt{10}} = 41.6 \text{ hours},$$

a margin of error twice as large. If the sample size were doubled, then ε would be:

$$\varepsilon = 1.645 \frac{40}{\sqrt{2 \cdot 10}} = 14.7 \text{ hours}$$

Therefore, doubling the sample size does not cut the margin of error in half, rather it cuts it by the factor $\sqrt{2}$.

Sample Size

The development of this approach suggests a way to answer one of the most common and important questions that comes up when a statistical investigation gets under way: "How large a sample size do I need?" To answer that question, we respond that three other questions need to be answered first and then an answer to the question can be given quite easily. The three preliminary questions are: (1) "What level of confidence do you want?" (and therefore what is $Z_{1-\alpha/2}$?); (2) "What is the largest margin of error ε you can tolerate?" ; and (3) "What is your best guess as to the size of the population standard deviation σ?" With those questions answered, you can see from example 3 that a formula can be set up that is a function of the sample size n. The equation is:

$$\varepsilon = Z_{1-\alpha/2} \frac{\sigma}{\sqrt{n}} .$$

This is an equation with only unknown being n. Solving for n we get:

$$n = \left(Z_{1-\alpha/2} \frac{\sigma}{\varepsilon} \right)^2$$

Therefore, if we want to compute the sample size for estimating the mean height of male adults within 1/2 an inch with a 95% confidence level, knowing that the standard deviation of the population is about 3 inches, then $Z_{1-\alpha/2} = 1.96$, $\varepsilon = 0.5$ inches, and $\sigma = 3$ inches. The sample size calculation is:

$$n = \left(Z_{1-\alpha/2} \frac{\sigma}{\varepsilon} \right)^2 = \left(\frac{1.96 \cdot 3}{0.5} \right)^2 = 138.3$$

So, we would need a sample of about 140 adult males to provide an estimate of the mean height under the specified conditions. (In most instances, we only have a rough guess of the size of σ, therefore, this formula only gives a rough guess as to the size of the sample. Thus, the reason for "rounding off the sample size" upward to 140.)

EXERCISES

1. Employees in a large corporation were given the option of investing a certain percentage of their monthly pay in a savings and investment program, which the employer would match, up to 3% of their salary. A random sample of 20 employees in a particular wage bracket was selected with the purpose of estimating the average amount being saved in this plan each month with $\bar{x} = \$79$. If it is assumed that the amounts saved are normally distributed with a standard deviation of $50, what would be your estimate of the average monthly savings? What is the size of the "error of estimate?"

2. A quality inspector at a food processing plant was in charge of inspecting truckloads of pears that were to be canned. She would take a sample of pears from the load. For each pear she would cut out the bad spots and remove the core. She would weigh the waste and core material and compare to the total weight of the pair and produce a ratio of usable fruit to total weight. The following sample values were produced as percentages for a sample of 24 pears. What is your estimate of overall percentage for the entire load? Provide a value for the margin of error for a 95% confidence interval, assuming the sample is random and that the percentages are normally distributed with a standard deviation of 3%.

    65.59 65.53 65.71 59.93 65.67 63.54 66.21 65.20 69.50 63.20 64.45 59.43 65.61
    69.38 69.67 65.47 64.69 66.51 61.64 64.71 59.96 66.38 65.22 64.54

3. A public utility company was considering changing the rate structure by increasing the price per kilowatt hour during peak-load periods and reducing the price during non-peak periods. Before implementing this program, a random sample of 50 households was sampled and monitored to assess the the potential affect of changing the rate structure. Would it tend to increase the average monthly bill, decrease the average monthly bill, or leave things unchanged? Assume the monthly bills are normally distributed with a standard deviation of $7.50. Construct a 95% confidence interval on the mean monthly bill if the sample of 50 households produced a sample mean of $90.80 when the new rate structure was applied.

4. In a study a state taxing agency wanted to assess the average deduction taken on personal tax returns with a margin of error of $250. What sample size would be necessary to produce such a margin of error associated with a 95% confidence interval if the standard deviation is $1850?

5. Suppose in general, for a $100(1 - \alpha)\%$ confidence interval, the margin of error desired is the value d and the population has a standard deviation, σ. Find an equation that would allow you to compute the sample size to meet these conditions.

6. The breaking strength of a certain type of thread is known to be approximately normally distributed with a standard deviation of 3 pounds. A sample of 25 lengths of string is selected and the mean breaking strength for the sample is determined to be $\bar{x} = 17$ pounds. Compute a 95% confidence interval for the mean breaking strength of the thread type.

7. The daily electrical power consumption is believed to follow a normal distribution. After converting to more convenient units, a sample of 11 randomly selected days produced the following consumption units. What is a 99% confidence interval on the mean per day?

Data: 7.1   4.4   9.1   5.4   6.3   6.8   5.9   8.7   7.2   7.3   8.5

8. A study of the income tax is proposed by a state legislature and one of the issues that is primarily of interest is the average amount taken in deductions. How large should the sample size be if the standard deviation is believed to be about $2000 and a 95% confidence interval is wanted with a margin of error of $100?

9. A university is interested in the total number of credit hours, on the average, its graduating seniors have upon graduation. It is assumed the number is distributed somewhat bell-shaped and the minimum number required for graduation is 128 hours. After asking around it is found that the maximum is about 152. What would be your guess as to the size of the standard deviation for this bell-shaped distribution? [Hint: Assume that range from 128 to 152 is a 6 standard deviation range.]

10. Using the standard deviation found in exercise 9, what size of sample would you recommend in order to have a 90% confidence interval with a margin of error of ± 2 credit hours?

11. Recall that the standard error of the sample mean is $\frac{\sigma}{\sqrt{n}}$. Suppose a population has a known standard deviation σ and we want the estimate of the population mean to have a standard error equal to δ. Form a simple equation to represent this and solve for n to show what size of sample is needed.

12. A normally distributed population has a standard deviation of 40. What size of sample is needed to produce a standard error of the mean equal to 5?

7.6   INTERVAL ESTIMATORS OF THE DIFFERENCE IN MEANS

Introduction

In this section we address the problem of comparing the means from two different populations such as the mean salary of men versus women, or the average production per plant of treated versus untreated tomato plants, etc. Specifically, we want to estimate the difference in means of the two populations.

The theory we have developed thus far, allows us to derive confidence intervals for two different cases; (1) The case of sampling normally distributed populations with known standard deviations, and (2) the case of sampling normally distributed populations with unknown but equal standard deviations. We explore these two cases in what follows.

In this section and the ones that follow, each case and its assumptions will be given and then examples of its application will follow immediately.

Methodology

Case 1: Estimating the Difference in Means for Independent Samples of Two Normally Distributed Populations, Known Variances

Background Assumptions: Suppose $X_1, ..., X_n$ and $Y_1, ..., Y_m$ are two independent random samples from normal distributions having unknown means μ_x and μ_y and known variances σ_x^2 and σ_y^2. Our objective is to provide an interval estimate of the difference in the means, $\mu_x - \mu_y$.

1. As an estimate of the difference in means, $\mu_x - \mu_y$, the difference in sample means can be used. Under the assumptions given above, the difference in sample means $\bar{X} - \bar{Y}$ is normally distributed with

$$E(\bar{X} - \bar{Y}) = \mu_x - \mu_y,$$

and

$$V(\bar{X} - \bar{Y}) = \left(\frac{\sigma_x^2}{n} + \frac{\sigma_y^2}{m}\right)$$

(These results can be proved using the moment generating function technique.)

2. The pivotal quantity we use is

$$\frac{(\bar{X} - \bar{Y}) - (\mu_x - \mu_y)}{\sqrt{\left(\frac{\sigma_x^2}{n} + \frac{\sigma_y^2}{m}\right)}}$$

which is distributed as a standard normal distribution.

3. The probability statement is

$$P\left(-Z_{1-\alpha/2} < \frac{(\bar{x} - \bar{y}) - (\mu_x - \mu_y)}{\sqrt{\left(\frac{\sigma_x^2}{n} + \frac{\sigma_y^2}{m}\right)}} < Z_{1-\alpha/2}\right) = 1 - \alpha$$

4. The confidence interval, after manipulating this expression as before, is:

$$(\bar{x} - \bar{y}) \pm Z_{1-\alpha/2} \sqrt{\left(\frac{\sigma_x^2}{n} + \frac{\sigma_y^2}{m}\right)}$$

The point estimate is the difference $(\bar{x} - \bar{y})$, the margin of error is $Z_{1-\alpha/2} \sqrt{\left(\frac{\sigma_x^2}{n} + \frac{\sigma_y^2}{m}\right)}$ and the confidence coefficient is $1 - \alpha$.

Case 2: Estimating the Difference in Means for Independent Samples of Two Normally Distributed Populations, Equal but Unknown Variances.

Background Assumptions: Suppose $X_1, ..., X_n$ and $Y_1, ..., Y_m$ are two independent random samples from normal distributions having unknown means μ_x and μ_y and unknown but equal variances $\sigma_x^2 = \sigma_y^2 = \sigma^2$. Our desire is to provide an interval estimate of the difference in the means, $\mu_x - \mu_y$.

1.   As an estimate of the difference in means, $\mu_x - \mu_y$, the difference in sample means can be used. Under the assumptions given above, the difference in sample means is normally distributed with mean, $\mu_x - \mu_y$, and variance, $\sigma^2(\frac{1}{n} + \frac{1}{m})$.

2.   The pivotal quantity we use is

$$\frac{(\bar{X} - \bar{Y}) - (\mu_x - \mu_y)}{\sqrt{s_p^2 \left(\frac{1}{n} + \frac{1}{m}\right)}}$$

which is distributed as a t- distribution with $n + m - 2$ degrees of freedom, and where

$$s_p^2 = \frac{(n-1)s_x^2 + (m-1)s_y^2}{(n-1) + (m-1)}$$

is called the "pooled estimate of the variance."

3.   The probability statement is

$$P(-t_{1-\alpha/2} < \frac{(\bar{X} - \bar{Y}) - (\mu_x - \mu_y)}{\sqrt{s_p^2 \left(\frac{1}{n} + \frac{1}{m}\right)}} < t_{1-\alpha/2}) = 1 - \alpha$$

4.   The confidence interval, after manipulating this expression as before, is:

$$(\bar{x} - \bar{y}) \pm t_{1-\alpha/2} \sqrt{s_p^2 \left(\frac{1}{n} + \frac{1}{m}\right)}$$

The point estimate is the difference in sample means $(\bar{x} - \bar{y})$, the estimated margin of error is $t_{1-\alpha/2} \sqrt{s_p^2 (\frac{1}{n} + \frac{1}{m})}$, and the confidence coefficient is $1 - \alpha$.

DISCUSSION

More often than not when a statistical investigation is planned, a comparison of two or more population means is contemplated. If the comparison involves the means of the two then it is logical to ask how different the means are. The approach of this section is appropriate to address such a subject. We start out by assuming that we have two normally distributed random variables X and Y with their means and variances denoted by μ_x, μ_y, σ_x^2, and σ_y^2, respectively. A random sample of size n is selected from the X population and a random sample of size m is selected from the Y population. Two cases in particular result.

Case I:

Suppose the variances σ_x^2 and σ_y^2 are known. Then the random variable

$$\frac{(\bar{X}-\bar{Y}) - (\mu_x - \mu_y)}{\sqrt{\left(\frac{\sigma_x^2}{n} + \frac{\sigma_y^2}{m}\right)}}$$

will be normally distributed and can be used to form a $(1-\alpha)\bullet 100 \%$ confidence interval of the form

$$(\bar{x} - \bar{y}) \pm Z_{1-\alpha/2} \sqrt{\left(\frac{\sigma_x^2}{n} + \frac{\sigma_y^2}{m}\right)}$$

However, it is not usual to know the variance; they must be estimated from the sample. This leads to case II.

Case II:

Suppose the variances σ_x^2 and σ_y^2 are unknown. We need to find a "pivotal quantity" whose distribution is known and whose only unknown elements are the means. From Chapter 6 an expression was developed that possessed a t distribution under the assumption that σ_x^2 and σ_y^2 are unknown but equal; i.e., $\sigma_x^2 = \sigma_y^2 = \sigma^2$. The random variable in this case was

$$\frac{(\bar{X}-\bar{Y}) - (\mu_x - \mu_y)}{\sqrt{s_p^2 \left(\frac{1}{n} + \frac{1}{m}\right)}}$$

which is distributed as a t-distribution with $n + m - 2$ degrees of freedom, and where

$$s_p^2 = \frac{(n-1)s_x^2 + (m-1)s_y^2}{(n-1) + (m-1)}$$

This results in a $(1 - \alpha) \cdot 100 \%$ confidence interval of the form

$$(\bar{x} - \bar{y}) \pm t_{1-\alpha/2} \sqrt{s_p^2(\frac{1}{n} + \frac{1}{m})}$$

EXAMPLES

EXAMPLE 1: A marketing research study was conducted among businesses from large metropolitan cities which were to be compared to businesses from smaller cities and rural areas. A service was to be offered and the number of times per week the service would be used was asked. For 70 businesses from the large cities interviewed, the estimated usage was an average of 3.7 times per week with a standard deviation of 0.75. For 50 businesses from smaller cities, the estimated usage averaged 2.9 times per week with a standard deviation of 0.66 times per week. Estimate the difference in usage between large and small cities.

SOLUTION: Since the sample size is large we shall first work this problem assuming the sample standard deviations are essentially the same as the population standard deviations. Therefore, we use the following equation:

$$(\bar{x} - \bar{y}) \pm Z_{1-\alpha/2} \sqrt{\left(\frac{\sigma_x^2}{n} + \frac{\sigma_y^2}{m}\right)}$$

where $m = 70$ and $n = 50$. We shall assume that we want a 95% confidence interval, so that the Z-value is 1.96. Therefore, we obtain the following results.

$$(3.7 - 2.9) \pm 1.96 \sqrt{\frac{(0.75)^2}{70} + \frac{(0.66)^2}{50}}$$

or

$$0.8 \quad \pm \quad 0.25$$

This gives the 95% confidence interval (0.55, 1.05). Thus we are 95% confident that the difference in usage per week is between 0.55 and 1.05 times.

EXAMPLE 2: Work the problem of example 1 assuming the distributions have an unknown but common variance.

SOLUTION: In this case we use the formula

$$(\bar{x} - \bar{y}) \pm t_{1-\alpha/2} \sqrt{s_p^2(\frac{1}{n} + \frac{1}{m})}$$

where

$$s_p^2 = \frac{(n-1)s_x^2 + (m-1)s_y^2}{(n-1) + (m-1)}$$

As a result, we get for the pooled variance

$$s_p^2 = \frac{(n-1)s_x^2 + (m-1)s_y^2}{(n-1) + (m-1)} = 0.5098$$

which has $69 + 49 = 118$ degrees of freedom (df). The t value for a 95% confidence interval is $t_{0.975,118} = 1.98$, resulting in the confidence interval expression

$$(3.7 - 2.9) \pm 1.98 \sqrt{0.5098 \left(\frac{1}{70} + \frac{1}{50} \right)}$$

or

$$0.80 \pm 0.26$$

which gives the confidence interval (0.54, 1.06), a slight and probably not important difference, when compared to the results of example 1. However, under the assumptions the interval of example 1 is only approximate while that of example 2 is exact.

EXAMPLE 3: List the assumptions made to justify the solutions given in example 1 and 2.

SOLUTION: The assumptions of example 1 are that: (1) usage values reported by the various businesses interviewed, in both large and small cities, follow a normal distribution with unknown means and standard deviations, (2) two independent, random samples are selected from the two normally distributed populations, and (3) with a large sample the sample standard deviation is sufficiently close to the value of the true standard deviation σ to justify its substitution in the formula.

The assumptions of example 2 are that: (1) usage values reported by the various businesses interviewed, in both large and small cities, follow a normal distribution with unknown means and unknown but equal standard deviations, d small cities, follow a normal distribution with unknown means and standard deviations, and (2) two independent, random samples are selected from the two normally distributed populations.

Since the margin of error is slightly larger for the approach of example 2 than for example 1, the conservative individual might feel more comfortable with the approach of example 2 rather than 1.

EXERCISES

1. A study was conducted to compare the nicotine content of two different brands of cigarettes. Ten cigarettes each of Brand X and Brand Y were tested. The following data were obtained. Compute a 95% confidence interval on the difference in average nicotine content. Measurements are in milligrams.

	Brand X	Brand Y
Sample Size	10	10
Mean	3.2	2.8
St. Dev.	0.5	0.7

2. A study of two kinds of photocopying equipment shows that, of 50 failures on one kind of machine, it took on the average 85.7 minutes to repair with a standard deviation of 20.4 minutes. On the other type of machine, it took an average of 92.5 minutes with a standard deviation of 18.5 minutes from a total of 55 failures. Construct a 90% confidence interval on the difference in repair time for the two machines.

3. An orchardist wanted to compare the production from two equal sized plots of peach trees planted in different types of soil. From the 15 trees in the one plots the average production per tree was 188 pounds with a standard deviation of 12 pounds. From the 15 trees in the other plot, the average production per tree was 160 pounds with a standard deviation of 18 pounds. Estimate the difference in the production per tree for the two plots involved and place a 95% confidence interval on your estimate.

4. The rate of oxygen consumption by runners of different levels of training is indicative of the effect of the training. College males using two different training methods are to be compared relative to their oxygen consumption. Part of the males involved regular training on a daily basis and the other part involved intermittent training over the same period of time. Measurements of oxygen use are in milliliters per kilogram-minute, and the sample sizes, means and standard deviations are given below.

	Continuous Training	Intermittent Training
Sample Size	9	7
Means	42.72	38.28
Standard deviations	6.12	6.88

Place a 95% confidence interval on the difference in oxygen use.

5. In two different controlled settings, psychologists observe two equally sized groups of children and score them on a scale of aggressive behavior. Construct a 95% confidence interval for the difference between the means scores for the two groups, assume equal variances. The following information is given.

	Group 1	Group 2
Sample Size	50	50
Means	63.5	70
Standard Deviations	5.998	3.277

6. For the data of exercise 5, use the procedure shown in Example 1 to find the confidence interval on the difference of means.

7. Show that $\sqrt{s_p^2 \left(\frac{1}{n} + \frac{1}{m}\right)} = \sqrt{\left(\frac{s_x^2}{n} + \frac{s_y^2}{m}\right)}$ when $n = m$. Discuss this implication as it applies to exercises 5 and 6.

8. A potato-chip company examines two machines that package potato chips to estimate the mean difference in the weight of packages filled by each machine. The data (in ounces) from each packaging machine are given below. Compute a 98% confidence interval for the mean difference in weight.

Machine A:	16.05	15.98	15.97	15.99	16.11	16.03	16.10	16.00	14.96
Machine B:	14.89	16.07	16.01	16.00	14.92	14.97	16.00	14.89	16.01

9. A confidence interval on the difference of means $\mu_x - \mu_y$ could have upper and lower limits that are either both positive, both negative or opposite in sign thus overlapping zero. What are the implications of these possibilities relative to the question as to whether $\mu_x = \mu_y$?

10. Suppose a confidence interval is constructed for $\mu_x - \mu_y$ and an upper and lower confidence limit is obtained. Using the same data, suppose a confidence interval is constructed for $\mu_y - \mu_x$. Will the upper and lower limits be the same or different, different in sign, or what?

7.7   INTERVAL ESTIMATORS FOR A VARIANCE OR RATIO OF VARIANCES

Introduction

In this section we treat the problem of constructing confidence intervals on the variance of normally distributed populations. We will examine two cases again. The first case is when we have a sample from a single population and desire a confidence interval on the variance, σ^2. The second case is when we have samples from two normally distributed populations and want a confidence interval on the ratio of the two variances,

$$\frac{\sigma_x^2}{\sigma_y^2}$$

In the first case, we recognize that the sample variance is related to the chi-square distribution under appropriate normality assumptions. In the second case, we use the relationship that the ratio of sample variance follows the F-distribution under normality assumptions.

Methodology

Case 1: Estimating the Variance of a Normally Distributed Population

> **Background Assumptions:** Let $X_1, ..., X_n$ be a random sample from a normally distributed population with unknown mean and variance. We desire to produce an interval estimate of the variance σ^2 of the population.

1.   As an estimate of the variance, σ^2, the sample variance s^2 is used. Under the assumptions given above, the term

$$\frac{(n-1)s^2}{\sigma^2}$$

is distributed as a chi-square random variable with $v = (n-1)$.

2.   Therefore, the pivotal quantity we use is

$$\frac{(n-1)s^2}{\sigma^2}$$

3.   The probability statement is

$$P(\chi^2_{\alpha/2} < \frac{(n-1)s^2}{\sigma^2} < \chi^2_{1-\alpha/2}) = 1 - \alpha$$

4.   The confidence interval, after manipulating this expression to isolate the variance σ^2, is:

$$P\left(\frac{(n-1)s^2}{\chi^2_{\alpha/2}} > \sigma^2 > \frac{(n-1)s^2}{\chi^2_{1-\alpha/2}} \right) = 1 - \alpha$$

where $\dfrac{(n-1)s^2}{\chi^2_{1-\alpha/2}}$ is the lower confidence limit and $\dfrac{(n-1)s^2}{\chi^2_{\alpha/2}}$ is the upper confidence limit and the chi-square values are from a chi-square table with $n-1$ degrees of freedom. The confidence coefficient is $1 - \alpha$.

(A margin of error is not given in this case because the limits around the parameter are not symmetric (equidistant) and thus there is no simple way to decide on what to call a margin of error.)

Case 2: Estimating the Ratio of Two Variances When Sampling Two Independent Normal Distributions.

Background Assumptions: Let $X_1, ..., X_n$ and $Y_1, ..., Y_m$ be two independent samples from two normally distributed populations having unknown variances σ_x^2 and σ_y^2. We want to produce a confidence interval on the ratio of the two variances, $\dfrac{\sigma_x^2}{\sigma_y^2}$.

1. As an estimate of the ratio of variances $\dfrac{\sigma_x^2}{\sigma_y^2}$ we use the ratio of the sample variances. Under the normality assumptions given above, the ratio

$$\frac{(s_x^2/\sigma_x^2)}{(s_y^2/\sigma_y^2)} = [\frac{s_x^2}{s_y^2}] \cdot [\frac{\sigma_y^2}{\sigma_x^2}]$$

is distributed as a random variable from the F-distribution with numerator and denominator degrees of freedom of $n-1$ and $m-1$ respectively.

2. Therefore, the pivotal quantity we use is

$$\frac{(s_x^2/\sigma_x^2)}{(s_y^2/\sigma_y^2)} = [\frac{s_x^2}{s_y^2}] \cdot [\frac{\sigma_y^2}{\sigma_x^2}]$$

which is distributed as a random variable from the F-distribution.

3.   The probability statement is

$$P\left(F_{\alpha/2} < \frac{(s_x^2/\sigma_x^2)}{(s_y^2/\sigma_y^2)} < F_{1-\alpha/2} \right) = 1 - \alpha.$$

4.   The confidence interval, after manipulating this expression, is:

$$\left[\frac{s_x^2}{s_y^2}\right] \frac{1}{F_{\alpha/2;n-1;m-1}} > \frac{\sigma_x^2}{\sigma_y^2} > \left[\frac{s_x^2}{s_y^2}\right] \frac{1}{F_{1-\alpha/2;n-1;m-1}}$$

or, equivalently

$$\left[\frac{s_x^2}{s_y^2}\right] F_{1-\alpha/2;m-1;n-1} > \frac{\sigma_x^2}{\sigma_y^2} > \left[\frac{s_x^2}{s_y^2}\right] \frac{1}{F_{1-\alpha/2;n-1;m-1}}$$

The point estimate is the ratio of sample variances, $\dfrac{s_x^2}{s_y^2}$ and the confidence coefficient is $1-\alpha$. (Again, we cite no margin of error since the confidence interval is typically non-symmetric.)

DISCUSSION

Another parameter in a normally distributed population is the variance σ^2. To construct a confidence interval on a variance we use the sample variance s^2 and the pivotal quantity $\dfrac{(n-1)s^2}{\sigma^2}$ which we know to be a chi-square random variable when the population is normally distributed. After manipulating the probability expression given in the Definitions, the following expression results for the confidence interval on the variance.

$$\frac{(n-1)s^2}{\chi_{\alpha/2}^2} > \sigma^2 > \frac{(n-1)s^2}{\chi_{1-\alpha/2}^2}$$

Because the chi-square distribution is not symmetric, the upper and lower confidence limits are not equidistant from σ^2. Therefore, we cannot simply compute the margin of error as half the confidence interval width. Rather, we will not formally identify a margin of error. Instead, we simply report the upper and lower confidence limits.

If we have two independent populations and we want to compare their variances, we will do this by examining their ratio. If the two variances are equal, then their ratio should equal 1. Under the assumption of normally distributed random variables X and Y and two independent samples of size n and m respectively, then the ratio

$$\frac{(s_x^2/\sigma_x^2)}{(s_y^2/\sigma_y^2)}$$

is distributed as an F random variable with $n-1$ and $m-1$ degrees of freedom for numerator and denominator. Manipulating this expression as we have before we get the confidence interval expression that follows.

$$\left[\frac{s_x^2}{s_y^2}\right]F_{1-\alpha/2;m-1;n-1} > \frac{\sigma_x^2}{\sigma_y^2} > \left[\frac{s_x^2}{s_y^2}\right]\frac{1}{F_{1-\alpha/2;n-1;m-1}}$$

The F distribution is not symmetric either, so we handle the problem of citing a margin of error by choosing not to.

We demonstrate these concepts, now, with a few examples.

EXAMPLES

EXAMPLE 1: A quality control inspector in a cannery knows that the amount of product in each can will vary from can to can. The mean "fill" per can is important, but of equal importance is the variation in fill represented by the variance σ^2. If the variance is too large then the product is unsatisfactory because some cans will be too full while others will be seriously under-filled. To estimate the variation in fill of a particular machine, a sample of 15 cans was selected and the mean and standard deviation of the weights obtained producing the values, 7.92 ounces and 0.05 ounces respectively. Produce a 95% confidence interval for the variance, σ^2.

SOLUTION: To obtain the appropriate confidence interval, we have to get the appropriate values from the chi-square table and insert them into the expressions

$$\frac{(n-1)s^2}{\chi_{1-\alpha/2}^2}$$

and

$$\frac{(n-1)s^2}{\chi_{\alpha/2}^2}$$

This gives the following values

$$\frac{(15-1)(0.05^2)}{26.1190} \le \sigma^2 \le \frac{(15-1)(0.05^2)}{5.6276}$$

or

$$0.00134 \le \sigma^2 \le 0.00622 .$$

EXAMPLE 2: A manufacturing plant is trying to streamline its production line and increase it product control. Several production line configurations were tested which were finally reduced to one of two different configurations. It was suggested that the configuration that produced the smaller variation in the number of finished units be used. Two independent samples produced the following results.

	Configuration 1	Configuration 2
Sample size	8 days	8 days
Sample variance	1,556	3,242

Compute a 90% confidence interval on the ratio of variances.

SOLUTION: The formula for a confidence interval on a ratio is given above as

$$[\frac{s_x^2}{s_y^2}]F_{1-\alpha/2;m-1;n-1} > \frac{\sigma_x^2}{\sigma_y^2} > [\frac{s_x^2}{s_y^2}]\frac{1}{F_{1-\alpha/2;n-1;m-1}}$$

Using this expression, we get

$$(\frac{1556}{3242})\,3.79 > \frac{\sigma_x^2}{\sigma_y^2} > (\frac{1556}{3242})\frac{1}{3.79}$$

or

$$1.819 > \frac{\sigma_x^2}{\sigma_y^2} > 0.1266$$

Thus we are 90% confident that the ratio of variances of configuration 1 to configuration 2 is between 0.1266 and 1.819.

EXAMPLE 3: Given that the confidence interval for the ratio of two intervals encloses the value 1 as given in example 2, what can you conclude about the possible equality of the variances, σ_x^2 and σ_y^2?

SOLUTION: Since the 90% confidence interval on the ratio $\frac{\sigma_x^2}{\sigma_y^2}$ encloses 1, it is therefore possible that $\frac{\sigma_x^2}{\sigma_y^2} = 1$ and as a result it is possible that $\sigma_x^2 = \sigma_y^2$. That is a possibility we at least have to admit based on the data.

EXERCISES

1. List the assumptions that are needed to justify computing a confidence interval on a variance as given in this section.

2. List the assumptions that are needed to justify computing a confidence interval on the ratio of variances as given in this section.

3. What is the relationship between the table values for F_α and $F_{1-\alpha}$ when both numerator and denominator degrees of freedom are the same?

4. A state traffic engineer was investigating the effect of tinted versus non-tinted windshields on reaction time to peripheral stimuli. Thirty subjects from students enrolled in a drivers education program were randomly divided up into two groups of 15 each. One group of fifteen were assigned to cars with tinted windshields, and the other 15 were assigned to cars without tinting. One aspect of this investigation, required examination of the variation of reaction times. A 90% confidence interval on the ratio of variances in reaction times was desired. The following data were obtained.

	Tinted	Not tinted
Sample Sizes	15	15
Variances	9.61	4.84

5. Based on the confidence interval obtained for the ratio in exercise 4, what conclusion would you make about that possibility that the true variances are equal?

6. Using the data of exercise 4 construct a 90% confidence interval on each variance obtained, that is the variance of the tinted group and the variance of the not-tinted group.

7. Examine the confidence intervals in exercise 6 and note whether they overlap to any degree? If they did overlap what would you be inclined to say about the possibility that the true variances are equal? If they don't overlap what would you be inclined to say?

8. A university computer center wants to estimate the variability in the number of jobs it processes each day. A random sample of 20 days during the past 6 months was selected and the number of jobs processed each day was recorded. The sample variance turned out to be 32761. Find a 95% confidence interval on the true variance. Take the square root of these limits and explain the meaning of these numbers.

9. In constructing a confidence interval on the difference of two means (see Section 6), an assumption about the equality of variances is made to justify the use of the pooled variance s_p^2 in the case when the variances are unknown. Use the data of exercise 1 in section 7.6 to investigate this possibility. (Is it possible that the variances for Brand X and Brand Y are equal, based on the confidence interval for the ratio of variances?) The data are given again below.

	Brand X	Brand Y
Sample Size	10	10
Mean	3.2	2.8
St. Dev.	0.5	0.7

10. Use the data of section 7.6, exercise 4 to construct a confidence interval on the ratio of variances. Note whether the confidence interval encloses the value 1 in order to justify the assumption of equal variances. The data are given below.

	Continuous Training	Intermittent Training
Sample Size	9	7
Means	42.72	38.28
Standard deviations	6.12	6.88

7.8   INTERVAL ESTIMATORS FOR THE PROPORTION OF A BERNOULLI DISTRIBUTION

Introduction

In this section we present a method for a large sample confidence interval for the Bernoulli parameter p. We assume that we have a set of n independent Bernoulli observations and that the sample is large enough for the central limit theorem to give a good approximation to the binomial probabilities. We also present a method for a confidence interval on the difference $p_x - p_y$ of two Bernoulli parameters, p_x and p_y.

A need for these techniques are encountered frequently. For instance, suppose we are trying to estimate the percentage of voters that favor a particular political candidate. Therefore, we are interested in the Bernoulli parameter p, the proportion of voters favoring a particular candidate.

Or suppose that we have two different shifts in a production operation and we want to estimate the difference in the proportion of defective products produced by the two shifts of workers. In this case we want to estimate the proportion of defective produced by one shift compared to the proportion produced by the other shift; or more precisely look at the difference in proportions. This problem can be modelled by two Bernoulli distributions with the difference in the Bernoulli parameters p_x and p_y being of interest.

Methodology

Case 1: Estimating the Proportion p in *n* Repeated Bernoulli Trials

Background Assumptions: Suppose a random sample of n independent and identically distributed Bernoulli random variables $X_1, ..., X_n$, is obtained with n being large. A confidence interval about the unknown parameter p is desired.

1. As an estimate of the proportion p, the sample proportion $\hat{p} = \bar{x}$ can be used. (Note that $\bar{x}$ is the mean of zeros and ones.) Under the assumptions given above, the estimator $\hat{p}$ will be binomially distributed with parameters n and p and, if the sample size n is sufficiently large, the statistic $\dfrac{\hat{p} - p}{\sqrt{\dfrac{p(1-p)}{n}}}$ will be approximately standard normal.

2. The pivotal quantity we use is

$$\frac{\hat{p} - p}{\sqrt{\dfrac{p(1-p)}{n}}}$$

which is approximately distributed as a standard normal distribution.

3. The probability statement is

$$P\left(-Z_{1-\alpha/2} < \frac{\hat{p} - p}{\sqrt{\dfrac{p(1-p)}{n}}} < Z_{1-\alpha/2}\right) = 1 - \alpha.$$

4. The confidence interval, after manipulation, is:

$$\hat{p} \pm Z_{1-\alpha/2} \sqrt{\frac{p(1-p)}{n}}$$

But since p under the radical is unknown , we replace it with our estimate $\hat{p}$ which gives us the expression:

$$\hat{p} \pm Z_{1-\alpha/2} \sqrt{\frac{\hat{p}(1-\hat{p})}{n}}$$

The point estimate is the proportion $\hat{p}$, the margin of error is $Z_{1-\alpha/2} \sqrt{\frac{\hat{p}(1-\hat{p})}{n}}$, and the confidence level is $1 - \alpha$.

Case 2: Estimating the Difference in Proportions from Two Independent Bernoulli Distributions

Background Assumptions: Suppose a random sample of n independent and identically distributed random variables $X_1, ..., X_n$, is obtained from a Bernoulli distribution with parameter p_x. A second independent sample of size m, $Y_1, ..., Y_m$ is selected from a different Bernoulli distribution with parameter p_y.   A confidence interval about the difference in proportions, $p_x - p_y$, is desired.

1.   As an estimate of the difference in proportions, $p_x - p_y$, the difference in sample proportions,

$$\hat{p}_x - \hat{p}_y = \frac{\Sigma x}{n} - \frac{\Sigma y}{m}$$

can be used. Under the assumptions given above, if the sample sizes, n and m, are sufficiently large, the statistic

$$\frac{(\hat{p}_x - \hat{p}_y) - (p_x - p_y)}{\sqrt{\frac{p_x(1-p_x)}{n} + \frac{p_y(1-p_y)}{m}}}$$

will be approximately standard normal.

2.   The pivotal quantity we then use is

$$\frac{(\hat{p}_x - \hat{p}_y) - (p_x - p_y)}{\sqrt{\frac{p_x(1-p_x)}{n} + \frac{p_y(1-p_y)}{m}}}$$

which is approximately distributed as a standard normal distribution.

3.   The probability statement is

$$P\left(-Z_{1-\alpha/2} < \frac{(\hat{p}_x - \hat{p}_y) - (p_x - p_y)}{\sqrt{\frac{p_x(1-p_x)}{n} + \frac{p_y(1-p_y)}{m}}} < Z_{1-\alpha/2}\right) = 1 - \alpha.$$

4.   The resulting confidence interval is (using estimates in place of the parameters):

$$\hat{p}_x - \hat{p}_y \pm Z_{1-\alpha/2} \sqrt{\frac{\hat{p}_x(1-\hat{p}_x)}{n} + \frac{\hat{p}_y(1-\hat{p}_y)}{m}}$$

The point estimate is $\hat{p}_x - \hat{p}_y$, the margin of error is

$$Z_{1-\alpha/2} \sqrt{\frac{\hat{p}_x(1-\hat{p}_x)}{n} + \frac{\hat{p}_y(1-\hat{p}_y)}{m}}$$ and the confidence level is $1 - \alpha$.

DISCUSSION

A lot of the data that is collected in research investigations is nominal data or categorical data, where the unit under study either possess a given characteristic or it doesn't. For instance, a manufactured part has a defect or it does not; it meets specifications or it doesn't, the paint has a ripple in it or it doesn't, the edges have been adequately buffed or they have not. A respondent is either married or single, male or female, in agreement with the President's stand or not, either voted in the last election, watched the Thursday night move, has a new or used car, etc. There seems to be no limit to the number of issues that can be treated this way and therefore modelled by a Bernoulli random variable.

We have found that the estimator of the Bernoulli parameter p is the mean of the Bernoulli random variables, which is also the sample proportion since the Bernoulli random variable is either a 0 or a 1. We would like to find an interval estimate for the parameter p. Techniques are available to develop such estimates using the table of the binomial distribution, but we will not present them here. Rather, we present a large sample, central limit theorem based confidence interval using the fact that

$\hat{p} = \bar{x}$ has an expectation of p and a variance of $\dfrac{p(1-p)}{n}$ Therefore, in large samples $\dfrac{\hat{p} - p}{\sqrt{\dfrac{p(1-p)}{n}}}$ will

be approximately standard normal. This provides the basis for giving the confidence interval formula on p as

$$\hat{p} \pm Z_{1-\alpha/2} \sqrt{\frac{\hat{p}(1-\hat{p})}{n}}$$

If we have two separate Bernoulli parameters and two independent samples, the large sample confidence interval on the difference can be written as

$$\hat{p}_x - \hat{p}_y \pm Z_{1-\alpha/2} \sqrt{\frac{\hat{p}_x(1-\hat{p}_x)}{n} + \frac{\hat{p}_y(1-\hat{p}_y)}{m}}$$

Examples showing the use of both of these formulas are given below.

EXAMPLES

EXAMPLE 1: A telephone survey of households was interested in obtaining opinions and attitudes about agreement or disagreement with the president's military stance relative to the middle east. A question was raised about the male-female ratio of the sample. Of a total of 580 respondents, 340 were female and the remaining were male. Estimate the proportion p of females with a 95% confidence interval. Does this interval enclose the 50-50 split a balanced sample would be expected to have?

SOLUTION: The equation to use for this confidence interval is

$$\hat{p} \pm Z_{1-\alpha/2} \sqrt{\frac{\hat{p}(1-\hat{p})}{n}}$$

Inserting the appropriate numbers gives

$$\frac{340}{580} \pm 1.96 \sqrt{\frac{\frac{340}{580}(1-\frac{340}{580})}{580}}$$

or

$$0.59 \pm 0.04$$

Therefore we can say we are 95% confident that the female percentage is between 55% and 63%. Since this interval doesn't include 50% it would seem that there are more females in the sample than chance would provide if the population were really split 50-50.

EXAMPLE 2: The same telephone survey of example 1 was conducted in two different congressional districts in such a way that the sample in the two districts could be considered independent. In the last general election district I had been heavily republican while district II had been slightly democratic. A comparison of the two districts relative to the proportion of respondents that supported the president's tax proposal was desired. The sample sizes out of the two districts was 300 and 280 respectively. And the number supporting the president's tax package in each district was 220 and 190 respectively. Construct a 95% confidence interval on the difference in the two proportions.

SOLUTION: The two proportions were

$$\hat{p}_x = \frac{220}{300} = 0.7333$$

and

$$\hat{p}_y = \frac{190}{280} = 0.6786.$$

To construct a 95% confidence interval we use the following formula with $Z_{1-\alpha/2} = 1.96$.

$$\hat{p}_x - \hat{p}_y \pm Z_{1-\alpha/2} \sqrt{\frac{\hat{p}_x(1-\hat{p}_x)}{n} + \frac{\hat{p}_y(1-\hat{p}_y)}{m}}$$

Substituting the values in the expression gives:

$$(0.7333 - 0.6786) \pm 1.96 \sqrt{\frac{(0.7333)(1-0.7333)}{300} + \frac{(0.6786)(1-0.6786)}{280}}$$

or

$$0.0547 \pm 0.0741$$

Therefore, the 95% confidence interval on the difference in proportions for the two districts is from −0.0194 to 0.1288. Since this interval overlaps zero, it is conceivable that there is no difference in the districts on the tax package issue since the difference $p_x - p_y$ could equal zero.

EXERCISES

1. Stainless steel is frequently used in handling corrosive chemicals. However, it is still susceptible to what is called "stress corrosion cracking". In a sample of 280 steel alloy failures in oil refineries, it was found that 41.2% were caused by stress corrosion cracking. Construct a 90% confidence interval on the true proportion.

2. In a large university a survey was conducted to assess the overall educational perspective of the students. Of a sample of 200 freshmen, 90 reported that they had no writing experience in any of their classes during the last year. Of 185 seniors, 65 reported that they had no writing experience in any of their classes. Estimate the true difference in these proportions. Place a 95% confidence interval on this estimate.

3. An insurance company examined the claims of a sample of 300 male drivers and a comparable sample of 310 female drivers. The proportion of claims involving a police citation turned out to be 65% for the males and 58% for the females. Place a 95% confidence interval on each proportion respectively and then on the difference of the male and female proportions. What conclusions can you make in comparing the male and female drivers?

4. A department store wants to estimate the proportion of all charge accounts that are paid by the fifteenth of the month. A random sample of 500 credit accounts shows that 432 were paid before the fifteenth. Construct a 90% confidence interval on the true proportion of all charge accounts that are paid by the fifteenth.

5. A control chart on the proportion of defective nails produced in an iron works is to be set up. (A defective is one where the head is not uniform, or the point has "burrs" on it.) A sample of 100 nails is selected every half hour and the proportion recorded. Using the large sample concept of this section what is the upper and lower control limits if the true proportion where 15%? 10%? 5%?

6. Suppose we want the margin of error ε for an estimate of the proportion to be controlled at a given value with $(1-\alpha) \cdot 100\%$ confidence. Develop an equation that would show how to find the sample size to meet these conditions. [Hint: Remember that the margin of error is $Z_{1-\alpha/2}\sqrt{\frac{p(1-p)}{n}}$. Therefore, set up the equation $\varepsilon = Z_{1-\alpha/2}\sqrt{\frac{p(1-p)}{n}}$ and solve for n.]

7. Use the results of exercise 6 to find the sample size n when $\varepsilon = 0.05$ with 95% confidence (let $Z_{1-\alpha/2} = 2$) when p takes on the following values (fill in the following table):

p	$n = (\frac{2}{.05})^2 p(1-p)$
0.1	
0.3	
0.5	
0.7	
0.9	

What do your computations suggest about the maximum sample size needed to meet the conditions imposed? What parameter value p corresponds to this condition?

8. A pre-election poll is to be undertaken a few days before election day. What sample size is needed for a 95% confidence level and a margin of error of 2%? How would the sample size change if a 98% confidence level were wanted?

9. What is the maximum margin of error for a random sample used to estimate a proportion with 95% confidence when the sample size is 600?

10. What size of sample would be needed to cut the margin of error in half under the circumstances described in exercise 9?

7.9   ANALYSIS OF THE CHAPTER EXERCISE

At a particular university a study was done among the students attending the university regarding non-educational costs: housing, food, utilities, entertainment and so forth. A random sample of 400 students was selected from the student files who reported the following information. (Figures are monthly figures.)

Item	Mean	Std. Dev.
Rent	$104	$52
Utilities	$ 12	$ 4
Food	$ 68	$25

Estimate the mean amount spent on rent, utilities, and food per month per student along with a measure of "error of estimation". Also estimate the total money entering the community assuming all students at the university (20,000) are living in university or community housing, paying utilities, and making food purchases there as well. (If you are not willing to assume these categories are independent of one another, what additional information is needed?)

SOLUTION: If we define the "error of estimation" relative to a 95% confidence interval then we can summarize the estimates and their errors in the following table.

Item	Estimate	Margin of Error*
Rent	$104	$ 5.10
Utilities	$ 12	$ 0.39
Food	$ 68	$ 2.45

*Margin of Error $= \dfrac{1.96s}{\sqrt{400}}$

To estimate the total amount spent in the community by the 20,000 students per month we would create the estimate $Y = (20{,}000)104 + (20{,}000)12 + (20{,}000)68 = \$3{,}680{,}000$. The variance of this estimate is the variance of a sum (a linear combination, Chapter 3, Section 11) which in general will require knowledge of the covariances of the 3 categories. If we assume these covariances are zero then the variance of Y would be estimated using the following expression

$$V(Y) = \left\{ \frac{[(20000)(52)]^2}{400} + \frac{[(20000)(4)]^2}{400} + \frac{[(20000)(25)]^2}{400} \right\}$$

Taking the square root of this expression and multiplying by 1.96 provides the error of estimate of $113,400. Thus we estimate that the monthly money coming into the community by these 20,000 students is $3,680,000 with an error of estimate of $113,400.

To associate the probability 95% with this error of estimate requires the assumption of normality or a large sample appeal to the central limit theorem. In this case, since the sample size is 400, it is not unreasonable to expect the central limit theorem to apply comfortably and allow us to claim a 95% confidence interval around the estimate using the error of estimation.

CHAPTER SUMMARY

In this chapter we have formally introduced one process for making a statistical inference: Using the data in a sample to estimate a population parameter and placing a confidence interval around the estimated parameter that reflects our confidence in the process. Prior to completing such a task we had to learn how to use a sample to estimate the population parameter. The two methods for doing this presented in this chapter, though there are other procedures, was the *method of moments* and the *method of maximum likelihood*. We also discussed the ways of evaluating how good the estimators were using various criteria such as their *mean square errors,* their comparative *efficiency,* whether they were *unbiased, minimum-variance unbiased, consistent, sufficient,*. and so forth.

After these concepts were presented, we then presented the method and resulting formulas for confidence intervals for various special cases assuming sampling from a normally distributed population. A summary of these formulas is given in Table 7.3 the follows:

Table 7.3 Confidence interval formulas for various distributions and parameters.

Parameter	Assumptions	Endpoints
μ	$N(\mu, \sigma^2)$ or n large σ^2 known	$\bar{x} \pm Z_{1-\alpha/2}\dfrac{\sigma}{\sqrt{n}}$
μ	$N(\mu, \sigma^2)$ σ^2 unknown	$\bar{x} \pm t_{1-\alpha/2}\dfrac{s}{\sqrt{n}}$
$\mu_x - \mu_y$	$N(\mu_x, \sigma_x^2)$ $N(\mu_y, \sigma_y^2)$ variances known	$(\bar{x}-\bar{y}) \pm Z_{1-\alpha/2}\sqrt{\dfrac{\sigma_x^2}{n}+\dfrac{\sigma_y^2}{m}}$
$\mu_x-\mu_y$	$N(\mu_x,\sigma_x^2)\quad N(\mu_y,\sigma_y^2)$ $\sigma_x^2 = \sigma_y^2 = \sigma^2$	$(\bar{x}-\bar{y}) \pm t_{1-\alpha/2}\sqrt{s_p^2\left(\dfrac{1}{n}+\dfrac{1}{m}\right)}$
σ^2	$N(\mu, \sigma^2)$	$\dfrac{(n-)s^2}{\chi_{\alpha/2}^2}, \dfrac{(n-1)s^2}{\chi_{1-\alpha/2}^2}$
σ_x^2 σ_y^2	$N(\mu_x,\sigma_x^2)$ $N(\mu_y,\sigma_y^2)$ Independent distributions	$\dfrac{\sigma_x^2}{\sigma_y^2}\dfrac{1}{F_{1-\alpha/2;n-1;m-1}}, \dfrac{\sigma_x^2}{\sigma_y^2}F_{1-\alpha/2;m-1;n-1}$
p	Bernoulli	$\hat{p} \pm Z_{1-\alpha/2}\sqrt{\dfrac{\hat{p}(1-\hat{p})}{n}}$
$p_x - p_y$	Two independent Bernoulli distributions	$\hat{p}_x - \hat{p}_y \pm Z_{1-\alpha/2}\sqrt{\dfrac{\hat{p}_x(1-\hat{p}_x)}{n}+\dfrac{\hat{p}_y(1-\hat{p}_y)}{m}}$

It should be noted that in Table 7.3 an assumption of a random sampling is necessary for all the cases considered.

CHAPTER EXERCISES

1. A garden shop sells seed potatoes in "2 lb. bags" using a scale they keep on the premises. Due to the size of the potatoes, it would be difficult to ensure that each bag had exactly 2 pounds of potatoes. Therefore, the owner gave instructions to the employees to always put more than 2 pounds in each bag whenever the weight was close or questionable. A representative from the department of weights and measures weighed 10 bags of potatoes with the following results:

2.2558	2.3749	1.7493	2.2132	2.2294
2.6502	2.2209	2.2363	2.9533	2.4142

The mean and standard deviation for this data are:

MEAN =   2.3297   ST.DEV. =   0.314

Construct a 95% confidence interval on the mean weight. Does it appear that the instructions for weighing the packages are being followed.?

2. In order to obtain federal funds for a neighborhood reclamation project certain information is needed about the neighborhood in question. One such question is the population size. A sample of sixty households from 300 in the neighborhood produced a mean of 1.53 and a standard deviation of .65. The estimate is of the form: (300 • mean). Compute a 98% confidence interval on the total number of people.

3. In a study of the incidence of dental caries among children 5 years-of-age between two communities, one having fluoridation and the other not having fluoridation, the following data were collected from dental records. (The measurement is the number of decayed, missing or filled surfaces per child.)

	Communities	
	Fluoridation	No Fluoridation
Sample size	50	170
Mean	0.51	0.55
Standard Deviation	0.20	0.31

Construct a 95% confidence interval on the difference of means between the two communities.

4. On the basis of the confidence interval you obtained in exercise 3, would you say that there is statistical evidence that would support the conclusion that the mean number of dental caries differs from community to community?

5. If you conclude there is a difference in the mean number of dental caries from the two communities described in exercises 3 and 4, can you conclude that it is due to fluoridation? (Are there any other factors that might explain the difference?)

6. In a local school district, the school superintendent needed to estimate the proportion of children in the elementary schools coming from single parent, households. In investigating this subject, it was noted that there seemed to be a drastic difference in the inner city schools versus the more suburban and rural schools. In a more comprehensive study it was found that in the inner-city schools, out of a sample of 320 children, 145 were from single parent households, while in the suburban-rural sample of 300 only 92 were from single parent households. Estimate the difference in proportions between the two school types and place a 95% confidence interval around it.

7. Based on the confidence interval computed in exercise 6, would you conclude that there was a difference in percentage of single parent households when comparing the urban versus rural households? What characteristic of the interval leads to the conclusion you make?

8. A study of wages paid professionals in the medical technology field was being done. It was of interest to measure the variability of wages in the field as well as the mean wage. It turned out that a sample of 25 produced a standard deviation of $272 on the basis of a monthly wage. Place a 99% confidence interval on the true variance.

9. Using the confidence interval of exercise 8, what would you report for a confidence interval on the standard deviation?

10. Suppose a comparison of the difference in two means is to be performed and an equal sample size is proposed from each population ($m = n$). How large a sample from each population would you recommend if the maximum margin of error tolerated is 5 and the standard deviation of each population is 15. A 95% confidence interval is desired for the specified margin of error. [Hint: You will need to set up your own equation, but you can follow the model found in this chapter with only minor variation.]

7.10 APPLICATION: SCIENTIFIC SURVEY SAMPLING

Objectives

From the outset of this text we claimed the selection of a random sample was a vitally essential component of any statistical investigation. Random sampling validates the assumptions of independence and identical distributions that are so important in many of the theorems that enable us to analyze data an a setting of uncertainty. In addition, it intuitively eliminates many of the sources of bias that would result if the sample were drawn more subjectively. The data collector might be tempted to interview the more socially attractive respondent, avoiding the homeless people of the street, or avoiding the households with large barking dogs, etc. A random sampling procedure would not allow for such arbitrary avoidances in selecting a sample.

In this section we briefly introduce the important designs used in survey sampling along with some mathematical formulas for their use. Specifically, we will examine: (1) simple random sampling, (2) stratified random sampling, (3) one-stage cluster sampling, and (4) systematic random sampling. Be aware that entire textbooks are devoted to these topics and that we only scratch the surface in this section.

Introduction

The Federal Government has been mandated to collect information on a regular basis to measure, among other things, the monthly unemployment rate, population increases, agricultural production, etc. To conduct a complete census every month, let alone every year or five years is exorbitantly expensive. Therefore, a monthly sample survey of households called the Current Population Survey has been commissioned which samples about 60,000 households every month. The question is, "How do you design a sampling scheme that will provide estimates with a specified margin of error for a large and complex population such as the mobile U.S. population?"

Solving such complex sampling problems comes within the realm of survey sampling. Much of the development in theory and methodology associated with this field is traced back to the work done at the Bureau of the Census within the last 30-40 years. Following the work of Jerzy Neyman in the 1930's, the Bureau first implemented aspects of sampling into their 1940 census. Coincident with this was the creation of the Sample Survey of Unemployment, the precursor of the Current Population Survey. (The Department of Agriculture was the first federal agency to lead out in the field of survey sampling with the Bureau of the Census later making the major thrust in developing methods and theory.)

Since that time, probability sampling methods have become the generally accepted model for collecting sample data from large populations. Techniques that are used are referred to as stratification, cluster sampling, sampling with probability proportional to size (pps), and others. The purpose is to obtain as much information as possible (estimate has a low variance) for a given expenditure (constant sample size) as possible. Statistical concepts used in this area are the application of estimation techniques in the absence of a known population model, finding the variance of complex functions, finite population sampling, and so forth. See Cochran, W.G. (1977). *Sampling Techniques*, John Wiley & Sons, Inc.

General Concepts and Notation

We first assume that the population from which we are to draw our sample is a finite population with N units in it. We denote the population mean by:

$$\mu_y = \sum_{i=1}^{N} \frac{Y_i}{N}$$

and the population variance by

$$\sigma_y^2 = \sum_{i=1}^{N} \frac{(Y_i - \mu_y)^2}{N}$$

(See Chapter 1, Sections 1.5 and 1.6.) The ratio of the sample size to the population size is called the sampling fraction and is denoted by $f = \dfrac{n}{N}$. Samples are selected without replacement; i.e., once an observation is selected from the population it is "set aside" and cannot be selected again. This is in contrast to the "independent, with-replacement sampling" assumed in the rest of this text. However, if n is much less than N, (usually if $f < 0.05$) the difference between sampling with replacement and without replacement, while of theoretical interest, will show little practical difference numerically.

Simple Random Sampling

A simple random sample is a sample drawn so that every element of the population has an equal chance of selection. Denoting the finite population size by N and the size of the sample by n, sampling without replacement means there are $\binom{N}{n}$ total possible samples that can be drawn each of which must be equally likely (see Chapter 2, Appendix B for an explanation of this combinatorial counting rule). A simple method for selecting a simple random sample is to label each member of the population with a number from 1 to N. Then, using a random number table, n random numbers are selected from the range of numbers from 1 to N, disallowing duplicate numbers. The n random numbers are matched to the population numbers and indicate which of the population elements are selected for the sample.

As a simple illustration, suppose we have a high school graduating class of 500 students and we want to sample 25 of them to measure their attitudes concerning a senior class gift to the school. From the school rolls, we number the students from 1 to 500. Then, using a random number generator we generate 25 numbers in the range of numbers from 1 to 500. The following 25 numbers represent one such sample selection that might result.

117	473	83	220	385
217	40	441	269	71
378	368	76	293	330
91	222	176	450	249
134	56	297	300	218

Therefore, the persons number 117, 473, 83, etc., would be the individuals who would be interviewed for this investigation.

You should realize that though a simple random sample is the easiest sampling scheme to define and describe, it is a very difficult procedure to implement in most real-world sampling problems. The primary problem is that there are very few sampling problems encountered where a list (called a sampling frame) exists that can be labeled as described above. There is no list of female heads of households, of people who have been exposed to AIDS, of teenagers who have a CD player. Thus, to sample such specialized populations other methods of sampling have to be implemented rather than simple random sampling.

Another problem with simple random sampling, as well as other sampling methods, is that most populations are constantly changing by "births" and "deaths." Therefore, a list that may be found for simple random sampling is probably out-of-date by the time it is used to select a sample. This needs to be taken into account when describing the characteristics of the population.

The population parameters we usually want to estimate are the population mean, and the population total. We use the sample mean which is an unbiased estimator of the population mean. Along with the estimate, we want the variance of the estimate so we can formulate an estimate of the margin of error of the estimate. We appeal to the central limit theorem to justify our formula for a margin of error since we usually have a large sample size in investigations of the type described here. Thus, we use a Z-score of 2 (1.96 rounded up) for a margin of error associated with a 95% confidence interval concept. These formulas are summarized in Table 7.4 below. The first column identifies the parameter; the second column the estimator of the parameter; and the third and fourth columns identify the variance of the estimator and the estimated margin of error for the estimator. (The margin of error equals twice the standard error: $\varepsilon = 2 \cdot S.E.$)

Table 7.4 Summary of formulas for estimators and their characteristics in simple random sampling.

Parameter	Estimator	Variance of Estimator	Estimated Margin of Error $\hat{\varepsilon}$
Mean: μ	$\bar{y}$	$V(\bar{y}) = \dfrac{\sigma^2(N-n)}{n\,(N-1)}$	$2 \cdot \sqrt{\dfrac{s^2}{n}(1-f)}$
Total: τ	$N\bar{y}$	$V(N\bar{y}) = N^2\left(\dfrac{\sigma^2(N-n)}{n\,(N-1)}\right)$	$2 \cdot N\sqrt{\dfrac{s^2}{n}(1-f)}$

In the above expressions for the variance, the term $\dfrac{(N-n)}{(N-1)}$ is referred to as the "finite population correction" factor or the "fpc." In the expressions for the estimated margin of error, the fpc becomes $\dfrac{(N-n)}{N} = 1 - \dfrac{n}{N} = 1 - f$. This term is the adjustment to the variance that comes about because we are sampling a finite population, without replacement. The term $1 - f \le 1$ for all $n \ge 1$. Thus the effect of sampling without replacement is to reduce the variance over that which it would be if the population were infinitely large. This term will show up in the variance expressions for other sampling designs besides simple random sampling.

EXAMPLES

EXAMPLE 1: A simple random sample of 9 customer charge accounts is selected to estimate the total amount of money due for the entire population of 690 active accounts the business holds. Estimate the total amount due and compute the margin of error for the estimate. The data are:

| 89.50 | 60.00 | 75.50 | 81.00 | 82.50 | 78.00 | 69.00 | 79.00 | 84.50 |

SOLUTION: In this problem $N = 690$, $n = 9$, $f = \dfrac{9}{690}$, and from the data we get $\bar{y} = \$77.67$ and $s = \$8.77$. Therefore, the estimates of total due and the margin of error are:

$$N\bar{y} = 690 \bullet \$77.67$$

$$= \$53,592.30$$

$$2 \bullet N\sqrt{\frac{s^2}{n}(1-f)} = 2 \bullet 690\sqrt{\frac{8.77^2}{9}(1-\frac{9}{690})}$$

$$= \$4007.80$$

Therefore, we estimate the amount due is about $53,600 with a margin of error of about $4000. (Notice, we could report all the the digits from the computations above, but it would give an illusion of more "precision" that is probably justified.)

EXAMPLE 2: In a study on the effectiveness of a new toothpaste in inhibiting the development of plaque, a group of 500 schoolchildren participated. After 4 months a sample of 16 randomly selected children were examined and the number of teeth showing obvious development of plaque was recorded. Provide an estimate of the mean number of teeth with plaque problems and its standard error. The data are:

| 0 | 4 | 2 | 3 | 2 | 0 | 3 | 4 |
| 1 | 1 | 3 | 3 | 1 | 0 | 2 | 1 |

SOLUTION: We want to estimate the mean and its estimated standard error. For the above data we have $N = 500$, $n = 16$ and $f = \dfrac{16}{500}$. Also,

$$\bar{y} = 1.875$$

and

$$2 \bullet \sqrt{\frac{s^2}{n}(1-f)} = 0.67$$

Therefore, the mean number of teeth showing indication of plaque is 1.875 with a corresponding margin of error of 0.67.

Stratified Random Sampling

We often find that a more efficient method of selecting a sample is to first divide the population in mutually exclusive groups of homogeneous elements per group. These groups are called strata and sampling in this context is referred to stratified sampling. If we take a simple random sample within each group or stratum, we call it stratified *random* sampling. (There may be other reasons for stratifying a population other than to make a more efficient estimator. These are often due to administrative considerations or because of specific interest in the individual strata. For instance, in a survey of employee retirement plans, we may have a large company with 5 subdivisions located at 5 different locations throughout the country. It may be administratively simpler to let each sub-division represent a stratum, with each subdivision taking responsibility for appropriately sampling employees within the subdivision. Thus, a stratified sample results with 5 strata being created. In another situation, we may be interested in a sample survey of electrical usage in a particular state with particular interest in the 3 largest counties in the state. We may therefore create a stratum each for the 3 largest counties and a 4th stratum for the remaining part of the state. Thus, we would have a stratified sample of the state consisting of a total of four strata)

Some additional notation is necessary for stratified random sampling. Let

L = number of strata

N_l = number of units in stratum i

N = number of units in the population = $N_1 + N_2 + \ldots + N_L$.

μ_i = mean of the N_i units in stratum i

σ_i^2 = variance of the N_i units in stratum i

$f_i = \dfrac{n_i}{N_i}$ = sampling fraction in i^{th} stratum

The formulas we use to estimate a mean and a total in stratified sampling are shown in Table 7.5 below.

Table 7.5 Summary of formulas for estimators and their characteristics in stratified random sampling.

Parameter	Estimator	Variance of Estimator	Estimated Margin of Error $\hat{\varepsilon}$
μ	$\hat{\mu}_{str} = \displaystyle\sum_{i=1}^{L} \frac{N\bar{y}_i}{N}$	$\displaystyle\frac{1}{N^2}\sum_{i=1}^{L} N_i^2 \frac{\sigma_i^2}{n_i}(1-f_i)$	$2\sqrt{\displaystyle\frac{1}{N^2}\sum_{i=1}^{L} N_i^2 \frac{s_i^2}{n_i}(1-f_i)}$
τ	$\displaystyle\sum_{i=1}^{L} N_i\bar{y}_i$	$\displaystyle\sum_{i=1}^{L} N_i^2 \frac{\sigma_i^2}{n_i}(1-f_i)$	$2\sqrt{\displaystyle\sum_{i=1}^{L} N_i^2 \frac{s_i^2}{n_i}(1-f_i)}$

The application of these formulas will be shown in an example below.

EXAMPLE

EXAMPLE 3:   A zoning commission is formed to estimate the average appraised value of homes in a suburb of a city. Two voting districts in the city were used because of separate lists of dwelling units available for each district. Estimate the average appraised value for all houses in the suburb along with a margin of error.

SOLUTION: The data and some of the necessary calculations are given in Table 7.6 below.

Table 7.6 Summarized data and calculations for estimating the mean appraised value of homes.

Stratum	N_i	n_i	$\bar{y}$	$N_i \cdot \bar{y}$	s_i^2	$N_i^2 \frac{s_i^2}{n_i}(1 - f_i)$
1	110	20	12,000	1320000	5263158	2605263104
2	168	30	14,000	2352000	482759	373076160
Total	278			3672000		2978339264

Therefore, the estimate of mean appraised value is:

$$\hat{\mu}_{str} = \sum_{i=1}^{L} \frac{N_i \bar{y}_i}{N} = \frac{3672000}{278} = 13208$$

with a margin of error of

$$2\sqrt{\frac{1}{N^2}\sum_{i=1}^{L} N_i^2 \frac{s_i^2}{n_i}(1 - f_i)} = 2(\frac{1}{278})\sqrt{2978339264} = 392.620.$$

Therefore, we can say that estimate of the mean appraised value is approximately $13,200 with a margin of error of about $400.

One-stage Cluster Sampling

Cluster sampling is usually implemented for sampling convenience and economy rather than an expectation that better estimators will result. One-stage cluster sampling is implemented by first forming the elements in the population into groups (usually relatively small groups) or clusters. A simple random sample of these clusters is selected and measurements are obtained from every element in the clusters chosen. (If a *random sample* of elements is selected from each cluster selected rather than taking all elements, then a two-stage cluster sample results. Coverage of two-stage cluster sampling is more than we want to cover in this brief presentation.)

For example, suppose a manufacturing plant wants to assess the attitude of residents in a rural neighborhood about their positive and negative impact on the community life-style. They intend to restrict their investigation to heads of households, and spouses in a well-defined geographical area around the plant. There is no list of heads of households that is conveniently available, so they decide to let the households in the area represent clusters. An aerial photo is made of the geographical area and the households are numbered. A random sample of households is then taken from the aerial photo and interviews of both husband and wife are conducted.

The notation we use in one-stage cluster sampling is:

N = the number of clusters

n = the number of clusters selected in a sample

M_i = the number of elements in cluster i

M_o = the total number of elements in the population $(M_o = \sum\limits_{i=1}^{N} M_i)$

y_{ij} = the observed value of the j th element within the i th cluster

y_i = the unit total $(y_i = \sum\limits_{i=1}^{M_i} y_{ij})$

$\bar{Y} = \sum\limits_{i=1}^{M_i} \dfrac{y_i}{N}$ = the mean of the cluster totals

The formulas for estimates and margins of error for a one-stage cluster sample are given in Table 7.7 below.

Table 7.7 Summary of formulas for estimators and their characteristics in one-stage cluster sampling.

Parameter	Estimator	Variance of Estimator	Estimated Margin of Error $\hat{\varepsilon}$
Mean	$\dfrac{N}{nM_o}\sum\limits_{i=1}^{n} y_i$	$\dfrac{N^2(1-f)}{nM_o^2} \dfrac{\sum\limits_{i=1}^{N}(y_i - \bar{Y})^2}{N-1}$	$2\sqrt{\dfrac{N^2(1-f)}{nM_o^2} \dfrac{\sum\limits_{i=1}^{n}(y_i - \bar{y})^2}{n-1}}$
Total	$\dfrac{N}{n}\sum\limits_{i=1}^{n} y_i$	$\dfrac{N^2(1-f)}{n} \dfrac{\sum\limits_{i=1}^{N}(y_i - \bar{Y})^2}{N-1}$	$2\sqrt{\dfrac{N^2(1-f)}{n} \dfrac{\sum\limits_{i=1}^{n}(y_i - \bar{y})^2}{n-1}}$

An example will illustrate the application of these formulas.

EXAMPLE

EXAMPLE 4: The employees of a company are given the option in their benefit package of buying varying increments of supplemental life insurance. The company would like to estimate the total life insurance obligation its insurer would have at the present time. A random sample of 7 departments is selected and the records of all the employees in each of the departments are examined. The total amount of supplemental life insurance selected is recorded in units of $1000. Find the estimate of total life insurance obligation knowing there are 28 departments in the company. The data and some of the computations are shown in the following solution.

SOLUTION: The totals for the 7 departments in the sample are:

| y_i | 61300 | 59670 | 72350 | 65490 | 55750 | 67320 | 70510 |

$$\text{Total} = \frac{N}{n} \sum_{i=1}^{n} y_i = \frac{28}{7}(452390) = \$1809560 \ (1000\text{'s})$$

$$2\sqrt{\frac{N^2(1-f)}{n} \cdot \frac{\sum_{i=1}^{n}(y_i-\bar{y})^2}{n-1}} = 2\sqrt{\frac{28^2(1-0.25)}{7}(36115556)} = \$110158 \ (1000\text{'s})$$

Therefore, we can say that the total insurance liability is $1,810,000,000 with a margin of error of $110,200,000.

EXAMPLE 5: Using the information from example 3, and knowing that there are 6277 employees in the company, what is the estimate of the mean amount of supplemental life insurance that employees are selecting?

SOLUTION: The solution is very simple using the computations of example 3. To find the estimate of the mean we compute

$$\frac{\$1,809,560,000}{6277} = \$288,284 \ (\text{or } \$288,000)$$

with a margin of error of

$$\frac{\$110,158,000}{6277} = \$17,549 \ (\text{or } \$17,500).$$

Systematic Random Sampling

In many instances, conditions suggest such that sampling using a "system" would be much easier to implement and with a certain intuitive appeal as well. For instance, suppose we want to sample shoppers of a particular grocery chain. It would be impossible to find a list of shoppers, but it would be very easy to stand outside the store on a given day and interview every 5th shopper coming out of the store. In addition, this system would spread the shoppers out over the whole day getting early, late and mid-day shoppers in the sample.

The process of implementing a systematic random sample in its ideal setting is to determine the size of sample and then compute $k = \frac{N}{n}$ (hopefully an integer value). We then choose a "random" starting point between 1 and k and starting with that element on the list, sample that one and every kth one as we proceed through the list. This should produce a total of n units in the sample. (Obviously, in most cases k will not be an integer, so depending on whether we round up or round down we could end up with $n - 1$ or $n + 1$ observations in the sample depending on the starting point that randomly turns up. We won't regard this as a serious problem in most cases since n will typically be fairly large anyway.)

There are advantages and disadvantages to systematic sampling. Among the advantages are its characteristic of spreading the sample out across the entire population. If our list has any natural stratification in it, the systematic sample will mimic a stratified sample. Systematic sampling is usually easier and less costly to implement than a simple random sample, and in some cases it may be the only logical way to select a sample.

For instance, some years ago I was associated with a study of state tax returns. The sample was to be selected as they were mailed in so that the sample would be complete shortly after April 15th. There is an expectation that those tax returns coming in early will receive refunds, while those coming in late will owe taxes. Thus, two strata through time are suggested. A systematic random sample of returns was selected producing a sample evenly spread out according to the number of returns coming in per time period. A simple random sample would not be practical in this case because it would require that we wait until all the returns were in before drawing the sample.

A major disadvantage of systematic sampling is the absence of a pure estimate of the standard error of the estimate. A systematic sample is equivalent to a cluster sample made up of k clusters. We select one cluster at random and you cannot get a measure of variability of the between-cluster variance with only one cluster in the sample.

Another disadvantage is the risk that there may be natural periodicities that match the sampling interval k. If so we could get very poor estimates and estimates of the standard error could be very misleading. For instance, suppose we sample every 6th can of tomatoes coming out of a machine with the purpose of determining whether the acid content is acceptable. Suppose that coincidentally the machine consists of a turret of 6 cans capacity that fills the cans. Thus, our systematic sample would be examining the cans coming from only one filling position from the machine. Thus, if that particular position has a tendency to overfill or underfill, our measurements would reflect that and would not be representative of all cans coming off the machine.

The usual technique for estimating means and totals and their standard errors and margins of error is to assume that the list is in random order. Thus, the formulas of Table 7.4 will apply when finding estimates of the mean and total. If the list is not in random order, the estimates could be very good or very bad depending on what patterns may exist in the list relative to the characteristic being measured. In any event, the estimate of the margin of error would most likely not be reliable for a systematic random sample of a non-randomly ordered list.

EXAMPLE

EXAMPLE 6: In an orchard of young cherry trees we want to estimate the total harvest in pounds. Rather than take a simple random sample it is easier to instruct the workers to select every 20th tree after which an experienced picker then estimates the weight of the crop on the tree. A total of 70 trees is examined in this fashion. Find an estimate of the total weight in pounds for the harvest and compute its margin of error. The following summary computations have resulted. (There are 1400 trees in the orchard.)

$$\sum_{i=1}^{70} y_i = 5480 \qquad \sum_{i=1}^{70} y_i^2 = 576780$$

SOLUTION: An estimate of total and its standard error assuming the yield is distributed randomly through the orchard is:

$$N\bar{y} = 1400 \cdot \frac{17,066}{70} = 109,600 \text{ pounds}$$

and

$$2 \cdot N\sqrt{\frac{s^2}{n}(1-f)} = 2 \cdot 1400\sqrt{\frac{2141.66}{70}(1-\frac{70}{1400})} = 15095.45 \text{ pounds.}$$

Thus, we would estimate the total yield to be about 110,000 pounds with a margin of error of about 15,100 pounds. (An assumption of random order is probably not too reasonable in this situation. There are probably sections of the orchard which have more fertile soil, or that get better water drainage, and so forth. This argues against "random" order. The systematic sample would probably produce results closer to a stratified sample in this case.)

EXERCISES

1. The manager of a machine shop wishes to estimate the average time an operator needs to complete a simple task. The shop has 98 operators. Eight operators are selected at random and the time to completion of the task is recorded. Estimate the average time to completion for the task and compute the margin of error. Write one or two sentences to explain your estimates for the manager of the machine shop.

Time (in minutes)	
4.3	5.2
4.9	4.7
7.8	5.0
3.6	4.2

2. A tape and record company, as part of its quality control program, samples cassette tapes to check for defects before packaging them. A total of 30 tapes is selected and the following results were produced. (Here we let $y_i = 0$ if no defect is found, and $y_i = 1$ if a defect is found.) Compute the sample mean and interpret it. Also compute the sample variance and compare it to the expression y (Notice how close the form of this expression is to the variance of a Bernoulli random variable, especially since $\hat{p} = \bar{y}$.) The data are given below.

Cassette Number	Response, y_i
1	1
2	0
3	1
4	0
5	0
.	.
.	.
30	0
	$\displaystyle\sum_{i=1}^{L} y_i = 2$

3. Use the data as it is found in exercise 2 to estimate the proportion (mean) of cassettes that are defective. Place a margin of error around your estimate.

4. A forester wants to estimate the total number of farm acres planted in trees for a state. The number of acres of trees varies with the size of the farm so she created four strata. A stratified random sample of 40 farms out of the 240 in the state was selected with the following data reported for the number of acres of trees per farm selected. Estimate the total number of acres of trees and find the margin of error.

Stratum	N_h	n_h	Data
1	86	14	97 67 42 125 25 92 105 86 27 43 45 59 53 21
2	72	12	125 155 67 96 256 47 310 236 220 352 142 190
3	52	9	142 256 310 440 495 510 320 396 196
4	30	5	167 655 220 540 780

5. Take the data of exercise 4 and merge it together into one sample of size 40. If this were a simple random sample of size 40 estimate the total number of acres of trees and find the margin of error. How does this margin of error compare to the margin of error of exercise 4.

6. Accountants frequently obtain cost inventories by sampling rather than doing a complete examination. A random sample is often difficult to implement, but with items arranged on shelves, a random sample of shelves is relatively easy to select. A sample of 10 of 48 shelves was randomly selected with the results shown in the following table. Estimate the total dollar value of the items on the shelves and compute an estimate of the margin of error.

Shelf (cluster)	Number of items	Total dollar amount
1	42	83
2	27	62
3	38	45
4	63	112
5	72	96
6	12	58
7	24	75
8	14	58
9	32	67
10	41	80

7. Use the data of exercise 6 to estimate the total number of items in the inventory. Provide an estimate of the margin of error.

8. List all possible samples of size $n = 2$ that can be selected from the population $\{1, 2, 3, 4, 5\}$. Calculate σ^2 for the population and the mean for each sample. Then calculate the variance of these means by the formula

$$\sigma^2 = \sum_{i=1}^{K} \frac{(\bar{y}_i - \mu_{\bar{y}})^2}{K}$$

where $K = \binom{N}{n}$, the number of possible samples. Thus, verify the formula $V(\bar{y}) = \frac{\sigma^2(N-n)}{n\,(N-1)}$.

9. Consider a population consisting of the elements {1, 1, 2, 2, 3, 3}. If three strata were to be constructed with one observation to be taken from each strata, what would you think is the best way to stratify this population? If two clusters were to be formed and one cluster is randomly selected, what is the best way to form the clusters?

10. Suppose a systematic random sample of size 3 is to be selected from the population of exercise 9, with the order of the population given in the list above. Is this type of sample most like a stratified sample or cluster sample?

8

TESTING HYPOTHESES I: NORMALLY DISTRIBUTED POPULATIONS

OBJECTIVES

Upon completing this chapter you should understand the vocabulary of tests of hypotheses and be able to relate that vocabulary to a practical problem. You should be able to apply the methods of this chapter to a variety of settings, such as testing for means, or variances for both one or two samples. Most importantly you should be able to list the background assumptions for each test used in this section to be valid. In addition, the scientific method as applied to a statistical problem should be understood.

You will also receive an introduction to the concept of a likelihood ratio test and the Neyman-Pearson Lemma which defines the "best" critical region for a simple hypothesis.

INTRODUCTION

Many problems of a statistical nature present themselves naturally as a "test of hypothesis." A tire salesman claims that a radial tire improves gas mileage. The candidate for senator claims that a majority of the constituents in the state agree with the position the senator has taken on foreign aid. An educational psychologist claims that grade school students show appreciable differences on math test scores in classes that average 22 students per class compared to classes averaging 25 students per class or less.

In each case presented here, a statement, claim, or assertion has been made which begs to be proved or disproved. One way to validate or invalidate the claim is to select a sample from the appropriate populations described and look at the data. If the data support the claim then we may decide the claim is true; if the data are inconsistent with what we would expect if the claim is true, then we may decide the claim is false. Or the data may be inconclusive and so we don't know what to decide. Another concern that we have to deal with is that the data may be "bad" and thus we end up supporting one conclusion when in fact we should make a different conclusion.

In this chapter, we intend to develop a logical approach to dealing with these types of problems and questions. In doing so, a new set of vocabulary terms will also be introduced. You will find that there are many similarities between the SIM of this chapter and the SIM of chapter 7. However, it will be important for you to make a clear distinction in your mind between what we do here and what was done in chapter 7. Then, after you have sorted everything out and are feeling comfortable with the concepts and methods presented here, you can then go back and start to examine the similarities and differences between "tests of hypotheses" and "confidence intervals."

Where We Are Going in This Chapter

The topic of this chapter deals with the very important statistical method called "tests of hypotheses". In reality this is an application of a procedure generally referred to as the "scientific method". When describing the scientific method, typically about 5 steps are defined: (1) Stating the problem; (2) Formulating the hypothesis; (3) Experimentation and data collection; (4) Evaluation and interpretation of the data; and, (5) Drawing conclusions on the basis of the data evaluation. Conceivably, this 5-step process could be "never-ending" since step 5 could lead to a clearer statement of the problem, a new hypothesis, more experimentation, etc. As we develop the methods of this chapter these steps should be remembered and noted as we formalize the techniques of statistical tests of hypotheses.

In the context of our course and terminology, when we formulate a hypothesis, we will state it by specifying a value or values of the parameters of an assumed pdf $f(x, \omega)$. The experimentation process will consist of selecting a a random sample from the population defined by the theoretical model. See Figure 8.1.

Analyzing the data involves computing various sample statistics of interest and of computing the probabilities of getting the experimental results under the model assumptions. The conclusion will amount to deciding whether the observed data are consistent with the model in a probability sense or whether the data and the model are inconsistent with one another. See Figure 8.1 which displays these various components.

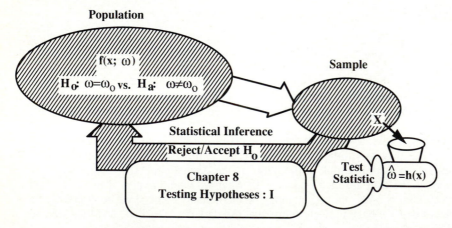

Figure 8.1 SIM—rejecting or accepting a hypothesis using a sample from a population.

A large part of this chapter will require that we establish new terminology and notation along with clarifying the logic of our application of the scientific method. You will see many parallels with the foundations we dealt with in the last chapter on estimation, but you must make a very strong effort to catalog and separate the logic and terminology of the two techniques or you will intermix the two inappropriately.

Chapter Exercise

With oil and gas prices showing substantial increases over the last decade, more and more interest has developed concerning methods to improve engine efficiency, gas mileage, and so forth. In an attempt to lure more customers to their franchise stations, suppose a major oil company has developed a gas additive that is supposed to increase gas mileage in subcompact cars. However, before marketing the new additive, the company conducts the following experiment: One hundred subcompact cars are randomly selected and divided into two groups of 50 cars each. The gasoline additive is dispensed into the tanks of the cars in one group but not the other. The miles per gallon obtained by each car in the study is then recorded. The data are summarized in the the accompanying table. Is there sufficient evidence for the oil company to claim that the average gas mileage obtained by subcompact cars with the additive is greater than without the additive?

	Without Additive	With Additive
Sample Size	50	50
Mean (mpg)	27.6	31.9
St.Dev.(mpg)	9.3	11.3

8.1 TESTS OF HYPOTHESES AND THE SCIENTIFIC METHOD

Introduction

In an experiment to assess the affect of cocaine on experimental rats, a miniature infusion pump was surgically implanted in each rat from a set of experimental rats. Some rats were then chosen randomly to receive a regular infusion of cocaine; others to receive an infusion of a saline solution. It was anticipated that the cocaine would have a depressing affect on the activity of the animals. This belief represents the hypothesis to be investigated in the experiment.

To translate this into a workable statistical problem, a measurement was defined that represented the number of times a rat broke a light beam in the course of an hour. Thus, a more active animal will break the light beam more times. We might state, as a "statistical hypothesis," that the mean number of times the beam is broken per hour by "cocaine infused" rats is less than the mean times per hour for the "saline infused" rats. Symbolically we write: $\mu_{cocaine} < \mu_{saline}$ which is called the "research hypothesis" or alternative hypothesis H_a.

Our starting position for investigating this hypothesis is to assume that there is no difference; i.e., that $\mu_{cocaine} = \mu_{saline}$ which is called the "null hypothesis" denoted by H_o.

To objectively investigate this question, a set of rules for conducting the experiment has to be set up (sometimes called a protocol) and a rule established to decide when the data is strong enough to support the research hypothesis in contrast to the null hypothesis. Such a set of rules is formally called a "test of hypothesis." There is an obvious connection between these rules and the investigative procedures commonly referred to as the *scientific method*.

In this section we introduce some basic definitions and notation in order to get started. Then the scientific method, as implemented in the discipline of statistics, is introduced. Several examples are presented to illustrate the concepts presented and to make the notation more explicit. It would be useful to read through the section completely including the examples. Once this first reading is completed, then go back and reread the definitions and the description of the scientific method, referring often to the worked-out examples. This will help solidify all the ideas presented in this section.

Definitions

DEFINITION 8.1: A *statistical hypothesis* is an assertion about the distribution of one or more random variables. If the statistical hypothesis completely specifies the distribution, including all parameter values, it is referred to as a *simple hypothesis*, otherwise it is referred to as a *composite hypothesis*.

DEFINITION 8.2: A *research hypothesis* or *alternative hypothesis*, H_a, is a statement that typically represents the position to be proved by the researcher. In most cases it defines the "new thinking," the "new theory," the "new explanation" of things. In most examples, the alternative hypothesis is a composite hypothesis.

DEFINITION 8.3: The *null hypothesis*, H_o, is the complementary condition of the "alternative" hypothesis. It denotes the "status quo," the "current thinking" on the subject, the position of "no change." We usually will write the null hypothesis as a "simple hypothesis."

DEFINITION 8.4: A *test statistic* $TS(\hat{\omega},\omega)$ is a function of an estimator $\hat{\omega}$ and the parameter ω. It is used to decide whether to reject or accept the null hypothesis. It is essential that the sampling distribution of $TS(\hat{\omega},\omega)$ be known under the assumed pdf, $f(x;\omega)$ in order that we may calculate probabilities of correct or incorrect decisions. (In most cases in this chapter we assume that $f(x;\omega)$ is a normal distribution).

DEFINITION 8.5: A *decision rule* is a rule that specifies the conditions under which the test statistic leads to rejecting the null hypothesis. The decision rule produces two regions on the real line called the rejection region, and the acceptance region. If the test statistic produces a value in the rejection region, then the null hypothesis is to be rejected. If the value of the test statistic falls in the acceptance region, then we have "failed to reject H_o."

DEFINITION 8.6: The *level of significance*, denoted by α, is the probability selected to classify an event as "rare."

DISCUSSION

In Chapter 3 and in succeeding chapters the concept of a "rare event" was introduced. The argument presented there and demonstrated in exercises and examples in succeeding chapters was that if an event occurred which would be classified as a rare event—an event with a very small probability of occurring under a given set of assumptions—wisdom would suggest that we should examine the assumptions to see if one or more has been violated which violation would explain the occurrence of the rare event. This simple line of reasoning will be explored in much more detail in this chapter with an introduction of methods and vocabulary of the principles of "tests of hypotheses." The relationship of these principles to the scientific method will also be closely noted.

In the following, the probability chosen to classify an event as a "rare event" is called the *level of significance* and is denoted by the Greek symbol α. Typically α will be small, on the order of 0.10, 0.05 or less. Consider, now, the following example.

One of the largest experiments ever conducted was the field test of the Salk poliomyelitis vaccine. Over a million children were involved in the experiment and, because of the significance of the potential findings, elaborate precautions were taken to ensure sound scientific results. The children involved in the experiment were randomly divided into one of two groups: the experimental group who received the Salk vaccine, and the control group who received a placebo of a simple salt water solution.   Neither the physicians administering the vaccine and later assessing the health status of the participants nor the children themselves knew who had received the vaccine and who had received the placebo. The method for random assignment of each child to the experimental or control group, however, guaranteed that each participating child was as likely to receive the vaccine as the placebo. (There are other political, and health related issues that had to be dealt with in this study but are not pertinent to our discussion here. For a more complete description of the experiment see *Statistics: A Guide to the Unknown* by Tanur, J.M. and others, Holden-Day, Inc.)

After the specified passage of time, data on the incidence of polio in both the experimental and control groups were examined and compared, with the expectation that, if the the vaccine had any affect, the experimental group would have a lower incidence of polio per 100,000 (the rate) than the control group. This difference in incidence had to take into account the ordinary differences that "chance" would produce; i.e., was the difference in incidence rates large enough to be classified as a "rare event?"

This important experiment is a classical example of the approach and methods of the scientific method and of testing a hypothesis, and can be used to illustrate the various definitions presented in this section.

The Scientific Method

The "scientific method" as a process is described in different ways depending on the person whose description you read. Though there is no single formal definition, it appears that all definitions have four components in common: (1) the process of exploring the known base of knowledge in order to narrow the list of possible hypotheses to those that justify further investigation, (2) the postulation of a specific hypothesis in sufficient detail so that the data that needs to be collected and the measurements that need to be made are clearly understood, (3) the collection of data from experimentation or observation that provides "reasonable" persons the basis of evaluating the truth of the proposed hypothesis, and (4) the preparation and writing of the conclusion based on the interpretation of the data to indicate whether the hypothesis has been confirmed or not and the direction that further study should take.

Step 4 of this method can be seen as merging with step 1 to set you on a new cycle through the entire sequence. This process continues and in its wake "scientific truth" is obtained.

In this chapter, and this section specifically, we examine the application of the scientific method to research under the formal heading of "tests of hypothesis." You should be able to identify the four components of the scientific method described above in the discussion and examples that follow.

Tests of Hypotheses

Let's reconsider the example of the Salk vaccine experimental trial that introduced this section. Prior to the field trial that was used on human subjects, there was substantial prior study and laboratory research. Finally, it was determined that the evidence gathered over time justified the implementation of a large-scale study on human subjects.

The hypothesis to be tested was that the true rate of polio for those who receive the vaccine would be lower than those who do not receive the vaccine. This statement is the *research* hypothesis and is called the *alternative* hypothesis in this text. The *null* hypothesis is that there is no difference in the rates.

To frame this null and alternative hypothesis in a form more consistent with our approach in this text, we argue as follows: Since each person either contracts the disease or does not, we can use the Bernoulli distribution $f(x;p) = p^x(1-p)^{1-x}$ as a model for this experiment. Then, letting the Bernoulli probability, p, denote the probability of getting the disease, we express the null and alternative hypothesis in terms of the parameters as:

$$H_o: p_S - p_P = 0 \text{ vs. } H_a: p_S - p_P < 0$$

where p_S = "the probability of contracting poliomyelitis with the Salk vaccine", and p_P = "the probability of contracting it with the placebo." This form of the null hypothesis $H_o: p_S - p_P = 0$ is called a "simple hypothesis"—we have specified that $p_S - p_P = 0$ which gives a complete specification of the distribution. The alternative is a "composite hypothesis"—it is composed of many "simple hypotheses."

Note that we have taken a general statement of an alternative and null hypothesis and assumed a specific model $f(x,\omega)$ involving some parameter ω. The null and alternative hypothesis were then expressed in terms of the parameter(s) of the distribution.

Next, an experiment was defined with sufficient controls and "blinds" to protect the validity of the conclusions to be made. Involved in this decision is the choice of the size of the sample, and a choice of the level of significance α.

Next we choose a "test statistic" as a function of the sample values that would provide a logical basis to decide whether the alternative hypothesis was true or not. In the polio example we would look most likely at the proportion of children in the experimental group who contract the disease, and compare it to the proportion in the control group who contract the disease. (We can compare these two proportions by looking at the arithmetic difference of the sample proportions $\hat{p}_s -$ $\hat{p}_p$. If the null hypothesis is true, this difference should be close to zero. If the alternative hypothesis is true and the null hypothesis is false, then the difference ought to be negative.)

Finally, the experiment defined earlier would be implemented, the data collected and analyzed and a decision as to whether differences observed where more consistent with the alternative hypothesis or the null hypothesis. If a "rare event" occurred then we would reject the null hypothesis in favor of the alternative hypothesis. In this particular illustration, the difference in polio rates between the vaccine group and the placebo group was sufficiently large that chance was not a reasonable explanation of the difference (a "rare event" was observed.) Therefore, the null hypothesis of no real difference between groups was rejected in favor of the alternative hypothesis that the group receiving the Salk vaccine has a lower rate of polio than the placebo group.

We have illustrated the concept of a null and alternative hypothesis and a test statistic with a specific illustration. In general, we need the procedures of "testing hypothe___" whenever individuals are involved in basic research. As the "researcher" becomes more knowledgeable about the area of investigation, ideas about the underlying causes and effects in the system begin to develop. This leads to new explanations, new understanding, new theories and thus new hypotheses intended to better explain relationships than the present theories do.

Thus, the researcher proposes a hypothesis that is more "enlightening" than the current theories. This claim or hypothesis is called the research or alternative hypothesis, (it is an "alternative" hypothesis in the sense that it provides an alternative to the current thinking). We will denote the alternative hypothesis by the symbol H_a. The null hypothesis, denoted by the symbol H_o is the complement of the alternative hypothesis.

At this point the null and alternative hypothesis may be expressed in very general verbal terms such as "this vaccine will reduce the growth of cancer cells or it will not" or "adding a gasoline additive increases the life of the engine or it will not," and so forth. However, before we can pursue a "test"of the hypotheses for the purposes of this course, we must first find a probability model $f(x;\omega)$ to model the population that would be investigated and then translate these general hypotheses into a null and alternative hypothesis about a probability density function $f(x;\omega)$ or its parameters. Such a null and alternative hypothesis is called a *statistical* hypothesis.

Once the null and alternative hypotheses are clearly stated and understood, a method of proof is developed. In this text the method of proof will involve designing some experiment or investigation that will produce a set of data (by random sampling from the population modelled by $f(x;\omega)$). The data are used to compute the value of a test statistic whose probability properties (the sampling distribution) are known under the null hypothesis. We then compute the likelihood of producing the particular sample that we got, or anything more extreme, favoring H_a assuming the null hypothesis is true. If this probability is small, then we have evidence that would suggest that the null hypothesis is incorrect in favor of believing the alternative hypothesis.

This cycle of steps to coming to scientific "truth" is the essence of the scientific method. This is summarized in the steps given below:

The Scientific Method Applied to Statistical Tests of Hypotheses

Step 1:   Formulate a null and alternative hypothesis concerning the parameter of interest from the assumed pdf $f(x;\omega)$. (Most often our assumed pdf will be the normal distribution and the parameters to be examined will be the mean μ, and the variance σ^2. It is possible to examine them simultaneously, but we will restrict our examination to considering only one at a time.)

Step 2:   Design an experiment or method of collecting data that reasonable persons would regard as providing proof concerning the truth of the alternative hypothesis. Choose the size of sample for the experiment so that risks of errors are acceptably small; that is, choose a small value for the level of significance α.

Step 3:   Using the concepts of sampling distributions, decide on a meaningful test statistic (TS), whose distribution you know when assuming that you sample from the distribution identified in step 1 above, $f(x;\omega)$. This test statistic will involve the parameter ω in some fashion along with a sample statistic. (In the cases shown in this chapter, the test statistic, $TS(\hat{\omega},\omega)$, is identical in form to one of the "pivotal quantities" covered in chapter 7 on estimation.) Establish a decision rule that identifies what values of the test statistic lead to rejecting the null hypothesis H_o.

Step 4:   Conduct the experiment, collect the data and compute the value of the test statistic $TS(\hat{\omega},\omega)$ defined above (call it ts_{cal}). Then calculate the probability of getting the value

ts_{cal} or any value more extreme, assuming the null hypothesis H_o is true. This probability value is often called the *p*-value for the test.

$$p = P(\text{TS}(\overset{\wedge}{\omega},\omega) = ts_{cal} \text{ or values more extreme} \mid H_o\colon \omega = \omega_o).$$

Step 5:   If the *p*-value is small ($p \le \alpha$) it signifies that the probability of getting that particular sample statistic is remote when the null hypothesis is true. Therefore, "reasonable " persons would reject the null hypothesis in favor of the alternative. (What value of α constitutes "small" is left to the choice of the individual researcher, but common practice uses the guideline of 0.05 or less as signifying that condition.) It should be noted that a small *p*-value does not necessarily mean that the null hypothesis is false, rather it may indicate that insufficient care was used in selecting a sample and thus the sample is at fault, or the model $f(x;\omega)$ may be an incorrect choice of models *or* it may actually be that the null hypothesis is true and a very unlikely outcome has occurred. Stranger things have happened. However, if we assume: (1) the model to be correct, and (2) that we have spent the necessary time and money to insure a defensible sample then a small *p*-value would then logically call the validity of the null hypothesis into question.

There should be a note of caution and clarification inserted at this point. The philosophy of a test of hypothesis utilizes a "proof by contradiction" approach to making decisions. The null hypothesis is assumed to be true unless a contradiction to it shows up in the data. If such a contradiction to the null hypothesis shows up, supported by a small p-value, then we conclude the null hypothesis is false and we "reject the null hypothesis." If the sample fails to contradict the null hypothesis, we have not "proved" it to be true. We simply have insufficient evidence to contradict it. Thus, our language is not to "accept H_o" but to report that we have failed to disprove it. These 5 steps are displayed in Figure 8.2 below:

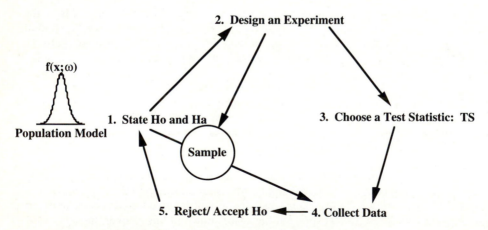

Figure 8.2 5-Step procedure for testing a statistical hypothesis.

EXAMPLES

EXAMPLE 1:   Suppose that we have reason to suspect the balance of an ordinary coin and believe that it favors heads more than tails. Conduct a test of this hypothesis.

SOLUTION: We will follow the 5 steps described above to test the hypothesis:

Step 1:   A single toss of the coin is modelled by a Bernoulli trial; that is

$$f(x;p) = p^x(1-p)^{1-x}.$$

Therefore, to evaluate the the balance of the coin we formulate a hypothesis about the Bernoulli parameter, p. That is,

$$H_o: p = 1/2 \quad \text{vs. } H_a: p > 1/2.$$

Step 2:   A logical method of conducting an experiment to test the above hypothesis would be to toss the coin several times and observe the number of heads that occur. To be specific, suppose we let $n = 20$ tosses and $\alpha = 0.10$.

Step 3:   A meaningful test statistic would be to observe the number of heads Y that occurs in the twenty tosses. Under the assumption of 20 independent Bernoulli trials, then the test statistic Y, the total number of heads that occur, follows the Binomial distribution with $n = 20$ with p being the same as the Bernoulli p. In this case, the test statistic TS is Y, the number of heads in 20 tosses which follows a binomial distribution with $n = 20$ and $p = 1/2$ if H_o is true.

Step 4:   Now suppose we have tossed the coin 20 times and observed 14 heads. The concept of an extreme outcome in this case would be those values that could occur in the experimental process that would tend to contradict H_o and favor H_a. Such a condition would occur if we tended to get a large number of heads. Therefore, we want to compute the probability

$$P(Y \geq 14 \mid p = 1/2)$$

(If $Y = 14$ is extreme, then $Y > 14$ is more extreme.) This probability is the same as $1 - F(13)$ where $F(13) = 0.9423$ which is obtained from the binomial tables. Therefore the p-value for this test is

$$p = 1 - 0.9423 = 0.0577.$$

Step 5:   This p-value is less than $\alpha = 0.10$ so that we conclude on the basis of these results that the coin is not balanced and in fact favors head. (We reject H_o in favor of H_a.)

EXAMPLE 2: Suppose that historically the average on a math competency exam has been 65.50 with a known standard deviation of 7 points. A math instructor using computer aided instruction (CAI) has developed methods claimed to improve student performance so that the mean score on the test will be larger than 65.50 (the standard deviation is assumed to remain the same). Write the 5 steps of the hypothesis testing procedure?

SOLUTION: The 5 steps are:

Step 1:   We decide to model this population with a normal distribution with unknown mean and standard deviation of 7, that is we assume X is $N(\mu,7^2)$. The hypothesis to test is

$$H_o: \mu = 65.50 \qquad \text{vs.} \qquad H_a: \mu > 65.50$$

(The null hypothesis is a simple hypothesis; the alternative is a composite hypothesis.)

Step 2:   A method of conducting an experiment would be to take a random sample of students who are taking the course and use the CAI procedure on them. At the conclusion of the training, they would be given a standard test for evaluation. (In reality, it is typically not possible to randomly assign students to various experimental programs. Rather, one simply takes the students available, making the assumption that they are a representative group.) Suppose we have a class of 25 students to work with. Let $\alpha = 0.02$ be the level of significance for this test.

Step 3:   The test statistic (TS), to be used in this case is

$$Z = \frac{(\bar{x} - \mu)}{\sigma/\sqrt{n}} = \frac{(\bar{x} - 65.5)}{7/\sqrt{25}}$$

which will be distributed as a standard normal variable under the assumptions of step 1.

Step 4:   Suppose the training and testing have been carried out and the sample mean is 68.80. In view of our alternative hypothesis, "extreme" outcomes would be associated with sample means larger that 65.50. Therefore, we will evaluate the probability

$$p = P(\bar{x} \geq 68.8 \mid \mu = 65.50)$$

$$= P(Z \geq 2.36) = 1 - F(2.36) = 0.0091$$

Step 5:   This p-value is smaller than $\alpha = 0.02$ so that chance is ruled out as the explanation of the higher sample mean. Therefore, we reject H_o and conclude that the training (other things being equal) explained the higher average test scores.

EXAMPLE 3: A pain inhibiting drug is being tested to see if it does in fact raise the pain threshold. Previous studies for standard groups of subjects showed a pain threshold of 7.5 units on a standard test with a standard deviation (σ) of 0.5 units. A group of 10 volunteer subjects, receiving the drug are tested. The data are summarized below. Is there reason to believe that the drug does, in fact, increase the pain threshold?

Summarized Data: $\bar{x} = 7.83,$ $s = 0.59$

SOLUTION: The 5 steps to test the hypothesis are:

Step 1: The data are assumed to be normally distributed and we assume that the volunteer subjects behave as a random sample. Therefore,

$$H_o: \mu = 7.5 \text{ vs. } H_a: \mu > 7.5$$

Step 2: The sample size is $n = 10$ subjects and we compute the average "pain threshold." Let the level of significance α be set at 0.05.

Step 3: Let the Test Statistic (TS) equal $\bar{X}$, the mean pain threshold of the sample of subjects.

Under our assumptions, $\overline{X}$ will be normally distributed with a mean of 7.5 and a standard deviation of $\dfrac{0.5}{\sqrt{10}} = 0.158$.

Step 4: The observed value of the TS is $\overline{x} = 7.83$. Is this a rare outcome? If so, then any value greater than 7.83 would be more rare. Therefore,

$$p = P(\overline{x} \geq 7.83 | \mu = 7.5, \sigma = 0.5) = P(Z \geq \frac{7.83 - 7.5}{0.5/\sqrt{10}} = 2.09) = 0.0183$$

Step 5: Since the p-value of 0.0183 < 0.05, the level of significance, reject the null hypothesis and conclude that the average pain threshold is higher than 7.5.

EXAMPLE 4: In an elementary school, the average number of children taking "hot lunch" each day over the last few years has been 125 children per day with a standard deviation of 4 children per day. That is, $\mu = 125$, $\sigma = 4$. Because of changes in economic conditions, school boundaries, and the need to hire or release cooks, buy more or fewer supplies, etc., the question is whether the mean is still the same.

Assuming the number of children taking hot lunch each day is reasonably modelled by a normal distribution, and that the number taking hot lunch during the second week of school is a random sample of days of the year, test the hypothesis that the mean is unchanged on the basis of the results of the "second week sample."

SOLUTION: The 5 steps are:

Step 1: The data are assumed to be normally distributed and we assume that we have a random sample. Therefore,

$$H_o: \mu = 125 \qquad \text{vs.} \qquad H_a: \mu \neq 125$$

Step 2: The sample size is $n = 5$ days and we compute the "average number of children taking hot lunch for those 5 days." Let $\alpha = 0.05$.

Step 3: Let the TS be $Y = \overline{X}$, the average number of children taking hot lunch.   Under our assumptions, Y will be normally distributed with a mean of 125 and a standard deviation of $\dfrac{4}{\sqrt{5}} = 1.79$.

Step 4: Suppose we observed $\overline{X} = 129$ students per day. Is this event rare? (Note that if we regard 129 students as rare compared to 125, then anything larger would be more rare. However, since we are looking for any departure from the mean of 125, in either direction, if 129 or more is rare compared to 125, then 121 or less would be just as rare by an argument of symmetry.) Therefore, the probability of a "rare event" is

$$p = P(\overline{X} \geq 129 \mid H_o \text{ is true}; \mu = 125) + P(\overline{X} \leq 121 | H_o \text{ is true}; \mu = 125)$$

$$= 2 \cdot P(\overline{X} \geq 129 \mid H_o \text{ is true}; \mu = 125) = 2 \cdot P(Z \geq \frac{129 - 125}{4/\sqrt{5}}) = 2.24)$$

$$= 2 \cdot (0.0125) = 0.025 \quad \text{(This is called a two-tailed test.)}$$

Step 5: Since $p = 0.025 \leq \alpha = 0.05,$ I would suggest the null hypothesis be rejected and conclude that the mean is unequal to 125. (The data suggest that the mean is greater than 125.)

EXERCISES

1. A random number generator on a pocket calculator was to be evaluated for "randomness." One simple characteristic to be studied was to see if the proportion of even and odd numbers was the same. (The first digit to the right of the decimal was examined to see if it was even or odd.) Fill in the first three steps of the test of hypothesis procedure for this problem.

2. For the problem defined in exercise 1, suppose, in a random sample of 25 trials, 16 odd numbers were generated compared to 9 even. Compute the p-value for this sample outcome and make a conclusion with the level of significance set at 10%.

3. For the same random number generator described in exercise 1, the 25 digits were grouped into sets of 5 and the mean of each group was then computed. Fill in the first three steps of the test of hypothesis procedure with an appropriate hypothesis about the mean μ of the uniform distribution being made. Recall what the mean and standard deviation of a uniform–discrete random variable is theoretically when its possible values are 0–9.

4. Using the principle of the central limit theorem, the 5 means obtained would be approximately normally distributed with $\mu = 4.5$ and $\sigma = 1.2845.$ Verify these results by applying basic principles about the sampling distribution of the mean.

5. For exercises 3 and 4 above, it was found that of the actual mean of the 5 "means" was 6.13. Compute the p-value for this sample outcome and make a conclusion. Suppose the level of significance was set at 4%. Notice the approach of Example 4 for an "not-equal"–type alternative hypothesis when calculating the p-value.

6. Suppose that Y is binomial with $n = 100$ and $p.$ We want to know if $p > 0.8.$ Set up the first three steps of the test of hypothesis procedure.

7. Suppose that the random variable Y in exercise 6 produced 85 success in the 100 trials. What is an approximate p-value for this situation using a normal approximation to the binomial probability presented in Chapter 6.

8. Let p denote the probability that for a particular tennis player, the first serve is good. From records kept over time it was found that p was approximately 0.40. The tennis player took lessons from a professional for a period of 3 months. Records of the number of good first-serves was kept for 50 trials. Set up the firs three steps of the test of hypothesis procedure.

9. In exercise 8, the 50 trials produced 25 successes (good first-serves). Complete steps 4 and 5 of the test of hypothesis routine using $\alpha = 0.06.$

10. Assume that SAT mathematics scores are N(530,8100). A small liberal arts college has received criticism that it is weak in the area of mathematics. A sample of 36 students was found to have a mean of 515. Fill in the 5 steps of the test of hypothesis for this case letting $\alpha = 0.05.$

8.2   ERRORS AND THEIR PROBABILITIES

Introduction

Any decision we have to make, has a risk of unacceptable consequences associated with it. If we decide to drive to work rather than taking the bus, we run the risk of an accident with property damage at the least and loss of life at the most. If I decide to jog today, I may get rained on, and so forth. This is an uncertain world and we generally try to go through life making decisions that minimize the risks in the anticipation that the benefits we gain offset the potential risks that are associated with it.

It is no different when testing a statistical hypothesis. We are placed in a situation where we must decide whether the null hypothesis is true or false. The only information we have to go on is a sample drawn from a population, incomplete information at best. Therefore, when we make a decision about the population using the sample data, we may make an error. We may decide that the null hypothesis is true when, in fact, it is false; or we may decide that the null hypothesis is false when in fact it is true.

For example, when a new drug is developed by a pharmaceutical firm, it must pass stringent tests and evaluations. Some of these tests are first performed on experimental animals. If the drug passes these tests, then tests on human subjects are performed. From these tests, the data may lead investigators to conclude that the drug is effective for a particular disorder, when, in fact, it has only slight benefits ... and it may even have serious side-effects. Alternatively, the data may suggest that the drug under evaluation doesn't do that much good when, in fact, it may have long-range, useful benefits.

In this illustration, we have described two types of errors that can occur: the error of marketing a useless and potentially harmful drug, and the error of overlooking a drug that may provide definite benefits. There is no way to insure that decisions are always correct short of examining the entire population and obtaining error-free measurements.   The best we can hope for is to control the probability of their occurrence and to keep this probability small.

In this section we formally define the two types of errors mentioned above. In addition, we show how to evaluate the probability of occurrence of each of the error types.

Definitions

DEFINITION 8.7: A *type I error* occurs whenever H_o is rejected (concluded to be false) when it is really true. The probability of a type I error is denoted by α. That is,

$$\alpha = P(\text{Reject } H_o \mid H_o \text{ is true})$$

(The probability of a type I error α is also called the *level of significance.*)

DEFINITION 8.8: A *type II error* occurs when H_o is accepted (concluded to be true) when it is really false.   The probability of a type II error is denoted by β.

$$\beta = P(\text{Accept } H_o \mid H_o \text{ is false})$$

DEFINITION 8.9: The *critical value of the test statistic* TS_{cv} or the *boundary value of the rejection region* is that value of the test statistic that separates the acceptance and rejection regions.

DISCUSSION

Whenever a decision is made on less than complete and perfect information, the decision could be wrong. Such is the case in deciding whether to reject or accept a null hypothesis in favor of an alternative hypothesis. To understand this examine Figure 8.3.

In this figure, we assume that the population parameter takes on only one of two possible values (the true state of nature). The row headings show the two possible conditions for ω, i.e., $H_0: \omega = \omega_o$ or $H_a: \omega = \omega_1$. The column headings denote the two possible decisions we can make. We can decide to accept H_0 or we can decide to reject H_0. Depending upon our decision, as the figure shows, we may make a correct decision, or an error.

		Decision	
		Reject H_0	Accept H_0
STATE	$H_0: \omega = \omega_0$	Type I Error	Correct Decision
OF			
NATURE	$H_a: \omega = \omega_1$	Correct Decision	Type II Error

Figure 8.3 Type I and type II errors.

In Figure 8.4, this decision process is demonstrated in the case of a jury making a legal decision about the guilt or innocence of a person accused of a crime. The testimony given in court is analogous to statistical evidence that we collect from an unknown population. The jury can make one of two errors depending on whether the person is guilty or innocent. If the person is innocent and the jury gives a verdict of guilty, then a type I error has been made. If the person is innocent and the jury finds the person innocent, then the correct decision has been made. (See the first row in the figure.)

			Decision: Jury Says	
			Guilty	*Innocent*
			Reject H_0	Accept H_0
STATE	$H_0:$	Person is Innocent	Type I Error	Correct Decision
OF				
NATURE	$H_a:$	Person is Guilty	Correct Decision	Type II Error

Figure 8.4 Type I and II errors—court room setting.

On the other hand, if the person is guilty and the jury finds the person innocent, then a type II error is made. If the jury finds the guilty person guilty then a correct decision is made. (See the second row of the figure.)

The challenge we face is to learn how to calculate the probabilities of the type one and type two errors under certain model assumptions; that is, assuming the data comes from a particular pdf, $f(x;\omega)$. This process will be carefully demonstrated in the following examples and can be generalized to other cases.

EXAMPLES

EXAMPLE 1: Suppose an instructor has created several normal distributions for demonstration in the classroom. In a rush to get to class, the instructor picks up one of the distributions, but fails to note which one it is. The instructor knows that the mean is either 50 or 75 and decides to challenge the class to decide which value it is on the basis of a sample of 16 observations drawn from the population. The class arbitrarily decides that if the mean of the sample exceeds 65, they will choose 75 as their best guess of the mean. If the sample mean is less than 65, they will choose 50 as their best guess of the true mean.   The standard deviation is known to be 40 units.   What are the probabilities of errors in this example?

SOLUTION:

Step 1: Suppose we state the null hypothesis and alternative as follows:

$$H_o{:}\mu = 50 \text{ vs. } H_a{:}\mu = 75.$$

Step 2: To investigate the hypothesis we select a random sample of 16 elements from the population. In this case, α is to be determined.

Step 3: Our TS is the sample mean $\bar{x}$.

Step 4: We have a decision rule as follows:

$$\text{Reject } H_o \text{ if } \bar{x} \geq 65$$
$$\text{Accept } H_o \text{ if } \bar{x} < 65$$

This decision rule is displayed in Figure 8.5.

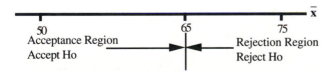

Figure 8.5 Acceptance and rejection regions for the null hypothesis.

The Probability of the Type I Error:

To calculate the probability of the type I error we need to calculate the probability that the sample mean will be in the rejection region ($\bar{x} \geq 65$) when the true mean is $\mu = 50$. In this case, the sample mean is normally distributed with a mean of 50 and a standard deviation of $40/\sqrt{16} = 10$. Therefore,

$$\alpha = P(\text{Reject } H_o \mid H_o \text{ is true})$$

$$= P(\bar{x} \geq 65 \mid \mu = 50) = P(Z \geq \frac{(65 - 50)}{40/\sqrt{16}})$$

$$= P(Z \geq 1.50) = 1 - F_Z(1.50) = 1 - 0.9332 = 0.0668.$$

This result is displayed in Figure 8.6 below.

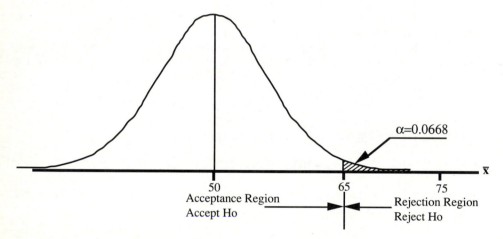

Figure 8.6 Region where type I error probability is 0.0668.

Probability of the Type II Error:

To get the probability of the type II error we need to calculate the probability of the sample mean falling in the acceptance region (< 65) when the true mean is 75. Therefore,

$$\beta = P(\text{Accept } H_o \mid H_o \text{ is false})$$

$$= P(\bar{x} < 65 \mid \mu = 75) = P\left(\frac{(\bar{x} - 75)}{40/\sqrt{16}} < \frac{(65 - 75)}{40/\sqrt{16}}\right)$$

$$= P(Z < -1.0) = F_Z(-1.0) = 0.1587.$$

See Figure 8.7 for a picture of the area denoting β.

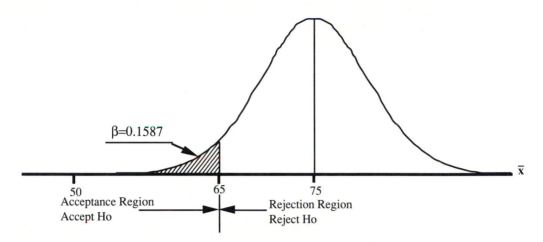

Figure 8.7 Region of type II error probability of 0.1587.

Figures 8.6 and 8.7 are merged in Figure 8.8 to show the relative sizes of α and β. Notice that α is calculated under the assumption that the normal distribution is centered at 50 while β is calculated under the assumption that the normal distribution is centered at 75.

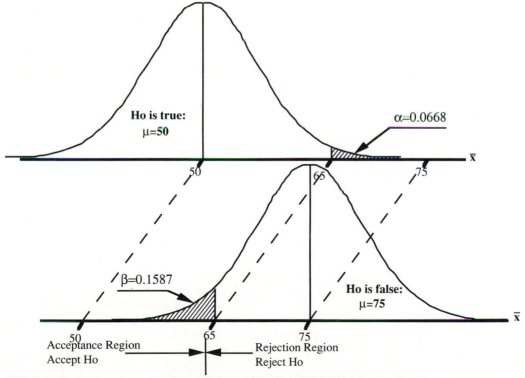

Figure 8.8 Composite graph showing relationship of type I and type II errors.

From these calculations, the likelihood of incorrectly concluding that the population the instructor has brought to class has a mean of 75, when in fact it has a mean of 50, is only 6.68% (the probability of a type I error). Similarly, the probability of incorrectly concluding that the population has a mean of 50 when in fact it has a mean of 75, is 0.1587, a slightly higher probability—the type II probability. By looking at Figure 8.8, we see that we can change these values by shifting the boundary that separates the rejection and acceptance regions. Shifting the boundary to the right will reduce the type I error while increasing the type II error. Likewise, shifting the boundary to the left would increase the value of α while decreasing the value of β. The only way to reduce both errors would be to increase the sample size which would reduce the standard deviation of both curves, leaving smaller areas in the respective tails.

From this example it can be seen that:

$$\alpha = P(\overline{x} \geq 65 \mid \mu = 50) = 0.0668$$

and

$$\beta = P(\overline{x} < 65 \mid \mu = 75) = 0.1587$$

The value 65 is called the *critical value of the test* or the *boundary value of the rejection region*. We can also say that α is the "size" or area of the rejection region when H_o is true and β is the "size" or area of the acceptance region when H_o is false.

This leads to the following generalization to be used in computing a value for α and β. First, let C denote the "rejection region" and C', the complement of C, denote the "acceptance region." Our decision rule is: If the test statistic $TS(\hat{\omega},\omega)$ belongs to C, we reject H_o; if $TS(\hat{\omega},\omega)$ belongs to C' we "fail to reject H_o." Therefore,

$$\alpha = P(TS(\hat{\omega},\omega) \in C \mid \omega \in H_o)$$

and

$$\beta = P(TS(\hat{\omega},\omega) \in C' \mid \omega \in H_a)$$

In addition, for the examples and assumptions made in this course the rejection region C is continuous with either an lower boundary value or an upper boundary value or some combination of the two. The "boundary" value(s) is called the critical value of the test statistic.

The preceding suggests a slight variation in the way we set up the 5 steps in hypothesis testing.

An Alternative to Using P-Values in Testing Hypotheses

In the procedure for testing hypotheses described in section 10.1, the logic was to compare the results obtained in a sample with what we would expect to occur assuming the null hypothesis were true. If the sample results were unlikely under the null hypothesis, then we would say we have reason to reject the null hypothesis. This probability or likelihood was reported as a *p*-value. If a small *p*-value occurred, we decided to reject H_o in favor of H_a. We reject the null hypothesis if $p < \alpha$, where α is a small number like 0.05, etc. In this case $\alpha = P(\text{Rejecting } H_o \mid H_o \text{ is true}) = P(\text{Type I error})$. Whatever choice for α we make determines the type I error risk for that problem. For instance, if an investigator says that any *p*-value less than 0.025 leads to rejection, then $\alpha = 0.025 = P(\text{Type I error})$ for that investigator. It can also be argued that whenever $p < \alpha$, then equivalently the test statistic takes on a value in a region commonly called the *rejection region*..

The connection between α and a *p*-value is demonstrated graphically in Figure 8.9. For example, suppose we test the null hypothesis that $\mu = \mu_o$, where large sample means would tend to lead us to reject H_o. We let $\overline{X}_{cv}$ denote the point on the curve so that the area in the tail is equal to

α (the gray region).   We call the region on the axis to the right of X_{cv} the *rejection region*, and the region to the left is called the *acceptance region*. Next suppose that the sample mean takes on the value $\overline{X}_o$. The area to the right of $\overline{X}_o$ is the *p*-value (the black region). From this figure, we can see that if the sample mean is found anywhere to the right of $\overline{X}_{cv}$, the *p*-value will be less than α, and thus we should reject the null hypothesis. The value, $\overline{X}_{cv}$, is the *critical value* or the *boundary value* of the rejection region. Therefore, an alternative but equivalent 5-step procedure for testing hypotheses can be formulated as follows.

Step 1: Formulate H_o and H_a.

Step 2: Design an experiment or data collection procedure. Pre-select a value for the type one error risk, α and choose the sample size n.

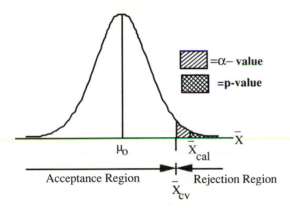

Figure 8.9 Relationship of *p*-value to α.

Step 3: Decide on an appropriate test statistic, TS, and form the decision rule:

Reject H_o if TS_{cal} (calculated value of the test statistic) is more extreme than TS_{cv} (the critical value of the test statistic), for the chosen value of α. Note that $p < \alpha$ implies TS_{cal} is more extreme that TS_{cv}.

Step 4: Collect the data and compute the value of the test statistic, TS_{cal}.

Step 5: Apply the decision rule and make a decision.

See Figure 8.10 for a pictorial representation of the decision rule.

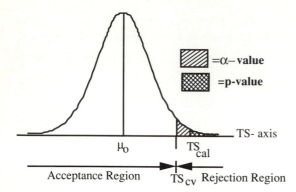

Figure 8.10 Relationship of p-value to α—general case.

EXAMPLE 2: Consider Example 2, section 8.1 concerning students taking a competency exam in math after using Computer Assisted Instructional (CAI) methods. Identify the rejection region using a type I error probability of $\alpha = 0.025$ and test the hypothesis that the CAI methods improve test scores.

Step 1: H_o: $\mu \leq 65.50$ vs H_a: $\mu > 65.50$

Step 2: Draw a sample of 25 students and let $\alpha = 0.025$.

Step 3: Let the test statistic be

$$Z = \frac{(\bar{x}-\mu)}{\sigma/\sqrt{n}} = \frac{(\bar{x}-65.5)}{7/\sqrt{25}}$$

Let the decision rule be: Reject H_o if $Z_c \geq Z_{cv} = Z_{0.975} = 1.96$ in order for the Type I error to be $\alpha = 0.025$.

Step 4: It was found that the sample mean was 68.8. Therefore

$$Z_{cal} = \frac{(68.8 - 65.5)}{7/\sqrt{25}} = 2.36.$$

Step 5: On the basis of the experimental evidence, reject H_o and conclude that the CAI instruction improves performance on the competency exam at the 0.025 level of significance.

In working this example in section 10.1, we decided that large average scores would provide reason to reject H_o in favor of H_a. We computed a p-value of 0.0091 which for $\alpha = 0.025$ leads to a conclusion to reject H_o and conclude that H_a is true. Under the conditions given in the problem, ($n = 25$ and $\sigma = 7$) a type I error rate of 0.025 can be shown to be associated with a rejection region of $Z_{cv} = 1.96$ or equivalently $\bar{X}_{cv} > 68.24$, (Note that $68.24 = 65.5 + 1.96(\frac{7}{\sqrt{25}})$.)

The number, 68.24, is also known as the critical value or boundary value of the rejection region. This relationship is shown in Figure 8.11.

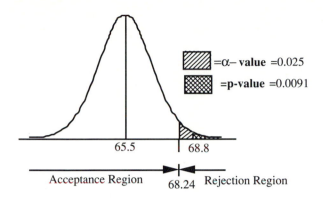

Figure 8.11 Relationship of p-value to α, Example 2.

This gives three different, but equivalent, decision rules for making a decision in Example 2 . This logic can be applied to any other problem of this type.

Reject H_o if :
(1) $p < \alpha = 0.025$ (We prefer this approach and will use it exclusively in later sections of this chapter.)

(2) $Z_{cal} > Z_{cv} = 1.96$

(3) $\bar{X}_{cal} > \bar{X}_{cv} = 68.24$

Obviously in (1) above, the type I error probability is $\alpha = 0.025$. Likewise, for (2) and (3), it is easy to show that the type I error probability is also $\alpha = 0.025$.

EXAMPLE 3: Suppose that an investigator has selected a type I error probability of 0.025 as in Example 2 concerning test scores after training with CAI. If the true mean μ of the newly trained group has really shifted to 70, what is the probability that such a shift will not be detected by the decision procedure defined? That is, what is the probability of a type II error, given that $\mu = 70$ with $n = 25$ and $\sigma = 7$?

SOLUTION: To obtain the value of the type II error, the following computations are required:

$$\beta = P(\text{Accepting } H_o \,|\, H_o \text{ is false})$$

$$= P(\bar{x} \text{ is in acceptance region} \,|\, \mu = 70.0)$$

$$= P(\bar{x} < 68.24 \,|\, \mu = 70.0)$$

$$= F(68.24 \,|\, \mu = 70.0)] = F_z\!\left(\frac{68.24 - 70}{7/\sqrt{25}} = -1.26\right) = 0.1038$$

This result is displayed graphically in Figure 8.12 below.

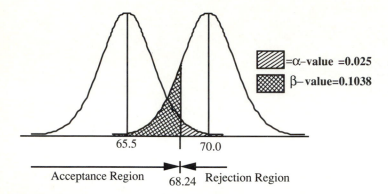

Figure 8.12 Probability of type II error, $\alpha = 0.025$.

Thus, there is a 10.38% chance of not detecting an improvement of 5 units in the average test score (from a mean of 65 to a mean of 70).

EXERCISES

1. Example 1 stated: "Suppose an instructor has created several normal distributions for demonstration in the classroom. In a rush to get to class, the instructor fails to note which population has been picked up. The instructor knows that the mean is either 50 or 75. The class is challenged to decide which value it is on the basis of a sample of $n = 16$ observations drawn from the population. The class arbitrarily decides that if the mean of the sample exceeds 65, they will choose 75 as their best guess of the mean. If the sample mean is less than 65, they will choose 50 as their best guess of the true mean. The standard deviation is known to be 40 units. What are the probability of the type I and type II errors in this example?" Change the critical value from 65 to 60 and calculate the type I and type II errors. Compare them to the results obtained in the Example 1.

2. Under the problem description of exercise 1, (Example 1), find the rejection region boundary for $\alpha = 0.05$. Determine the corresponding value of β when $\mu = 75$.

3. Again using exercise 1, find the rejection region boundary for $\beta = 0.10$. Determine the corresponding value of α.

4. For the problem description given in exercise 1 and amended in exercise 2 find the value of β if $n = 36$. That is find β for $H_o:\mu = 50$ *vs.* $H_a:\mu = 75$, $\alpha = 0.05$ and $n = 36$.

5. Rework exercise 4 for $n = 49$. What do you notice about the size of the critical value and the size of β as we have increased n from 16 to 36 to 49? Draw a picture of these three cases to help your understanding.

6. A machine fills cereal boxes with a popular brand of breakfast cereal. The mean weight of the boxes after filling should be 25 ounces. A quality control analyst knows from past experience that the standard deviation of the weight of the boxes is 0.5 ounces. The hypothesis is: $H_o: \mu = 25$ ounces *vs.* $H_a:\mu \neq 25$ ounces. Suppose a sample of $n = 25$ is selected and the level of

significance (the probability of the type I error) is set at 0.05. This problem has two critical values, one above $\mu = 25$ and one below since we are trying to detect changes in the machine either towards over-filling or under-filling. Find the two critical values corresponding to putting $\alpha/2 = 0.025$ in each of the rejection regions. [Draw a picture with a normal curve centered at $\mu = 25$ and with 0.025 area in each end (or tail) of the distribution.]

7. For the problem described in exercise 6, find the value of β if $\mu = 24.95$, that is the machine is slightly underfilling the boxes on the average? [Use the diagram drawn in exercise 6 with $\mu = 25$ and clearly indication the location of the two critical values. Add another normal curve centered at 24.95 and notice that the area in the lower tail has increased and the area in the upper tail has decreased. Find the area between these two points.]

8. Suppose we want to test the null hypothesis that $H_o{:}\mu = \mu_0$ versus $H_a{:}\ \mu = \mu_1$ where $\mu_1 < \mu_o$ for a normally distributed population with standard deviation of σ. For some value of α write the expression for $\overline{X}_{cv}$ in terms of μ_o, σ, and n. Is $\overline{X}_{cv}$ to the right or left of the mean μ_o?

9. Suppose we want to test the null hypothesis that $H_o{:}\mu = \mu_o$ versus $H_a{:}\ \mu = \mu_1$ where $\mu_1 < \mu_o$ for a normally distributed population with standard deviation of σ. For some value of β write the expression for $\overline{X}_{cv}$ in terms of μ_1, σ, and n. Is $\overline{X}_{cv}$ to the right or left of the mean μ_1?

10. Looking at exercises 8 and 9 and realizing that $\overline{X}_{cv}$ is at the same location in both cases, set the two equations for $\overline{X}_{cv}$ equal to one another and solve for n in terms of the Z's, μ's, and σ. Use this equation to determine the sample size to test a hypothesis when $\mu_o = 50$, $\mu_1 = 75$, $\alpha = 0.05$, $\beta = 0.10$ and $\sigma = 40$.

8.3   THE POWER FUNCTION AND SAMPLE SIZE

Introduction

In section 8.2 we examined the type I and type II errors and their probabilities for a simple null hypothesis and a simple alternative hypothesis; that is $H_0{:}\omega = \omega_0$ vs. $H_a{:}\omega = \omega_1$ where $\omega_o \neq \omega_1$ and ω represents the parameter in the distribution . However, the type of null and alternative hypotheses most often encountered are composite and are of the form $H_0{:}\omega \leq \omega_o$ vs. $H_a{:}\omega > \omega_o$, etc. In this case, there are an infinite number of possible values for ω that satisfy H_0 and an infinite number of possible values for ω when H_a is true. Therefore, there are an infinite number of values of α and also of β. In such a case, we find it more acceptable to create a *curve* of type one error values or type two error values. These curves are called power curves or operating charcteristic curves (OC curves), depending on which curve we produce.

    In this section we define what is meant by a power function or power curve and the operating characteristic (OC) curve for composite hypotheses. The derivation of the points on the curves don't require any new computations beyond that covered in the last section; we simply do more computations of the same kind.

In addition, it becomes obvious in this section that there is a direct relationship between α, β, and the sample size n. This relationship will be developed in this section.

Definitions

DEFINITION 8.10: A *one-tailed test* is a test of hypothesis having the rejection region in only one "tail" of the distribution defined by H_o. The alternative hypotheses $\mu > \mu_o$ and $\mu < \mu_o$ produce "one-tailed" tests for test about means.

DEFINITION 8.11: A *two-tailed test* is a test of hypothesis having the rejection region in two "tails" of the distribution defined by H_o. The alternative hypothesis $\mu \neq \mu_o$ produces a "two-tailed" test.

DEFINITION 8.12: The *power function* is the probability that H_o is rejected given a specific value of ω. If we denote the power function by $P(\omega)$, then

$$P(\omega) = \begin{cases} \alpha & \text{for each } \omega \text{ in } H_o \\ 1 - \beta & \text{for each } \omega \text{ in } H_a \end{cases}$$

DEFINITION 8.13: The *operating characteristic curve* (OC curve) is a curve of probabilities that represent the probabilities of acceptance for the various values of ω under consideration. That is,

$$OC = 1 - P(\omega).$$

DISCUSSION

One-Tailed versus Two-Tailed Tests

The distinction between a one-tailed vs. two-tailed test is determined by the particular alternative hypothesis we have in a given test of hypothesis. For instance, if we test $H_o : \mu = \mu_o$ against the alternative $H_a : \mu > \mu_o$, then sample means which are to the right of μ_o lead to rejection of H_o, and therefore the rejection region is in the right tail of the normal distribution defined by H_o.

Similarly, if the alternative is of the form $H_a : \mu < \mu_o$, then values of the sample mean to the left of μ_o lead to rejection and thus the rejection region is in the left tail of the null hypothesis distribution. If the alternative is simply $H_a : \mu \neq \mu_o$ then means which are either to the left or the right could lead to rejection of H_o, and thus we have a two-tailed test.

These relationships are depicted in Figure 8.13 below, with the rules to use in determining the p-value under the three alternative hypotheses. Recognition of the correct rejection region is necessary in order to correctly compute power or OC curves.

Computing the Power Function

The power function provides a curve of probabilities that represent the probability of rejecting H_o when it should be rejected. In this chapter, it represents the area in the rejection region under an assumed normal distribution. Thus, the computation of these probabilities is very much like the determination of the probability of a type I error. The steps involved in obtaining the power function are explained below and then illustrated with an example.

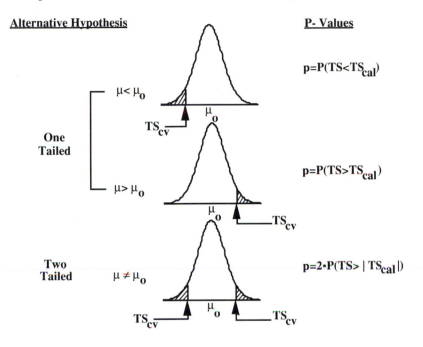

Alternative Hypothesis

One
Tailed

$\mu < \mu_o$

$\mu > \mu_o$

Two
Tailed $\mu \neq \mu_o$

P- Values

$p = P(TS < TS_{cal})$

$p = P(TS > TS_{cal})$

$p = 2 \cdot P(TS > |TS_{cal}|)$

Figure 8.13 Diagrams showing one- vs. two-tailed tests.

1.   For a series of values of ω in the region defined by the alternative hypothesis, compute $P(\text{Rejecting } H_o)$.
      That is:
      $$P(\omega_o) = P(TS \text{ in Rejection Region} \mid \omega = \omega_o)$$

      For values of ω in H_o compute the probabilities the same way. The values thus obtained represent the probability of a Type I error, however.

2.   For each probability value just obtained plot the probability on the vertical axis against the corresponding value of ω on the horizontal axis. This produces a curve called the power function and represents the probability of rejecting a false hypothesis.

EXAMPLE 1: Compute the power function for the decision rule associated with the investigation concerning the improvement of competency exam scores after training using CAI.

SOLUTION: Referring to Example 2 of both sections 8.1 and 8.2 of this chapter, for $\alpha = 0.025$, the decision rule is given as: Reject H_o if $\bar{x} \geq 68.24$. Therefore, to compute the power we have the expression:

$$P(\mu_1) = P(\bar{x} \geq 68.24 \mid \mu = \mu_1, \mu_1 > 65.50)$$

We shall compute this probability for μ equal to 66, 68, 70, and 72 in order to get the general form of the power function curve. Doing so we get:

$$P(66) = P(\bar{x} \geq 68.24 \,|\mu = 66)$$

$$= 1 - F_z(1.60) = 0.0548$$

$$P(68) = P(\bar{x} \geq 68.24 \,|\mu = 68)$$

$$= 1 - F_z(0.17) = 0.4325$$

$$P(70) = 1 - F_z(-1.26) = 0.8962$$

$$P(72) = 1 - F_z(-2.69) = 0.9964$$

The power curve is displayed in Figure 8.14 below. In this figure, for each value of μ on the horizontal axis, the point on the curve represents the corresponding power at that given value of μ.

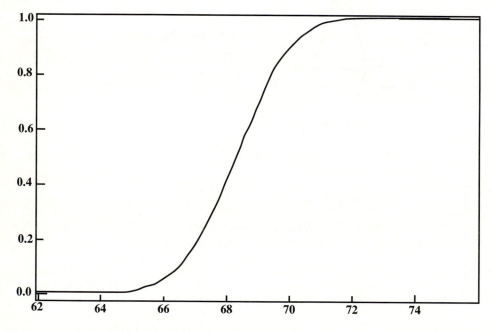

Figure 8.14 Power curve for math competency exams.

A general solution to the problem of calculating the power can be solved. We have to treat each alternative hypothesis separately to derive the results. The formula for a one-tailed test is different than a that for a two-tailed test. Also, the formula for a right-tailed test is different than for a left-tailed test. The approach is similar, however, in all cases: we calculate the area of the rejection region for a given value of the mean, say μ_1. In order to do this we have to find the rejection region boundary. The details are given next.

Power Calculations for Hypotheses about μ in the Normal Distribution

First, let $\bar{X}_{cv}$ denote the boundary value for the rejection region with type I error probability set at α. Therefore,

$$\bar{X}_{cv} = \mu_0 + Z\frac{\sigma}{\sqrt{n}}$$

where $H_o:\mu = \mu_0$ is the null hypothesis and Z is positive or negative depending on the alternative H_a. We have 3 cases to consider: (1) when the alternative leads to a two-tailed rejection region, (2) when the rejection region is one-tailed in the right tail of the distribution, and (3) when the rejection region is one-tailed in the left tail of the distribution.

Case 1: Two-Tailed Test

By definition the power is the area in both tails of the distribution. Therefore,

$$\text{Power} = P(\mu_1) = P(\bar{X} > \bar{X}_{cvu} \text{ or } \bar{X} < \bar{X}_{cvl} \mid \mu = \mu_1)$$

$$= P(\bar{X} > \mu_o + Z_{1-\frac{\alpha}{2}} \frac{\sigma}{\sqrt{n}} \mid \mu = \mu_1) + P(\bar{X} < \mu_o - Z_{1-\frac{\alpha}{2}} \frac{\sigma}{\sqrt{n}} \mid \mu = \mu_1)$$

where $\bar{X}_{cvu}$ is the critical value of the upper rejection region and $\bar{X}_{cvl}$ is the critical value of the lower rejection region. See Figure 8.15 below to identify these two values.

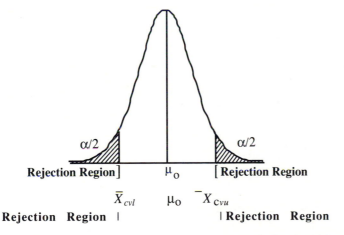

Figure 8.15 Figure showing location of rejection region boundaries for two-tailed test.

After standardizing, we get:

$$\text{Power} = P\left(Z > \frac{\mu_o - \mu_1}{\sigma/\sqrt{n}} + Z_{1-\frac{\alpha}{2}}\right) + P\left(Z < \frac{\mu_o - \mu_1}{\sigma/\sqrt{n}} - Z_{1-\frac{\alpha}{2}}\right)$$

$$= 1 - F_Z\!\left(\frac{\mu_0 - \mu_1}{\sigma/\sqrt{n}} + Z_{1-\frac{\alpha}{2}}\right) + F_Z\!\left(\frac{\mu_0 - \mu_1}{\sigma/\sqrt{n}} - Z_{1-\frac{\alpha}{2}}\right)$$

Case 2: One-Tailed Test—Right-Tailed Rejection Region

If the rejection region is in the right tail, we need to find the area to the right of $\bar{X}_{cvu}$ given that $\mu = \mu_1$. This regions is shown in as diagram (a) in Figure 8.16. (Diagram (b) shows the location of the rejection region for a one-tailed left tail test.) The calculations are shown next.

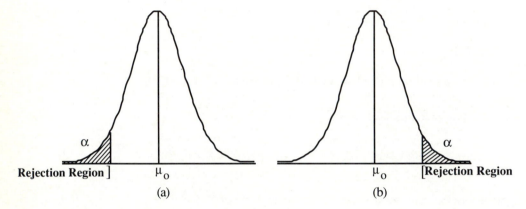

Figure 8.16 Location of rejection regions for one-tailed tests of means.

$$\text{Power} = P(\mu_1) = P(\bar{X} > \bar{X}_{cvu} \mid \mu = \mu_1)$$

$$= P(\bar{X} > \mu_o + Z_{1-\alpha}\,\frac{\sigma}{\sqrt{n}} \mid \mu = \mu_1)$$

After standardizing:

$$\text{Power} = P\!\left(Z > \frac{\mu_o - \mu_1}{\sigma/\sqrt{n}} + Z_{1-\alpha}\right)$$

$$= 1 - F_Z\!\left(\frac{\mu_o - \mu_1}{\sigma/\sqrt{n}} + Z_{1-\alpha}\right)$$

Case 3: One-Tailed Test—Left-Tailed Rejection Region

If the rejection region is in the left tail (see diagram (b) in Figure 8.16) we do the following computations.

$$\text{Power} = P(\mu_1) = P(\bar{X} < \bar{X}_{cvl} \mid \mu = \mu_1)$$

$$= P(\bar{X} < \mu_o - Z_{1-\alpha}\,\frac{\sigma}{\sqrt{n}} \mid \mu = \mu_1)$$

$$= F_Z\left(\frac{\mu_o - \mu_1}{\sigma/\sqrt{n}} - Z_{1-\alpha}\right)$$

EXAMPLE 2: Repeat the calculations for Example 1 assuming the sample size from the normal distribution is changed from $n = 25$ to $n = 36$. Plot the power curve on the same axis and compare the two curves.

SOLUTION: Let us use the formulas we have derived above to expedite the calculations. In this case we have a one-tailed, right-tailed test so the formula to use would be:

$$P(\mu_1) = 1 - F_Z\left(\frac{\mu_o - \mu_1}{\sigma/\sqrt{n}} + Z_{1-\alpha}\right)$$

where $\mu_o = 65.5$, $\sigma = 7$, $n = 36$, $Z_{1-\alpha} = Z_{0.975} = 1.96$, and μ_1 takes on the values greater than 65.5. We get:

$$P(66) = 1 - F_Z\left(\frac{65.5 - 66}{7/\sqrt{36}} + 1.96\right) = 1 - F(1.5314) = 0.0628$$

$$P(68) = 1 - F_Z\left(\frac{65.5 - 68}{7/\sqrt{36}} + 1.96\right) = 1 - F(-0.1829) = 0.5725$$

$$\text{etc.}$$

The complete power curve superimposed on the diagram of Figure 8.14 is shown as Figure 8.17 below.

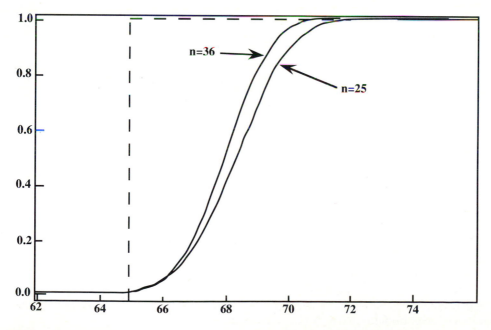

Figure 8.17 Power curve for math competency exams for $n = 25$ and $n = 36$.

Notice the power curve for $n = 36$ is steeper and above that for $n = 25$. Thus for a given value of μ on the horizontal axis the power is larger for $n = 36$ than it is for $n = 25$. The dotted line shows the "ideal" power which would be 0 for $\mu \leq 65.5$ and would be 1.0 for $\mu > 65.5$. As the sample size increases we would expect the power curve to approach more closely this "ideal."

The power curve gives us an evaluation on the strength of our decision procedure. If the power curve is not steep enough to make the probabilities of risks acceptably small then we should consider an increase in the sample size or a change in the value of α. Thus, the power curve is a "pre-experimental" device to help us decide if the experimental procedure we propose to carry out will do a satisfactory job. Another implication in all of this is that we should be able to find a unique sample size to produce an acceptable power curve. We consider the approach to this problem next.

The Sample Size

EXAMPLE 3: Suppose H_o: $\mu = \mu_0$ vs. H_a: $\mu > \mu_o$. What sample size is needed to test the hypothesis concerning the mean of a normally distributed population with Type I error probability α and Type II error probability when $\mu = \mu_1$ is β?

SOLUTION: Let $\bar{X}_{cv}$ denote the critical value for the rejection region such that

$$P(\bar{X} > \bar{X}_{cv} \mid \mu = \mu_0) = \alpha$$

and

$$P(\bar{X} > \bar{X}_{cv} \mid \mu = \mu_1) = 1 - \beta$$

These properties are shown in Figure 8.18. (The vertical dotted line indicates the location of the critical value $\bar{X}_{cv}$.) If we standardize $\bar{X}_{cv}$ relative to μ_o and μ_1 respectively we get the following two results:

$$\frac{\bar{X}_{cv} - \mu_o}{\sigma/\sqrt{n}} = Z_{1-\alpha}$$

and

$$\frac{\bar{X}_{cv} - \mu_1}{\sigma/\sqrt{n}} = Z_{\beta}$$

We solve for $\bar{X}_{cv}$ in both equations and set them equal to one another. This gives the following equation:

$$\mu_o + Z_{1-\alpha}\frac{\sigma}{\sqrt{n}} = \mu_1 + Z_{\beta}\frac{\sigma}{\sqrt{n}}$$

Solving for n we get

$$n = \frac{\sigma^2(Z_\beta - Z_{1-\alpha})^2}{(\mu_o - \mu_1)^2}.$$

Thus, we have an equation that allows us to determine the sample size for a given choice of α, β, and deviation shift in the mean from μo to $\mu 1$. It is assumed that σ is known.

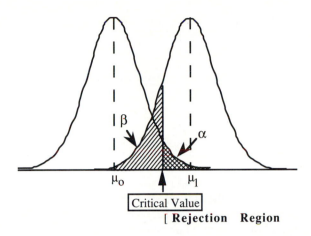

Figure 8.18 Location of $\bar{X}_{cv}$ and its relationship to α and β.

EXAMPLE

EXAMPLE 4: Suppose we want to test the hypothesis H_o:$\mu = 50$ versus H_a: $\mu > 50$ with the specific requirement that $\alpha = 0.05$ when $\mu = 50$ and $\beta = 0.10$ when $\mu = 75$. The standard deviation σ is assume to be 40. What size of sample is needed if the populations is normally distributed?

SOLUTION: Given: $\sigma = 40$, $\mu_o = 50$, $\mu_1 = 75$, $Z_{.95} = 1.645$, $Z_{.10} = -1.28$. Solving for n we get:

$$n = \frac{\sigma^2(Z_\beta - Z_{1-\alpha})^2}{(\mu_o - \mu_1)^2} = \frac{40^2(--1.28 - 1.645)^2}{(50 - 75)^2}$$

$$= 21.90$$

Obviously we can't select of sample of 21.9 units; our sample size would have to be 21 or 22. Choosing n = 22 would give us a value of α and β slightly smaller than requested. (Practically, our use of $\sigma = 40$ is probably on a guess as to the value of σ. Therefore, we might more realistically anticipate that we need a sample size of 20–30 units. Budget constraints might then help revise our sample size decision. At least we know that we don't need a sample of 50 to 100 or that we couldn't get by with a sample of 10 and meet our conditions.)

EXERCISES

1. In establishing federal EPA standards for miles-per-gallon for a particular type of automobile, the hypothesis to be tested is H_o: $\mu = 28$ vs. H_a:$\mu \neq 28$ mpg. If it is known that the standard deviation is 6 miles per gallon and the sample size consists of 25 automobiles:
    a.   For $\alpha = 0.05$, determine the critical values for the sample mean for this test.
    b.   Sketch the complete power curve after obtaining at least 4 points on the curve. For instance, use $\mu_1 = 28.5, 29, 29.5$ and 30. (What would happen if you were to use 27.5, 27, 26.5 and 26 instead? Note that this is a two-tailed test.)

2. From the test of hypothesis proposed in exercise 1, suppose the study were carried out and the sample mean were 29 mpg. What is the *p*-value for this test of hypothesis and what is your decision?

3. Suppose in exercise 1 above, the sample size were increased to 100 with α remaining the same.
   a. Determine the new critical values.
   b. Sketch, on the same graph as in exercise 1, part b, the new power curve.

4. For the revision of the problem of exercise 1 with an increase in sample size to 100, (see exercise 3) suppose the sample mean were 29 mpg. What is the *p*-value for this test of hypothesis and what is your decision?

5. For a two-tailed test of means, assuming σ is known, and α and β are specified, derive a formula for the sample size needed to detect a shift in the mean from μ_o to μ_1. (Draw a picture for this case similar to Figure 8.18, and follow the logic of the derivation given in this section.)

6. Find the sample size for a two-tailed test of hypothesis of $H_0{:}\mu = 28$ versus $H_a{:}\mu \neq 28$ with $\alpha = 0.05$ when a shift of 1 unit is to be detected with a probability of 0.85. Assume $\sigma = 6$. (Remember the connection between power and β.) Use the formula derived in exercise 5.

7. Test the following one-tailed hypothesis and sketch a power curve for it using $\alpha = .025$, a sample size of 16, and a known standard deviation of 10.

$$H_0{:}\ \mu = 50 \text{ vs. } H_a{:}\ \mu < 50.$$

8. Derive a sample size formula for a one-tailed test, left-tail rejection region in terms of a given value for σ, α, β, and $\delta = \mu_o - \mu_1$. (Draw a picture for this case similar to Figure 8.18, and follow the logic of the derivation given in this section.)

9. Sketch a power curve for a two-tailed test and compare it to the curves that would result with a one-tailed test with (a) a left-tail rejection region, and (b) a right-tail rejection region.

10. Suppose we have the null and alternative of $H_0{:}\ \mu = 50$ vs. $H_a{:}\ \mu > 50$. Assess the impact on the sample size if:
   (a) The standard deviation were to double.
   (b) We allowed the type I error to double.
   (c) The shift that we want to detect $\delta = |\mu_o - \mu_1|$ were to be cut in half.

8.4 THE ONE-SAMPLE TEST OF MEANS IN NORMALLY DISTRIBUTED POPULATIONS

Introduction

In a manufacturing process it has been found that the average time to assemble a particular piece of electronics for installation in a computer is 10.5 minutes. An industrial engineer studying the assembly line proposes a change in the process of assembly that she feels will reduce the time it takes to assemble the electronic component. Therefore, the conjectures is that mean time to assembly will be less than 10.5 minutes. If we assume the assembly times are normally distributed we want to devise an experiment and a test of hypothesis procedure to test the hypothesis:

$$H_o: \mu = 10.5\text{minutes} \qquad vs. \qquad H_a: \mu < 10.5 \text{ minutes.}$$

Many details of the experiment would have to be worked out such as training employees in the new assembly procedures, controlling as much as possible for other factors that might influence the outcome, and so forth. Ultimately, the assembly times for a sample of components would be used to test the above hypothesis. The preceding sections have demonstrated the process we would follow to conduct such a test for a set of data.

In this section we formalize the procedures of the previous sections into the steps needed for testing a specific hypothesis about a single mean from a normally distributed population. We present the steps for two cases: (1) when the standard deviation is known, and (2) when the standard deviation is unknown. You should notice that the test statistic used in these cases is identical to the pivotal quantity encountered in the last chapter on confidence intervals.

Methodology

Case 1:   Normal Distribution, Known Standard Deviation

Background assumptions: Assume that we have a random sample from a normal distribution with an unknown mean μ and known standard deviation, σ. We want to test a hypothesis concerning the unknown mean μ.

Step 1: The Hypotheses:

Null Hypothesis   Possible Alternative Hypotheses

$$H_o: \mu = \mu_o \qquad H_a: \begin{cases} \mu \neq \mu_o \\ \mu > \mu_o \\ \mu < \mu_o \end{cases}$$

Step 2: Select a random sample of size n from the assumed normal distribution. Determine the level of significance α.

Step 3: The test statistic is

$$Z = \frac{(\bar{x} - \mu_o)}{\sigma/\sqrt{n}}$$

which follows a standard normal distribution.

Step 4: Compute the *p*-value as follows:

Alternative Hypothesis	p-value Computation
$\mu \neq \mu_o$	$p = 2P(Z > \lvert Z\rvert_c)$
$\mu > \mu_o$	$p = P(Z > Z_c)$
$\mu < \mu_o$	$p = P(Z < Z_c)$

where

$$Z_c = \frac{(\bar{x} - \mu_o)}{\sigma/\sqrt{n}}$$

Step 5: If $p < \alpha$, then reject H_0.

Case 2: Normal Distribution, Unknown Standard Deviation

Background assumptions: Assume that we have a random sample of n observations from a normal distribution with an unknown mean μ and an unknown standard deviation, σ. We want to test a hypothesis concerning the unknown mean μ.

Step 1: The Hypotheses:

Null Hypothesis   Possible Alternative Hypotheses

$$H_0: \mu = \mu_o \qquad\qquad H_a: \begin{cases} \mu \neq \mu_o \\ \mu > \mu_o \\ \mu < \mu_o \end{cases}$$

Step 2: Select a random sample of size n from the assumed normal distribution. In addition, let the selected probability of a type I error be α.

Step 3: The test statistic is

$$t = \frac{(\bar{x} - \mu_o)}{s/\sqrt{n}}$$

which follows the t-distribution with $df = n - 1$.

Step 4: Compute the p-value as follows

Alternative Hypothesis	p-value Computation
$\mu \neq \mu_o$	$p = 2P(t > \lvert t\rvert_c)$
$\mu > \mu_o$	$p = P(t > t_c)$
$\mu < \mu_o$	$p = P(t < t_c)$

where

$$t_c = \frac{(\bar{x} - \mu_o)}{s/\sqrt{n}}$$

Step 5: If $p < \alpha$, then reject H_0.

DISCUSSION

From this point on and through the next chapter, the sections are "catalogs" of methods for testing hypotheses under various conditions. Therefore, we introduce a standard format of presentation that will persist through these sections. First, we identify, by title, the distribution model assumed. Next, we identify the "background" assumptions that are needed to validate the procedure presented. Then, the 5-step procedure for testing hypotheses is presented. Examples of each method are also provided.

In this section we assume explicitly that we have a random sample from a normally distributed population and we have formulated a hypothesis about the mean of the population. The research hypothesis may be one-tailed or two-tailed. Regardless of which it is, the test statistic has the same form and is either a Z-score or a t-score. If we know the standard deviation of the population, the test statistic is

$$Z = \frac{(\bar{x} - \mu_o)}{\sigma/\sqrt{n}}$$

If the standard deviation σ is unknown, then we use its estimator s and the test statistic is

$$t = \frac{(\bar{x} - \mu_o)}{s/\sqrt{n}}$$

We recommend the p-value approach to making a decision since the publication of p-values is frequently the method of reporting results in research journals. This allows anyone reading the research to select their own value for α, the probability of the type I error, and decide whether they would reject the null hypothesis or not by comparing it with the published p-value. A table that indicates the alternative (research) hypothesis and a "generic" approach to computing a p-value is given as Table 8.1.

Table 8.1 The p-value computation for a given alternative hypothesis.

	Alternative Hypothesis		p-value Computation*
(a)		$\mu \neq \mu_o$	$p = 2P(TS > |TS|)$
(b)		$\mu > \mu_o$	$p = P(TS > TS_c)$
(c)		$\mu < \mu_o$	$p = P(TS < TS_c)$

*TS stands for "test statistic" and will usually represent a Z, t, χ^2, or F random variable

A few examples illustrate the specifics of this approach.

EXAMPLES

EXAMPLE 1: The EPA sets a limit of 5 parts per million (ppm) on PCB, a poisonous chemical pollutant, in river water. A major manufacturing firm producing PCB for insulation of electrical cables discharges a small amount through its waste water system. Building in a safety factor, federal instructions require the plant to halt production if a statistical test of hypothesis indicates the mean level exceeds 3 ppm using $\alpha = 0.01$. Experience has shown the standard deviation of measurements made under normal conditions to be about 0.5 ppm.

A random sampling of fifty water specimens yielded a mean of 3.1 ppm. At $\alpha = 0.01$, do the data present strong enough evidence for a shutdown ?

SOLUTION:

Step 1: $H_o: \mu = 3$ ppm   vs. $H_a: \mu > 3$ ppm

Step 2: A random sample of 50 water samples is selected and the Type I error is to be $\alpha = 0.01$.

Step 3: The test statistic is

$$Z = \frac{(\bar{x} - 3)}{0.5/\sqrt{50}}$$

Step 4: The actual value of the test statistic is

$$Z_c = \frac{(3.1 - 3)}{0.5/\sqrt{50}}$$

which has a p-value of 0.0793. That is $p = P(Z > 1.41) = 0.0793$.

Step 5:   For the given value of α, we have insufficient statistical evidence to reject H_o, ($p > 0.01$) and therefore conclude that there is insufficient proof that the manufacturer has exceeded government standards.

EXAMPLE 2: Computer response time is the time a user has to wait for access to information on the disk. A computer center wants the computer response time to average 60 milliseconds or less. A random monitoring of 15 response times was made and the mean and standard deviation were found to be 68 and 18 milliseconds respectively. Does this provide statistical evidence that the response time exceeds 60 milliseconds? Or could chance provide a reasonable explanation for the sample outcome?

SOLUTION:

Step 1: $H_o: \mu = 60$ vs. $H_a: \mu > 60$

Step 2: A random sample of 15 response times will be collected.

Step 3: The test statistic is

$$t = \frac{(\bar{x} - 60)}{s/\sqrt{15}}$$

which, based on our background assumptions, is distributed as a t random variable with $df = 14$.

Step 4: The sample of fifteen response times yielded a mean of 68 milliseconds and a standard deviation of 18 milliseconds. The calculated value of the test statistic is

$$t_c = \frac{(68 - 60)}{18/\sqrt{15}} = 1.72$$

Step 5:   The p-value from the t-table for this value of t is: $0.05 < p < 0.10$. (An exact calculation of the p-value is 0.0536). On the basis of this p-value, convention ($\alpha = 0.05$) would suggest we do not reject, however, common sense suggests that we are seeing a very unlikely event and at least further study might be warranted.

EXERCISES

1.   In a taste-testing experiment for a favorite soft drink, 10 persons were randomly selected and asked to rank their preference for the taste on a scale from 1 to 10. Experience suggests that the mean score would have to exceed 6.0 on the scale in order for the drink to be competitive. Test this hypothesis using an α of 0.05. The mean of the scores reported was $\bar{x} = 6.8$ with a standard deviation of $s = 1.2$. Comment on the validity of assumptions in this problem and note any that are questionable.

2.   A real estate firm followed the market very closely and divided the city in which they did business into 225 zones. Each month they would take a random sample of 25 of these zones and record the number of houses listed for sale in the zone. The mean per zone for all 225 zones the previous year was 8.3 houses for sale with a standard deviation of 2.7 houses. For the same month this year they obtained a sample mean of 6.5 houses. Test the hypothesis that the real estate listings per zone is the same versus it is different. Use $\alpha = .01$. State the assumptions that you must make and respond to their validity in this case.

3.   A mining firm regularly takes ore samples from their drilling sites. A site is regarded with potential as long as the average number of ounces per sample of a particular type of mineral exceeds 4 oz. From 16 ore samples from a given site, the mean per sample of the mineral was 5.1 with a standard deviation of 0.8. Test the hypothesis of interest in this problem using a significance level of $\alpha = 0.01$.

4.   A cannery processing tuna fish in cans is required to test the fish for various contaminants. If the level of mercury, for instance exceeds 8 ppm, then the fish is regarded as unacceptable for human consumption. The quality control inspector regularly takes a sample of 8 fish from the processing line and subjects them to chemical analysis to detect and measure the mercury content. The latest sample produced a mean of 8.3 ppm with a standard deviation of 0.27 ppm. Test the hypothesis of interest with $\alpha = 0.005$.

5.   Suppose the problem of exercise 4 were stated as: "If the level of mercury is less than 8 ppm the fish is regarded as being safe for human consumption." Which of these statements of the hypothesis would the cannery propose? Which would the USDA (United States Department of Agriculture) propose, as advocates of the consumer?

6.   After being stored for a week, newly made cinder blocks are tested for moisture content. If the mean moisture content is above 5.0, the block cannot be used for building material, but must be allowed to dry more. From a sample of 12 cinder blocks the sample mean was found to be $\bar{x} = 5.98$. From experience it is known that $\sigma = 1.20$. Test the hypothesis for $\alpha = 0.05$.

7.   From the data of exercise 6 it was found that $s = 1.15$. Compute t_c for this data and compare with Z_c for exercise 6. Compute the p-values for both tests and comment on the reason for the difference you see in them.

8. Highway engineers testing the reflective characteristics of a certain type of paint take 64 samples. The results recorded were $\bar{x} = 136$ and $s = 66$. Is there evidence to indicate that the mean reflective characteristic is more than 115? Let $\alpha = 0.10$.

9. Let X be normally distributed with an unknown mean μ and a standard deviation $\sigma = 4$. Test the hypothesis $H_o:\mu = 0$ against $H_a:\mu \neq 0$ with a random sample of size $n = 25$ that produced a sample mean $\bar{x} = 0.28$. Do we reject H_o with $\alpha = 0.10$ or not. Does the 90% confidence interval contain $\mu = 0$?

10. Examine the hypothesis of exercise 9 and the confidence interval computed. Could you have a result occur where H_o is "accepted" and you find $\mu = 0$ is not in the confidence interval when $\alpha = 0.10$ and the confidence level is 90%. Explain why or why not.

8.5   TWO SAMPLE TESTS OF MEANS FOR NORMALLY DISTRIBUTED POPULATIONS

Introduction

Most tests of hypothesis about normally distributed populations will involve more than a single population. For instance, we want to compare male salaries with female salaries. We wonder if married students do better than unmarried students. We want to compare an experimental method with a control group.

In each of these situations, we are comparing one population of measurements with another. If we are willing to assume the populations of measurements are normally distributed, or that we have large enough samples to justify the large sample properties of the central limit theorem, then we need a method to compare the two populations involved. If our purpose is to compare the means of the two populations, then the results of this section are appropriate. If the purpose is to compare the variances then section 8.6 should be consulted.

As before, the test statistics, but not the approach, depends upon whether we know the standard deviations or not. The first method assumes known standard deviations.

Methodology

Case 1: Two Normal Distributions, Known Standard Deviations

Background assumptions: Assume that we sample independently from two normal distributions with unknown means μ_x and μ_y, and known standard deviations, σ_x and σ_y. We want to test a hypothesis concerning the difference of the unknown means, $\mu_x - \mu_y$.

Step 1: The Hypotheses:

Null Hypothesis   Possible Alternative Hypotheses

$$H_o: \mu_x = \mu_y \qquad\qquad H_a: \begin{cases} \mu_x \neq \mu_y \\ \mu_x > \mu_y \\ \mu_x < \mu_y \end{cases}$$

Step 2: Select a random sample of size m and n from the assumed normal distributions. In addition, let the selected probability of a type I error be α.

Step 3: The test statistic is

$$Z = \frac{(\bar{x} - \bar{y}) - (\mu_x - \mu_y)}{\sqrt{(\frac{\sigma_x^2}{n} + \frac{\sigma_y^2}{m})}}$$

which follows a standard normal distribution.

Step 4: Compute the p-value as follows

Alternative Hypothesis   p-value Computation

$\mu_x \neq \mu_y$ $p = 2P(Z > |Z_c|)$

$\mu_x > \mu_y$ $p = P(Z > Z_c)$

$\mu_x < \mu_y$ $p = P(Z < Z_c)$

where Z_c is

$$Z_c = \frac{(\bar{x} - \bar{y}) - (\mu_x - \mu_y)}{\sqrt{(\frac{\sigma_x^2}{n} + \frac{\sigma_y^2}{m})}}$$

Step 5: If $p < \alpha$, then reject H_o.

Case 2: Two Normal Distributions, Unknown But Equal Variances

Background assumptions: Assume that we are sampling from two normal distributions with unknown means μ_x and μ_y and unknown but equal variances, i.e., $\sigma_x^2 = \sigma_y^2 = \sigma^2$. We want to test the hypothesis about the difference in the unknown means, $\mu_x - \mu_y$.

Step 1: The Hypothesis:

Null Hypothesis   Possible Alternative Hypotheses

$$H_o: \mu_x = \mu_y \qquad\qquad H_a: \begin{cases} (a)\ \mu_x \neq \mu_y \\ (b)\ \mu_x > \mu_y \\ (c)\ \mu_x < \mu_y \end{cases}$$

Step 2:   Select a random sample size m and n from the assumed normal distributions respectively. In addition, let the selected probability of a type I error be α.

Step 3:   The test statistic is

$$t = \frac{(\bar{x} - \bar{y}) - (\mu_x - \mu_y)}{\sqrt{s_p^2(\frac{1}{n} + \frac{1}{m})}}$$

where

$$s_p^2 = \frac{(n-1)s_x^2 + (m-1)s_y^2}{(n-1) + (m-1)}.$$

which follows a t-distribution with $m + n - 2$ df.

Step 4:   Compute the p-value as follows

Alternative Hypothesis	p-value Computation
$\mu_x \neq \mu_y$	$p = 2P(t > |t_c|)$
$\mu_x > \mu_y$	$p = P(t > t_c)$
$\mu_x < \mu_y$	$p = P(t < t_c)$

where t_c is

$$t_c = \frac{(\bar{x} - \bar{y}) - (\mu_x - \mu_y)}{\sqrt{s_p^2(\frac{1}{n} + \frac{1}{m})}}$$

Step 5:   If $p < \alpha$, then reject H_o.

DISCUSSION

In this section we assume that we have two separate, normally distributed populations identified with the random variables X and Y, respectively. We are interested in constructing a hypothesis about the difference in the means, the most common hypothesis being that the are equal. Therefore, the null hypothesis will usually take on the form: H_o: $\mu_x = \mu_y$, or H_o: $\mu_x - \mu_y = 0$. (We could also hypothesis that the difference in the means is a specified value such that H_o: $\mu_x - \mu_y = \delta$. The test statistic we use will accommodate either situation.) We assume that a random sample of size n is selected from the X-population and a random sample of size m is selected from the Y-population. We assume either that the variances of the two populations, σ_x^2 and σ_y^2, are known or that they are unknown but equal. In the former case we can construct a Z-score for our test statistic of the form

$$Z = \frac{(\bar{x} - \bar{y}) - (\mu_x - \mu_y)}{\sqrt{(\frac{\sigma_x^2}{n} + \frac{\sigma_y^2}{m})}}$$

which will be standard normal in distribution.

In some instances when n and m are large, it is common practice to substitute the values of s_x^2 and s_y^2 in place of σ_x^2 and σ_y^2 in the above formula when the true variances are unknown, expecting that the sample variances would provide reliable estimates. If this is done, an approximate standard normal random variable would be obtained.

If we assume $\sigma_x^2 = \sigma_y^2 = \sigma^2$, then the test statistic can be written as

$$t = \frac{(\bar{x} - \bar{y}) - (\mu_x - \mu_y)}{\sqrt{s_p^2(\frac{1}{n} + \frac{1}{m})}}$$

which will follow a t distribution with $df = (n - 1) + (m - 1) = n + m - 2$.

Notice in both formulas above that H_o could be written as $\mu_x - \mu_y = 0$ or $\mu_x - \mu_y = \delta$. The test statistics as given will allow for either case to be tested. The justification for these results are presented in Chapter 6 if review is needed. The following examples illustrate their use.

EXAMPLES

EXAMPLE 1: In a study of the recreation habits of teenagers in an inner-city high school of a large metropolitan city, one question asked was the number of video games played per week. The population was divided into two age groups, 13-15 year olds and 16-18 year olds. In the study there were 170 in the 13-15 age group and 120 in the 16-18 age group. Test the hypothesis that the average number of games played per week is the same for the two different age groups. The following data were obtained. (Assume that due to the large sample size, the sample standard deviations are essentially equal to the population standard deviations.)

	Age Group	
	13-15 Years	16-18 Years
Sample size	170.0	120.0
Mean	18.5	14.8
St. Dev.	2.8	5.3

SOLUTION: Let X denote the measurements for the 13-15 year age group and Y denote the measurements for the 16-18 year age group. Therefore we have the following steps:

Step 1: The Hypotheses:

Null Hypothesis Alternative Hypothesis

H_o: $\mu_x = \mu_y$ H_a: $\mu_x \neq \mu_y$

Step 2: Random samples of size $m = 170$ and $n = 120$ from assumed normal distributions are selected.

Step 3: The test statistic is

$$Z = \frac{(\bar{x} - \bar{y}) - (\mu_x - \mu_y)}{\sqrt{(\frac{\sigma_x^2}{n} + \frac{\sigma_y^2}{m})}}$$

which follows a standard normal distribution approximately if the sample standard deviations are substituted for the true standard deviations.

Step 4: The computed value of the test statistic is:

$$Z_c = \frac{(\bar{x} - \bar{y}) - (\mu_x - \mu_y)}{\sqrt{(\frac{\sigma_x^2}{n} + \frac{\sigma_y^2}{m})}} = \frac{(18.5 - 14.8) - 0}{\sqrt{(\frac{(2.8)^2}{170} + \frac{(5.3)^2}{120})}} = 6.99$$

which has an approximate *p*-value of 0.0000.

Step 5: With a *p*-value this small, we reject the null hypothesis and conclude there is a difference between the means for the two different age groups and that the younger age group plays more games on the average than does the older age group.

EXAMPLE 2: Redo the computations for the test of hypothesis of Example 1 under an assumption of unknown but equal (common) variance. Compare to the results of Example 1.

SOLUTION: All the steps of the hypothesis would be the same except for the test statistic we would use and the table in which we would find the p-value. The computation of the test statistic is:

$$t_c = \frac{(\bar{x} - \bar{y}) - (\mu_x - \mu_y)}{\sqrt{s_p^2(\frac{1}{n} + \frac{1}{m})}} = \frac{(18.5 - 14.8) - 0}{\sqrt{16.2(\frac{1}{170} + \frac{1}{120})}} = 7.71$$

where

$$s_p^2 = \frac{(170 - 1)(2.8^2) + (120 - 1)(5.3^2)}{(170 - 1) + (120 - 1)} = 16.2$$

The *p*-value for this test is also 0.0000 which is the same as for Example 1 though they would differ in places beyond the four decimal-place accuracy reported here.

EXAMPLE 3: Many college and university professors have been accused of "grade inflation" in recent years. This means they assign higher average grades now than in the past even though the students' work is of the same caliber. To test the grade inflation theory at one university, a professor selected a random sample of 30 majors who were graduating and whose records and gpa's were available. However, the professor was only able to obtain the gpa's of 10 majors from 10 years ago. A summary of the results is given below. Using $\alpha = 0.10$, do the data support the accusation of grade inflation?

	10 YEARS AGO	PRESENT
Sample size	10	30
Sample mean	2.82	3.04
Sample standard deviation	0.43	0.38

(In addition, comment on the validity of background assumptions, along with any flaws in the approach to solving the problem.)

SOLUTION:

Step 1: H_o: $\mu_x = \mu_y$ H_a: $\mu_x \neq \mu_y$

Step 2:   A sample of 10 and 30 students is obtained from the two populations, respectively. The level of significance is chosen to be $\alpha = 0.10$.

Step 3:   The test statistic is

$$t = \frac{(\bar{x} - \bar{y}) - (\mu_x - \mu_y)}{\sqrt{s_p^2(\frac{1}{n} + \frac{1}{m})}} = \frac{(\bar{x} - \bar{y}) - (\mu_x - \mu_y)}{\sqrt{s_p^2(\frac{1}{10} + \frac{1}{30})}}$$

which, under our assumptions, possesses a t-distribution with 38 degrees of freedom.

Step 4:   As a preliminary computation, the pooled variance is:

$$s_p^2 = \frac{(10 - 1)(0.43^2) + (30 - 1)(0.38^2)}{(10 - 1) + (30 - 1)} = 0.154$$

The computed t-value is

$$t_c = \frac{(2.82 - 3.04) - 0}{\sqrt{.154\left(\frac{1}{10} + \frac{1}{30}\right)}} = -1.5353$$

which from the t table has a p-value such that $0.05 < p < 0.10$. (The actual p-value is 0.0665)

Step 5:   For $\alpha = 0.05$ we would not reject, but for $\alpha = 0.10$ we would. Since we chose $\alpha = 0.10$ as our level of significance, we conclude that there is evidence to statistically conclude that the mean gpa of the present students is higher than students of 10 years ago.

The assumptions we require to validate the analysis are assumptions of normality and equal variance. Since a g.p.a. is the average of numerous measurements, it would be reasonable to expect the distribution to be near normal. The assumption of equal variance is not unreasonable and could expected to be approximately true.

There are other questions that should be asked about the validity of the approach, however.

How equal were the courses of study of the individuals involved in the samples? Is it reasonable to regard the 10 records from 10 years ago as being random and representative of the records then? Are the instructors the same? What kind of controls were implemented to ensure that the difference, if found, is due to grade inflation and not other factors?

EXERCISES

1. Makers of two of the popular soft drinks in the United States have waged a battle concerning which one was the most preferred. An independent research firm conducted the following study. Twenty persons were asked to rate the taste of brand A cola on a scale from 1 to 10. An independent group of twenty persons were asked to rate Brand B cola in the same fashion. Test the hypothesis of equal ratings. Comment upon any assumptions you feel are questionable. Use the data in the following table and $\alpha = 0.05$.

Brand		
	A	B
Sample Size	20	20.0
Mean	7.3	6.2
St. Dev.	1.3	2.8

2. Analyze the data of exercise 1 assuming the sample sizes are large enough that the two sample standard deviations are "good" estimates of the true standard deviations and can therefore be used in place of the true standard deviations. Note the similarities in the computational results that you get compared to the outcome of exercise 1.

3. Show that $s_p^2(\frac{1}{n} + \frac{1}{m}) = \frac{s_x^2}{n} + \frac{s_y^2}{m}$ when $n = m$.

4. In exercises 1 and 2 compare the two p-values obtained and comment on the reasons for their being different.

5. Construct a confidence interval for the difference of means for exercise 1 using the same assumptions that you used in that exercise. Does the confidence interval enclose the value $\mu_x - \mu_y = 0$? Discuss how this notion compares to the conclusion of your test of hypothesis in exercise 1.

6. The effectiveness of two different formulations of fluoride mouth wash was being tested by a dental research laboratory under controlled conditions. Ten children were treated with one of the formulations while another ten, with similar dental histories were treated with the other. After a 2-year time period, the number of dental caries observed and treated were reported for the two groups. Test the hypothesis of equal effectiveness. Use $\alpha = 0.025$.

	Formulation	
	X	Y
Sample Size	10	10
Mean	4.1	5.3
St. Dev.	0.8	1.3

7. A manufacturer of insulation has developed a type of insulation that it feels is superior to their

current product. To test this hypothesis they found a housing subdivision being built that essentially was made up of the same size home and floor plan, the differences in units being aesthetic only. Arrangements were made to install their regular insulation in 14 homes as a control group and they then installed their new product in 18 identical homes. During a home show, where they had control over temperature settings, they kept track of the BTU's of heat burned in the two different groups of homes for a period of 2 weeks. Is there sufficient statistical evidence from this study to indicate that the new insulation is superior to the old? Comment on any other aspects of this experiment that causes you concern.

	Insulation Type	
	Old	New
Sample Size	14	18
Mean	89.0	76.0
St. Dev.	6.8	5.5

8. Construct a confidence interval on the difference of the two means using the data of of exercise 7, choosing the value of α that would provide an equivalent decision rule for testing the hypothesis that the difference in means is zero.

9. Using the data of exercise 7, test the hypothesis that the benefit of the new insulation is to reduce the average use of BTU's by more than 10 BTU's. (The null hypothesis for this case would be $\mu_x - \mu_y = 10$. Set up the appropriate alternative hypothesis and complete the test.)

10. An anthropologist, studying fertility and other characteristics of two different tribes in a remote country in South America, took a random sample of 50 married women from each tribe and recorded the number of pregnancies each women reported. (The women were chosen from those women who were no longer of child-bearing age.) The mean and standard deviations of the number of pregnancies reported, for each tribe were, respectively:

	Tribe	
	One	Two
Sample Size	50	50
Mean	5.1	6.2
St. Dev.	2.1	2.8

Test the hypothesis of equal means by assuming they come from populations with a common variance. Use $\alpha = 0.05$.

11. Analyze the data of exercise 10 assuming the sample sizes are large enough that the two sample standard deviations are "good" estimates of the true standard deviations and can therefore be used in place of the true standard deviations. Note the similarities in the computational results that you get compared to the outcome of exercise 10.

12. For samples of the sizes given in exercise 10, how different will the Z and t-score values be when making a decision rule based on them for $\alpha = 0.05$.

11. Construct a confidence interval on the difference of the two means of exercise 10 choosing the value of α that would provide an equivalent decision rule for testing the hypothesis that the difference in means is zero.

8.6 TEST OF MEANS FOR PAIRED OBSERVATIONS

Introduction

One of the objectives of good "statistics" and good "science" in general is to improve the soundness of investigations, to reduce the errors that might mislead the investigator. In statistics we try to accomplish this end by using good experimental designs—methods of conducting experiments and statistical investigations to minimize the errors that might obscure the truth. One way of doing this when comparing two population means is through the *paired comparison experiment*.

For example, suppose we want to compare the effectiveness of two methods of teaching first grade children to read. We first want to control for all the factors that could influence the learning rate of the children involved: the teachers, the environment, the time of day, etc. But even after controlling for all these factors there is still the basic difference of the students involved. It is possible that the observed difference in performance scores or whatever is used to evaluate the teaching methods may be attributed to differences in students rather than differences in teaching methods.

One way to remove the affect of student differences, though difficult and often costly to implement, is to use twin-pairs. One member of each pair would be assigned at random to the one method, the other receiving the other method. Thus, when the comparison is made, the observed difference is much more likely to be attributed to the teaching methods, rather than the individual differences in students since the individuals involved were as much alike as they could possibly be. However, in the analysis of this data, we should also take this characteristic of the experiment into account and work with the differences in scores from each twin pair, rather than the individual scores themselves.

Another common situation that dictates the same method of analysis, is the experiment using a "before/after" approach where the subject is measured before and after some "treatment" is applied. Here, obviously, the scores appear as pairs of data. If we let X denote the "before" score, and Y denote the "after" score, then we should examine the difference, $X - Y$, for each person in the study. If these differences are normally distributed, then the method of analysis described in this section is appropriate.

Methodology

Paired Observations, Normally Distributed Differences

Background Assumptions: Assume that we have a sample of n bivariate observations, $(X_1, Y_1), ..., (X_n, Y_n)$ such that the differences, $D_i = X_i - Y_i$ constitute a random sample from a normal distribution where $E(D_i) = \mu_D$ with an unknown standard deviation. We want to test the hypothesis that the unknown mean, μ_D, takes on a specific value.

Step 1: The Hypotheses:

Null Hypothesis   Possible Alternative Hypotheses

$$H_0: \mu_D = \mu_o \qquad\qquad H_a: \begin{cases} \mu_D \neq \mu_o \\ \mu_D > \mu_o \\ \mu_D < \mu_o \end{cases}$$

(Usually μ_o is specified to be 0.)

Step 2: Select a random sample size n from the bivariate population of pairs and the level of significance α.

Step 3: First, compute the n differences, $D_i = X_i - Y_i$. The test statistic is

$$t = \frac{(\bar{d} - \mu_o)}{s_d/\sqrt{n}}$$

where $\bar{d}$ is the mean of the n differences, and s_d is the standard deviation of the n differences. The test statistic follows a t distribution with $n - 1$ degrees of freedom.

Step 4: Compute the p-value as follows

Alternative Hypothesis	p-value Computation
$\mu_D \neq \mu_o$	$p = 2P(t > \lvert t_c \rvert)$
$\mu_D > \mu_o$	$p = P(t > t_c)$
$\mu_D < \mu_o$	$p = P(t < t_c)$

where t_c is

$$t_c = \frac{(\bar{d} - \mu_o)}{s_d/\sqrt{\text{I}}}$$

Step 5: If $p < \alpha$, then reject H_0.

DISCUSSION

One of the primary objectives of good statistical practice is to reduce the effect of uncertainty that occurs in so many problems. One simple way to do this when comparing the means of two populations is to use a "paired comparison" experiment. This requires that observations from the two populations be paired on the basis of common characteristics, thus, creating a strong correlation between the measurements obtained from each member of the pair. The analysis is performed on the differences in the pairs, with the effect that much of the source of variability will have been removed by the process of pairing.

For example, the use of identical twins in studies is one form of pairing. Another is to set up a "before–after" experiment using the same person for each "before–after" measurement. Less effective, but still helpful, would be to examine the subjects before experimentation and pair them up on the basis of common characteristics that would provide correlated measurement values.

The theoretical argument for such pairing is based on the notion of having a random variable X that is positively correlated with a random variable Y. If we are interested in the difference $D = X - Y$, then the variance of D is

$$V(D) = V(X - Y) = V(X) + V(Y) - 2\text{Cov}(X,Y) = V(X) + V(Y) - 2\rho\sqrt{V(X)} \cdot \sqrt{V(Y)}.$$

If we make the same assumption of equal variances that we made in section 8.5 we get, after simplification:

$$V(D) = \begin{cases} 2\sigma^2(1 - \rho) & \text{if } X \text{ and } Y \text{ are correlated} \\ 2\sigma^2 & \text{if } X \text{ and } Y \text{ are uncorrelated.} \end{cases}$$

If the correlation is positive, then $V(D)$ will be smaller than if X and Y are uncorrelated, and thus we have succeeded in removing some uncertainty by reducing the size of the variance by the process of pairing.

The application and implementation of these ideas are illustrated in the following examples.

EXAMPLES

EXAMPLE 1: A manufacturer is comparing two new models of a die stamping machine for average time to produce a part. If it can be shown that one is better (smaller average time to produce the part) than the other, they will be begin to phase out their current machines with the new, "superior" machine. It is recognized that one machine may be better doing certain kinds of tasks, while the other machine might be better doing other types of tasks. Therefore, over the last 3 months, 10 typical but different types of products were randomly selected and the time to produce a part was recorded for each machine. Is there evidence at $\alpha = 0.05$ to conclude that one machine is "better" than the other? The time it took to produce each of the 10 different parts by each machine is presented in the following table:

Time in Seconds										
Part Number										
Machine	1	2	3	4	5	6	7	8	9	10
X	6	24	15	5	16	50	18	16	24	37
Y	2	21	12	3	18	46	15	15	18	40
Difference	4	3	3	2	−2	4	3	1	6	−3

SOLUTION: The pairing or creation of correlated measurements was implemented by the choice of the different parts that each machine was required to produce. The time it takes for a given part produced by one machine should be correlated with the time it takes the other machine to produce the part. (If a part is complex, it will take both machines a "long" time to produce it; if a part is simple, it will take both machines a "short" time to complete it.) The analysis of the data proceeds as follows.

Step 1: $H_o: \mu_X = \mu_Y$ or $\mu_D = 0$ $H_a: \mu_X \neq \mu_Y$ or $\mu_D \neq 0$

Step 2:   The experimental procedure is described above and α was set at 0.05.

Step 3:   The test statistic is

$$t = \frac{(\bar{d} - \mu_o)}{s_d/\sqrt{n}}$$

which follows the t distribution with $df = 9$.

Step 4:   The mean and standard deviation of the differences are 2.1 and 2.767 respectively. Thus, the computed value of t is

$$t_c = \frac{(2.1 - 0)}{2.767\sqrt{10}} = 2.40$$

which, from the table, is associated with a p-value of $2(0.025) < p < 2(0.01)$. (The actual value is $p = 2 \cdot (0.0199) = 0.0399$.)

Step 5:   Since $p < \alpha = 0.05$, we reject H_o and conclude machine X gives statistically different mean production times when compared to machine Y. (The data would seem to suggest that Machine Y takes less time on the average than Machine X. Therefore, Machine Y would seem to be a faster machine.)

EXAMPLE 2: A study was conducted to determine the extent that alcohol blunts a person's thinking ability for performing a given task. Ten persons of varying characteristics were selected at random and asked to participate in the experiment. After suitable briefing, each person performed the task with no alcohol in the blood system. Then the task was performed again after each person had consumed enough alcohol to raise alcohol content in his or her system to 0.1%. Can one conclude at the $\alpha = 0.05$ level of significance that the average "before" time is less than the average "after" time? The data and some summarizing computations are given below.

Response times (in minutes) for completing a task before and after ingestion of alcohol

	Before	After	Diff = After–Before
	28	39	11
	22	45	23
	55	67	12
	45	61	16
	32	46	14
	35	58	23
	40	51	11
	25	34	9
	37	48	11
	20	30	10
Mean	33.9	47.9	14
Variance	118.7667	139.2111	26.44445

SOLUTION:

Step 1: $H_o: \mu_B = \mu_A$ or $\mu_D = 0$ $H_a: \mu_B < \mu_A$ or $\mu_A - \mu_B > 0$

Step 2:   The experimental procedure is described above and α was set at 0.05.

Step 3:   The test statistic is

$$t = \frac{(\bar{d} - 0)}{s_d/\sqrt{10}}$$

which follows the t distribution with $df = 9$.

Step 4:   The mean and standard deviation of the differences are 14 and $\sqrt{26.44445} = 5.1424$ respectively. Thus, the computed value of t is

$$t_c = \frac{(14 - 0)}{5.1424 /\sqrt{10}} = 8.609$$

which, from the table, is associated with a p-value of $p < 2(0.005) = 0.010$. (Actual value to 4 places is $p = 0.0000$.)

Step 5:   Since $p < \alpha = 0.05$, we reject H_o and that the mean response time with alcohol is greater than without alcohol.

EXAMPLE 3: Suppose the data in Example 2 were analyzed as if it were a two-independent sample test. How does the variance of the difference compare to the variance of the difference under the "paired comparison" test? Why are the test statistic values for the two tests different?

SOLUTION: The variance of the two-independent sample approach is estimated by

$$\left(\frac{s_x^2}{n} + \frac{s_y^2}{m}\right) = \left(\frac{118.7667}{10} + \frac{139.2111}{10}\right) = 25.7978$$

compared to

$$\frac{s_d^2}{n} = \frac{26.4445}{10} = 2.644445.$$

The value of the test statistic for the two-independent sample approach is

$$\frac{14 - 0}{\sqrt{25.7978}} = 2.756$$

compared to 8.609 for the paired-comparison test. The reason for the large difference is the difference in the sizes of the two variances, since the numerators are both the same.

EXERCISES

1.   In a comparison of tastes of two well known soft drinks a group of 10 "taste-testers" was used. Each person was given a taste of each drink in unmarked cups. In addition, the decision as to which soft drink was presented first was determined randomly. The taste tester was then asked to give each drink a score from 1 to 10. The following data were obtained. Test the hypothesis of equal mean scores between the two soft drinks. Use $\alpha = 0.05$.

Taste Tester	Soft Drink	
	X	Y
1	9	7
2	8	8
3	8	6
4	9	8
5	8	9
6	5	3
7	7	4
8	8	5
9	9	9
10	8	7

2. A psychologist conducted the following experiment in an introductory psychology course. Students were given a sequence of digits to memorize within 15 seconds. The digits were covered and the students were asked to list the digits in order. Each student then recorded the number of digits listed correctly starting to the left of the list. Next the instructor provided a short lesson on a memory technique. The exercise was repeated with a different string of digits. The purpose was to see if the memory exercise helped in the memorization process. There were 45 students in the class and the mean of the difference in scores (the improvement scores) was 2.4 with a standard deviation of .76. For $\alpha = 0.025$, is there statistical evidence for concluding the memory exercise helped with the memorization of a string of digits?

3. A manufacturer of wood stains wanted to compare a new stain against the standard stain currently being used. Since there is quite a difference in the way different woods accept stains, six different types of wood were selected and the staining surface was divided into two halves. One of the stains was randomly assigned to the left half, the right half receiving the other stain. A stain "expert" then compared the two halves and gave a score to the difference observed. The following difference scores were obtained. Test the hypothesis of no difference in stains for $\alpha = 0.05$. Difference scores: 3, 1, 0, 2, –1, 2

4. Prove that if we define $D = X - Y$ for the random variables X and Y, that from a random sample of n X,Y pairs that $\bar{d} = \bar{x} - \bar{y}$; that is, the mean of the differences equals the difference in means.

5. Find an expression that would indicate what the correlation must be between X and Y in order for $V(\bar{d})$ to be less than $V(\bar{x} - \bar{y})$ when X and Y are from independent samples. Assume that X and Y have a common variance.

6. What is the correlation between X and Y for the data given in exercise 1?

7. Suppose a set of pairs such as that given in exercise 1 are obtained from a paired comparison experiment but that the correlation between X and Y is actually zero. Suppose coincidentally a set of identical measurements were obtained from a "two-independent sample" experiment. Explain why the p-values of the two test statistics would be different even though the values of the test statistics would theoretically be the same? Which p-value would be smaller and thus which would be the "most powerful" experimental procedure under these conditions.

8. Suppose that we have an additive that is reported to increase gas mileage when added to regular unleaded gasoline. Five cars are chose for a test. Each is assigned a driver who will first test-drive the car without the additive and then test-drive it with the additive. Let $\alpha = 0.05$. Is there any statistical evidence to support the claim that gas mileage is increased? The data are given below.

	Miles per Gallon	
Car	Additive	No Additive
1	19.8	18.7
2	22.0	20.4
3	27.2	25.3
4	35.1	33.6
5	27.0	24.7

9. For the data of exercise 8 compute a 90% confidence interval on the mean difference. Does this confidence interval include 0? How does this confirm the conclusions of your test of hypothesis done in exercise 8.

10. Use the data of exercise 8 to test the hypothesis that the additive increases miles-per-gallon by more than 1 mpg. Let $\alpha = 0.05$. Does the confidence interval of exercise 9 confirm this conclusion?

8.7 TESTS OF VARIANCES IN NORMALLY DISTRIBUTED POPULATIONS

Introduction

In many situations that we encounter, it is the variability of the process of set of measurements about which we are concerned. When that is the case, we want a test of hypothesis about variances with an appropriate test statistic. In this section we examine both the one-sample and two-sample tests of hypotheses. In both cases we assume that the data are normally distributed with some unknown variance about which we want to conduct our test. In the two sample/population case, we again assume normality along with an assumption of independence of samples.

Methodology

Case 1: The One Sample Test of Variance

    Background assumptions: Assume that we have a random sample of n observations, $X_1, ..., X_n$ having come from a normal distribution with unknown mean and variance. We want to test a hypothesis concerning the unknown variance.

    Step 1: The Hypotheses:

Null Hypothesis   Possible Alternative Hypotheses

$$H_0: \sigma^2 = \sigma_o^2 \qquad\qquad H_a: \begin{cases} \sigma^2 \neq \sigma_o^2 \\ \sigma^2 > \sigma_o^2 \\ \sigma^2 < \sigma_o^2 \end{cases}$$

Step 2: Select a random sample size n from a normally distributed population and choose a value α for the level of significance.

Step 3: The test statistic is

$$\chi^2 = \frac{(n-1)s^2}{\sigma^2}$$

which follows a chi-square distribution with $df = (n-1)$.

Step 4: Compute the p-value as follow

Alternative p-value Computation
Hypothesis

$\sigma^2 \neq \sigma_o^2$

$$p = \begin{cases} 2P(\chi^2 > \chi_c^2) & \text{if } \chi_c^2 > (n-1) \\ 2P(\chi^2 < \chi_c^2) & \text{if } \chi_c^2 < (n-1) \end{cases}$$

$\sigma^2 > \sigma_o^2 \qquad\qquad p = P(\chi^2 > \chi_c^2)$

$\sigma^2 < \sigma_o^2 \qquad\qquad p = P(\chi^2 < \chi_c^2)$

where χ_c^2 is

$$\chi_c^2 = \frac{(n-1)s^2}{\sigma^2}$$

Step 5: If $p < \alpha$, then reject H_0.

Case 2: The Two-Sample Test of Equal Variances

Background assumptions: Assume that we have two independent random samples from different normally distributed populations with unknown means and variances. We want to test a hypothesis concerning the equality of the unknown variances.

Step 1: The Hypothesis:

Null Hypcthesis   Possible Alternative Hypotheses

$$H_0: \sigma_x^2 = \sigma_y^2 \qquad\qquad H_a: \begin{cases} \sigma_x^2 \neq \sigma_y^2 \\ \sigma_x^2 > \sigma_y^2 \\ \sigma_x^2 < \sigma_y^2 \end{cases}$$

Step 2: Select a random sample size n and m from two normally distributed populations identified with the random variables X and Y, respectively, and choose a value α for the level of significance.

Step 3: The test statistic is

$$F = \frac{s_x^2}{s_y^2}$$

which follows an F distribution with $n-1$ numerator and $m-1$ denominator degrees of freedom respectively.

Step 4: Let the larger of the sample variances represent the random variable X and then compute the p-value as follows

Alternative Hypothesis	p-value Computation
$\sigma_x^2 \neq \sigma_y^2$	$p = 2P(F > F_c)$
$\sigma_x^2 > \sigma_y^2$	$p = P(F > F_c)$
$\sigma_x^2 < \sigma_y^2$	$p = P(F < F_c)$

where F_c is the computed value of the TS in step 3 and is

$$F_c = \frac{s_x^2}{s_y^2}$$

Step 5: If $p < \alpha$, then reject H_o.

DISCUSSION

Suppose that the characteristic of a normally distributed population that we are concerned about is the variance σ^2. We know that the sample variance s^2 is an unbiased estimator of σ^2 and that the random variable $\frac{(n-1)s^2}{\sigma^2}$ follows a chi-square distribution with $n-1$ degrees of freedom. This expression becomes a logical test statistic. If the null hypothesis is $H_o: \sigma^2 = \sigma_o^2$, then $\frac{(n-1)s^2}{\sigma_o^2}$ will be approximately equal to $(n-1)$ since $\frac{s^2}{\sigma_o^2}$ will be approximately equal to one. Therefore, the larger the value of $\frac{(n-1)s^2}{\sigma_o^2}$ compared to $(n-1)$ the more likely we are to think that $\sigma^2 > \sigma_o^2$; the smaller it is compared to $(n-1)$ the more likely we are to think that $\sigma^2 < \sigma_o^2$. This leads very naturally to the five-steps of a test of hypothesis given in the methodology section above.

If we sample two normally distributed populations and we want to compare the two variances with a null hypothesis of equality; i.e., H_o: $\sigma_x^2 = \sigma_y^2$, the ratio of the sample variances, $\frac{s_x^2}{s_y^2}$, will follow an F distribution. If we select a sample of size n from the X-population and a sample of size m from the Y-population, then the numerator degrees of freedom will be $n - 1$ and the denominator degrees of freedom will be $m - 1$. If the null hypothesis is true, then this ratio will be close to one, if not then the ratio will be significantly different from 1.

A useful device to enable us to use the table of the F distribution in the appendix, which has values in the upper-tail of the F distribution only, is to wait until the variances are computed and then label the population with the larger sample variance as the X-population. By doing this, we will get values for the F-ratio that always exceed 1. This allows us to use the table as it is without having to convert to a lower tail value.   The five-step procedure given above shows how to calculate p-values for this situation for any particular alternative hypothesis.

EXAMPLES

EXAMPLE 1: A new gun-like apparatus has been devised to replace the needle in administering vaccines. The apparatus, which is connected to a large supply of vaccine, can be set to inject different amounts of the serum, but the variance in the dose must not be greater than 0.06 to meet FDA standards (and to prevent accidental overdoses or inadequate dosages). A random sample of 25 injections for a model of the vaccine gun produced a set of measurements whose variance was 0.135 cc's. Use $\alpha = 0.025$ to test the hypothesis indicated.

SOLUTION:
Step 1: H_o: $\sigma^2 = 0.06$ vs. H_a: $\sigma^2 > 0.06$

Step 2: A random sample of 25 measured injections is taken and α of 0.025 is selected. .

Step 3: The test statistic is

$$\chi^2 = \frac{(25 - 1)s^2}{0.06}$$

which under the background assumptions will follow the chi-square distribution with 24 degrees of freedom.

Step 4: The computed value of the test statistic is

$$\chi_c^2 = \frac{(24)0.135}{0.06} = 54.00.$$

Using the chi-square table with 24 degrees of freedom we would report the p-value as $p < 0.001$. (The actual value is 0.0004.)

Step 5: On the basis of the p-value and for $\alpha = 0.025$, we reject H_o. Therefore, we conclude that there is statistical evidence to indicate that the vaccine gun is exceeding FDA requirements in terms of its variability of dosage.

EXAMPLE 2: In section 8.5, Example 2, the hypothesis was to be tested concerning the possibility of "grade inflation." This hypothesis was tested using a two-sample test of means. The test procedure used requires an assumption of equal variances. Conduct a test of equal variances for the data in that example as a preliminary test to the test of equal means. Do this for $\alpha = 0.05$. The data are reproduced here.

	10 YEARS AGO	PRESENT
Sample size	10	30
Sample mean	2.82	3.04
Sample standard deviation	0.43	0.38

SOLUTION:

Step 1:　　　　　　　$H_o: \sigma_x^2 = \sigma_y^2$　　　　vs.　　　　$H_a: \sigma_x^2 \neq \sigma_y^2$

Step 2:　The experiment is defined in section 5 of this chapter. The level of significance was selected to be 0.05.

Step 3:　The test statistic is

$$F = \frac{s_x^2}{s_y^2}$$

which is distributed as an F-statistic with 9 and 29 degrees of freedom under the background assumptions.

Step 4:　The computed value of the test statistic is

$$F = \frac{(0.43)^2}{(0.38)^2} = 1.28$$

which is F with 9 and 29 degrees of freedom respectively. Using the F table the p-value for this outcome is

$$p > 2(0.10) = 0.20$$

(The actual value is $p = 0.5781$.)

Step 5:　On the basis of this p-value, we do not reject the null hypothesis, and therefore, conclude that an assumption of equal variances is not disproved. Therefore, there is no reason to question the validity of the test on means because on a concern for the assumption of equal variances.

　　　Some caution should be exercised on the claims made in implementing a procedure as described in Example 2. The test of means and the interpretation of the level of significance α, if preceded by a test of equal variances, is really conditional upon the outcome of the test of equal variances. Therefore, the true value of the type I error is a value larger than the level of significance chosen for the test of equal means if a test of equal variances precedes it. The actual value of the type I error depends in on the decision making strategies chosen by the research in such a case and involves more complicated arguments than we want to spend time on here. Simply be aware that caution should be exercised in claiming a value for the type I error probability when a "joint" test of equal variances and equal means is implemented.

EXERCISES

1. Males and females differ in many respects. One experimenter explored the variability in pain threshold between the two sexes. Test the hypothesis of no significant difference between the variances of men and women using the following data and $\alpha = 0.10$.

Pain Threshold Measurement Values	Males	Females
Sample Size	14	10
Mean	16.5	14.8
St. Dev.	11.9	26.4

2. Two of the popular soft drink manufacturers in the United States have waged a battle concerning which one was the most preferred.   An independent research firm conducted the following study. Two independent groups of twenty persons each were asked to rate brand A and brand B cola. In order to test the hypothesis of equal ratings, an assumption of equal variance is necessary. Test the hypothesis of equal variances as a preliminary test to the subsequent test of means.   Use the data in the following table and $\alpha = 0.05$.

Brand	A	B
Sample Size	20	20
Mean	7.3	6.2
St. Dev.	1.3	2.8

3. The ratings that produced the data for exercise 2 are scaled to have a variance equal to about 6. Test this hypothesis for the group tasting Brand A and then do a separate test for the group using brand B. Use $\alpha = 0.10$ in both tests.

4. The typical variation in diameter in a metal sleeve that fits over an axle in a high performance motor cycle is 0.055 cm^2.   A sample of 5 sleeves is randomly selected from the last hour's production and the variance in the measurements computed. The last sample of 5 produced a sample variance of 0.131 cm^2. For $\alpha = 0.05$, does this sample present sufficient reason to say that the variance in the machinery producing the sleeves has increased?

5. Use the data of exercise 4 to test a hypothesis that the variance does not equal 0.055 cm^2. Report the p-value you get for this computation.

6. Use the data of exercise 4 to test a hypothesis that the variance does not equal 0.150 cm^2. Report the p-value you get for this computation.

7. Use the data of exercise 4 to construct both a 90% and a 95% confidence interval on the variance. Do the intervals produced enclose 0.055. Which of the two intervals provide results and conclusions consistent with the test of hypothesis of exercise 4?

8. A manufacturer of insulation has developed a type of insulation that it feels is superior to their current product. To test this hypothesis they found a subdivision being built that essentially was made up of the same size home and floor plan, the differences in units being aesthetic only. They

made arrangements to install their regular insulation in 14 homes as a control group and they then installed their new product in 18 identical homes. During a home show, where they had control over temperature settings, they kept track of the BTU's of heat burned in the two different groups of homes for a period of 2 weeks. In addition to a question of equal means is the question of equal variability. The data are given below. Test the hypothesis of equal variances for this data using $\alpha = 0.10$.

Insulation Type		
	Old	New
Sample Size	14	18
Mean	89	76
St. Dev.	6.8	5.5

9. Use the data of exercise 8 to test that hypothesis that the old insulation has less variability than does the new. Report the proper *p*-value for this test.

10. Use the data of exercise 8 to compute a 90% confidence interval on the ratio of the variances. Does this interval include the value 1?   What is the significance of your answer relative to the test of hypothesis of equal variances?

8.8   THE NEYMAN-PEARSON LEMMA AND LIKELIHOOD RATIO TESTS—OPTIONAL

Introduction

Suppose for a normally distributed population we have a simple null and alternative as follows: H_o:$\mu = 10$ vs. H_a:$\mu = 15$. Using the methods of preceding sections we would select a random sample from the population, and if the sample mean is larger than some value k we would reject H_o in favor of H_a.

That is, the rejection region is $\bar{x} > k$. The specific value of k would depend on the choice of α that was selected. A question we might ask: "Is this the best α-level rejection region for testing the hypothesis? It is intuitively logical, but is there any other decision rule that could be better?"

In this section we use the Neyman-Pearson Lemma to provide sufficient conditions to show that the rejection region defined above is the most powerful rejection region; i.e., its power is a at least as large as any other rejection region having the same value α for its type I error probability. The Neyman-Pearson Lemma expresses these conditions in the context of the ratio of likelihood functions.

In this section, we introduce, in general, the concept of a "most powerful critical region," a "uniformly most powerful test."

Another principle that uses the ratio of likelihood functions is the likelihood ratio test. The likelihood ratio test provides a very general method to construct a procedure for testing a hypothesis when both the null and alternative, H_o and H_a are composite. This procedure assumes that we know the functional form of the pdf which depends on one or more unknown parameters. As it turns out, most of the tests of sections 8.4 through 8.7 are likelihood ratio tests, or modifications of them.

Theorems/Definitions

THEOREM 8.1 (THE NEYMAN-PEARSON LEMMA): Suppose that we have a simple null and alternative hypothesis about the parameter ω of the form $H_o : \omega = \omega_0$ versus $H_a : \omega = \omega_1$. If there exists a rejection region C of size α and a positive constant k such that

$$\frac{L_o(x_1, ..., x_n; \omega_o)}{L_1(x_1, ..., x_n; \omega_1)} \leq k \quad \text{inside } C$$

and

$$\frac{L_o(x_1, ..., x_n; \omega_o)}{L_1(x_1, ..., x_n; \omega_1)} \geq k \quad \text{outside } C,$$

then C is the *most powerful critical region of size* α for testing $H_o : \omega = \omega_0$ against $H_a : \omega = \omega_1$, where L_o and L_1 are the likelihood functions relative to H_o and H_a, respectively.

DEFINITION 8.14 (UNIFORMLY MOST POWERFUL TESTS): A test is a *uniformly most powerful test* if it is the most powerful test against every possible value in a composite alternative hypothesis.

DEFINITION 8.15 (GENERALIZED LIKELIHOOD RATIO TEST): Consider a random sample $X_1, X_2, ..., X_n$ from the probability density function $f(x; \omega)$ where $\omega \in \Omega$, the parameter space. Suppose we have the composite null and alternative hypothesis $H_o : \omega \in \Omega_0$ vs. $H_a : \Omega \in \Omega_1$ where $\Omega = \Omega_0 \cup \Omega_1$. The generalized likelihood ratio λ is defined as the expression

$$\lambda = \frac{L(\hat{\Omega}_o)}{L(\hat{\Omega})}$$

where $L(\hat{\Omega}_o)$ denotes the likelihood function with all unknown parameters replaced with their maximum-likelihood estimates subject to the restriction that $\omega \in \Omega_o$, and $L(\hat{\Omega})$ denotes the likelihood function with all unknown parameters replaced with their maximum-likelihood estimates over the entire parameter space; i.e., $\omega \in \Omega$. Then the critical region

$$\lambda \leq k$$

where $0 < k < 1$, defines a likelihood ratio test where k is chosen so the level of significance is of size α.

THEOREM 8.2: If certain very general regularity conditions hold, then $-2 \ln \lambda$ is distributed approximately as a chi-square with 1 degree of freedom.

DISCUSSION

The Neyman-Pearson Lemma

Consider a simple null and alternative hypothesis where

H_o: $f(x)$ has the following form

X	1	2	3	4
f(X)	$\frac{4}{10}$	$\frac{3}{10}$	$\frac{2}{10}$	$\frac{1}{10}$

and

H_a: $f(x)$ has the following form

X	1	2	3	4
f(X)	$\frac{1}{10}$	$\frac{3}{10}$	$\frac{3}{10}$	$\frac{3}{10}$

One observation is to be selected from the distribution $f(x)$ and we want $\alpha = 0.10$. Upon comparing the two distributions we decide to let the rejection region C consist of the sample point $x = 4$. For this case $P(x \in C|H_o \text{ is true}) = P(X = 4| H_o \text{ is true}) = \frac{1}{10} = \alpha$. It is also easy to verify that Power $= 1 - \beta = P(X = 4 |H_a \text{ is true}) = \frac{3}{10}$.

Notice that the likelihood under H_o is $L(X = 4, H_o) = \frac{1}{10}$ and under H_a is $L(X = 4,H_a) = \frac{3}{10}$ and thus the "ratio of the likelihoods" is

$$\frac{L(X = 4,H_o)}{L(X = 4,H_a)} = \frac{\frac{1}{10}}{\frac{3}{10}} = \frac{1}{3}$$

The ratio of the likelihoods for the other possible values can easily be seen to be

$$\frac{L(X = 3,H_o)}{L(X = 3,H_a)} = \frac{\frac{2}{10}}{\frac{3}{10}} = \frac{2}{3} \qquad \frac{L(X = 2,H_o)}{L(X = 2,H_a)} = \frac{\frac{3}{10}}{\frac{3}{10}} = \frac{3}{3} \qquad \frac{L(X = 1,H_o)}{L(X = 1,H_a)} = \frac{\frac{4}{10}}{\frac{1}{10}} = \frac{4}{1}$$

Looking at the pattern of the likelihoods is seems that there is a value k where if x is an element of the rejection region C then the ratio of the likelihoods is less than or equal to k and for the x's not in C, the ratio is greater than k. In this specific case, the value of k is $\frac{1}{3}$. Does this pattern persist for other cases? Can it be generalized in any way? The answer to these questions is the subject of the Neyman-Pearson Lemma.

In a more general setting with a more general notation, suppose we have a simple hypothesis about a parameter ω in a probability density function $f(x;\omega)$ with the null hypothesis of the form $H_o: \omega = \omega_o$ versus a simple alternative hypothesis $H_a: \omega = \omega_1$. A random sample of size n is taken from $f(x;\omega)$ resulting in the vector of observations $x = (x_1, x_2, ..., x_n)$. In addition, suppose we have found a rejection region or *critical region C* which is of size α. Using the notation of the power function from section 8.3 we can write:

$$\text{Power} = P(\omega) = P((x_1, x_2, ..., x_n) \in C \mid \omega) = P(x \in C \mid \omega)$$

so that

$$\alpha = P(\omega_o) = P(x \in C \mid \omega = \omega_o) = P(\text{Type I Error}) = \text{Size of Critical Region}$$

However, there are many possible critical (rejection) regions C which satisfies this property. We would like is to choose that single critical region C^* of size α that is "most powerful" when H_a is true; i.e., when $\omega = \omega_1$. Such a critical region is called a *most powerful critical region* of size α. The critical region C^* will have the property:

$$\text{Power} = P(\omega_1) = P((x_1, x_2, ..., x_n) \in C^* \mid \omega = \omega_1) \geq P((x_1, x_2, ..., x_n) \in C \mid \omega = \omega_1)$$

where C is any other critical region of size α.

The Neyman-Pearson Lemma gives us a method for deriving the most powerful critical region C^* under the conditions described above in terms of a ratio of likelihood functions. First, consider the likelihood functions under the two assumptions that $\omega = \omega_o$ and $\omega = \omega_1$ denoted by $L_o(x_1, ..., x_n; \omega_o)$ = and $L_1(x_1, ..., x_n; \omega_1)$, respectively. Now, suppose we have a sample of observation, $(x_1, ..., x_n)$ that represents a point in the rejection region C. If the null hypothesis is true the likelihood $L_o(x_1, ..., x_n; \omega_o)$ should be small; if the alternative hypothesis is true the likelihood should be large, so that for such a point $L_o(x_1, ..., x_n; \omega_o) \leq L_1(x_1, ..., x_n; \omega_1)$ and thus the ratio

$$\frac{L_o(x_1, ..., x_n; \omega_o)}{L_1(x_1, ..., x_n; \omega_1)}$$

will be relatively small for all such points.

Alternatively, suppose we have a sample point that falls in the acceptance region C' (the complement of C). Such a point would imply that we accept the null hypothesis. Therefore, an argument similar to the above suggests that under H_o the likelihood L_o will be high since we "should" accept H_o, and the likelihood L_1 would be low since we want a low probability of accepting when the alternative is true. Thus, the ratio

$$\frac{L_o(x_1, ..., x_n; \omega_o)}{L_1(x_1, ..., x_n; \omega_1)}$$

will be relatively large. These ideas could be summarized in a simple table such as Table 8.2 and then generalized in the Neyman-Pearson Lemma which follows with examples.

The Neyman-Pearson Lemma states that for some rejection region C of size α and for some value k if

$$\frac{L_o(x_1, ..., x_n; \omega_o)}{L_1(x_1, ..., x_n; \omega_1)} \leq k \qquad \text{inside } C$$

and

$$\frac{L_o(x_1, \ldots, x_n; \omega_o)}{L_1(x_1, \ldots, x_n; \omega_1)} \geq k \qquad \text{outside } C$$

then C is a "most powerful" critical region for testing $H_o: \omega = \omega_o$ versus $H_a: \omega = \omega_1$.

Table 8.2 Characteristics of likelihood functions for a sample point x and a rejection region C.

Sample Outcome $x = (x_1, x_2, \ldots, x_n)$	$H_o: \omega = \omega_o$	$H_a: \omega = \omega_1$	Ratio of Likelihoods
$x \in C$	$L_o(x_1, \ldots, x_n; \omega_o)$ is small	$L_1(x_1, \ldots, x_n; \omega_1)$ is large	$\dfrac{L_o(x_1, \ldots, x_n; \omega_o)}{L_1(x_1, \ldots, x_n; \omega_1)}$ is small
$x \in C'$	$L_o(x_1, \ldots, x_n; \omega_o)$ is large	$L_1(x_1, \ldots, x_n; \omega_1)$ is small	$\dfrac{L_o(x_1, \ldots, x_n; \omega_o)}{L_1(x_1, \ldots, x_n; \omega_1)}$ is large

EXAMPLES

EXAMPLE 1: Let $X_1, \ldots, X_n$ denote a random sample from an exponential distribution with unknown parameter θ. Determine the best critical region of size α for testing

$$H_o: \theta = \theta_o \qquad \text{vs. } H_a: \theta = \theta_1, \text{ with } \theta_1 > \theta_o$$

SOLUTION: (Realize that in the following solution, the purpose is not to a specific value of k; rather, it is to find an expression the defines the location of the "rejection region on the real line.) The likelihood ratios are formed to get

$$\frac{\left(\dfrac{1}{\theta_o}\right)^n e^{-\Sigma x_i/\theta_o}}{\left(\dfrac{1}{\theta_1}\right)^n e^{-\Sigma x_i/\theta_1}} \leq k$$

After taking the natural log and simplifying and gathering terms we get

$$\ln\left(\frac{\theta_1}{\theta_o}\right)^n \Sigma x_i \left(\frac{1}{\theta_1} - \frac{1}{\theta_o}\right) \leq k$$

After further manipulation, this inequality can be re-expressed as

$$\Sigma x \geq k_1$$

where k_1 is:

$$k_1 = \frac{\ln((\frac{\theta_o}{\theta_1})^n k)}{(\frac{1}{\theta_1} - \frac{1}{\theta_o})}$$

Notice that the inequality sign changes direction because of dividing by the term $\left(\frac{1}{\theta_1} - \frac{1}{\theta_o}\right)$ which is negative whenever $\theta_1 > \theta_o$. Therefore, the best rejection region is when the sum of the x's gets large. Since we know the sum of independent exponential random variables is distributed as a gamma distribution we could determine the value of k_1 so that the area to the right of it is α when $\theta = \theta_o$. This the same rejection region we would have set up intuitively for the same problem using our previous development. However, we now have theoretical justification for using such a critical region.

The next thing to notice is that the same rejection would result for any value of θ_1 so long as $\theta_1 > \theta_o$. Thus, if we have the composite alternative: $H_a: \theta > \theta_o$, then the critical region C would be most powerful for each value of θ where $\theta > \theta_o$ in the sense of a simple alternative. A test procedure that has this property for a composite hypothesis is called a *uniformly most powerful* test.

EXAMPLE 2: Let $X_1, \ldots, X_n$ be a random sample from the normal distribution $N(\mu, 1)$. We want to find the best critical region for testing the simple hypothesis $H_o: \mu = 5$ versus $H_a: \mu = 10$.

SOLUTION: The likelihood ratios are formed to get

$$\frac{(2\pi)^{-n/2} \exp\left[-(\frac{1}{2})\sum_1^n (x_i - 5)^2\right]}{(2\pi)^{-n/2} \exp\left[-(\frac{1}{2})\sum_1^n (x_i - 10)^2\right]} \leq k$$

Canceling, and taking the natural logs we get

$$-10\sum_1^n x_i - n5^2 + n10^2 \leq (2)\ln k.$$

Thus,

$$\frac{1}{n}\sum_1^n x_i \geq -\frac{1}{10n}[n5^2 + n10^2 + (2) \ln k]$$

or equivalently,

$$\bar{x} \geq k_1$$

Thus, the best critical region C is the region

$$C = \{(x_1, x_2, \ldots, x_n) \text{ such that } \bar{x} \geq k_1\}$$

so that

$$P((x_1, x_2, ..., x_n) \in C \mid \mu = 5) = \alpha.$$

Suppose, $n = 25$ and $\alpha = 0.05$, then the "most powerful rejection region" consists of those value of $\bar{x} \geq k_1$ such that

$$0.05 = P(\bar{X} \geq k_1 \mid \mu = 5) = P(\frac{\bar{X} - 5}{1/\sqrt{25}} \geq \frac{k_1 - 5}{1/\sqrt{25}} \mid \mu = 5)$$

The Z-score value from the standard normal tables that satisfies this condition is $Z = 1.645$. Therefore, $\frac{k_1 - 5}{1/\sqrt{25}} = 1.645$ implies that

$$k_1 = 5 + 1.645(\frac{1}{5}) = 5.33$$

EXAMPLE 3: For the situation described in Example 2 does a uniformly most powerful test result and what is it like?

SOLUTION: In Example 2 suppose we were testing $H_o:\mu = 5$ versus $H_a:\mu = 15$. The Neyman-Pearson Lemma would lead us through the same operations to the same conclusion as given in that example. The critical region defined by $k_1 = 5.33$ for $\alpha = 0.05$ would be the same. Therefore, for any $\mu > 5$, the same conclusion would result, and thus this test meets the conditions for a "uniformly most powerful test."

The Likelihood Ratio Test

The Neyman-Pearson approach to hypothesis testing provides a most powerful test and rejection region for a simple null versus simple alternative hypothesis in most of the cases we would consider in this course. In addition, we can identify uniformly most powerful conditions, usually associated with one-tailed critical regions. However, it does not always provide a solution when composite hypotheses are encountered. In addition, there may be distributions where there are nuisance parameters present; i.e., the distribution has more than one unknown parameter. We need an approach to handle these other conditions and the likelihood ratio test is one very useful approach available to us.

First, assume that $X_1, X_2, ..., X_n$ constitute a random sample of size n from a probability density function $f(x;\omega)$ where the *parameter space* is Ω. That is, Ω denotes the set of possible values that ω could take on. Suppose we partition Ω into two mutually exclusive and complementary sets Ω_0 and Ω_1 such that the null hypothesis is that $H_o:\omega \in \Omega_0$ versus $H_a:\Omega \in \Omega_1$. Let $L(\hat{\Omega}_o)$ denote the likelihood function with all unknown parameters replaced with their maximum-likelihood estimates, subject to the restriction that $\omega \in \Omega_0$. Similarly, let $L(\hat{\Omega})$ denote the likelihood function with all unknown parameters replaced with their maximum-likelihood estimates over the entire parameter space; i.e., $\omega \in \Omega$. Notice that these two likelihood expressions are positive and the unknown nuisance parameters are replaced by estimates from the sample. Therefore, expressions are obtained that would now reflect whether H_o or H_a is true without nuisance parameters clouding the picture.

The likelihood ratio is defined by

$$\lambda = \frac{L(\hat{\Omega}_o)}{L(\hat{\Omega})}$$

As the ratio of two positive numbers, λ will be positive as well and it can be shown that λ is bounded by 1. Therefore, $0 \le \lambda \le 1$.

Next, recall that both the numerator and denominator are functions of the actual sample values. If the sample is supportive of H_o then $L(\hat{\Omega}_o)$ ought to be large relative to $L(\hat{\Omega})$; if the sample is supportive of H_a, then $L(\hat{\Omega}_o)$ ought to be small relative to $L(\hat{\Omega})$. Therefore, a small value of λ would lead to rejection of H_o. This suggests the following for a *likelihood ratio test:*

The *critical region for the likelihood ratio test* is the set of points in the sample space for which

$$\lambda = \frac{L(\hat{\Omega}_o)}{L(\hat{\Omega})} \le k$$

where $0 < k < 1$ and k is selected so that the test has a significance level of α.

Hopefully, as we find the expression for λ, we find that λ is a function of the sample such that its distribution will be known. Then the theory we have already used will allow us to enter the appropriate tables and find the critical region C of size α. However, if the distribution cannot be found it is helpful, and true, that for large n, the distribution of $-2 \ln \lambda$ approaches the chi-square distribution with 1 degree of freedom. So that, at the least, we can find an approximate test and critical region for testing the hypothesis defined above.

EXAMPLES

EXAMPLE 4:   The weight X in ounces of "1-pound packages" of potato chips is assumed to be normally distributed with unknown mean and a variance of 0.36 (oz.)^2. Test the hypothesis H_o:$\mu = 16$ against H_a:$\mu \ne 16$. Use the likelihood ratio test.

SOLUTION: The parameter space is $\Omega = \{\mu: -\infty < \mu < \infty \}$ with $\Omega_o = \{16\}$ and Ω_1 is the complement of Ω_o with respect to Ω. The likelihood function in general is

$$L = (2(0.36)\pi)^{-n/2} \exp\left[-(\frac{1}{2(0.36)})\sum_1^n (x_i - \mu)^2 \right]$$

Under the null hypothesis, $\Omega_o = \{16\}$ and has only the one value. Therefore, the maximum of L is

$$L(\hat{\Omega}_o) = ((0.72)\pi)^{-n/2} \exp\left[-(\frac{1}{0.72})\sum_1^n (x_i - 16)^2 \right]$$

Also, the likelihood estimate of μ in Ω is the sample mean so that

$$L(\hat{\Omega}) = ((0.72)\pi)^{-n/2} \exp\left[-(\frac{1}{0.72})\sum_1^n (x_i - \bar{x})^2\right]$$

Therefore, the likelihood ratio is

$$\lambda = \frac{((0.72)\pi)^{-n/2} \exp\left[-(\frac{1}{0.72})\sum_1^n (x_i - 16)^2\right]}{((0.72)\pi)^{-n/2} \exp\left[-(\frac{1}{0.72})\sum_1^n (x_i - \bar{x})^2\right]}$$

which after algebraic simplification becomes

$$\lambda = \exp\left[-\frac{n}{(0.72)}(\bar{x} - 16)^2\right]$$

If $\bar{x}$ is close to 16, then λ would be close to 1 and would support the null hypothesis. On the other hand, the further away from 16 the sample mean $\bar{x}$ is, in either direction, the stronger the support for H_a, and λ gets smaller as this happens.

Therefore, there is a value k such that when $\lambda \le k$ we reject the null hypothesis. The value λ depends directly on $\bar{x}$, and under the assumptions of the problem $\bar{x}$ will be normally distributed with mean 16 under H_o. Therefore, as we move away from 16 symmetrically to find that value such that under the null hypothesis the rejection region has the significance level of α, we will find a value of k, after standardizing such that the critical region is

$$C = \left\{\bar{x}: \frac{|\bar{x} - 16|}{\sigma/\sqrt{n}} \ge Z_{1-\alpha/2}\right\}$$

EXAMPLE 5: Suppose that $X_1, X_2, ..., X_n$ is a random sample from a normal distribution with an unknown mean and variance. We want to test $H_o:\mu = \mu_o$ against $H_a:\mu > \mu_o$. Determine the likelihood ratio test.

SOLUTION: First, note that $\Omega_o = \{\mu_o\}$ and $\Omega_1 = \{\mu > \mu_o\}$ and therefore, $\Omega = \{\mu \ge \mu_o\}$. To find $L(\hat{\Omega}_o)$ under the restriction of the null hypothesis means that we don't have to estimate μ; we set it equal to μ_o. However, we must estimate σ^2 for the case where $\mu = \mu_o$. The principle of maximum likelihood applied to this case gives the estimate

$$\hat{\sigma}_o^2 = \frac{\sum_{i=1}^n (X_i - \mu_o)^2}{n}$$

The unrestricted maximum likelihood estimator of μ is $\bar{x}$. Therefore, over the unrestricted space the maximum likelihood estimator is the larger of $\bar{x}$ or μ_o. The maximum likelihood estimator of σ^2 is

$$\hat{\sigma}^2 = \frac{\sum_{i=1}^{n}(X_i - \hat{\mu})^2}{n}$$

Following the lead of Example 4 we get $L(\hat{\Omega}_o)$ by replacing μ with μ_o and σ^2 by $\hat{\sigma}_o^2$, which gives

$$L(\hat{\Omega}_o) = \left(\frac{1}{\sqrt{2\pi}}\right)^n \left(\frac{1}{\hat{\sigma}_o^2}\right)^{n/2} e^{-n/2}$$

(The terms in the exponent of e cancel out except for the n and 2.) We get $L(\hat{\Omega})$ by replacing μ with $\hat{\mu}$ and σ^2 by $\hat{\sigma}^2$, which gives

$$L(\hat{\Omega}) = \left(\frac{1}{\sqrt{2\pi}}\right)^n \left(\frac{1}{\hat{\sigma}^2}\right)^{n/2} e^{-n/2}$$

Thus, forming the ratio of these two terms and cancelling common expressions we get

$$\lambda = \frac{L(\hat{\Omega}_o)}{L(\hat{\Omega})} = \left(\frac{\hat{\sigma}^2}{\hat{\sigma}_o^2}\right)^{n/2} = \begin{cases} \left[\dfrac{\sum_{i=1}^{n}(X_i - \bar{X})^2}{\sum_{i=1}^{n}(X_i - \mu_o)^2}\right]^{n/2} & \text{if } \bar{X} > \mu_o \\ \\ 1 & \text{if } \bar{X} \leq \mu_o \end{cases}$$

Next, because

$$\sum_{i=1}^{n}(X_i - \mu_o)^2 = \sum_{i=1}^{n}[(X_i - \bar{X}) + (\bar{X} - \mu_o)]^2 = \sum_{i=1}^{n}(X_i - \bar{X})^2 + n(\bar{X} - \mu_o)^2$$

the rejection region defined by $\lambda \leq k$ is equivalent to

$$\frac{\displaystyle\sum_{i=1}^{n}(X_i - \bar{X})^2}{\displaystyle\sum_{i=1}^{n}(X_i - \mu_o)^2} < k^{2/n} = k_1$$

or

$$\frac{\displaystyle\sum_{i=1}^{n}(X_i - \bar{X})^2}{\displaystyle\sum_{i=1}^{n}(X_i - \bar{X})^2 + n(\bar{X} - \mu_o)^2} < k_1$$

or

$$\frac{1}{1 + \dfrac{n(\bar{X} - \mu_o)^2}{\displaystyle\sum_{i=1}^{n}(X_i - \bar{X})^2}} < k_1$$

Clearing out the 1's in this expression and then multiplying by $\dfrac{n-1}{n-1}$, with another constant k_2, defined we get

$$\frac{n(\bar{X} - \mu_o)^2}{\displaystyle\sum_{i=1}^{n}\frac{(X_i - \bar{X})^2}{n-1}} > k_2$$

or

$$\frac{(\bar{X} - \mu_o)}{s/\sqrt{n}} > \sqrt{k_2}$$

This last step is true as long as $\bar{X} > \mu_o$ which is an assumption we made many steps before this. Now, note that this last expression is the expression for a t statistic that was used for testing this hypothesis in section 8.4 of this chapter.

EXAMPLE 6: In the problem described in Example 5, what is an approximate chi-square test?

SOLUTION: Looking at the solution for Example 5 we find an expression for λ under the restriction that $\bar{X} > \mu_o$ of the following form:

$$\lambda = \left[\frac{\displaystyle\sum_{i=1}^{n} (X_i - \bar{X})^2}{\displaystyle\sum_{i=1}^{n} (X_i - \mu_o)^2} \right]^{n/2}$$

which is strictly a function of sample values and the hypothesized mean with $\lambda < 1$. By a previous claim, $-2 \ln \lambda$ will then be approximately chi-square with 1 degree of freedom for large n. We would reject the null hypothesis at the α-level of significance in favor of the alternative if

$$-2 \ln \lambda > \chi^2_{1-\alpha}$$

where the chi-square value is obtained from the table with 1 degree of freedom.

Final Note: As you compare the tests that result in the above examples, you should note their similarity to those tests presented in sections 8.4 through sections 8.7. It can be concluded that the tests of those sections are either likelihood ratio tests or slight modifications of those procedures. This will be true for most tests involving normal distribution assumptions.

EXERCISES

1. Suppose a random sample of 20 observations is taken from a normal distribution with an unknown mean and known variance of $\sigma^2 = 10$. We want to test $H_o : \mu = 15$ versus $H_a : \mu < 15$.

    (a) Find the uniformly most powerful test with significance level of 0.05.
    (b) For the test of part (a) find the power against the alternatives: $\mu = 14$, and $\mu = 12$.

2. Suppose a random sample of n observations is selected from a normal distribution with a *known* mean μ and an unknown variance σ^2. Find the most powerful α-level test of $H_o : \sigma^2 = \sigma^2_o$ versus $H_a : \sigma^2 = \sigma^2_1$ $(\sigma^2_1 > \sigma^2_o)$. Show that this test is equivalent to a χ_2 test. Is the test uniformly most powerful for $H_a : \sigma^2 > \sigma^2_o$?

3. The number of successes X in n Bernoulli trials is to be used to test the null hypothesis that the parameter p of a binomial population equals $\frac{1}{2}$ against the alternative that it does not equal $\frac{1}{2}$.

    (a0 Find an expression for the likelihood ratio statistic.
    (b) Use the result of part(a) to show that the critical region of the likelihood ratio test can be written as

    $$x \bullet \ln x + (n - x) \bullet \ln(n - x) \geq k$$

    where x is the number of successes.

4. A random sample of size n from an exponential distribution is used to test $H_o : \theta = \theta_o$ versus $H_a : \theta \neq \theta_o$. Find an expression for the likelihood ratio statistic.

5. Use the results of exercise 4 to show that the critical region of the likelihood ratio test can be written as

$$\overline{x} \bullet e^{-\overline{x}/\theta_o} \leq k$$

6. For the test of exercise 4, suppose that $\overline{x} = 25$, and $\theta_o = 20$ with $n = 15$. Compute $-2 \ln \lambda$ and complete the approximate chi-square test for $\alpha = 0.10$.

7. Let s_1^2 and s_2^2 denote, respectively, the variances of two independent random samples of size n and m from normal distributions with means μ_1 and μ_2 and common variance σ^2. If the two means are unknown, construct a likelihood ratio test of $H_o: \sigma^2 = \sigma_o^2$ versus $H_a: \sigma^2 = \sigma_1^2$ $(\sigma_1^2 > \sigma_o^2)$.

8. Suppose we have a random sample of size n and we want to test $H_o:X$ is Uniform $(0,1)$ against $H_a: X$ is Exponential $(\theta = 1)$. What is the most powerful critical region for such a test?

9. Find the likelihood ratio test for testing $H_o: \lambda = \lambda_o$ versus $H_a: \lambda = \lambda_1$ where $\lambda_1 > \lambda_o$ in a random sample of size n from a Poisson distribution.

10. Suppose we have a random sample of size n from a Poisson distribution with parameter λ_1 and an independent random sample of size m from a Poisson distribution with parameter λ_2. Test H_o: $\lambda_1 = \lambda_2$ versus $H_a: \lambda_1 \neq \lambda_2$ by the likelihood ratio method with the level of significance of size α. [Hint: Write the likelihood of the *joint* sample first and then recognize that Ω_o consists of one point, namely, $\Omega_0 = \{\lambda_1 = \lambda_2 = \lambda\}$.]

11. Repeat the exercise 10. However, assume the distributions sampled are exponential distributions instead of Poisson distributions.

8.9   ANALYSIS OF THE CHAPTER EXERCISE

With oil and gas prices showing substantial increases over the last decade, more and more interest has developed concerning methods to improve engine efficiency, gas mileage, and so forth. In an attempt to lure more customers to their franchise stations, suppose a major oil company has developed a gas additive that is supposed to increase gas mileage in subcompact cars. However, before marketing the new additive, the company conducts the following experiment: One hundred subcompact cars are randomly selected and divided into two groups of 50 cars each. The gasoline additive is dispensed into the tanks of the cars in one group but not the other. The miles per gallon obtained by each car in the study is then recorded. The data are summarized in the the accompanying table. Is there sufficient evidence for the oil company to claim that the average gas mileage obtained by subcompact cars with the additive is greater than without the additive?

SOLUTION:

	Without Additive	With Additive
Sample Size	50	50
Mean (mpg)	27.6	31.9
St.Dev.(mpg)	9.3	11.3

To address the issue of the effectiveness of the gasoline additive we propose to conduct a test of hypothesis of equal means for two independent samples. However, such a test of hypothesis requires an assumption of normally distributed data and an assumption of a common variance.

We shall assume that the data are normally distributed at the best, and at the worst, the sample size is large enough for the Central Limit Theorem to apply to the distribution of means. Therefore, the only assumption of concern is the assumption of equal variances. If we can confirm this assumption, then we will proceed with a test on the means.

Test of Hypothesis for Equal Variances:

Step 1: $H_o: \sigma_x^2 = \sigma_y^2$ $H_a: \sigma_x^2 \neq \sigma_y^2$

Step 2: The experiment is defined above.

Step 3: The test statistic is

$$F = \frac{s_x^2}{s_y^2}$$

which is distributed as an F-statistic with 49 and 49 degrees of freedom under the background assumptions.

Step 4: The computed value of the test statistic is

$$F = \frac{(11.3)^2}{(9.3)^2} = 1.48$$

which is F with 49 and 49 degrees of freedom respectively. The p-value for this outcome is: $p > 2(0.10) = 0.20$.

Step 5: On the basis of this p-value, do not reject the null hypothesis. We conclude an assumption of equal variances is not disproved. Therefore, there is no reason to question the validity of the test on means because of concern for the assumption of equal variances.

Test of Hypothesis of Equal Means, Independent Samples:

Step 1: $H_o: \mu_x = \mu_y$ $H_a: \mu_x < \mu_y$

Step 2: A sample of 50 automobiles is obtained from each of the two populations, respectively.

Step 3: The test statistic is

$$t = \frac{(\bar{x} - \bar{y}) - (\mu_x - \mu_y)}{\sqrt{s_p^2(\frac{1}{n} + \frac{1}{m})}} = \frac{(\bar{x} - \bar{y}) - 0}{\sqrt{s_p^2(\frac{1}{50} + \frac{1}{50})}}$$

which, under our assumptions, possesses a t distribution with 98 degrees of freedom.

Step 4:   To compute the t-value we first obtain the pooled variance as

$$s_p^2 = \frac{(50 - 1)(9.3^2) + (50 - 1)(11.3^2)}{(50 - 1) + (50 - 1)} = 107.09$$

which is inserted into t_c to produce

$$t_c = \frac{(27.6 - 31.9) - 0}{\sqrt{107.09(\frac{1}{50} + \frac{1}{50})}} = -2.08$$

which has a p-value such that

$$p < 0.02.$$

Step 5:   On the basis of this p-value we reject the null hypothesis for any $\alpha > 0.02$. Therefore we recommend rejecting the null hypothesis and conclude the the cars with the additive had a higher mpg rating than those without the additive.

CHAPTER SUMMARY

In this chapter we have introduced the vocabulary and concepts of testing hypotheses. You should know how to write a null hypothesis and alternative hypothesis concerning means and variances of normally distributed random variables, what a Type I and Type II error is, and how to interpret their probabilities α and β. With sufficient information, you should be able to compute the "power" of a test procedure to evaluate how it would compare to others test procedures, a procedure changed by making either a different choice of α or a different sample size. You should also be able to determine the size of sample necessary to meet a given Type I and Type II error probability.

After these preliminary ideas were presented, a series of special cases were examined within a 5-step process for testing hypotheses. In all cases the assumptions made were random sampling from normally distributed distributions with hypotheses formed about means or variances. The results are summarized in Table 8.3 below which specifies whether the sample is drawn from one or two populations, whether the null hypothesis concerns means or variances, and the test statistic to be used for each particular case.

Table 8.3 Summary of the null hypothesis and the test statistic for various cases: normally distributed observations.

Type of Population	Null Hypothesis	Test Statistic
One Population: Sample of Size n	$\mu = \mu_o$	$Z = \dfrac{(\bar{x} - \mu_o)}{\sigma/\sqrt{n}}$ σ known $t = \dfrac{(\bar{x} - \mu_o)}{s/\sqrt{n}}$ σ unknown $df = n - 1$
Two Independently sampled populations: Sample sizes of n & m, respectively	$\mu_x = \mu_y$	$Z = \dfrac{(\bar{x} - \bar{y}) - (\mu_x - \mu_y)}{\sqrt{\left(\dfrac{\sigma_x^2}{n} + \dfrac{\sigma_y^2}{m}\right)}}$ known variances $t = \dfrac{(\bar{x} - \bar{y}) - (\mu_x - \mu_y)}{\sqrt{s_p^2\left(\dfrac{1}{n} + \dfrac{1}{m}\right)}}$ unknown variances* $*df = n + m - 2$
Population of pairs, or matched samples: Sample of Size n	$\mu_D = \mu_o$	$t = \dfrac{(\bar{d} - \mu_o)}{s_d/\sqrt{n}}$ $df = n - 1$
One Population: Sample of Size n	$\sigma^2 = \sigma_o^2$	$\chi^2 = \dfrac{(n-1)s^2}{\sigma^2}$
Two Independently sampled populations: Sample sizes of n & m, respectively	$\sigma_x^2 = \sigma_y^2$	$F = \dfrac{s_x^2}{s_y^2}$

    The objective of this chapter, though distinct from the objective of a confidence interval, can be realized by using the notion of a confidence interval computation. You should be careful, however, not to negligently interchange the vocabulary. It is very easy to say the wrong thing and confuse the concepts. Briefly, however, if the a test of hypothesis has a level of significance of α, a table value from the standard normal, t, chi-square or F can be pulled against which the test statistic is compared. If the same table value is used to construct a confidence interval, then the confidence interval will overlap H_o if the null hypothesis should not be rejected. If the test of hypothesis procedure leads to rejecting the null hypothesis, then the confidence interval constructed should not overlap the null hypothesis value. This relationship is straight-forward for two-tailed alternatives. For one-tailed alternatives, you have to be careful in selecting the correct table value to construct the confidence interval. Work out enough exercises to gain the experience that you need to understand this relationship.

This chapter concluded with a presentation of the Neyman-Pearson Lemma and an introduction to likelihood ratio tests. These two concepts provide justification for the kind of rejection regions we used in the previous sections of this chapter. This allowed us to define tests as being most powerful or uniformly most powerful under appropriate conditions.

The likelihood ratio procedure gave us a way to solve many problems of a more general nature and provides a means of finding a test of hypothesis for the unusual cases that may be encountered. It was noted that under large sample conditions, $-2 \ln \lambda$ is approximately chi-square with 1 degree of freedom, so we can always get an approximate test for a likelihood ratio expression if the regularity conditions for a likelihood ratio test are met.

CHAPTER EXERCISES

1. Consider the following data for conducting a two sample test of means.

#	Sample 1	Sample 2
1	40	45
2	36	37
3	85	91
4	65	68
5	21	28
6	58	58
7	49	51
8	72	75
9	14	19

    Perform a "two independent sample" t-test of equal vs. unequal means on this data. Note specifically the difference in the means of the two samples and the size of the standard deviations.

2. Using the data in exercise 1, regard the data as a set of 9 pairs. Compute the differences of the nine pairs of measurements and perform a "paired comparison" test of equal means. Again note the numerical value of the mean of the differences. Note, also, the size of the standard deviation of the difference. Why is there such a dramatic difference in the two t-values and their corresponding p-values?

3. Compute the covariance of the pairs of data points in exercise 1, above. If we let $D = X - Y$, then show that the variance of the differences is equal to the sum of the variance of sample 1 plus the variance of sample 2 minus twice the covariance, i.e.,

$$s_d^2 = s_x^2 + s_y^2 - 2\, s_{xy}$$

Discuss how this explains why the two-independent sample test is so much different from the paired comparison test of exercises 1 and 2.

4. Using the same data of exercise 1, test the hypothesis of equal variances and note your conclusions based on the p-value obtained.

5. Suppose an engineer wishes to compare the number of complaints per week filed by union members for two different shifts at a manufacturing plant. Fifty independent observations on the number of complaints per week for shift 1 produced a mean of $\bar{x} = 25$ for shift 1 and $\bar{y} = 28$ for shift 2. Assume that the number of complaints per week has a Poisson distribution with parameters λ_1 and λ_2, respectively, for the two different shifts. Test $H_o: \lambda_1 = \lambda_2$ versus $H_a: \lambda_1 = \lambda_2$ by the likelihood ratio method with the level of significance of size $\alpha = 0.01$.

6. Suppose a box contains four marbles, ω white ones and $4 - \omega$ red ones. Test $H_o: \omega = 3$ versus $H_a: \omega \neq 3$ by doing the following: Select two marbles with replacement and reject H_o if both marbles are the same color; otherwise do not reject.

   (a) Compute the probability of a Type I error.
   (b) Compute the probability of a Type II error under all possible values for ω.

7. Rework all parts of exercise 6 assuming the sampling is done without replacement. What conclusions does this suggest about sampling with replacement versus sampling without replacement.

8. Let a random sample of size n be selected from the following probability density function:

$$f(x;\omega) = \begin{cases} \left(\dfrac{1}{\omega}\right) mx^{m-1} e^{-x^m/\omega} & x > 0 \\ \\ 0 & \text{elsewhere} \end{cases}$$

where m denotes a known constant.

   (a) Find the uniformly most powerful test of $H_o: \omega = \omega_o$ against $H_a: \omega > \omega_o$.
   (b) Find the likelihood ratio test for this same hypothesis.

9. Let a random sample of size n be selected from the following probability density function:

$$f(x;\omega) = \begin{cases} \left(\dfrac{1}{\omega_1}\right) e^{-(x-\omega_2)/\omega_1} & x > \omega_1 \\ \\ 0 & \text{elsewhere} \end{cases}$$

Find the likelihood ratio test for testing $H_o: \omega_1 = 5$ versus $H_a: \omega_1 > 5$ with ω_2 unknown. Next, rewrite this test procedure for any hypothesized value ω_0.

10. Referring to exercise 9, find the likelihood ratio test for testing $H_o: \omega_2 = 5$ versus $H_a: \omega_2 > 5$ with ω_1 unknown, then generalize it for other values than 5.

11. A health officer is asked to give a city swimming pool a chlorine safety check and test the null hypothesis that it is safe to swim. Explain the practical meaning of a Type I error versus a Type II error in this situation.

12. Let X_1 and X_2 be a random sample of size 2 from the population given by

$$f(x;\omega) = \begin{cases} \omega x^{(\omega-1)} & 0 < x < 1 \\ 0 & \text{elsewhere} \end{cases}$$

Suppose we define the critical region C as $C = \{x_1, x_2: x_1 \bullet x_2 \geq \frac{3}{4}\}$ to test the hypothesis that $\omega = 1$ versus the alternative that $\omega = 2$. What is the power of this test at $\omega = 2$?

9

TESTING HYPOTHESES II: LARGE SAMPLE APPROXIMATE TESTS

OBJECTIVES

In Chapter 1, you were introduced to the various levels of measurement, and it was observed that certain operations that were appropriate for interval-ratio data made no sense for ordinal or nominal measurement. (Nominal measurement produces data often called categorical data. It comes from a simple operation of classifying responses into one or more unordered categories, such as when recording the marital status of an individual, or their race, or hair color, etc. See section 1.2, for a complete discussion.)

In this chapter we present a variety of test procedures that apply to nominal measurement. Upon completing this chapter, you should be able to apply the tests described here to the types of problems defined, and you should be able to decide when a particular kind of analysis is appropriate. Each test procedure presented has certain assumptions upon which it is based and you should be able to list the assumptions associated with each of the procedures presented in this chapter. You should also be able to verify the appropriateness of the assumptions when applied to a given problem.

INTRODUCTION

Much of the data that is collected in research are at the nominal level and are often simply dichotomous (they take on one of two values, such as yes-no, heads-tails, success-failure, etc.). For instance, in a study on the affect of certain school policies on the attendance of students, the only measurement obtained was whether a given student was present or absent on a particular day. Drug studies often are only able to report whether the experimental animal survives or not, whether a cancer develops or doesn't, and so forth. For such data, the simplest model to be proposed is the binomial distribution. Thus in testing hypotheses or constructing confidence intervals, usual normal distribution theory results are not valid. Other techniques are necessary, adapted to the special nature of the data and the distributional assumptions appropriate for them.

The presentation of those specialized techniques are often reserved for a course in non-parametric statistics. However, if the sample size is fairly large, the Central Limit Theorem provides a large-sample set of procedures that are approximately normal for several commonly encountered types of problems. A few of these test procedures are presented in this chapter. (Note, the procedures presented here are acceptable for data which are at most nominal data. Data which are ordinal or higher can be treated with more powerful techniques using rank tests, some of which are mentioned briefly in Chapter 10.)

Where We Are Going in This Chapter

In Chapter 8 we introduced the basic principles of tests of hypotheses. In that chapter the typical assumptions made were those of independent and *normally* distributed observations. The data were interval or ratio measurements. However, much data from random samples, possessing the properties of independence, fail to meet the assumption of normally distributed observations. One reason for such a failure is that the measurement level for the data is nominal, or ordinal at most.

In this chapter we examine the chi-square goodness-of-fit test, a test procedure to test how well a particular probability model fits a set of data. This procedure is related to the binomial (more appropriately the multinomial) distribution but in its large sample form the test statistic is distributed approximately as a chi-square random variable. Next, we examine two procedures for testing hypotheses using data from one or two binomial distributions. We test hypotheses concerning the Bernoulli parameter p.

Finally, we conclude with a test of homogeneity of r populations and a test of independence for nominal, bivariate data whose test statistic through large sample assumptions is also approximately chi-square in distribution. The data for these two tests are summarized in r by c contingency tables or cross-tabulations.

The SIM diagram of Chapter 8 will apply here without modification (see Figure 9.1), since we are still testing hypotheses. The "statistical inference" is to decide whether to reject the null hypothesis or not. We will still work within the 5-step test of hypothesis procedure, we simply have different types of test statistics defined for each of the methods presented.

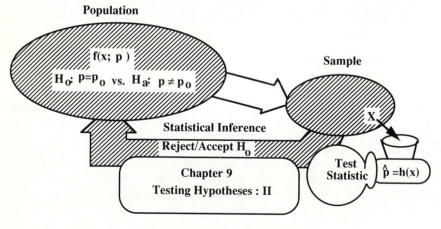

Figure 9.1 SIM—Testing hypotheses about Bernoulli parameter p.

9.1   A LARGE SAMPLE, χ^2 GOODNESS-OF-FIT TEST

Introduction

We have frequently encountered the problem of trying to decide which population model best describes a given set of data. The approach taken thus far has been to plot the data, make a histogram and look at the data, examine the theoretical underpinnings of the data, all with the hope of being able, by experience or good judgment and luck, to choose a distribution that adequately fits the data.

Once a distribution has been chosen as a candidate for consideration, then a more objective procedure needs to be used to determine if the choice is a good one or a bad one. That is the purpose of a goodness-of-fit test. In this chapter we will examine one such test called the chi-square goodness-of-fit test. Other procedures are available.

Methodology

The χ^2 Goodness-of-fit Test

Background Assumptions: Let a random sample of $X_1, ..., X_n$ be selected from some unknown distribution with distribution function $F(x)$. The sample is divided up into k mutually exclusive categories (as when constructing a frequency table or histogram). Let n_i denote the number of observations in the sample falling into category i, where N_i is a random variable that takes on the value n_i. We want to test whether $F(x) = F_o(x)$, where $F_o(x)$ is the distribution function for some specified pdf.

Step 1: $H_o: F(x) = F_o(x)$ vs. H_a: $F_o(x)$ is not the distribution function of $F(x)$.

Step 2: Select a random sample of observations from the unknown distribution and categorize them into k mutually exclusive classes. Choose a level of significance α.

Step 3:   The test statistic is

$$\chi^2 = \sum_{i=1}^{k} \frac{(n_i - np_i)^2}{np_i},$$

the sum being over all k classes. In this formula, p_i denotes the probability of an observation falling in class i under the assumption of the null hypothesis.   Thus, under the null hypothesis, $E(N_i) = np_i$. This statistic will be distributed approximately as a χ^2 random variable with $k - 1$ degrees of freedom for large n. (See the explanation that follows for an argument justifying the large sample chi-square distribution.)

Step 4: The p-value is:

$$p = P(\chi^2 \geq \chi_c^2)$$

where $df = k - 1$. A useful table to use to display the summarized data and computations is the following:

Class or Class Interval	Observed n_i	Probability p_i	np_i	$\dfrac{(n_i - np_i)^2}{np_i}$
1	n_1	p_1	np_1	$\dfrac{(n_1 - np_1)^2}{np_1}$
2	n_2	p_2	np_2	$\dfrac{(n_2 - np_2)^2}{np_2}$
$\vdots$	$\vdots$	$\vdots$	$\vdots$	$\vdots$
k	n_k	p_k	np_k	$\dfrac{(n_k - np_k)^2}{np_k}$
Total	n	1.00	n	$\sum\dfrac{(n_i - np_i)^2}{np_i} = \chi_c^2$

Step 5: If $p \leq \alpha$, then reject the null hypothesis.

DISCUSSION

The Multinomial Distribution

Suppose a random variable can take on any of k possible outcomes with probabilities $p_1, p_2, \ldots, p_k$, and a random sample of n of these random variables is selected. If we now define k new random variables $X_1, X_2, \ldots, X_k$ to represent the total number of times each of the k possible outcomes occurred, then the random variables are said to follow the multinomial distribution, a generalization of the binomial distribution (where $k = 2$), and has the following probability density function.

$$f(x_1, x_2, \ldots, x_k; p_1, p_2, \ldots, p_k) = \binom{n}{x_1, x_2, \ldots, x_k} p_1{}^{x_1} p_2{}^{x_2} \ldots p_k{}^{x_k}$$

$$= \frac{n!}{x_1! x_2! \ldots x_k!} p_1{}^{x_1} p_2{}^{x_2} \ldots p_k{}^{x_k}$$

where $x_1 + x_2 + \ldots + x_k = n$ and $p_1 + p_2 + \ldots + p_k = 1$ and the x's take on positive, integer values. Notice that knowing any k-1 values determines the last; i.e., knowing $x_1, x_2, \ldots, x_{k-1}$, then x_k is determined since $x_1 + x_2 + \ldots + x_k = n$. Similarly, knowing $p_1, p_2, \ldots, p_{k-1}$ automatically determines what p_k will have to be.

As an example of a multinomial experiment, suppose we roll a single balanced die 30 times, recording the number of times each face turns up. This experiment follows the multinomial distribution above with $n = 30$, $k = 6$, and $p_1 = p_2 = \ldots = p_6 = \dfrac{1}{6}$. The probability that we would get an equal number of each face turning up would be:

$$f(X_1 = 5, X_2 = 5, \ldots, X_6 = 5; p_1 = p_2 = \ldots = p_6 = \tfrac{1}{6}) = \binom{30}{5, 5, \ldots, 5}(\tfrac{1}{6})^5(\tfrac{1}{6})^5\ldots(\tfrac{1}{6})^5$$

$$= \frac{30!}{5!\,5!\,\ldots\,5!}(\tfrac{1}{6})^{30}$$

The multinomial probability density function serves as the model for the next test of hypothesis.

Goodness-of-Fit Test

A major perspective that we have taken in this course, starting with Chapter 2, was to make certain model assumptions about the population from which a set of data have come or will come. But up to this point we have had no objective method of evaluating how well a particular model "fits" a data set. This section remedies that problem as we present the chi-square goodness-of-fit test. With this test we will be able to assess whether a normal distribution is a good model for heights of adult males, whether "waiting time" data are exponentially distributed, whether the number of heads when actually tossing a coin 5 times is modelled by a binomial distribution with $n = 5$ and $p = 0.5$.

Notice as the method is presented that the 5–step hypothesis testing procedure is used. In addition, note that the distribution of the test statistic, though in large samples is approximately distributed as a chi-square random variable, is fundamentally constructed on a Bernoulli concept, and thus is related to the binomial distribution (actually the multinomial distribution of above). That is, we note how many observations fall into each of k particular intervals or range of values. The Bernoulli property arises since a given observation either is in the interval in question or it is not.

Consider a very simple case where observations or measurements fall into one of two classes so that in the notation above $k = 2$. Therefore, we divide the sample observations into 1 of 2 categories. Let p_1 represent the theoretical probability under the null hypothesis of an observation falling into category 1, and p_2 the probability of an observation falling into category 2, $(p_1 + p_2 = 1)$. Then we let n_1 denote the number of observations in the sample falling into category 1, then n_1 is a *binomial* random variable with parameters n and p_1 and $1 - p_1 = p_2$. As a result, for large n, it has been shown that

$$Z = \frac{n_1 - np_1}{\sqrt{np_1(1 - p_1)}}$$

will be approximately standard normal and therefore, Z^2 would be chi-square with 1 degree of freedom.

Next it is possible to show algebraically, that

$$Z^2 = \left(\frac{n_1 - np_1}{\sqrt{np_1(1 - p_1)}}\right)^2 = \frac{(n_1 - np_1)^2}{np_1(1 - p_1)} = \frac{(n_1 - np_1)^2}{np_1} + \frac{(n_2 - p_2)^2}{np_2} = \sum_{i=1}^{2}\frac{(n_i - np_i)^2}{np_i}$$

by letting $n_1 = (n - n_2)$ and $p_1 = 1 - p_2$. Generalizing to k classes, we get

$$\chi^2 = \sum_{i=1}^{k} \frac{(n_i - np_i)^2}{np_i}$$

which is approximately a chi-square random variable with $k - 1$ degrees of freedom. This test statistic with its approximate distributional properties was first proposed by Karl Pearson around 1900.

EXAMPLES

EXAMPLE 1: In an examination of employees reporting sick during a week, managers wondered whether the number reporting sick were evenly distributed over the 5 working days, or not. The following data were obtained from office records.

Day	No. Reporting Sick by Day of Week				
	Mon.	Tues.	Wed.	Thurs.	Fri.
# Sick	49	35	32	39	45

Do the data support a conclusion of an equal distribution throughout the week? (That is, does the discrete uniform distribution "fit" the data?)

SOLUTION: The 5-steps are:

Step 1: $H_0: p_m = p_t = \dots = p_f = \frac{1}{5}$ vs. H_a: At least one $p_i \neq \frac{1}{5}$

Step 2: A random sample of 200 sick-leave records is selected and the day of the week an illness was reported was noted.

Step 3: The test statistic is

$$\chi^2 = \sum_{i=1}^{5} \frac{(n_i - np_i)^2}{np_i}$$

which has a chi-square distribution with $df = 5 - 1$ for large samples. In this expression, each p_i equals $\frac{1}{5}$.

Step 4: The computations are as indicated next:

Class	Observed n_i # Sick	Prob: p_i	np_i	$\frac{(n_i - np_i)^2}{np_i}$
Monday	49	1/5	40	81/40
Tuesday	35	1/5	40	25/40
Wednesday	32	1/5	40	64/40
Thursday	39	1/5	40	1/40
Friday	45	1/5	40	25/40
Total	200	1.00	200	196/40 = 4.9

Therefore the value of the test statistic is $\chi^2 = 4.9$ which is approximately chi-square with 4 degrees of freedom. The *p*-value is: $0.1 < p < 0.9$

Step 5: Do not reject H_o; there is insufficient evidence to reject the conclusion that sick days are evenly distributed throughout the week. (The model provides an adequate fit to the data.)

Observations

There are some special conditions that can occur in the goodness-of-fit test that may require a modification of the procedure as shown above. Three conditions of particular importance are as follows:

1. If any parameters of the distribution have to be estimated using the data—either by the principle of maximum likelihood or by the method of moments—one degree of freedom is subtracted from the total for each parameter that is estimated. Therefore the degrees of freedom in general can be written as $df = k - 1 - r$ where k is the number of classes created and r the number of parameters estimated.

2. The decision to "reject" H_o should be exercised with caution if $np_i \leq 5$ for 20% or more of the categories or if $np_i \leq 1$ for any cell. When this happens, we can get an artificially large value for the chi-square, leading to a possible incorrect decision to reject . A typical solution to this problem is to collapse two or more adjacent cells together so the the "expected" number will exceed 5.

3. For very large samples, we have the tendency to always reject due to the sensitivity of the test. Therefore, we may reject the hypothesis of a "good" fit due to slight but unimportant deviations from the exact model. Beware of this temptation!

EXAMPLE 2: Suppose in performing a chi-square calculation category j has an observed value of 95 and an expected value of 100; and category h has an observed value of 0 and an expected value of 5. Compute the chi-square "pieces" for these two categories and comment on the effect of the small "expected" value.

SOLUTION: For category j we have the computation as

$$\frac{(O_j - E_j)^2}{E_j} = \frac{(95 - 100)^2}{100} = \frac{25}{100} = 0.25.$$

For category h we get:

$$\frac{(O_h - E_h)^2}{E_h} = \frac{(0 - 5)^2}{1} = \frac{25}{1} = 25.$$

There is a dramatic difference in the sizes of the two computations even though the numerator is the same for each. The difference is in the size of the denominators, the "expected value." Thus, a small "expected value" could produce a very large chi-square contribution, contributing to a large overall chi-square value and leading to the decision to reject, possibly when one shouldn't reject. This could lead to a higher likelihood for a type I error than expected.

EXAMPLE 3: A biologist studied the growth of bacterial colonies in samples of milk film due to a concern about the proper storage temperature of milk. Four hundred observations of bacterial colonies within the field of a microscope were made and the following table resulted. Test the hypothesis that the data fit the Poisson distribution.

Number of Colonies per Field x_i	Frequency of Observation (f_i)
0	56
1	104
2	80
3	62
4	42
5	27
6	9
7	9
8	5
9	3
10	2
11	0
12	1
Total	400

SOLUTION: Before laying out the steps of the hypothesis test we should note that the parameter λ of the Poisson is not specified and thus must be estimated. The maximum likelihood estimate is the sample mean which gives $\hat{\lambda} = \dfrac{\sum x_i f_i}{400} = 2.422$ colonies per field.

Step 1: H_o: The Poisson model fits the data vs. H_a:The Poisson model does not fit the data

Step 2:   A random sample of 400 microscope fields is observed and the number of bacterial colonies per field is recorded. Let $\alpha = 0.10$.

Step 3:   The test statistic is

$$\chi^2 = \sum_{i=1}^{13} \frac{(n_i - np_i)^2}{np_i}$$

which has a chi-square distribution with $df = 13 - 1 - 1$ for large samples. (We have to estimate one parameter.) In this expression, each p_i is obtained from a Poisson distribution with $\lambda = 2.4$ using

$$f(x_i) = F(x_i) - F(x_i - 1)$$

Step 4:   The resulting data occurred with computations performed as indicated:

x	(n_i)	p_i	np_i
0	56	.0907	36.28
1	104	.2177	87.08
2	80	.2613	104.52
3	62	.2090	83.60
4	42	.1254	50.16
5	27	.0602	24.08
6	9	.0241	9.64
7	9	.0083	3.32
8	5	.0025	1.00
9	3	.0007	.28
10	2	.0002	.08
11	0	.0000	.00
12	1	.0000	.00
Total	400	1.0000	400.00

Note that the categories from $x = 7$ to $x = 12$ are such that $np_i \le 5$. Therefore, we propose that the categories from 6 to 12 be combined together into one category. The resulting table and computations are:

x_i	n_i	p_i	np_i	$\dfrac{(n_i - np_i)^2}{np_i}$
0	56	.0907	36.28	10.72
1	104	.2177	87.08	3.29
2	80	.2613	104.52	5.75
3	62	.2090	83.60	5.58
4	42	.1254	50.16	1.32
5	27	.0602	24.08	.35
6+	29	.0358	14.32	15.05
Total	400	1.0000	400.04	42.06

Therefore the value of the test statistic is

$$\chi^2 = 42.06$$

which is approximately chi-square with $7 - 1 - 1 = 5$ degrees of freedom. The p-value from the chi-square table is $p \ll 0.005$ (where $\ll$ means "much less than.")

Step 5:   Reject H_o. The Poisson does not provide an adequate fit to the data.

One last observation about this test is needed. In Chapter 8 the researcher collected data with the hope that it would lead to rejection of the null hypothesis. If we reject H_o, then the Type I error was controlled by a choice for α that was acceptably small. In the goodness-of-fit test, our hope is *not to reject* but rather to accept the null hypothesis that the given distribution "fits" the data. However, the decision to accept leads to the possibility of a Type II error occurring, not a Type I error, and we have not pre-selected a value for the Type II error. To moderate this conflict, we can select a value for α that is larger than ordinary, say of 0.10 or 0.15. This, as we have seen previously, will reduce the Type II error probability by making the rejection region larger and the acceptance region smaller. We still will not know the value of β, but we can have some reassurance that it is smaller than it would have been if the size of α were set as small as we usually do.

EXAMPLE 4: The following 40 observations was generated from a gamma distribution with $\alpha = 2$ and $\theta = 10$. Apply the goodness-of-fit test to this data using a level of significance of 10%.

49.0662	15.2768	23.3803	2.2905	13.7978
37.7604	33.9633	22.8145	16.0873	43.4627
40.6222	10.2427	36.0867	26.9065	20.2895
16.0575	5.6385	14.9561	36.7405	13.6624
17.3373	13.9306	30.7167	34.9067	27.5957
20.0064	40.2106	64.0425	6.9909	50.5328
20.7794	3.8205	7.1160	2.7182	31.7776
7.6585	9.3867	6.0724	5.3690	22.0052

SOLUTION: To begin with, the data were organized into a frequency table consisting of 13 intervals. The mean and standard deviation of the data are also given.

Class Interval	Frequency
$0.0 < x \le 5.0$	3
$5.0 < x \le 10.0$	7
$10.0 < x \le 15.0$	5
$15.0 < x \le 20.0$	4
$20.0 < x \le 25.0$	6
$25.0 < x \le 30.0$	2
$30.0 < x \le 35.0$	4
$35.0 < x \le 50.0$	3
$40.0 < x \le 45.0$	3
$45.0 < x \le 50.0$	1
$50.0 < x \le 55.0$	1
$55.0 < x \le 60.0$	0
$60.0 < x$	1

$$n = 40$$
$$\text{Mean} = 22.552$$
$$\text{St. Dev.} = 15.1$$

Since the data are theoretically continuous, $p_i = F(x_i) - F(x_{i-1})$ where p_i is the probability of an observation falling in the ith class interval, x_i refers to the end point of the ith class interval and $F(x_i)$ is the cumulative probability density function for a gamma distribution. Recall in this case, that $F(x_i;\alpha,\theta) = 1 - F_P(\alpha - 1;\lambda = \frac{x_i}{\theta}) = 1 - F_P(1;\lambda = x/10)$. The following tables show some of the computations when applying the goodness-of-fit test.

Class Interval End Point, x_i	f_i	$F(x)$	p_i	$E_i = 40 \cdot p_i$
5.0	3	0.0902	0.0902	3.608
10.0	7	0.2642	0.1740	6.960
15.0	5	0.4422	0.1780	7.120
20.0	4	0.5940	0.1518	6.070
25.0	6	0.7127	0.1187	4.648
30.0	2	0.8009	0.0882	3.528
35.0	4	0.8641	0.0632	2.528
40.0	3	0.9084	0.0443	1.772
45.0	3	0.9389	0.0305	1.220
50.0	1			
55.0	1	* The "expected values" from here down are too small,		
60.0	0	so the rows and are collapsed together. See the next		
65.0+	1	table.		

After collapsing the class intervals together from 45 on up, we get:

Class Interval End Point, x_i	n_i	$F(x)$	p_i	np_i	$\dfrac{(n_i - np^*)^2}{np^*}$
5.0	3	0.0902	0.0902	3.608	0.1025
10.0	7	0.2642	0.1740	6.960	0.0002
15.0	5	0.4422	0.1780	7.120	0.6312
20.0	4	0.5940	0.1518	6.070	0.7059
25.0	6	0.7127	0.1187	4.648	0.3933
30.0	2	0.8009	0.0882	3.528	0.6618
35.0	4	0.8641	0.0632	2.528	0.8571
40.0	3	0.9084	0.0443	1.772	0.8510
45.0	3	0.9389	0.0305	1.220	2.5970
45.0+	3	1.0000	0.0611	2.444	0.1265
	40.0		1.0000	39.898	6.9265

This gives a Chi-square value of 6.9265 which is to be compared to a tabled value having $10-1 = 9$ degrees of freedom. The p-value for this test is > 0.05, (it is actually 0.6448) therefore, we would accept the null hypothesis that the gamma distribution "fits" the data.

EXAMPLE 5:   If the null hypothesis had been "The data come from a gamma distribution" what would you have been done differently in Example 4 to address the problem when phrased this way?

SOLUTION: The difference between this problem and that of Example 4 is that the parameters values are not specified. Therefore, we would first need to estimate the parameters for the gamma distribution. Much of the work would have been the same from that point on. The probabilities in the p_i column would have been different and so a different value of chi-square would have been obtained. In addition, the degrees of freedom would now be: $df = 10 - 1 - 2 = 7$.

EXERCISES

1. In a pharmaceutical laboratory, rats were housed with 4 rats per cage. Accidently, 160 cages of rats were exposed to a disease carried in their water. It was speculated that the likelihood of an individual rat contracting the disease was 0.5. The number of rats per cage contracting the disease was reported and is listed below. Does a binomial distribution with $n = 4$ fit this data?

Number with Disease/Cage	Frequency
0	12
1	35
2	60
3	45
4	8

2. An avid Monopoly player wanted to test if the die she was using was true. She proceeded to roll the die 60 times with the following results. Does the uniform discrete distribution provide an adequate fit for this data?

Number on face	Number of Outcomes
1	8
2	11
3	15
4	6
5	9
6	11

3. A class of 164 students reported their heights to the nearest inch and the data, reported as a frequency table, is shown below. Does a normal distribution with mean of 70 inches and standard deviation of 3 inches provide an adequate fit to this data?

Interval in Inches	Frequency
61-64	10
65-68	21
69-72	83
73-76	49
77-80	1

4. Fill in the steps to prove the following equality for the multinomial (binomial) distribution when $k = 2$ categories.

$$Z^2 = \left(\frac{n_1 - np_1}{\sqrt{np_1(1 - p_1)}} \right)^2 = \frac{(n_1 - np_1)^2}{np_1} + \frac{(n_2 - np_2)^2}{np_2} = \sum_{i=1}^{2} \frac{(n_i - np_i)^2}{np_i}$$

5. In a cannery, the machinery that fills the cans has 8 nozzles mounted on a circulating system. Each can is filled from one of the 8 nozzles. A regular monitoring process is implemented to check for a tendency for a particular nozzle to overfill or underfill the cans. Cans are selected randomly through the day with approximately equal numbers coming from each nozzle. It is noted whether the cans are underfilled or overfilled beyond that which is acceptable variation in

fill. The following data resulted. Is there any reason to think that a particular nozzle has a tendency to over or underfill the cans? Let $\alpha = 0.15$.

Nozzle Position	# of Cans Over/Under Filled
1	22
2	18
3	25
4	30
5	19
6	24
7	15
8	20
Total	173

6. While testing a used tape for bad records, a computer operator counted the number of flaws per 100 feet of tape. Let X equal this random variable. Test the null hypothesis that X has a Poisson distribution with a mean of $\lambda = 2.4$ given that 40 observations of X yielded 5 zeros, 7 ones, 12 twos, 9 threes, 5 fours, 1 five, and 1 six. Choose your own value for α.

7. The following data were generated from a normal distribution with $\mu = 50$ and $\sigma = 10$. In this case $n = 50$. Create 5 classes with equal class interval widths and use the goodness-of-fit test with the level of significance set at $\alpha = 0.10$.

52.19	59.05	53.08	56.01	56.44	66.32	41.81	38.47	53.90	50.36
54.62	47.08	64.69	51.00	23.50	53.12	29.06	57.82	47.91	49.84
41.54	34.94	51.48	37.52	71.26	43.03	49.85	56.68	50.59	56.30
44.15	46.65	57.47	60.80	52.49	43.13	45.24	60.33	40.19	28.90
42.73	48.21	65.99	58.51	52.39	46.58	39.58	46.07	33.33	47.81

8. For the data of exercise 7, perform the same test of hypothesis using 5 classes with an equal number of observations per class. Use the same level of significance as used in exercise 7.

9. Suppose you use the data of exercise 7 to estimate the mean and standard deviation to perform the test of hypothesis. Use 5 classes with an equal number of observations per class.

10. Consider the test proposed in exercise 9. Does this procedure raise a question in your mind that you have "favored" the null hypothesis by using the data to estimate the parameters? What error would you be concerned about by using the data to estimate the parameters, the type I or the type II error?

9.2   LARGE SAMPLE TESTS OF PROPORTIONS: ONE AND TWO POPULATIONS

Introduction

When the underlying distribution of data can be modelled by a Bernoulli distribution, the resulting test statistic for a test of hypothesis about the Bernoulli parameter p will be a binomial random variable. However, in this chapter we will rely on the large sample properties of the binomial random variable which will be approximately normally distributed according to the central limit theorem. Thus, by an appropriate transformation, the test statistics presented in this section will be z-scores which will be approximately standard normal in distribution.

Specifically, in this section we examine two cases. The first case assumes we have a random sample of n Bernoulli trials and we test the hypothesis about the parameter p. The second case assumes we have two independent samples, constituting two different Bernoulli populations. We test the hypothesis about the difference in the two parameters, $p_x = p_y$.

Methodology

One Bernoulli Population, Test of One Proportion

Background assumptions: Suppose we have a random sample of n Bernoulli random variables with unknown probability of success p. We want to test the hypothesis about the parameter p.

Step 1: The Hypotheses:

Null Hypothesis	Possible Alternative Hypotheses
$H_0: p = p_o$	$H_a: \begin{cases} p \neq p_o \\ p > p_o \\ p < p_o \end{cases}$

Step 2: Select a random sample of size n from a Bernoulli distribution with unknown parameter, p. Determine the level of significance α.

Step 3: The test statistic is

$$Z = \frac{(\hat{p} - p)}{\sqrt{\dfrac{p_o(1 - p_o)}{n}}}$$

which is distributed approximately as a standard normal random variable.

Step 4: Compute the *p*-value as follows:

Alternative Hypothesis	*p*-value Computation
$p \neq p_o$	$p = 2P(Z > |Z_c|)$
$p > p_o$	$p = P(Z > Z_c)$
$p < p_o$	$p = P(Z < Z_c)$

where

$$Z_c = \frac{(\hat{p} - p)}{\sqrt{\dfrac{p_o(1 - p_o)}{n}}}$$

Step 5: If $p < \alpha$, then reject H_o.

Two Bernoulli Populations, Tests of Two Proportions

Background assumptions: Suppose we have two different Bernoulli distributions with parameters p_x and p_y respectively. We take a random sample of n and m respectively from the two distributions and we want to test the hypothesis about the difference in proportions, $p_x - p_y$.

Step 1:

Null Hypothesis	Possible Alternative Hypotheses
$H_o: p_x = p_y$	$H_a: \begin{cases} p_x \neq p_y \\ p_x > p_y \\ p_x < p_y \end{cases}$

Step 2: Select independent random samples of size *n* and *m* from two different Bernoulli distributions with unknown parameters, p_x and p_y, respectively. In addition, let the selected probability of a type I error be α.

Step 3: The test statistic is

$$Z = \frac{(\hat{p}_x - \hat{p}_y) - 0}{\sqrt{\hat{p}_o(1 - \hat{p}_o)(\dfrac{1}{n} + \dfrac{1}{m})}}$$

where we let *x* denote the total number of successes in the sample of size *n*, and *y* denote the total number of successes in the sample of size *m*. Therefore, p_x is estimated by $\hat{p}_x = \dfrac{x}{n}$, p_y by $\hat{p}_y = \dfrac{y}{m}$, the difference $p_x - p_y$ by $\hat{p}_x - \hat{p}_y = \dfrac{x}{n} - \dfrac{y}{m}$ and the common proportion under $H_o: p_x = p_y = p_o$ is estimated by $\hat{p}_o = \dfrac{(x + y)}{(n + m)}$. The test statistic *Z* will be approximately standard normal by the central limit theorem with large *n* and *m*.

Step 4: Compute the p-value as follows

Alternative Hypothesis	p-value Computation
$p_x \neq p_y$	$p = 2P(Z > |Z_c|)$
$p_x > p_y$	$p = P(Z > Z_c)$
$p_x < p_y$	$p = P(Z < Z_c)$

where Z_c is

$$Z_c = \frac{(\hat{p}_x - \hat{p}_y) - 0}{\sqrt{\hat{p}_o(1 - \hat{p}_o)(\frac{1}{n} + \frac{1}{m})}}$$

Step 5: If $p < \alpha$, then reject H_o.

DISCUSSION

The One-Sample Problem

The classical Bernoulli experiment is the coin tossing experiment. If we want to test whether a given coin is balanced or not, we would set up a hypothesis about the parameter p. Such a test would be called a test of one proportion. Tests such as this are very common and can arise in a variety of settings. For instance, suppose we want to compare the proportion of male and female respondents in a telephone survey with the proportion known to be present in the population, or suppose we want to compare the use of smokeless tobacco in a particular teenage population with the nationwide average. Both of these hypothetical situations would lead to a test of hypothesis about the proportion p in a Bernoulli distribution. There are numerous other possibilities that can be set up this way.

The logical test statistic in this case would be Y the number of successes where $Y = \sum_{i=1}^{n} X_i$ such that each X_i is a Bernoulli random variable. Under the null hypothesis and assuming a random sample of independent and identically distributed Bernoulli observations, Y will be distributed as a binomial random variable with parameters n and p_o (H_o:$p = p_o$). The rejection region would be associated with one or the other tails of the binomial distribution when the alternative hypothesis leads to a one-tailed test, or if a two-tailed test is appropriate, then the rejection region would be in both tails of the binomial distribution. The critical values could be found from the binomial tables. However, we will be restricted to the values of n and p of Table A. To allow for a more general treatment we assume a large sample and then appeal to the central limit theorem (see section 6.10 of Chapter 6). This allows us to use the standard normal distribution to find p-values where the test statistic is

$$Z_c = \frac{Y - np}{\sqrt{np_o(1 - p_o)}} = \frac{(\hat{p} - p_o)}{\sqrt{\frac{p_o(1 - p_o)}{n}}}$$

The estimate of p is $\hat{p} = \frac{Y}{n}$, the proportion of successes in the sample.

The Two-Sample Problem

In other situations, comparisons will often involve proportions from two or more populations. For instance, suppose we want to compare the proportions of heavy smoking males compared to females, the proportion of 6th grade students to 11th grade students who are absent on a given day, the proportion of germinating tomato seeds from one supplier compared to another, and so forth. Such a test would involve comparing two Bernoulli populations and their respective parameters p_x and p_y. In this section this is called a test of two proportions. We assume that $X_1, X_2, ..., X_n$ is a random sample of size n from one population associated with a Bernoulli random variable X, and $Y_1, Y_2, ..., Y_m$ is an independent random sample of size m from another Bernoulli population associated with the random variable Y. This leads to two random variables to consider: $X = \sum_{i=1}^{n} X_i$

and $Y = \sum_{i=1}^{m} Y_i$ each of which will be binomial random variables. To adjust for the difference in

sample sizes we propose a test statistic based on the proportions $\hat{p}_x = \dfrac{\sum X_i}{n}$ and $\hat{p}_y = \dfrac{\sum Y_i}{m}$. If both

sample sizes are relatively large then a central limit theorem test statistic that will be approximately normally distributed is

$$Z = \frac{(\hat{p}_x - \hat{p}_y) - (p_x - p_y)}{\sqrt{V(\hat{p}_x - \hat{p}_y)}}$$

If the null hypothesis is $H_o: p_x = p_y = p_o$, then $(p_x - p_y) = 0$, and because of independent samples

$$V(\hat{p}_x - \hat{p}_y) = V(\hat{p}_x) + V(\hat{p}_y)$$

$$= \frac{p_o(1 - p_o)}{n} + \frac{p_o(1 - p_o)}{m}$$

$$= p_o(1 - p_o)(\frac{1}{n} + \frac{1}{m})$$

Therefore, under the null hypothesis the Z-score becomes

$$Z = \frac{(\hat{p}_x - \hat{p}_y) - 0}{\sqrt{p_o(1 - p_o)(\frac{1}{n} + \frac{1}{m})}}$$

with p_o is the *proportion of successes in the combined sample* and is estimated by $\hat{p}_o = \dfrac{(\sum X_i + \sum Y_i)}{(n + m)}$

If we let x denote the number of successes in the sample of size n from population X, and y denote the number of successes in the sample of size m from population Y, then we can write

$$\hat{p}_o = \frac{(x + y)}{(n + m)}$$

so the resulting test statistic is:

$$Z = \frac{(\hat{p}_x - \hat{p}_y) - 0}{\sqrt{\hat{p}_o(1 - \hat{p}_o)(\frac{1}{n} + \frac{1}{m})}}$$

which will be approximately standard normal under a large sample size assumption.

The data for such a test of hypothesis is conveniently arranged in two-by-two table as follows:

Population	Successes	Failures	Sample size
X	x	$n - x$	n
Y	y	$m - y$	m
Total	$x + y$	$(n + m) - (x + y)$	$n + m$

EXAMPLES

EXAMPLE 1: In a recent sociological survey concerning the attitude toward seat belts of 400 respondents, the male-female ratio in the sample was computed in order to assess whether a representative sample with regard to sex was obtained. In the population at large, it was known from census figures that the proportion of females is 51% to 49% for the males. In the sample there were 220 females. Are there statistical reasons to conclude the male-female ratio in the sample is inconsistent with the population?

SOLUTION: This is a test of one proportion. The steps are as follows:

Step 1: $H_o: p_f = 0.51$ vs. $H_a: p_f \neq 0.51$

Step 2:   A random sample of 400 persons was selected and the sex of each individual was recorded.

Step 3:   The test statistic is

$$Z = \frac{(\hat{p} - 0.51)}{\sqrt{\frac{0.51(1 - 0.51)}{400}}}$$

which for large samples will be approximately standard normal by the central limit theorem. The expression, $\hat{p}$, is the maximum likelihood/method of moments estimator of the unknown parameter p and is the proportion of females in the sample.

Step 4:   The calculated value of Z is

$$Z_c = \frac{\left(\frac{220}{400} - 0.51\right)}{\sqrt{\frac{0.51(1 - 0.51)}{400}}} = 1.60$$

Since this is a two-tailed test, this has a p-value of

$$p = 2(0.0548) = 0.1096$$

Step 5:   On the basis of the statistical evidence it would seem unreasonable to conclude that we have a non-representative sample relative to the sex ratio.

EXAMPLE 2:   In a recent exit poll, the question was raised as to whether the proportion of voters in Congressional District X voting for the incumbent president was different than the proportion in Congressional District Y. For a sample of size 760 in District X, it was found that 562 (74%) of them voted for the incumbent, while of 745 voters interviewed in District Y, 521 (70%) of them voted for the incumbent. Does this suggest that the proportion is different in the two districts? This data are arranged in the following table for convenience.

District	For	Against	Sample size
X	562	198	760
Y	521	224	745
Total	1083	422	1505

SOLUTION:
Step 1: $H_o : p_x = p_y$ vs. $H_a : p_x \neq p_y$

Step 2:   A random sample of voters was selected from each district and were asked if they voted for the incumbent president.

Step 3:   The test statistic is

$$Z = \frac{(\hat{p}_x - \hat{p}_y) - 0}{\sqrt{\hat{p}_o(1 - \hat{p}_o)(\frac{1}{n} + \frac{1}{m})}}$$

where $\hat{p}_o = \dfrac{(x + y)}{(m + n)}$ and x denotes the number of voters in District X favoring the incumbent, while y denotes the number of voters in District Y favoring the incumbent.

Step 4:   The values to be inserted into the test statistic are: $n = 760$, $m = 745$, with $x = 562$ voters from District X voting for the incumbent and $y = 521$ voters from District Y. Thus,

$$\hat{p}_o = \frac{(562 + 521)}{(760 + 745)} = \frac{1083}{1505} = 0.72 \ .$$

This leads to the following Z score.

$$Z_c = \frac{\left(\frac{562}{760} - \frac{521}{745}\right) - 0}{\sqrt{\frac{1083}{1505}\left(1 - \frac{1083}{1505}\right)\left(\frac{1}{760} + \frac{1}{745}\right)}} = 1.73$$

and $P(Z > 1.73) = 0.0418$. Since this is a two-tailed test, this gives us a p-value of $p = 2 \cdot (0.0418) = 0.0836$.

Step 5:   On the basis of the statistical evidence it would seem unreasonable to conclude that there is a difference in District X and District Y relative to their preference for the incumbent president. (This is the appropriate decision for any value of $\alpha < 0.0836$.)

EXAMPLE 3: Recognizing that a squared Z-score is a chi-square random variable with 1 degree of freedom, use this fact to test the same hypothesis as given in Example 2.

SOLUTION: Since the alternative hypothesis in Example 2 is two-tailed, then either large positive or large negative values lead to a decision to reject the null hypothesis.   Therefore, squaring these "large" Z-scores will produce large chi-square values. Therefore, we would reject the null hypothesis if $Z^2 = \chi^2$ is large or if the p-value is less than α. For this particular case then $Z^2 = (1.73)^2 = 2.9929$ and the p-value is

$$p = P(\chi_1^2 > 2.9929) = 0.0836$$

the same p-value we obtained in Example 2.   Remember that the p-value of Example 3 is only approximate, so the approach of this example is also only approximate as is the resulting p-value.

EXERCISES

1.   In an unofficial taste-test conducted in a supermarket, two popular soft drinks were being compared with one another. In unmarked cups, customers passing by were asked to compare the two soft drinks and indicate which one they preferred. Out of 150 customers trying the two soft drinks, 90 preferred brand A. Is this sufficient evidence to indicate a clear preference for one brand over another, or could this have occurred by chance under an assumption of equal preference?

2.   In the setting of exercise 1 above, age was also noted by simply classifying the "taste-tester" as being "under 30" or "over 30" with the intention of getting approximately equal numbers in each group. (The classification by age was a subjective evaluation by the person distributing the drinks.) However, the sample sizes of the two groups didn't turn out even because of the clientele of the supermarket. The sample resulted in 110 in the "under 30" category and 40 in the "over 30" category. Of the 110 "under 30" group, seventy reported a preference for brand A. In the "over 30" group, 20 reported a preference for brand A. Does this suggest a difference in preference due to age group?

3.   Use the approach of Example 3 to test the hypothesis of exercise 2 above.

4.   Describe why caution should be used in applying the approach of Example 3 if the rejection region is "one-tailed."

5.   Psychologists have noted a tendency that mothers who have immediate access to their babies after birth, prefer to hold their babies on the left side, while those without immediate access prefer to hold their babies on the right side. (Speculation on this preference centers around the mother's heartbeat which is more pronounced on the mother's left side.) To substantiate this perception, the following data were obtained. Do the data support this contention. (Note that this table is "on its side" compared to the layout used in the section discussion.)

Mother's Preference	Type of Contact	
	Immediately after birth	No early contact
Left side	75	40
Right side	25	60
Total	100	100

6. Television viewing habits of children were a concern of parents and PTA officers in a particular community. One school in the community was using a pilot program to encourage more reading at home in contrast to TV viewing, so it was decided to compare it with another school that had no particular reading program. Eighty children in the same grade were randomly selected from each school. The number of children watching more than 5 hours of TV per day versus less than 5 were noted and is reported in the following table. Does there tend to be a difference between the two schools in amount of TV watched?

No. of Hours/Day	School	
	With Reading Program	Without Reading Program
More than 5	9	20
Less than 5	71	60
Total	80	80

7. Indicate why the "squared Z-score" approach of Example 3 would or would not be appropriate for exercises 5 and 6. To answer this question, think in terms of whether the alternative hypothesis is one-tailed or two-tailed and its implications.

8. A state that has a mandatory seat-belt law still only showed about 21% of front-seat occupants in personal vehicles using a seat belt or other restraining device. A campaign on TV as well as by the highway patrol was implemented and after two months, of 480 drivers observed, 135 were using some restraining device. Does this indicate that the use of restraining devices had increased? Use $\alpha = 0.05$ for this test.

9. In exercise 8 suppose a 95% confidence interval on p were used. Would this confidence interval exclude the value $p = 0.21$ with the same probability as chosen for α in that exercise?

10. Under what conditions would a confidence interval and test of hypothesis about one proportion become equivalent procedures if both used the same value for α?

9.3   TWO-WAY CONTINGENCY TABLES: A LARGE SAMPLE χ^2 TEST

Introduction

One of the more interesting questions to ask in many statistical investigations is about the presence or absence of a correlation between two random variables. For instance, we may want to know if there is a correlation between the processing temperature of canned corn and its shelf life. Is there a correlation between violence among preschoolers and the type of TV programs they watch? Is there reason to expect the "baby boomers" to be more liberal or conservative than the rest of the population?

One problem encountered in the examination of the correlation between two variables is that the variables may be nominal variables. If this is the case the concepts of correlation and covariance encountered in Chapters 1 and 3 do not carry over in a straightforward way. Therefore, we have to develop other ways of treating the concept of association and correlation when dealing with nominal measurements.

In this section we examine one method for addressing the presence of a correlation. First, the data are summarized in a two-way table called a contingency table, or a cross tab. A method of testing the hypothesis of association or no association is presented. The method of analyzing the data in such a 2-way table is fairly simple and easily computerized and is called a test of *independence*.

In addition, an extension of the 2 population test for Bernoulli proportions of section 9.2 to the case of r populations is presented. This same procedure extends the multinomial test of section 9.1 to a test of equlity of r multinomial distributions. These two extensions fall under the general heading of a test of *homogeneity* and result in the same test statistic as the test of independence referred to above.

Methodology

Test of Homogeneity

Background assumptions: Suppose a random sample is obtained independently from r different populations, and the size of each sample is $n_1, n_2, ..., n_r$. From each element sampled, a nominal measurement Y is made where the measurement can take on any of c categories. The data are summarized in an r by c, 2-way table called a contingency table, shown as Table 9.1 below.

Table 9.1 Example of an r by c contingency table-test of homogeneity.

Population	Y 1	2	...	c	Total
1	n_{11}	n_{12}	...	n_{1c}	n_1
2	n_{21}	n_{22}	...	n_{2c}	n_2
.	.	.		.	.
.	.	.		.	.
.	.	.		.	.
r	n_{r1}	n_{r2}	...	n_{rc}	n_r
Total	$n_{.1}$	$n_{.2}$	...	$n_{.c}$	n

Note that $n_1 + n_2 + ... + n_r = n$, the overall sample size and each n_i is a fixed sample size (but not necessarily equal.). Also note that $n_{.j}$ denotes the sum of the jth column in the table.

5 Steps of Test of Hypothesis:

Step 1: Let p_{ij} denote the probability that an observation from population i will fall in category j of the c possible categories associated with the random variable Y. It will be true that

$$\sum_{j=1}^{c} p_{ij} = 1$$

Then, the null and alternative hypotheses are:

H_o: $p_{11} = p_{21} = ... = p_{r1}$, and $p_{12} = p_{22} = ... = p_{r2}$, and. .., $p_{1c} = p_{2c} = ... = p_{rc}$

H_a: At least one of these equalities fails.

(The null hypothesis is equivalent to saying the distribution of probabilities across the c categories of Y is the same for every one of the r populations sampled.)

Step 2: A random sample of fixed size is selected from each of r populations. These sample sizes are denoted by n_1, n_2, ..., and n_r. Next a categorical measurement Y is made on each member of the various samples which results in any of c discrete values. The frequencies of measurements are recorded in an $r \times c$ contingency table.

Step 3: Under the null hypothesis, the common probabilities $p_{11} = p_{21} = ... = p_{r1}$ are estimated by $\frac{n_{1.}}{n}$; $p_{12} = p_{22} = ... = p_{r2}$ by $\frac{n_{.2}}{n}$; and finally $p_{1c} = p_{2c} = ... = p_{rc}$ by $\frac{n_{.c}}{n}$. The test statistic used is

$$\chi^2 = \sum_{i=1}^{r} \sum_{i=1}^{c} \frac{(O_{ij} - E_{ij})^2}{E_{ij}}$$

where O_{ij} is the observed frequency in cell ij, and E_{ij} is the expected frequency in cell ij and is such that

$$O_{ij} = n_{ij} \text{ and } E_{ij} = \frac{n_i n_{.j}}{n}$$

This test statistic is approximately chi-square with $df = (r - 1)(c - 1)$.

Step 4: Compute the value of the test statistic and find p as

$$p = P(\chi^2 \geq \chi_c^2)$$

Step 5: If $p \leq \alpha$ then reject H_o.

Test of Independence

Background assumptions: Suppose a random sample of n observations is selected and bivariate, nominal measurements are made on each observation producing the pair of measurements (X_i, Y_i), $i = 1, ..., n$. Suppose the nominal measurement X has c categories and the measurement Y has r categories which are summarized in Table 9.2 called a two-way table or a *contingency table*.

Table 9.2 Example of an r by c contingency table-test of independence.

X	1	2	...	c	Total
			Y		
1	n_{11}	n_{12}	...	n_{1c}	$n_{1.}$
2	n_{21}	n_{22}	...	n_{2c}	$n_{2.}$
.	.	.		.	.
.	.	.		.	.
.	.	.		.	.
r	n_{r1}	n_{r2}	...	n_{rc}	$n_{r.}$
Total	$n_{.1}$	$n_{.2}$	...	$n_{.c}$	n

The notation in the table is such that the numbers in each of the cells represent the total number of measurement pairs having that particular combination of measurement values. The totals using the "dot" notation are simply the row and column totals respectively where the dot represents the subscript over which the sum made.

5 Steps of Test of Hypothesis:

Step 1:   If p_{ij} denotes the probability of an observation falling in the cell associated with row i and column j, and $p_{i.}$ and $p_{.j}$ denote the marginal probabilities of an observation being classified into row i and column j respectively, then a hypothesis of independence would be:

$$H_o: p_{ij} = p_{i.}p_{j} \quad \text{vs.} \quad H_a: p_{ij} \neq p_{i.}p_{j}$$

that is, the null hypothesis is that measurements are independent, without association. The alternative hypothesis indicates a dependence or association of the random variables X and Y.

Step 2:   A random sample of n bivariate measurements is selected and the frequency of measurements in each cell of the r by c contingency table is recorded.

Step 3:   If the true marginal probabilities are known, a test statistic that is approximately chi-square with $rc - 1$ degrees of freedom would be used. However, the marginal probabilities are generally not known and therefore must be estimated using the maximum likelihood estimates:

$$\hat{p}_{i.} = \frac{n_{i.}}{n} \quad \text{and} \quad \hat{p}_{j} = \frac{n_{.j}}{n}$$

As a result we get a Pearson-type the test statistic

$$\chi^2 = \sum_{i=1}^{r} \sum_{i=1}^{c} \frac{(n_{ij} - \frac{n_{i.}n_{.j}}{n})^2}{\frac{n_{i.}n_{.j}}{n}}$$

or

$$\chi^2 = \sum_{i=1}^{r} \sum_{i=1}^{c} \frac{(O_{ij} - E_{ij})^2}{E_{ij}}$$

where O_{ij} is the observed frequency in cell ij, and E_{ij} is the expected frequency in cell ij and is such that

$$O_{ij} = n_{ij} \qquad \text{and} \qquad E_{ij} = \frac{n_{i.}n_{.j}}{n}$$

The distribution of this test statistic if the sample size is large is approximately chi-square with $df = (rc - 1) - (c - 1) - (r - 1) = (r - 1)(c - 1)$.   [The degrees of freedom expression results from having rc categories and $(c - 1) + (r - 1)$ marginal probabilities to estimate.]

Step 4:   Compute the value of the test statistic and find p as

$$p = P(\chi^2 \geq \chi^2_{cal})$$

Step 5:   If $p \leq \alpha$ then reject H_o.

DISCUSSION

Test of Homogeneity

Suppose we have r Bernoulli populations which are sampled independently. If we want to know whether the probability of a "success" is the same for each of the populations we can summarize the data in an $r \times 2$ contingency table where the r rows denote the r populations and the 2 columns represent "success" and "failure." The hypothesis of equality of populations relative to probability of a success and is called a test of homogeneity.

On the other hand, suppose we have r random variables each of which is a multinomial with c categories or classifications. A test of equality of multinomial distributions results in an $r \times c$ contingency table, each row representing a population and the c columns representing the classifications for the multinomial. This also is called a test of homogeneity. We want to test whether the response to a question is consistent or homogeneous across the r populations.

For example, a study on tobacco use was carried out by sampling a fixed number of residents from each of three different cities. We want to know whether tobacco use is the same from city to city, specifically their use of chewing tobacco also called "smokeless tobacco." This is a test of homogeneity of tobacco use across the three cities. (See Table 9.3.)

Table 9.3 Test of homogeneity of tobacco use across 3 cities.

Question: "Have you used smokeless tobacco in the last 10 days?"

City	Yes	No	Total
Spring Glen	15	41	56
Farm Cove	11	54	65
Hooper	7	52	59
Total	33	147	180

The theoretical table defining the parameters of the 3 populations that would correspond to this case is shown as Table 9.4.

Table 9.4 Contingency table of parameters for 3 cities.

Question: "Have you used smokeless tobacco in the last 10 days?"

City	Yes	No	Total
Spring Glen	p_{11}	p_{12}	1.0
Farm Cove	p_{21}	p_{22}	1.0
Hooper	p_{31}	p_{32}	1.0

The null hypothesis is that the probabilities in the first column are equal and the probabilities in the second column are equal (H_o:$p_{11} = p_{21} = p_{31}$ and $p_{12} = p_{22} = p_{32}$. However, since the rows sum to one, $p_{11} + p_{12} = 1$ and so on, the second part of the null hypothesis, $p_{12} = p_{22} = p_{32}$, is redundant.) Under this hypothesis, an estimate of the common value of $p_{11} = p_{21} = p_{31}$ from Table 9.3 would be $\dfrac{33}{180}$ and the expected number in cell (1,1) would be: $56 \cdot \left(\dfrac{33}{180}\right) = \dfrac{56 \cdot 33}{180}$. "Expected values" can be obtained for the other cells of the table and thus a chi-square test statistic can be formed as:

$$\chi^2 = \sum_{i=1}^{3} \sum_{i=1}^{2} \frac{(O_{ij} - E_{ij})^2}{E_{ij}}$$

And for a general $r \times c$ table the test statistic would be:

$$\chi^2 = \sum_{i=1}^{r} \sum_{i=1}^{c} \frac{(O_{ij} - E_{ij})^2}{E_{ij}}$$

The degrees of freedom for such a table are obtained by the following argument. Each population has $c - 1$ categories or $c - 1$ degrees of freedom, therefore for the r populations there are $r(c - 1)$

categories. From the null hypothesis statements there are $(c-1)$ independent estimates to make since all c parameters are probabilities and must sum to 1. Therefore,

$$df = (\text{\# of categories}) - 1 - (\text{\# of estimates}) = r(c-1) - [(c-1)] = (r-1) \cdot (c-1)$$

The example above can be viewed as an extension of the test of 2 Bernoulli populations to 3 Bernoulli populations. It can also be viewed as a test of equality of 3 multinomial distributions where each multinomial has only 2 categories (and each multinomial is therefore a binomial random variable). In either case, the test statistic and 5-step setup for testing the hypothesis of homogeneity is the same and can be generalized to r populations or random variables each of which has measurements that fit into one of c classifications or categories.

Test of Independence

In a recent study on attitudes and behavior concerning use of both smokeless (chewing tobacco) and non-smokeless tobacco, a wide variety of information was gathered which included several demographic variables, such as age, level of education, sex, marital status, and so forth.   Some interesting questions to be investigated were about the presence or absence of association between variables. Is there a correlation between age and tobacco use? What about level of education; does it have any association with the use of tobacco?

In answering such questions, the data are summarized in a two-way table called a contingency table. This will typically be the case when the data are nominal or ordinal measurements. A typical contingency table format is shown as Table 9.5 below. Here, the row headings indicate the sex of the respondent and the column headings indicate whether the respondent has used smokeless tobacco within the last 10 days. This is called a *2 x 2*, two-way contingency table or a "cross-tabulation" (a "cross-tab" for short).

Table 9.5 Two-way table on sex and use of smokeless tobacco.

Question: "Have you used smokeless tobacco in the last 10 days?"

Sex	Yes	No	Total
Male	15	41	56
Female	7	52	59
Total	22	93	115

The entries in the table are the frequencies associated with a set of bivariate measurements which for this illustration are Bernoulli-type random variables made on each unit in a sample of size n. For instance if X is 0 for Male and 1 for Female, and Y is 0 for Yes and 1 for No to the question, then from a sample of 115 individuals we have 15 bivariate measurements that are (0,0), 41 measurements that are (0,1) and so forth.   In this context, it is reasonable to ask, "Are X and Y correlated?" However, for nominal data, it does not make sense to use the usual computations of covariance as they were introduced in Chapters 1 and 3. Instead, other approaches to the problem have been developed.

In this section, we formally present a method of testing the hypothesis of *no association* vs. *association*.   This method assumes that we have a random sample upon which bivariate measurements X and Y have been made. We formally test whether X and Y are independent random variables. The assumption of independence implies that the probability that

$$P(X = x, Y = y) = P(X = x) \cdot P(Y = y).$$

For the above example this implies that the probability that a randomly selected person will be "male" and respond with a "yes" to the question about smokeless tobacco (the entry in row 1, column 1 of Table 9.5 called cell (1,1)) is the product of the probability that a person is "male" and a person says "Yes." That is,

$$p_{11} = P(\text{"Male \& Yes"}) = P(\text{"Male"}) \cdot P(\text{"Yes"}) = p_{1.}p_{.1}$$

where $p_{1.}$ is the sum of the probabilities for row 1 and $p_{.1}$ is the sum of the probabilities in column 1. This relationship is a direct application of the definition of independent events from Chapter 2 and the ideas of a joint probability distribution as described in Chapter 3, section 3.7. Table 9.6 shows this joint distribution table with its marginal probabilities.

Table 9.6 Joint probability of two Bernoulli random variables X and Y.

		Y		
		0	1	
X	Sex	Yes	No	Total
0	Male	p_{11}	p_{12}	$p_{1.}$
1	Female	p_{21}	p_{22}	$p_{2.}$
	Total	$p_{.1}$	$p_{.2}$	1.0

We propose to use a Pearson-type test statistic and therefore need the "expected frequency" for each cell. For instance, the "expected frequency" of cell (1,1) if n randomly chosen persons are selected from a population of persons, is

$$E_{11} = \text{"Expected Number in Cell [1,1]"} = n \cdot p_{11}$$

If X and Y are independent this expected value is also:

$$E_{11} = n \cdot (p_{1.}p_{.1}) = n \cdot P(\text{"Male"}) \cdot P(\text{"Yes"})$$

However, the parameters $p_{1.}$ and $p_{.1}$ are unknown to us, but a maximum likelihood estimate of them from the observed data would be:

$$\hat{p}_{1.} = \frac{n_{1.}}{n} = \frac{56}{115} \quad \text{and} \quad \hat{p}_{.1} = \frac{n_{.1}}{n} = \frac{22}{115}$$

Therefore, under the hypothesis of "independence," E_{11} would be estimated by

$$E_{11} = n\hat{p}_1.\hat{p}_{.1} = 115(\frac{56}{115})(\frac{22}{115})$$

A similar argument could be made for each of the other 3 cells in the table as well. The following computation produces an approximate chi-square random variable with $(r-1)\bullet(c-1) = (2-1)\bullet(2-1) = 1$ degrees of freedom.

$$\chi^2 = \sum_{i=1}^{r} \sum_{i=1}^{c} \frac{(O_{ij}-E_{ij})^2}{E_{ij}} = \sum_{i=1}^{2} \sum_{i=1}^{2} \frac{(O_{ij}-E_{ij})^2}{E_{ij}}$$

This provides a test statistic with which a test of the hypothesis of independence could be made. If the hypothesis of independence is valid, then $(O_{ij}-E_{ij})$ will be close to zero as the "observed" and "expected" agree with one another throughout the table and the overall chi-square computation will be small relative to the values of a chi-square distribution. Therefore, the rejection region of the test falls in the right-tail of the chi-square distribution.

This argument can be easily extended to a larger table than a 2 by 2 table. The generalization allows us to set up an r by c table where the random variable X can take on any of r categories and the random variable Y can take on any of c categories. Table 9.2 above shows us such a table and the test statistic becomes:

$$\chi^2 = \sum_{i=1}^{r} \sum_{i=1}^{c} \frac{(O_{ij}-E_{ij})^2}{E_{ij}}$$

In this statistic there are $r \bullet c$ classifications with r estimates of the type $\hat{p}_{i.}$ and with c estimates of the type $\hat{p}_{.j}$. However, since the sum of $\hat{p}_{i.}$ equals 1 as does the sum of $\hat{p}_{.j}$, knowing $(r-1)$ of the estimates determines the rth and knowing $(c-1)$ of the estimates determines the cth. Therefore, the degrees of freedom are:

$$df = (\text{\# of categories }) - 1 - (\text{\# of estimates}) = (rc-1) - [(r-1)-(c-1)] = (r-1)\bullet(c-1)$$

A five-step test of hypothesis procedure can be easily formulated for this situation.

Upon examination, the test of homogeneity and the test of independence have identical computations and have test statistics with identical large-sample distributional properties. The difference between them is in the sampling set-up and the resulting formulation of the null hypothesis. In the test of independence, a random sample of n bivariate pairs (X, Y) is selected and a hypothesis about independence of random variables is proposed. In the test of homogeneity, r different populations are sampled with a pre-determined sample size being selected from each of the r populations. A common nominal measurement X is then made on every member of the combined samples. The hypothesis is one of common distributions across the r different populations. Examples of the application of these two procedures follow.

EXAMPLES

EXAMPLE 1: A production process uses five machines in its three-shift operation. A random sample of 164 breakdowns was classified according to the machine that produced it and the shift in which the breakdown occurred. The data are given in Table 9.7 below. Based on this information, is there reason to doubt the independence of shift and machine breakdown? Use $\alpha = 0.05$.

SOLUTION:

Step 1: $H_o: p_{ij} = p_i.p_{.j}$ vs. H_a: The observations are dependent—association is present between shift and breakdowns.

Step 2: A random sample of 164 bivariate measurements is selected and the measurements are classified into a 3 by 5 contingency table given below.

Table 9.7 Cross-tabulation of shift versus type of breakdown.

Shift	A	B	C	D	E	Totals ($n_i.$)
1	10	12	8	14	8	52
2	15	8	13	8	11	55
3	12	9	14	12	10	57
Total ($n_{.j}$)	37	29	35	34	29	164 = n

Step 3: The test statistic is

$$\chi^2 = \sum_{i=1}^{r} \sum_{i=1}^{c} \frac{\left(n_{ij} - \frac{n_i.n_{.j}}{n}\right)^2}{\frac{n_i.n_{.j}}{n}}$$

or

$$\chi^2 = \sum_{i=1}^{r} \sum_{i=1}^{c} \frac{(O_{ij} - E_{ij})^2}{E_{ij}}$$

where $r = 3$ and $c = 5$. The distribution of this test statistic is approximately chi-square with $df = (r - 1)(c - 1) = (3 - 1)(5 - 1) = 8$.

Step 4: The expected value for each of the 8 cells must be computed. After which, the value of the test statistic is:

$$\chi_c^2 = \frac{(10 - 11.73)^2}{11.73} + \frac{(12 - 9.20)^2}{9.20} + \dots + \frac{(10 - 10.08)^2}{10.08} = 5.709.$$

Therefore, $p = P(\chi^2 \geq \chi_c^2 = 5.709) > 0.10$.

Step 5: Since $p > 0.05$, we do not reject H_o; there is insufficient statistical evidence to conclude a relationship exists between the number of breakdowns and the work shift.

EXAMPLE 2: In a study of drug abuse in a local high school, the school counselor selected 100 sophomores, 100 juniors and 100 seniors randomly from the respective rolls for each grade. Each student was then asked if they used a particular drug frequently, seldom or never. The data are summarized in the table given below. Is there evidence to suggest that the frequency of drug use is the same across the three different grades?

(Population) Grade	Frequency of Drug Use			Total
	Frequently	Seldom	Never	
Sophomore	15	30	55	100
Junior	20	35	45	100
Senior	25	35	40	100
Total	60	100	140	300

SOLUTION: This is a test of homogeneity since a fixed sample of size 100 is selected from each of three different populations. The measurement, frequency of drug use, has 3 possible categories. Therefore, a 3 by 3 contingency table is constructed to summarize the data. The test is summarized by the following steps.

Step 1: Let p_{ij} denote the probability of an observation from grade level i will fall in category j of the 3 possible categories associated with frequency of drug use. Then, the null and alternative hypotheses are:

$$H_o: p_{11} = p_{21} = p_{31}, \ p_{21} = p_{22} = p_{32}, \text{ and } p_{13} = p_{23} = p_{33}.$$

$$H_a: \text{At least one of these equalities fail.}$$

(The null hypothesis is equivalent to saying the distribution of probabilities across the 3 categories of drug use is the same for every one of the 3 grade levels (populations) sampled.)

Step 2: A random sample of 100 is selected from each of 3 grade level populations.

Step 3: Under the null hypothesis, the common probabilities $p_{11} = p_{21} = p_{31}$ are estimated by

$\frac{n_{.1}}{n}$; $p_{21} = p_{22} = p_{32}$ by $\frac{n_{.2}}{n}$; and $p_{13} = p_{23} = p_{33}$ by $\frac{n_{.3}}{n}$. The test statistic used is

$$\chi^2 = \sum_{i=1}^{r} \sum_{i=1}^{c} \frac{(O_{ij} - E_{ij})^2}{E_{ij}}$$

where O_{ij} is the observed frequency in cell ij, and E_{ij} is the expected frequency in cell ij and is such that $O_{ij} = n_{ij}$ and $E_{ij} = n_i(\frac{n_{.j}}{n})$. The distribution of this test statistic is approximately chi-square with $df = (3 - 1)(3 - 1) = 4$.

Step 4: The computed value of the test statistic is

$$\chi^2 = \frac{(15 - 20)^2}{20} + \frac{(30 - 33.33)^2}{33.33} + \frac{(55 - 46.67)^2}{46.67} + \dots + \frac{(40 - 46.67)^2}{46.67} = 5.5$$

which has an approximate chi-square distribution with 4 degrees of freedom. The p-value is

$$p = P(\chi^2 \geq 5.5) > 0.10$$

Step 5:   On the basis of this p-value we would not reject H_0. Thus, we conclude there is no important difference between the three grade levels relative to frequency of use of drugs.

EXERCISES

1. A publisher is trying to determine the best of three cover designs for a paperback book to be placed on the market. Four hundred people in California, Indiana and Massachusetts are interviewed and asked which of the three covers they preferred. The data are reported below. Does there seem to be regional differences in preference between the three covers?

State	First Cover	Second Cover	Third Cover
California	81	78	241
Indiana	60	93	247
Massachusetts	182	95	123

2. Describe carefully whether the test of hypothesis of exercise 1 is a test of homogeneity or a test of independence.

3. Preference for a new disposable diaper to be marketed is tested in two separate test markets. The results of the test are given below. Does there seem to be a difference in preference between the two markets?

Market	Prefer Diaper	Don't prefer Diaper
1	105	195
2	140	160

4. Three hundred people were questioned about their political preference and their opinions on nuclear disarmament. Does there appear to be an association between the two?

Political Party	Don't Favor Disarming	Favor Disarming
Republican	50	86
Democrat	94	70

5. Describe carefully whether the test of hypothesis of exercise 4 is a test of homogeneity or a test of independence.

6. A large university used high school GPA and ACT test scores to determine admissions to the university. However, a sociologist proposed that living conditions were a stronger determinant of success in school. The sociologist selected 200 freshmen at random and assigned a rank to the the roommate of the student from 1 to 5 where 1 indicates a roommate who was difficult to live with and was not serious about school and a 5 was a roommate who was easy to live with and encouraged study and scholarship. The GPA of the selected student was also recorded and the following 5 by 4 contingency table resulted. Test the hypothesis of no association between roommate and GPA with $\alpha = 0.05$.

Rank of Roommate	Grade Point Average (GPA)				
	Under 2.00	2.00-2.69	2.70-3.19	3.20-4.00	Totals
1	8	9	10	4	31
2	5	11	15	11	42
3	6	7	20	14	47
4	3	5	22	23	53
5	1	3	11	12	27
Totals	23	35	78	64	200

7. Notice in the data of exercise 6 that the roommate rank is ordinal measure as is the grade point interval classification. If there were a strong "linear" relationship between rank and GPA, what pattern would you expect to see in the table? Is that pattern present in the table of exercise 6?

8. A certain brand of children's chewable vitamins comes in five animal shapes and five fruit flavors. The combinations of shapes and flavors are supposed to be approximately equal and the same frequency is expected to be found in each bottle. A 100-tablet bottle contained the distribution found in the following table. Does this table support the null hypothesis of independence of shape and flavor? Use $\alpha = 0.05$.

Flavor	Shape					
	Hawk	Squirrel	Turtle	Penguin	Dog	Totals
Grape	5	4	3	10	2	24
Cherry	3	3	7	3	2	18
Orange	2	4	3	1	3	13
Apple	5	2	10	4	4	25
Lemon	5	6	3	5	1	20
Totals	20	19	26	23	12	100

9. Rework exercise 8 using the following argument and a goodness-of-fit test. There are $5 \cdot 5 = 25$ combinations of shapes and flavors. Therefore, if each of the twenty-five combinations are to have equal frequency, then in 100 tablets, each flavor-shape combination would be expected to have 4 tablets. Use $\alpha = 0.05$.

10. Comment on the "small expected" frequency problem in exercise 9 and make a recommendation to correct for it.

CHAPTER SUMMARY

In this chapter we have presented several large sample approximate tests when dealing with nominal data. For the most part, the original populations can be modeled by a Bernoulli distribution, or its extension, the multinomial distribution of which the binomial is a special case. Some of the results are summarized in Table 9.8 which follows.

Table 9.8 Summary of the null hypothesis and the test statistic for cases of Chapter 9.

Type of Population	Null Hypothesis	Test Statistic
Multinomial	$F(x) = F_o(x)$ or $p_1 = p_{1o}$, $p_2 = p_{2o}$, ... $p_k = p_{ko}$	$$\chi^2 = \sum_{i=1}^{k} \frac{(n_i - np_i)^2}{np_i}$$ $df = k - 1$
One Bernoulli Population	$p = p_o$	$$Z = \frac{(\hat{p} - p)}{\sqrt{\dfrac{p_o(1 - p_o)}{n}}}$$
Two Bernoulli Populations	$p_x = p_y$	$$Z = \frac{(\hat{p}_x - \hat{p}_y) - 0}{\sqrt{\hat{p}_o(1 - \hat{p}_o)(\dfrac{1}{n} + \dfrac{1}{m})}}$$
Test of Homogeneity: r Multinomials	$p_{11} = p_{21} = ... = p_{r1}$, and $p_{12} = p_{22} = ... =$ p_{r2}, and ..., $p_{1c} = p_{2c}$ $= ... = p_{rc}$	$$\chi^2 = \sum_{i=1}^{r} \sum_{i=1}^{c} \frac{(O_{ij} - E_{ij})^2}{E_{ij}}$$ where O_{ij} is the observed frequency in cell ij, E_{ij} is the expected frequency in cell ij and is such that $O_{ij} = n_{ij}$ and $E_{ij} = \dfrac{n_i.n_j}{n}$ $df = (r - 1)(c - 1)$
Test of Independence: One Sample of n Pairs of Nominal Random Variables X and Y	$p_{ij} = p_i.p_j$	$$\chi^2 = \sum_{i=1}^{r} \sum_{i=1}^{c} \frac{(n_{ij} - \dfrac{n_i.n_j}{n})^2}{\dfrac{n_i.n_j}{n}}$$ See notation and degrees of freedom under *Test of Homogeneity*

CHAPTER EXERCISES

1. Consider the following three 3 by 3 contingency tables. Compute the chi-square test statistic for each table along with the corresponding *p*-values. Explain the implication of the presence or absence of association as may be the case in the three tables.

Table 1
Education Level

		High School	College	Post Graduate	Total
	Disagree	30	0	0	30
Opinion	Neutral	0	30	0	30
	Agree	0	0	30	30
	Total	30	30	30	90

Table 2
Education Level

		High School	College	Post Graduate	Total
	Disagree	20	10	0	30
Opinion	Neutral	10	10	10	30
	Agree	0	10	20	30
	Total	30	30	30	90

Table 3
Education Level

		High School	College	Post Graduate	Total
	Disagree	10	10	10	30
Opinion	Neutral	10	10	10	30
	Agree	10	10	10	30
	Total	30	30	30	90

2. Examine Table 1 of exercise 1 and indicate what opinion you would predict a person with a high school education to have. Next, do the same for Table 3. Relate this ability to "predict" in Table 1 and your uncertainty to be able to predict in Table 3 to the conclusion as to whether there is an association between the two variables as found in exercise 1.

3. Explain why it does not make sense to report a "correlation" between −1 and +1 for nominal data.

4. A measure of correlation for nominal data is often simply scaled between 0 and 1 where 0 represents "no predictability" (see Table 3 of exercise 1) and a 1 indicates a "perfect predictability" (see Table 1 of exercise 1). A common measure of correlation proposed is Cramer's contingency coefficient defined by

$$C = \frac{\chi^2}{n(q-1)}$$

where χ^2 is the computed chi-square value, n is the total number of observations, and q is the smaller of r or c, the number of rows and columns in the contingency table. Compute this contingency coefficient for the three tables of exercise 1 and note whether it seems consistent with your claims for association or no association as computed in exercise 1.

5. A publisher is trying to determine the best of three cover designs for a paperback book to be placed on the market. Four hundred people in California, Indiana and Massachusetts are interviewed and asked which of the three covers they preferred. The data are reported below. Do the covers seem to be equally preferred or not. If not, which cover would you recommend? Would this recommendation be the same for each region of the country as represented by the various states?

State	First Cover	Second Cover	Third Cover
California	81	78	241
Indiana	60	93	247
Massachusetts	182	95	123

6. In section 9.1 the following problem was described. "Psychologists have noted a tendency that mothers who have immediate access to their babies after birth, prefer to hold their babies on the left side, while those without immediate access prefer to hold their babies on the right side. (Speculation on this preference centers around the mothers heartbeat which is more pronounced on the mother's left side.) To substantiate this perception, the following data were obtained. Do the data support this contention."

	Type of Contact	
Mother's Preference	Immediately after birth	No early contact
Left side	75	40
Right side	25	60
Total	100	100

Analyze this data using both a two-sample test of proportions and also as a 2-way contingency table. Recall, that $Z^2 = \chi^2$ with 1 degree of freedom. Check to see if this connection holds for the two analyses.

7. The number of defective items found each day by the inspectors in a manufacturing plant is recorded.

Number of Defectives per Day	Times Observed
0-10	6
11-15	11
16-20	16
21-25	28
26-30	22
31-35	19
36-40	11
41-45	4

Is it reasonable to conclude that these data come from a normal distribution?

8. The following data were generated from a normal distribution with a mean of 50 and a standard deviation of 10. Perform a goodness-of-fit test using $\alpha = 0.10$. Create intervals 10 units wide going from 20-30, 30-40, etc.

50.14	40.65	53.70	40.24	40.38	46.73	49.90	69.73	53.45	50.15
70.07	64.35	70.77	61.56	52.53	43.72	29.90	36.38	50.99	41.23
52.92	65.08	44.98	61.67	53.91	55.81	39.02	54.52	63.82	62.25
42.57	50.54	51.75	45.83	49.70	69.02	53.22	51.62	37.56	52.79
44.54	43.88	59.66	47.25	61.19	46.26	64.06	42.53	51.30	28.56
53.26	29.77	50.15	62.80	41.48	68.44	57.07	43.80	44.06	69.06
58.44	75.57	47.88	48.92	65.24	64.92	42.58	41.42	46.83	68.29
44.46	57.17	49.28	53.40	48.43	72.31	50.49	62.21	47.53	38.23
42.92	44.00	47.85	58.02	47.96	63.19	42.31	44.34	59.68	35.87
30.27	58.06	43.10	59.59	52.83	58.84	40.05	53.03	65.75	45.84

9. Repeat the test of exercise 8; however, use the estimated mean and standard deviation from the data and write the corresponding hypothesis. Compare the chi-square obtained, or the p-value with that of exercise 8.

10. Repeat the test of exercise 8 and 9; however, estimate the mean and the standard deviation from the "grouped" data. That is, let x_i denote the midpoint of class interval i and f_i the number (frequency) of observations in group i, then compute

$$\bar{x} = \frac{\sum\limits_{i=1}^{k} f_i x_i}{100} \quad \text{and} \quad s^2 = \frac{\sum\limits_{i=1}^{k} f_i(x_i - \bar{x})^2}{\sum\limits_{i=1}^{k} f_i}$$

Write the corresponding hypothesis for this case. Compare the chi-square obtained, or the p-value with that of exercise 8 and exercise 9. (Note: This procedure is recommended to increase the power of the test.)

10

OTHER APPLICATIONS

OBJECTIVES

When you complete this chapter you should be able to analyze data and test hypotheses for data coming from more than two populations. The computations of analysis of variance on data from a completely randomized design and a randomized block design are presented. You should be able to use some of the basic terminology of designing experiments and the analysis of variance.

In addition you should be able to apply a simple sign test, signed rank test, the Mann-Whitney test and the Kruskall-Wallis test to data that fail the assumption of normality or that are ordinal data at the most.

Lastly, you should be able to obtain a simple linear regression equation for data that satisfy the assumptions and know how to set up the computations for a more general linear regression model.

INTRODUCTION

As you look back on the topics we have covered in this text you can note that the overall driving force was associated with making inferences about a population based on a sample from the population. Our purpose in the preceding chapters has been to develop a theoretical base for making sound inferences by developing a model that explains the variability that seems to be inherent in the data. Ideally, we would like to explain away all of the variability, but realistically we never achieve that goal. However, the smaller we can make the "unexplained variability" the more meaningful will be the information we have about what is going on.

For instance, quality control inspectors may observe variation in the smoothness of a finish on metal filing cabinets. Some of that variation may be attributed to differences in the baking temperature when the finish was applied, to differences in the thickness of the finish coat, or to differences in the sealing coat after the finish was "baked" on. Corrections and adjustments to those factors may eliminate much of the problem. However, even when these factors have been taken into account, there may still be variation that remains. It may be possible to find other explanations that would further reduce the amount of "unexplained" variation, but a point will be reached in practice where we simply have to attribute the remaining variation to "chance."

We then want to develop a mathematical/statistical model that links the measure of smoothness of finish to the "explanatory" variables upon which it depends. Once this model is developed we can test to see whether there is really a relationship of one or all of the variables to our "smoothness" measurement; we may try to determine which factors have the strongest relationship, and so forth.

To perform this kind of analysis, we need to collect the data carefully so that the variables under consideration can be investigated. Therefore, we need to carefully design an experiment that will allow measurements of correlation and patterns to be observed. In setting up the experiment we want to minimize the error or unexplained variability as much as possible so that it doesn't cloud the information that is being sought. The unexplained variability can be thought of as "noise" or "static" that, if loud enough, will completely obscure the signal (information) that may be present.

In this chapter we will discuss some of the terminology and techniques of "experimental design" along with some methods of analyzing data for a few simple designs that are in common use. If the data can be assumed to be normally distributed, then methods of analysis that fall under the heading of "analysis of variance" can be applied. If the data are not normally distributed or if the data are only ordinal data, then distribution-free methods can be applied. We discuss the methods of analysis for such situations in sections 10.1 through 10.6.

In section 10.7 we introduce the concept of least squares linear regression. This topic addresses a similar set of objectives but takes the approach of trying to find an equation that would allow us to predict a response when certain values of other related variables may be set or observed.

In all these sections you need to remember that only the "tip of the iceberg" is being examined—just enough for you to have some familiarity with the language and terminology and with the type of problem being addressed. To become really knowledgeable about these topics requires extensive study to understand all of the potential that the techniques represent. There are numerous good references you can find in the library for each of the main areas that are only touched on in the chapter—design and analysis of variance, nonparametric methods, and linear regression methods.

10.1 EXPERIMENTAL DESIGN, THE ONE- FACTOR EXPERIMENT, AND THE RANDOMIZED BLOCK DESIGN

Introduction

In this section some of the vocabulary of experimental design is presented along with its objectives. Next a completely randomized design or experiment is described along with a method of analyzing the data that come from such a design. The method of analysis is commonly called an analysis of variance (ANOVA). Formulas for computation are provided and a convenient summary table called an ANOVA table is given.

Definitions

DEFINITION 10.1: The measurement of primary interest in an investigation is the *dependent variable*. Other variables which we believe to affect the measurements obtained on the dependent variable are called *independent* or *explanatory* or *predictor variables*. In this context we say that the dependent variable is explained by the independent variables.

DEFINITION 10.2: In an experiment, each of the explanatory variables can be used as a *factor* in the experiment. The different values or settings of the factor are called the *factor levels*.

DEFINITION 10.3: A *treatment* or *treatment combination* is represented by a single level of a factor or a combination of factor levels being used simultaneously in conducting the experiment.

DEFINITION 10.4: An *experimental unit* is the basic element to which a given treatment or treatment combination is applied and from which a measurement on the dependent variable is obtained.

DEFINITION 10.5: The difference in measurements for the dependent variable between two identical experiment units receiving the same treatment or treatment combination is called *experimental error*. If two or more identical experimental units are subjected to the same treatment, the experiment is said to be *replicated*.

DEFINITION 10.6: A o*ne-factor experiment* or *completely randomized design* is an experiment where a set of *n* experimental units are assigned randomly to one of *k* treatments or levels of a factor in an experimental setting. If the number assigned to each treatment is the same we say the design is balanced. If it differs from one treatment to another we say the design is unbalanced.

DEFINITION 10.7: A *randomized block design* describes an experiment wherein a set of *k* similar experimental units, called a block, are randomly assigned to *k* different levels of a factor, or to *k* different treatments. This assignment of units to treatments is repeated for a set of *r* different blocks. It is assumed that the *k* experimental units in each block are very similar or "homogeneous" but the blocks may differ from block to block (heterogeneity of blocks).

DISCUSSION

The Design of Experiments

The objective of experimental design is to design the way an experiment is done so that maximum information is obtained at a minimum of cost. This objective is so important that many texts have been written on the subject of experimental design. We can only touch the surface of the topic in this text.

We have already encountered one simple application of "experimental design philosophy." In Chapter 8, the two independent sample t-test was presented to test the hypothesis that the means of two different populations were the same or different in some way. Under an assumption of normally distributed observations and common variance for the two populations a test statistic that followed a *t* distribution was formulated.

However, in the next section of Chapter 8 it was pointed out that a more "powerful" procedure for making such a test could be possible if a "pairing" of units could be performed. It was important that one unit of each pair belonged to one population and the other unit of the pair belonged to the other. For instance we may be interested whether an adjustment on the lathe speed will make an improvement in a product that requires turning on the lathe. If there are 10 different lathes used in manufacturing the product we could conduct an experiment where 5 of the machines are randomly selected to produce products with the normal lathe speed being used. The other 5 machines are adjusted to the proposed speed. However, it is possible in this case to have all 10 lathes produce products at the normal speed and then each is adjusted to the new speed and a set of products are produced. The first situation would be analogous to a "two-independent sample test" whereas the latter would correspond to a pair-comparison test.

In the two-independent sample test, if a difference in the products produced by the two groups is found, it could be due to the different lathe speeds or it could be a difference in the two groups of machines that were selected. In the paired comparison test, if a difference in the products before and after adjustment were observed, it would be most likely attributed to the speed. There is no "group" factor to cloud the picture.

From a more theoretical point-of-view, the variability in the two-independent sample test is measured by $V(\overline{X} - \overline{Y}) = V(\overline{X}) + V(\overline{Y})$ because of independence of samples. In the paired-comparison case the variability is measured by $V(\overline{X} - \overline{Y}) = V(\overline{X}) + V(\overline{Y}) - 2\text{Cov}(\overline{X},\overline{Y})$ which will be smaller if the covariance between X and Y in each pair is positive; i.e., $V(\overline{X}) + V(\overline{Y}) - 2\text{Cov}(\overline{X},\overline{Y}) \leq V(\overline{X}) + V(\overline{Y})$ if $\text{Cov}(\overline{X},\overline{Y}) \geq 0$. Thus, we hope to reduce variability in a paired-comparison experiment over a two-independent sample test by wisely pairing to create pairs that are positively correlated. We can then take advantage of that pairing to devise a test statistic that has a smaller variance, and thus a more powerful test than we might otherwise have.

In this section, we examine logical extensions of the two-independent sample t test called a completely randomized design or one-factor experiment, and the extension of the paired comparison t test called the randomized block experiment. These two designs represent only a brief introduction to the body of knowledge and methods of analysis that fall under the heading of *design of experiments* and *analysis of variance*.

The One-Factor Experiment or Completely Randomized Design

In some experimental situations we desire to know whether a set of k different treatments are the same or different. For instance, we may want to know whether a particular type of plastic pipe shows different brittleness characteristics when subjected to temperatures between 0 and 20 degrees Fahrenheit. We may choose to randomly assign 10 pieces of pipe from a total of 30 pieces of pipe to each of three temperatures levels—0, 10 and 20 degrees—for a fixed length of time. The brittleness measure for the three groups of 10 pipes is obtained and an analysis is needed to help decide whether there is a difference in brittleness for the three different temperatures. This is an example of a one-factor experiment, the factor being temperature. This is also referred to as a completely randomized design and can be thought of as taking independent random samples from each of k (in this case 3) populations. The development of a method of analysis with a resulting test statistic is argued in the following paragraphs.

First, suppose we have k normally distributed population each with an unknown mean μ_i but common variance σ^2 (that is, the variance is assumed to be the same for each of the k populations). A random sample of n observations is selected from each population and the sample means are computed, resulting in k means, $\overline{X}_1, \overline{X}_2, ..., \overline{X}_k$. While the individual observations are $N(\mu_i,\sigma^2)$, the k sample means will be independent and normally distributed with mean μ_i and variance σ^2/n, i.e., each $\overline{X}_i$ is $N(\mu_i,\sigma^2/n))$. These results and a typical notation are summarized in Table 10.1 that follows.

Table 10.1 Notation for individual sample values and summary statistics for k independent samples from the same population.

Observation	Population 1	Population 2	...	Population k	Overall
1	X_{11}	X_{12}	...	X_{1k}	
2	X_{21}	X_{22}	...	X_{2k}	
3	X_{31}	X_{32}	...	X_{3k}	
...	...	...	...	...	
n	X_{n1}	X_{n2}	...	X_{nk}	
Mean	$\bar{X}_1$	$\bar{X}_2$	...	$\bar{X}_k$	$\bar{\bar{X}}$

Now, consider one of two alternatives: either the k populations have the same mean such that $\mu_1 = \mu_2 = \ldots = \mu_k$, or at least one of the means is different. We want to develop a test statistic that will test this hypothesis: H_o: $\mu_1 = \mu_2 = \ldots = \mu_k = \mu$, with μ unspecified vs. H_a: At least one mean is unequal to μ.

It is obvious that if the null hypothesis is true, then the sample means $\bar{X}_1, \bar{X}_2, \ldots, \bar{X}_k$ would be very similar in numerical value. If the alternative is true, at least one of the means would be expected to differ from the others. We want a test that would reflect which of these two situations were true. In the presentation here the objective is to present an intuitive and logical test statistic without getting too bogged down with formulas and notation.

First, note that the variance of the means, computed by the formula

$$s_{\bar{x}}^2 = \frac{\displaystyle\sum_{i=1}^{k} (\bar{X}_i - \bar{\bar{X}})^2}{k-1}$$

is an unbiased estimator of $\dfrac{\sigma^2}{n}$ if the null hypothesis is true. Therefore, $ns_{\bar{x}}^2$ will be an unbiased estimator of σ^2. However, if the alternative is true, it can be shown that $ns_{\bar{x}}^2$ will *overestimate* σ^2, and thus is expected to be larger than it would be if the null hypothesis were true. (Obviously, the sample means $\bar{X}_1, \bar{X}_2, \ldots, \bar{X}_k$ will show more variability if the population means are different than they would if the population means were all equal.)

Next, the variance of each sample

$$s_i^2 = \frac{\displaystyle\sum_{i=1}^{n}(X_{ij} - \bar{X}_i)^2}{n-1}$$

will be an unbiased estimator of σ^2. Therefore, the mean of these k sample variances will also be an unbiased estimator of σ^2.   That is, $E\left[\frac{1}{k}\sum_{i=1}^{k}s_i^2\right]=\sigma^2$. Then, we claim (without proof) that both $s_{\bar{x}}^2$ and $\dfrac{1}{k}\displaystyle\sum_{i=1}^{k}s_i^2$ are independent, and after an appropriate transformation, are chi-square random variables. Therefore, by theorem 6.10, their ratio will follow the F distribution under the null hypothesis. Specifically, we form the ratio

$$\frac{s_{\bar{x}}^2}{\dfrac{1}{k}\displaystyle\sum_{i=1}^{k} s_i^2}$$

which will be an F-ratio if the null hypothesis is true and will be larger than an F random variable if the null hypothesis is false. Therefore, we have a "test statistic" that we can use to test the null hypothesis versus the alternative hypothesis. Large values for the ratio (compared to values from the F-distribution) indicates that the null hypothesis is false and should be rejected in favor of the altnernative hypothesis that at least one of the means of the k populations is not equal to the others.

Before concluding this discussion we have yet to deal with several important issues: (1) How do we adjust the formulas to handle different sample sizes selected from each of the k populations? (2) Is there a simpler computational algorithm to obtain the above F-ratio? (3) What are the appropriate degrees of freedom to use for the F-ratio that we obtain? We will address these questions in order.

To adapt the equations to the case of unequal sample sizes for each of the k independent samples suppose, for instance, we have selected random samples of size $n_1, n_2,\ldots, n_k$ from each of the k populations with $n = n_1 + n_2 + \ldots + n_k$. The distributional arguments are the same as those used above, with a simple revision of the formulas being required. Each population is $N(\mu_i, \sigma^2)$ $i = 1, 2, \ldots,$ k and we denote the k sample means by $\bar{X}_1, \bar{X}_2, \ldots, \bar{X}_k$.

Next suppose we compute the overall sum of squares about the grand mean of all $n = n_1 + n_2 + \ldots + n_k$. individual observations. We can write this and its expansion as:

$$\sum_{i=1}^{k}\sum_{j=1}^{n_i}(X_{ij}-\bar{\bar{X}})^2 = \sum_{i=1}^{k}\sum_{j=1}^{n_i}(X_{ij}-\bar{X}_i+\bar{X}_i-\bar{\bar{X}})^2$$

$$= \sum_{i=1}^{k}\sum_{j=1}^{n_i}(X_{ij}-\bar{X}_i)^2 + \sum_{i=1}^{k}\sum_{j=1}^{n_i}(\bar{X}_i-\bar{\bar{X}})^2 + 2\sum_{i=1}^{k}\sum_{j=1}^{n_i}(X_{ij}-\bar{X}_i)(\bar{X}_i-\bar{\bar{X}})$$

The last term of the above expression—the cross-product term—can be shown to sum to zero. Also, the term preceding the cross-product term can be simplified so that we get

$$\sum_{i=1}^{k}\sum_{j=1}^{n_i}(X_{ij}-\bar{\bar{X}})^2 = \sum_{i=1}^{k}\sum_{j=1}^{n_i}(X_{ij}-\bar{X}_i)^2 + \sum_{i=1}^{k}n_i(\bar{X}_i-\bar{\bar{X}})^2$$

or, in *analysis of variance* terminology (letting SS denote "Sum of Squares")

$$SS(Total) = SS(Within\ Populations) + SS(Between\ Populations)$$

$$SS(Total) = SS(Error) + SS(Treatments) = SS(E) + SS(T)$$

since the different populations are usually associated with a different "treatment" being applied to each "population" and the variation observed among observations receiving the same treatment is called "error."

Using results from section 6.3 of and specifically an argument based on Theorem 6.5 we claim that under the null hypothesis the three terms above, properly transformed, are chi-square random variables where the term SS(Total) has $n-1$ degrees of freedom, SS(T) has $k-1$ degrees of freedom and SS(E) has $n-k$ degrees of freedom. [For computational convenience, it is usually easier to compute SS(Total) and SS(T) and then find SS(E) by taking the difference: SS(E) = SS(Total) − SS(T).]

If we let T_i denote the sum of all of the observations associated with treatment i and GT (the grand total) denote the sum of *all* the observations, then these formulas further simplify as shown in the following box.

Analysis of Variance Formulas—One Factor Experiment

$$GT = \sum_{i=1}^{k} \sum_{j=1}^{n_i} X_{ij} \qquad\qquad \text{Grand Total}$$

$$SS(T) = \sum_{i=1}^{k} \frac{T_i^2}{n_i} - \frac{(GT)^2}{n} \qquad\qquad \text{Sums of Squares for Treatments}$$

$$SS(Total) = \sum_{i=1}^{k} \sum_{j=1}^{n_i} X_{ij}^2 - \frac{(GT)^2}{n} \qquad\qquad \text{Total Sums of Squares}$$

$$SS(E) = SS(Total) - SS(T) \qquad\qquad \text{Sums of Squares for Error}$$

These computations, with their degrees of freedom and the appropriate F-ratio are summarized in a table called an Analysis of Variance Table or ANOVA Table, given as Table 10.2

Table 10.2 Typical layout of an ANOVA table.

Source	Sum of Squares	Degrees of Freedom	Mean Square (MS)	F-Ratio
Treatment	SS(T)	$k-1$	$MS(T) = \dfrac{SS(T)}{k-1}$	$\dfrac{MS(T)}{MS(E)}$
Error	SS(E)	$n-k$	$MS(E) = \dfrac{SS(E)}{n-k}$	
Total	SS(Total)	$n-1$		

The computations and interpretation of an analysis of variance table are best illustrated in applying them to an example.

EXAMPLE 1: An experiment was conducted to compare the wearing qualities of three types of paint when subjected to the abrasive action of a slowly rotating cloth-surfaced wheel. Ten paint specimens were tested for each paint type and the number of hours until visible abrasion was apparent was recorded. Using the data in Table 10.3 below is there evidence to indicate a difference in the three paint types?

Table 10.3 Abrasion resistant measures for three types of paint.

	Paint Type		
	1	2	3
	148	513	335
	76	264	643
	393	433	216
	520	94	536
	236	535	128
	134	327	723
	55	214	258
	166	135	380
	415	280	594
	153	304	465
Totals, T_i	2296	3099	4278

SOLUTION: To illustrate the computations we get:

$$GT = 2296 + 3099 + 4278 = 9673$$

$$SS(Total) = \sum_{i=1}^{k} \sum_{j=1}^{n_i} X_{ij}^2 - \frac{(GT)^2}{n}$$

$$= 148^2 + 76^2 + 393^2 + \ldots + 380^2 + 594^2 + 465^2 - \frac{9673^2}{30} = 9\,69,443$$

which has $30 - 1 = 29$ degrees of freedom, and

$$SS(T) = \sum_{i=1}^{k} \frac{T_i^2}{n_i} - \frac{(GT)^2}{n} = \frac{2296^2}{10} + \frac{3099^2}{10} + \frac{4278^2}{10} - \frac{9673^2}{30} = 198,772$$

which has $3 - 1 = 2$ degrees of freedom. Therefore,

$$SS(E) = 969,443 - 198,772 = 770,671$$

which has $29 - 2 = 27$ degrees of freedom. The numbers are summarized in the analysis of variance table below, Table 10.4.

Table 10.4 Analysis of variance table for paint data.

Source	df	Sum of Squares	Mean Square	F-ratio	p-Value
Type	2	198772	99386.2	3.48	0.0452
Error	27	770671	28543.4		
Total	29	969443			

From this table, we can see that the F ratio of 3.48 has a p-value of 0.0452. Therefore, we would reject the null hypothesis and conclude that the means of the abrasion measurements for the three paints are not the same. A boxplot of the data shows what seems to be going on. This boxplot is shown in Figure 10.1 that follows.

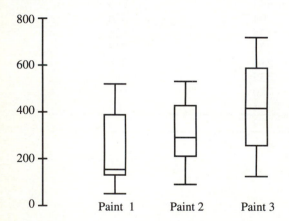

Figure 10.1 Boxplot of paint abrasion measures.

From the boxplot you can see that the medians show an upward shift from Paint 1 to Paint 3. The analysis of variance indicates at least one of the means is different from the other. Therefore, we might be tempted to conclude from the boxplot that all three means are different from each other. (An objective procedure to resolve this question–called a multiple comparison test—is reserved for a lengthier presentation of the subject of analysis of variance and is not addressed in this book.)

EXAMPLE 2: A company psychologist was concerned about reducing the stress level of the workers in an assembly line. Three different methods for helping workers deal with the stress were under investigation. A standard test (HST) was designed to measure the degree of stress workers had with a high score on the test indicating high stress. Eleven workers who scored high on the test were used in an experiment. Four were selected at random and treated with Method 1. Four were assigned to Method 2 and the remaining three were assigned to Method 3. All treatments were continued for a quarter of the work year with each worker taking the HST test again at the conclusion of the quarter. Perform an analysis of variance to determine if there is a difference in the three methods. The data are reported in Table 10.5 below.

Table 10.5 HST test scores for measuring stress.

	Method		
	1	2	3
	80	70	63
	92	81	76
	87	78	70
	83	74	
Totals, T_i	342	303	209

SOLUTION: In this example, $n_1 = 4$, $n_2 = 4$, and $n_3 = 3$. To illustrate the computations we get:

GT $= 342 + 303 + 209 = 854$

$$SS(Total) = \sum_{i=1}^{k} \sum_{j=1}^{n_i} X_{ij}^2 - \frac{(GT)^2}{n} = 80^2 + 70^2 + 63^2 + \ldots + 83^2 + 74^2 - \frac{854^2}{11} = 686.55$$

which has $11 - 1 = 10$ degrees of freedom, and

$$SS(T) = \sum_{i=1}^{k} \frac{T_i^2}{n_i} - \frac{(GT)^2}{n} = \frac{342^2}{4} + \frac{303^2}{4} + \frac{209^2}{3} - \frac{854^2}{11} = 452.13$$

which has $3 - 1 = 2$ degrees of freedom. Therefore, $SS(E) = 686.55 - (452.13) = 234.42$ which has $10 - 2 = 8$ degrees of freedom. The numbers are summarized in the analysis of variance table below, Table 10.6

Table 10.6 Analysis of variance table for HST stress scores.

Source	df	Sum of Squares	Mean Square	F-ratio	p-Value
Method	2	452.13	226.07	7.72	0.0136
Error	8	234.42	29.30		
Total	10	686.55			

Therefore, we reject the equality of mean scores for the three methods and conclude that the methods are different.

The Randomized Block Design

The one-factor experiment and its analysis using the "analysis of variance" is an extension of the two-independent sample t test of Chapter 8.   As you recall the methods of testing hypotheses of Chapter 8, remember the "paired comparison" test of that chapter. As it was described there, a set of n bivariate pairs (X_i, Y_i) was selected, the differences $X_i - Y_i$ were computed and a t test was performed to see whether there was a difference in the mean of the X's compared to the Y's; i.e., H_o: $\mu_x - \mu_y = 0$. We argued that if there was a positive correlation between X and Y, we could take advantage of this fact and come up with a test procedure that was more powerful than the ordinary t test of two independent samples. The randomized block design is simply an extension of the paired comparison test to the case of k treatments with blocks serving as the "pairing" variable.

For example, suppose we want to compare 4 different brands of cake mixes for moisture content after baking for 25 minutes at 350° Fahrenheit. To conduct this experiment each of the four cake mixes are prepared and poured into a different 9x9 cake pan. These four cake mixes are then are placed in an oven preheated to 350°. After 25 minutes the four cakes are removed and the moisture content at the center of each cake is measured and reported. This same procedure is then repeated four more times for a total of five replicates of the basic experiment. The moisture content data could be recorded as shown in Table 10.7 below.

Table 10.7 Table of moisture content data for a randomized block experiment.

Block	A	B	C	D	Total
1	X_{11}	X_{12}	X_{13}	X_{14}	B_1
2	X_{21}	X_{22}	X_{23}	X_{24}	B_2
3	X_{31}	X_{32}	X_{33}	X_{34}	B_3
4	X_{41}	X_{42}	X_{43}	X_{44}	B_4
5	X_{51}	X_{52}	X_{53}	X_{54}	B_5
Total	T_1	T_2	T_3	T_4	GT

Each replicate of four cakes would represent a *block*, each having the same oven temperature and baking environment in common. Therefore, there is a potential correlation between the results in one block which would be independent of other blocks. (The position in the oven might be randomly assigned for each replicate if it is felt that the position in the oven might have some effect on the moisture content.)

The development of a test statistic for this situation would follow the same kind of argument as was used for the one-factor experiment, resulting in an F test for the hypothesis of "no difference in treatments." The computations also follow in a similar vein. Using a notation similar to Table 10.7 above, assuming that we have k treatments and r blocks, we would compute as the box shows.

Analysis of Variance Formulas–Randomized Block Design

$$GT = \sum_{i=1}^{k} \sum_{j=1}^{r} X_{ij} \quad \text{Grand Total}$$

$$SS(T) = \sum_{i=1}^{k} \frac{T_i^2}{r} - \frac{(GT)^2}{rk} \qquad \text{Sums of Squares for Treatments}$$

$$SS(B) = \sum_{j=1}^{r} \frac{B_j^2}{k} - \frac{(GT)^2}{rk} \qquad \text{Sums of Squares for Blocks}$$

$$SS(Total) = \sum_{i=1}^{k} \sum_{j=1}^{n_i} X_{ij}^2 - \frac{(GT)^2}{n} \qquad \text{Total Sums of Squares}$$

$$SS(E) = SS(Total) - SS(T) - SS(B) \qquad \text{Sums of Squares for Error}$$

These computations would be summarized in an analysis of variance table like that of Table 10.8 that follows next.

Table 10.8 Analysis of variance table for randomized block design.

Source	Sum of Squares	Degrees of Freedom	Mean Square (MS)	F-Ratio
Treatment	SS(T)	$k - 1$	$MS(T) = \dfrac{SS(T)}{k - 1}$	$\dfrac{MS(T)}{MS(E)}$
Blocks	SS(B)	$r - 1$	$MS(B) = \dfrac{SS(B)}{r - 1}$	$\dfrac{MS(B)}{MS(E)}$
Error	SS(E)	$n - k - r + 1$	$MS(E) = \dfrac{SS(E)}{n - k - r + 1}$	
Total	SS(Total)	$n - 1$		

Consider the following example to illustrate the computations.

EXAMPLE 3: A study is conducted to determine the best method of assembling a simple mechanism. Three different assembly methods are to be examined. Four different assemblers of varying skill are trained in the three assembly methods; the order in which the three different methods are applied is randomly set for each of the four assemblers. Therefore, the treatments are the three different methods of assembly, and the blocks are the assemblers (each assembler should yield homogeneous results for each of the three different methods.) The number of assemblies completed in one hour is observed and recorded in Table 10.9. For $\alpha = 0.01$, can we conclude that there is a difference among the methods. (The measurements are counts and cannot truly be normally distributed. Thus, the p-values obtained in the following are only approximate p-values.)

Table 10.9 Number of assemblies per hour.

Assembler	Method 1	Method 2	Method 3	Totals, B_i
1	3	4	6	13
2	5	7	8	20
3	7	9	11	27
4	6	5	9	20
Totals, T_i	21	25	34	80

SOLUTION: The computations are shown below and summarized in the analysis of variance table, Table 10.10 which follows.

$$GT = \sum_{i=1}^{k} \sum_{j=1}^{r} X_{ij} = 80$$

$$SS(T) = \sum_{i=1}^{k} \frac{T_i^2}{r} - \frac{(GT)^2}{rk} = \frac{21^2}{4} + \frac{25^2}{4} + \frac{34^2}{4} - \frac{80^2}{12} = 22.1667$$

$$SS(B) = \sum_{j=1}^{r} \frac{B_j^2}{k} - \frac{(GT)^2}{rk} = \frac{13^2}{3} + \frac{20^2}{3} + \frac{27^2}{3} + \frac{20^2}{3} - \frac{80^2}{12} = 32.6667$$

$$SS(Total) = \sum_{i=1}^{k} \sum_{j=1}^{n_i} X_{ij}^2 - \frac{(GT)^2}{n} = 3^2 + 4^2 + 6^2 + \ldots + 9^2 - \frac{80^2}{12} = 58.6667$$

and

$$SS(E) = SS(Total) - SS(T) - SS(B) = 58.6667 - (32.6667) - (22.1667) = 3.8333$$

Table 10.10 Analysis of variance for number of assemblies per hour—randomized block design.

Source	df	Sum of Squares	Mean Square	F-ratio	Prob
Assembler	3	32.6667	10.8889	17.0	0.0024
Method	2	22.1667	11.0833	17.3	0.0032
Error	6	3.83333	0.6389		
Total	11	58.6667			

The p-value for "methods" is very small indicating there is a difference in methods.

EXERCISES

1. Four brands of tires were tested under controlled conditions for wear resistance. Four tires of each brand were randomly selected from the production line and tested. Use the following data to test whether there is a difference in average wear resistance among brands. Use $\alpha = 0.05$ for your level of significance.

	Brand			
	1	2	3	4
	17	13	14	12
	12	11	14	8
	14	13	11	8
	13	10	12	9
Totals, T_i	56	47	51	37

2. Using the data of exercise 1, compute the mean wear resistance for each of the four brands of tires. Next, using $\sqrt{\frac{MSE}{4}}$ as the standard error, compute a 95% confidence interval on each mean and note which intervals overlap or don't overlap. What implications does this have for the test of hypothesis done in exercise 1.

3. Suppose the data in exercise 1 were collected by mounting a tire of each brand on an automobile, locations on the car being determined randomly. Suppose the first row of data is associated with car 1, the second row with car 3 and so forth. Since the cars have different drivers, are driven under different conditions, each car can be thought of as a "block" and the data analyzed as a randomized block experiment. Perform this analysis and compare your results to the analysis of exercise 1. Specifically, note the size of the MSE term from each analysis. Which is smaller?

4. We want to determine whether there is a significant difference between an Ace Company Product and the products of Ace's two chief competitors. We randomly select five identical units from each of the three companies and independently rate them on the basis of their performance. Do the data in the following table suggest that the products rate equally well? Use $\alpha = 0.05$.

Ace	Competitor I	Competitor 2
28	34	29
31	33	33
32	30	37
26	31	36
25	27	37

5. Suppose five different evaluators were involved in the ratings obtained in exercise 4. Suppose each row in the table containing the data referred to one of the five judges involved. Analyze the data from an assumption of a randomized block experiment with rows (evaluators) representing blocks and columns representing treatments. Use the same level of significance as was used in exercise 4.

6. A tire manufacturer wishes to begin production operations in a mid-western city, and is considering four cities in which to locate the new plant. An important feature to be considered in the placement of the plant is the reliability of the workers in similar plants already in operation in these cities. The manufacturer takes a random sample of six plants in each of the four cities and investigates the number of days absent per 1000 work-days. The results are listed in the table below. Is the average number of days absent that same from city to city? Use $\alpha = 0.05$.

City 1	City 2	City 3	City 4
124	126	144	183
131	144	141	189
157	151	127	145
163	129	1111	161
132	127	104	157
105	98	131	136

7. Construct a boxplot for each of the columns of data and plot them on the same scale size by side. Does the boxplot visually confirm the conclusions of your analysis of variance?

8. The data in the table below represent the output from each of four machines that manufacture the same product. The data were collected over five consecutive workdays. Do the data suggest a significant difference in output among the four machines? Let $\alpha = 0.05$.

Days	Machines			
	A	B	C	D
1	293	308	323	333
2	298	353	343	363
3	288	358	365	345
4	281	323	350	368
5	260	343	340	330

9. Ignoring the "days" feature of the data in exercise 8, make side-by-side boxplots of the data for the four machines. What does this suggestion about the equality of means for the 4 machines?

10. Show that the following expression equals zero.

$$\sum_{i=1}^{k}\sum_{j=1}^{n_i}(X_{ij}-\overline{X}_i)(\overline{X}_i-\overline{\overline{X}}).$$

11. Show that the following equation is true:

$$\sum_{i=1}^{k}\sum_{j=1}^{n_i}(\overline{X}_i-\overline{\overline{X}})^2 = \sum_{i=1}^{k}n_i(\overline{X}_i-\overline{\overline{X}})^2.$$

10.2 SOME NONPARAMETRIC / DISTRIBUTION-FREE TESTS: THE SIGN TEST AND THE WILCOXON-SIGNED RANK TEST

Introduction

In this section and the next section a few of the most common nonparametric tests are presented. Specifically, in this section the versatile sign test is presented, a test based on the binomial distribution. Numerous types of data can be constrained in ways so that the sign test or slight variations of it can be applied to test hypotheses of interest.

Along with the sign test we present the Wilcoxon-signed rank test, a test for paired data having more power than the sign test. Both of these tests could be considered as alternatives for some of the techniques presented in Chapter 8.

Definitions

DEFINITION 10.8: Assume that a set of n, independent, identically distributed ordered pairs is available, denoted by $(X_1, Y_1), \ldots, (X_n, Y_n)$. The *sign test* is a procedure to test the hypothesis that $P(X_i < Y_i) = P(X_i > Y_i)$. (We assume that X_i and Y_i are continuous and so $P(X_i = Y_i) = 0$.) If we record a "+" when the $X_i > Y_i$ and a "−" when $X_i < Y_i$, and if $p+$ denotes the probability of a "+", then the null hypothesis is $H_0: p+ = \frac{1}{2}$. The test statistic is the total number of pluses in the n pairs observed. This test statistic will follow a binomial distribution with n and $p = \frac{1}{2}$.

DEFINITION 10.9: The *Wilcoxon matched-pairs, signed-rank test* is a procedure for testing whether two treatments are different when applied to two related samples. The data are collected as pairs (X_i, Y_i); the first member of each pair being understood to be associated with treatment 1 and the second member with treatment 2. It is assumed that the magnitude of the difference between the measurements of each pair can be assessed, along with the direction of the difference (+ or −). The absolute values of the differences are ranked from small to large, the differences are next replaced by the ranks and the signs are then reattached. The test statistic is then based on the sum of the "signed ranks."

DISCUSSION

Distribution-Free Techniques

In the preceding sections and chapters, we made an assumption of normality whenever we wanted to test a hypothesis or construct a confidence interval. Or, we needed a sample size large enough to justify the use of the central limit theorem. Does this mean that all data are normally distributed? If not what do we do? The answer to the first question is obviously "no!" We will encounter data approximated by many types of distributions: exponential, gamma, log-normal, and other distributions we have not talked about in this textbook. In answer to the second question, the literature is extensive with procedures for testing hypotheses and constructing confidence intervals under small-sample conditions, and under assumptions that are much less restrictive than that of a specific underlying distribution, such as a normal distribution. These procedures are often broadly classified as *nonparametric procedures* or *distribution-free procedures*.

Another situation that leads to a need for these kinds of techniques is when we have data that are at best nominal data (also called categorical data) or ordinal data consisting of ranks. When the level of measurement is nominal or ordinal at best, the usual techniques of Chapters 7 and 8 are not strictly applicable and other alternatives must be found. The most powerful of these techniques are based on the properties of order statistics (see section 6.7).

In this and the following section we shall briefly consider techniques for testing hypotheses that represent nonparametric analogs to the following procedures: (1) the two-independent sample test, (2) a paired sample test (2 techniques are presented here), and (3) a one-factor completely randomized design. Specifically, in this section we present (1) the sign test, a test based on the properties of the binomial distribution and illustrated in the context of a paired experiment (the sign test is a very flexible test and can be used in many other contexts than simply the paired experiment demonstrated here), and (2) the Wilcoxon matched-pairs signed rank test. In section 10.3 we present (3) the Mann-Whitney test of two independent samples, and (4) the Kruskal-Wallis test for k independent samples.

There are hundreds of "nonparametric" tests in the literature. Most are tailor-made for a specific experimental situation. Those included in this chapter are some of the more common methods used as alternatives to the familiar normal-distribution based methods that are a part of this and other introductory textbooks. Further, rather than include a set of tables for each of these cases, we will present the large sample results of these tests only. Such large sample results lead to Z-score computations and a standard normal distribution table look-up. If tables are needed for small sample tests, there are plenty of books of tables available for your use. (See, for instance, W. J. Conover, *Practical Nonparametric Statistics*, 2nd Edition, John Wiley and Sons, Inc.) Our purpose here is merely to provide some insight into alternatives to the standard normal-distribution methods and to present some of the theoretical justification for their properties.

The Sign Test

As previously mentioned the sign test is a very versatile procedure to test hypotheses. Its use is limited only by the lack of ingenuity of the researcher. It can be used to test for trends over time, predictable patterns and so forth. However, we will illustrate its use in the following "paired-sample test."

Suppose that we have pairs of measurements such that the first measurement in a given pair, denoted by X, is associated with treatment 1 and the second measurement of the pair, denoted by Y, is associated with treatment 2. Our purpose is to decide whether the treatments are the same or different. The measurement associated with each pair is either a "+" or a "−" depending on whether, say treatment 1 is "better" (+) or "worse" (−) than treatment 2. We have n independent pairs of measurements available and we simply count the number of plus signs in the totality of n signs recorded. If U = the number of plus signs then U will be a binomial random variable. If we then formulate the null hypothesis as being that "there is no difference in treatments" then $P(+) = p + =$ $P(-) = p - = \frac{1}{2}$. Therefore, U will be binomial with $p = \frac{1}{2}$ and "p-values" for a test of hypothesis can be obtained from the table of the binomial distribution, Table A.

If the alternative hypothesis is that Treatment 1 is "better" than Treatment 2 then we would expect to see a lot of plus signs. If the number of plus signs recorded is k, then the "p-value" will be:

$$p\text{-value} = P(U \geq k \mid n, p = \frac{1}{2}).$$

If the alternative is that Treatment 1 is "worse" than Treatment 2, then we would expect to see very few plus signs and the p-value will be

$$p\text{-value} = P(U \leq k \mid n, p = \frac{1}{2}).$$

If the alternative is that the treatments are different, then

$$p = 2 \bullet P(X \leq k \mid n, p = \frac{1}{2}) \qquad \text{if } k \text{ is less than } \frac{n}{2}$$

or

$$p = 2 \bullet P(X \geq k \mid n, p = \frac{1}{2}) \qquad \text{if } k \text{ is greater than } \frac{n}{2}.$$

Since the sign test is very east to carry out and is based on the binomial distribution, we shall not dwell on it further here except to say that, if we have large samples, we can utilize the normal approximation to the binomial to provide an approximate test. (See Chapter 9, section 9.2.) The appropriate test statistic for this large-sample case under $H_o : p + = \frac{1}{2}$, after simplifying the expression, is:

$$Z = \frac{U - n(\frac{1}{2})}{\sqrt{n(\frac{1}{2})(\frac{1}{2})}} = \frac{U - \frac{n}{2}}{\frac{\sqrt{n}}{2}} = \frac{2U - n}{\sqrt{n}}$$

[**Caution:** The sign test is very versatile and for nominal data is often the most powerful test. However, if we have ordinal or interval-ratio data much more powerful tests are available. Among those more powerful tests are the *t* test for normally distributed data, and the Wilcoxon test presented later in this section.]

Consider the following examples of the use of the sign test.

EXAMPLES

EXAMPLE 1: Suppose that you are comparing a new wood finish with a standard varnish. The new finish compares favorably in hardness, cost, and other characteristics. To test its appearance, you finish 15 pairs of wood chips representing 15 different types of wood under identical procedures, with one chip of each pair randomly assigned to the new finish, and the other to the standard varnish. You then ask an observer to indicate which chip in each of the 15 pairs has the more attractive finish. In 13 of the 15 pairs, the observer prefers the new finish. At the $\alpha = 0.05$ level of significance, can we conclude that the new finish is superior in appearance to the standard varnish.

SOLUTION: Formulating a 5-step hypothesis test we have:

Step 1: $H_o: p+ = \dfrac{1}{2}$ vs. $H_a: p+ > \dfrac{1}{2}$ where $p+$ is the probability that the new finish is better than the old varnish.

Step 2: The sample size is 15 pairs of wood chips and $\alpha = 0.05$.

Step 3: The test statistic is $U =$ the number of times the new finish is preferred over the standard varnish.

Step 4: The value of the test statistic was $U = 13$. Therefore, for the alternative given in step 1 we can get the *p*-value from the binomial table as:

$$p = P(U \geq 13 \mid n = 15, p = \tfrac{1}{2}) = 1 - P(X \leq 12) = 1 - 0.9963 = 0.0037.$$

Step 5: Since $p = 0.0037 < 0.05$ we reject the null hypothesis and conclude that the new finish is noticeably preferable over the old varnish.

EXAMPLE 2: A study was conducted to determine the extent that alcohol blunts a person's thinking ability for performing a given task. Ten persons of varying characteristics were selected at random and asked to participate in the experiment. After suitable briefing, each person performed the task with no alcohol in the blood system. Then the task was performed again after each person had consumed enough alcohol to raise the alcohol content in his or her system to 0.1%. Can one conclude at the $\alpha = 0.05$ level of significance that the average "before" time is less than the average "after" time? The data and some summarizing computations are given below. (This is Example 2 in section 8.6. See Chapter 8.)

SOLUTION:

Step 1: $H_o: p+ = \dfrac{1}{2}$ vs. $H_a: p+ > \dfrac{1}{2}$ where $p+$ is the probability that the "after" measurement will be larger than the "before" measurement.

Step 2: The experimental procedure is described above and α was set at 0.05.

Step 3: The test statistic is $U =$ the number of plus signs, which follows a binomial distribution with $n = 10$ and $p = \dfrac{1}{2}$ under H_o.

Response times (in minutes) for completing a task before and after ingestion of alcohol

Before	After	Sign of Difference: After–Before
28	39	+
22	45	+
55	67	+
45	61	+
32	46	+
35	58	+
40	51	+
25	34	+
37	48	+
20	30	+
U = Number of pluses		10

Step 4: The value of U is 10, so that

$$p = P(U \geq 10 \mid n = 10, p = 1/2) = \left(\frac{1}{2}\right)^{10} = 0.00098.$$

Step 5: Since $p < \alpha = 0.05$, we reject H_o and conclude that the mean response time with alcohol is greater than without alcohol.

Note that the p-value for this problem obtained previously using a paired-comparison t statistic was $p = 0.0000$ to four places after the decimal. Both procedures lead clearly to a decision to reject, but the sign test uses only a "nominal" characteristic of the data after taking differences where the t test uses the "interval-ratio" level of measurement the data possess. Both procedures, however, lead to the same conclusion. However, there may be conditions when the sign test does not reject, where a more powerful test would. You need to become familiar with the power of the various tests that might be applied in a given test of hypothesis.

EXAMPLE 3: In a telephone survey, 100 households were obtained where the same question was asked of both spouses. The number of times that the husband and wife answered the question the same way (both answered the question with a Yes or both answered with a No) was 60 out of 100. Does this suggest a correlation between responses within married couples?

SOLUTION: If there is no correlation between husband and wife, we would expect the responses to agree as often as they disagree. Therefore, the probability of agreement equals the probability of disagreement equals $\frac{1}{2}$. The alternative hypothesis is : $H_a: p + \neq \frac{1}{2}$. The test statistic is (for a large sample.)

$$Z = \frac{2U - n}{\sqrt{n}} = 2.00$$

The p-value for this z score is:

$$p = 2P(Z > 2.00) = 0.0455.$$

This would be significant if $\alpha = 0.05$. Therefore, there is strong statistical evidence for concluding a correlation of responses. (This kind of hypothesis and data set-up is often referred to as McNemar's test.)

The Wilcoxon Signed-Ranks Test

Let us extend the paired-comparison test demonstrated above with the sign test to a more powerful test. Again assume we have data (X_1, Y_1), (X_2, Y_2), ..., (X_n, Y_n), occurring as correlated pairs. As before we examine the differences $D_i = Y_i - X_i$ but we maintain information on the size of the difference as well as the sign of the difference.

Next we take the absolute value of the differences denoted by $|D_i|$ and order them from small to large and assign their ranks to them. (The "smallest" absolute value is given the "rank" of 1, the next smallest the rank of 2, and so on, with the largest absolute value getting the rank n.) Lastly, we "reattach" the "signs" to the ranks of the absolute differences denoting these "signed ranks" by R_i. The test statistic that is used for testing the hypothesis that the median of the X's is the same as the median of the Y's is

$$\text{TS} = \frac{\sum\limits_{i=1}^{n} R_i}{\sqrt{\sum\limits_{i=1}^{n} R_i^2}} = \frac{\sum\limits_{i=1}^{n} R_i}{\sqrt{\dfrac{n(n+2)(2n+1)}{6}}} \qquad \text{(if no ties are present)}$$

(In the above formula, the denominator is re-expressed assuming that there are no "ties." A tie occurs when two or more of the differences are equal and, therefore, a unique rank can't be assigned. Usually what we do in those cases is to assign the "average rank" to the tied observations. For instance, suppose that D_i and D_j are equal to one another or are "tied." Suppose we add (or subtract) a small amount ε to D_i, thus breaking the tie. The ranks they now receive are k and $k+1$, then the "average rank" assigned to them is $\dfrac{k + (k+1)}{2}$. The same principle applies if more than two observations are tied. See the examples for an application of this procedure.)

The logic behind the test statistic is that if we assume the differencies are symmetric, then under H_o the point of symmetry would be zero. Then we would expect that

$$E(\sum_{i=1}^{n} R_i) = 0$$

since the sign on a particular rank is as likely to be positive as negative. So in the long run, the sum of the ranks would be equal to zero. In addition, since we use the ranks of the differences instead of the individual differences, for any given sample size n, we can list all of the possible sums that could occur and thus calculate probabilities.

For a more simple-minded approach, consider a set of n chips numbered from 1 to n with the number of the chip inscribed on both sides of the chip. In addition, a plus sign (+) is written on one side of the chip and a negative sign (–) is on the other. Next the n chips are placed in a cup which is then thoroughly mixed. The chips are then poured on a table and then sorted out from smallest to largest in absolute value. However, there is an equal probability that the plus sign will show as the negative sign. A specific outcome might look like $(+1, +2, -3, +4, -5, -6, ..., +n)$. Since each of the n positions could either be positive or negative, there are $(2)^n$ possible outcomes. One extreme case would be when all chips are positive $(+1, +2, +3, +4, +5, +6, \ldots, +n)$, in which case $\sum_{i=1}^{n} R_i = \frac{n(n+1)}{2}$.

The probability under the null hypothesis of any specific outcome of this experiment is $\frac{1}{(2)^n}$, all possible values can be listed, the sum $\sum_{i=1}^{n} R_i$ and thus the value of the test statistic enumerated, and the distribution of the test statistic, TS, tabled. Under H_0 the distribution of the test statistic TS would be symmetric around zero. If the median of the X's is larger than the Y's then the value of the test statistic would tend to be negative. If the median of the X's is smaller than the Y's then the value of the test statistic TS would tend to be positive. In addition, the large sample distribution of TS is approximately standard normal under H_0. Therefore, a 5-step test of hypothesis procedure can be developed that follows logically from these observations. Consider the following examples.

EXAMPLES

EXAMPLE 4: A team of psychologists uses ten sets of kindergarten-age twins in an experiment to assess the effect of certain kinds of psychological supports to the learning process. One member of each set of twins serves as a control. The other twin receives the treatment. After ten weeks, the psychologists give the twins a standard test. Using an 0.025 level of significance, can we conclude from the following scores that the psychological supports improved performance on the test?

SOLUTION: The following 5-step test of hypothesis is as follows:

Step 1: H_0:Med$(D_i) = 0$ vs. H_a: Med$(D_i) < 0$

Step 2: The sample size consists of $n = 10$ pairs of observations. Let $\alpha = 0.025$.

Step 3: The test statistic is

$$TS = \frac{\sum_{i=1}^{n} R_i}{\sqrt{\sum_{i=1}^{n} R_i^2}}$$

which is approximately standard normal in distribution.

The data and computations are shown below. Notice that the column of absolute differences, after ordering show two 15's which would be ranked 7 and 8 respectively. Therefore, both are given the average rank of 7.5.

Twin Set	Control, Y_i	Treatment,X_i	$D_i = Y_i - X_i$	Signed Rank R_i	R_i^2
1	62	83	−21	−9.	81
2	43	75	−32	−10.	100
3	75	90	−15	−7.5	56..25
4	38	36	2	1	1
5	50	65	−15	−7.5	56.25
6	42	36	6	3	9
7	36	46	−10	−5	25
8	65	72	−7	−4	16
9	79	75	4	2	4
10	35	47	−12	−6	36
Sum				−43	384.50

Ordered |Differences|	Ranks
2	1
4	2
6	3
7	4
10	5
12	6
15	7 (7.5)
15	8 (7.5)
21	9
32	10

Step 4:   The value of the test statistic from the computations shown above is

$$TS = \frac{-43}{\sqrt{384.500}} = -2.203$$

which has an approximate p-value equal to 0.0138 after consulting the standard normal table.

Step 5:   Because the p-value is less than 0.025, we reject H_o and conclude the psychological aids did improve test performance.

EXAMPLE 5: A study was conducted to determine the extent that alcohol blunts a person's thinking ability for performing a given task. Ten persons of varying characteristics were selected at random and asked to participate in the experiment. After suitable briefing, each person performed the task with no alcohol in the blood system. Then the task was performed again after each person had consumed enough alcohol to raise alcohol content in his or her system to 0.1%. Can one conclude at the $\alpha = 0.05$ level of significance that the average "before" time is less than the average "after" time? The data and some summarizing computations are given below. (This is Example 2 in Chapter 8, section 8.6 which is also used above in an illustration of the sign test.)

SOLUTION: The data and test of hypothesis follow.

Response times (in minutes) for completing a task before and after ingestion of alcohol

Before, X_i	After, Y_i	$Y_i - X_i$	Signed Ranks, R_i	R_i^2
28	39	11	4	16
22	45	23	9.5	90.25
55	67	12	6	36
45	61	16	8	64
32	46	14	7	49
35	58	23	9.5	90.25
40	51	11	4	16
25	34	9	1	1
37	48	11	4	16
20	30	10	2	4
			55.0	382.50

The 5-step test of hypothesis is:

Step 1: H_0:Med$(D_i) = 0$ vs. H_a: Med$(D_i) < 0$

Step 2:   The sample size consists of $n = 10$ pairs of observations. Let $\alpha = 0.05$.

Step 3:   The test statistic is

$$TS = \frac{\sum\limits_{i=1}^{n} R_i}{\sqrt{\sum\limits_{i=1}^{n} R_i^2}}$$

which is approximately standard normal in distribution.

Step 4:   The value of the test statistic from the computations shown above is

$$TS = \frac{55.0}{\sqrt{382.50}} = 2.81$$

which has an approximate p-value of 0.0025 using a standard normal distribution.

Step 5:   Because the p-value is less than 0.05 we reject H_0 and conclude the psychological aids did improve test performance.

Note that the p-value for this problem obtained previously using a paired-comparison t statistic was $p = 0.0000$ to four places after the decimal, and for the sign test the p-value was 0.0010, rounded to 4 places. The p-value above of 0.0025 is only an approximate p-value; the actual p-value is 0.0010, the same value as for the sign test since the probability in both cases is $\dfrac{1}{(2)^{10}}$.   All procedures lead clearly to a decision to reject, but the sign test uses only a "nominal" characteristic of the data after taking differences where the signed rank test uses the ordinal characteristics of the data and the t test uses the "interval-ratio" level of measurement the data possess.

In terms of "power," the sign test is the least powerful, the signed-rank test next most powerful and the t-test is most powerful of the three though it has the most restrictive assumptions as well. (See exercise 1 that follows in this section.) However, the signed-rank test is only slightly less powerful in the cases where the t-test is most powerful and because of its less restrictive assumptions should be considered as an alternative to the prevailing preference to use the t test.

EXERCISES

1. Consider Examples 2 and 5 of this section. Suppose the data show only 1 negative difference associated with the rank of 1. What is the value of $\sum_{i=1}^{n} R_i$ and what is the p-value for that outcome? Suppose the sign test were applied to that outcome; what is the p-value for such a sign test? Note the difference in sizes of the p-values. Remember in the signed rank test you are calculating the probability that $\sum_{i=1}^{n} R_i \geq 53$, and for the sign test you are calculating the probability of getting 9 or more plus signs in 10 trials.

2. In a consumer study, two different kinds of product packaging were shown to 25 persons. Each person was asked to indicate which packing was preferred. Using $\alpha = 0.025$, test the null hypothesis of no strong preference for either packaging for the case in which 19 of the 25 prefer packaging 2.

3. A department store is conducting a preference test on two perfumes. Fifteen customers are asked which of the two different fragrances they prefer. However, unknown to the customers, the perfumes are the same. The only difference is that one bottle is much more expensive-looking. The perfume buyer feels that the customers will tend to choose the fragrance that is in the more elegant bottle. It turns out that 10 customers choose the perfume from the more expensive-looking bottle. Use $\alpha = 0.05$ to test the hypothesis.

4. A manufacturing unit in a plant has been making changes over time that they feel has caused an improvement in quality, with fewer defects being produced. To test this hypothesis, two-years worth of data were collected and paired by month: January of this year with January of last, February of this with February of last, and so forth. It is argued that under the hypothesis proposed, the defect rate of this year should be less than the defect rate of a year ago. Do the following data support the research hypothesis? Use $\alpha = 0.05$. (Note: This illustrates the use of the sign test as a "test of trend.")

					Defects per 1000 Parts							
					MONTH							
Year	Jan	Feb	Mar	Apr	May	Jun	July	Aug	Sep	Oct	Nov	Dec
1	4	5	3	4	4	3	4	5	3	2	1	3
2	3	3	2	3	3	4	2	2	4	3	0	2

5. A company's Employee Safety Committee wants to implement a safety campaign to cut down on

industrial accidents. The committee collected data on the number of accidents over one years time before and after the implementation of the safety program. This was done using the eight divisions of the company. The following data resulted. What is the *p*-value for the test that the safety program was effective in reducing the number of accidents using the signed-rank test.

| | | | | Number of Accidents | | | | |
| | | | | Division | | | | |
	1	2	3	4	5	6	7	8
Before Program	75	92	57	63	72	47	81	87
After Program	63	86	58	51	53	50	60	80

6. Use the sign test to find the *p*-value for the data of exercise 5.

7. A consumer testing agency wishes to compare two meat tenderizers to determine whether there is a discernible difference in their ability to tenderize different cuts of meat. Six cuts of meat are divided into two pieces. Each piece receives one of the two meat tenderizers by random assignment. A experienced chef is then asked to rate the tenderness on a 1–20 scale. What is the *p*-value for the following data using the signed-rank test.

| | | | Meat Cut | | | |
	1	2	3	4	5	6
Tenderizer 1	10	13	8	9	16	11
Tenderizer 2	12	17	3	8	10	10

8. Use the sign test to find the *p*-value for the data of exercise 7.

9. An instructor in a political science course suspected there was a tendency for the political affiliation of students in her course to reflect the political affiliation of the head of household. She asked her 75 students to report the political affiliation of their parents and then to report their own political affiliation. Of the 75 students, 45 reported the same political party as the head of household. Does this suggest a relationship?

10. A new set of procedures was proposed to be used by engineers in evaluating the properties of a particular type of material. A workshop was conducted and a pretest was given to each participant and then after the instruction, a post-test was given. The data for 10 participants is presented. Is there indication using the sign test that the instruction improved the understanding of participants?

Participant	Pretest	Posttest
1	32	34
2	31	31
3	29	35
4	30	33
5	29	28
6	20	26
7	24	27
8	31	30
9	23	26
10	15	16

11. Using the data of exercise 10, apply the signed-rank test to it to assess the hypothesis described in exercise 10.

10.3 MORE NONPARAMETRIC TESTS: THE MANN-WHITNEY TEST AND THE KRUSKALL-WALLIS TEST

Introduction

In this section we continue our discussion of a few of the more common nonparametric tests. Specifically, a two-independent sample test comparable to the two-independent sample t test is presented. It is referred to as the Mann-Whitney test or the Mann-Whitney-Wilcoxon test in recognition of those who were its probable developers. It is a test of equal location with regard to the two populations from which the samples were drawn.

A generalization of the Mann-Whitney test to the case of k independent samples is also presented, called the Kruskall-Wallis test. This test is an alternative to the one-factor analysis of variance procedure described in section 10.1.

Definitions

DEFINITION 10.10:   The *Mann-Whitney test* is a procedure to test the hypothesis that two populations are identical except for a possible shift in location (such as a mean or median) relative to each other. A sample of n observations X_i, $i = 1,2,...,$ n are selected from the one population and an independent sample of m observations Y_j, $j = 1,2,...,m$ are selected from the other population. The two samples are merged into one and are ranked from smallest to largest. Next, their ranks from 1 to $N = n + m$ are assigned to each observation. The test statistic is based on the sum of the ranks of, say, the X population.

DEFINITION 10.11:   The *Kruskall-Wallis test* is a procedure to test the hypothesis that k populations are identical except for a possible difference in their locations.   A sample of n_1, n_2, ..., n_k are selected independently of one another from the k populations under consideration. The samples are merged into one and are ranked from smallest to largest. Next, their ranks from 1 to $N = n_1 + n_2 + ...+ n_k$ are assigned to each observation. The test statistic is based on the mean rank for the observations associated with the k different populations. The method of computation is very close to the computations associated with the analysis of variance computations.

DISCUSSION

The Mann-Whitney Two Independent Sample Test

Recall the two independent sample *t*-test of Chapter 8 for testing a hypothesis about the means of two populations from which independent samples had been taken. We assumed that we had n normally distributed observations on X and m normally distributed observations on Y. The assumption of normally distributed data with equal or common variance was made for the t-test to be valid. However, suppose that you are uncomfortable about an assumption of normality for such data. A common and very powerful alternative to consider in such a case is the Mann-Whitney two independent sample test.

The assumptions for a valid application of the test are:

*   Both samples are random samples from their respective populations.

*   The two samples are independent between each other as well as within each sample.

*   The level of measurement is at least ordinal.

*   If $F(x)$ denotes the cumulative distribution of one sample and $G(y)$ denotes the cumulative distribution of the other, then if there is a difference in the two populations it is only a "location" difference. That is either $F(x) = G(y)$ or there is some constant c such that $F(x) = G(y + c)$ and c is the size of the difference in locations of the two populations. (That is, the two populations are assumed to be identical except for a possible difference in their locations.)

Under these assumptions, the null hypothesis is that the mean (or median) of the two populations are equal or different by some specified amount. The application of the test requires that we merge the two samples into one and order the combined observations from small to large and then replace them with their ranks, 1, 2, 3, …, N where $N = n + m$. (That is we create the "order statistics" of the combined sample.) However, you must keep track of which ranks are associated with the random variable X and which are associated with Y. Next we form the sum of the ranks of the X population,

$$U = \sum_{i=1}^{n} R(X_i).$$

We can consider three possible scenarios.

(1) If the two populations are equal, then the X and Y values would tend to be "randomly" mixed and ranks of the X's would have no tendency to be larger or smaller than the ranks of the Y's .

(2) If the random variable X is associated with a population having a smaller mean than the random variable Y, then the ranks of the X observations would tend to be smaller and the sum U would tend to be "small," relatively.

(3) If the X population had a larger mean than the Y population, then the X's would tend to have larger ranks than would be expected.

(Of course, if the sample sizes are different this would have an effect on what the sum of the ranks would be under each of the 3 situations described above.)

The form of the test statistic we will use is

$$T_{M-W} = \frac{U - n\left(\frac{N+1}{2}\right)}{\sqrt{\frac{nm}{N(N-1)} \sum_{i=1}^{N} R_i^2 - \frac{nm(N+1)^2}{4(N-1)}}}$$

where $\sum_{i=1}^{N} R_i^2$ refers to the sum of the squares of all N ranks from both samples. The test statistic, denoted by T_{M-W}, is approximately standard normal in distribution and therefore the standard normal tables can be consulted for obtaining approximate p-values for testing the null hypothesis.

The exact distribution of U under the null hypothesis is found by assuming that the random variable X and the random variable Y are identically distributed. If this is the case, then every permutation of the Xs and Ys in the ordered combined sample is equally likely. (Strictly speaking, this observation requires proof but is intuitively obvious if you attempt to make an argument that some permutations are more likely than another.)

Under the assumption of identical distributions the ranks associated with the set of X's is equivalent to selecting n integers randomly from the set of integers from 1 to $n + m$. The distribution of U can be obtained by enumerating the sum all of the equally likelihood possibilities of selecting n integers from the total of $n + m$ integers under consideration.

The number of ways of selecting n integers from a total of $n + m$ integers is $\binom{n+m}{n}$ each of which is equally likely. Therefore, the probability of any particular outcome is

$$\frac{1}{\binom{n+m}{n}}$$

Therefore, the probability that $U = k$ is found by calculating all the different sets of n integers that sum to k and then divide by $\binom{n+m}{n}$. For instance, if $n = 3$ and $m = 3$, then

$$\frac{1}{\binom{n+m}{n}} = \frac{1}{\frac{6!}{3!3!}} = \frac{1}{20}$$

so that the probability of any particular selection, assuming random selection without replacement, is 1 chance in 20.

Suppose that the elements chosen resulted in the integers: 1, 2, and 3. In this case $U = 1 + 2 + 3 = 6$ and therefore, $P(U = 6) = \frac{1}{20}$. We can find the probability distribution for U by continuing in this fashion until all 20 possible samples have been accounted for. This would allow us to obtain exact p-values for the test statistic. This approach allows for the preparation of exact tables for the Mann-Whitney two sample test.

Because U is the sum of the ranks of the n X's, for large n and m, the central limit theorem can be applied to produce an approximate sampling distribution. It can be shown (see the exercises) that, when there are no ties,

$$E(U) = \frac{n(N+1)}{2} \qquad \text{and} \qquad V(U) = \frac{n(N+1)m}{12}$$

so that

$$Z = \frac{U - \frac{n(N+1)}{2}}{\sqrt{\frac{n(N+1)m}{12}}}$$

When there are no ties present this formula produces the same value that would be obtained by applying the test statistic, T_{M-W} on the preceding page. Consider the following examples of the use of the Mann-Whitney test.

EXAMPLES

EXAMPLE 1: To test the effectiveness of two different approaches to a speed-reading course, an instructor randomly assigns 9 persons out of a group of 17 to one course, the remaining 8 to the other. After the students complete the six-week course, the instructor administers a standard reading comprehension test. Can we assume that the students in the two courses performed differently on the basis of the following test scores? Use the Mann-Whitney test to obtain a p-value.

					Test Scores				
Course 1	82	84	68	66	76	88	62	81	
(Ranks)	(10)	(11)	(3.5)	(2)	(5.5)	(13)	(1)	(9)	Total = 55
Course 2	80	76	92	89	68	94	86	96	78
(Ranks)	(8)	(5.5)	(15)	(14)	(3.5)	(16)	(12)	(17)	(7)

SOLUTION: In the table below, the combined scores after ordering are reported with some of the summary computations indicated. The ranks are then inserted into the data table above.

Sorted Scores	Group	Raw Rank	Ranks adjusted for ties
62	1	1	1
66	1	2	2
68	1	3	3.5
68	2	4	3.5
76	1	5	5.5
76	2	6	5.5
78	2	7	7
80	2	8	8
81	1	9	9
82	1	10	10
84	1	11	11
86	2	12	12
88	1	13	13
89	2	14	14
92	2	15	15
94	2	16	16
96	2	17	17

The sum of the ranks from group 1 is $U = 55$ and the sum of the squares of the ranks is 1784. Therefore, the test statistic value is

$$T_{M-W} = \frac{U - n\left(\frac{N+1}{2}\right)}{\sqrt{\frac{nm}{N(N-1)}\sum_{i=1}^{N}R_i^2 - \frac{nm(N+1)^2}{4(N-1)}}}$$

$$= \frac{55 - 8(\frac{17+1}{2})}{\sqrt{\frac{8\cdot 9}{17(17-1)}(1784) - \frac{8\cdot 9(17+1)^2}{4(17-1)}}} = -1.6378$$

The approximate p-value for this Z score, since it is a two-tailed test, is

$$p = 2 \cdot P(Z > 1.6378) = 2 \cdot (0.0507) = 0.1015.$$

Therefore, the scores are not sufficiently different to indicate that the two courses were different.

The Kruskal-Wallis Test for Several Independent Samples

Suppose we have a completely randomized design (see section 10.1 of this chapter) but we are unwilling to make the assumption of normality typical of the analysis of variance with its resultant F test as shown in section 10.1. In this section we present a "nonparametric" alternative for the hypothesis described in section 10.1.

In review, we assume that we have k independent samples with possibly different sample sizes within each. The data could be summarized as in Table 10.11 below.

Table 10.11 Table of notation for individual sample values and summary statistics for k independent samples from the same population.

Sample Number			
1	2	...	k
X_{11}	X_{12}	...	X_{1k}
X_{21}	X_{22}	...	X_{2k}
X_{31}	X_{32}	...	X_{3k}
...	...	...	...
$X_{n_1 1}$	...	...	...
	...	...	$X_{n_k k}$
	$X_{n_2 2}$		

We assume that the measurement scale is at least ordinal, and that the populations are either identical (the null hypothesis assumption) or that they differ only in location. (These assumptions, compared to the assumptions of the Mann-Whitney test, show that this is a generalization of the Mann-Whitney assumptions from 2 populations to k populations. Thus, the Mann-Whitney test could be viewed as a special case of the Kruskall-Wallis procedure that is described here.)

Letting N denote the total number of observations ($N = n_1 + n_2 + \ldots + n_k$,) first pool the N observations together into one large sample of N and order them from small to large. Assign rank 1 to the smallest, continuing until rank N is assigned to the largest. (Ties are handled as described previously.) Next, sort the "ranks" back into the k populations, letting $R(X_{ij})$ represent the ranks assigned to X_{ij}. Then let R_j be the sum of the ranks associated with jth sample. This can be shown in Table 10.12 below.

The test statistic used for the Kruskall-Wallis test is

$$TS_{K-W} = \frac{1}{S^2}\left(\sum_{j=1}^{k} \frac{R_j^2}{n_j} - N\frac{(N+1)^2}{4} \right)$$

where N and R_j have been defined previously and where

$$S^2 = \frac{1}{N-1}\left(\sum_{\text{all } ij} R(X_{ij})^2 - N\frac{(N+1)^2}{4} \right)$$

Table 10.12 Table of notation for individual sample values and summary statistics for k independent samples from the same population.

	Sample Number		
1	2	...	k
$R(X_{11})$	$R(X_{12})$	...	$R(X_{1k})$
$R(X_{21})$	$R(X_{22})$	...	$R(X_{2k})$
$R(X_{31})$	$R(X_{32})$	...	$R(X_{3k})$
...	...	...	...
$R(X_{n_1 1})$	...	...	$R(X_{n_k k})$
	$R(X_{n_2 2})$		
R_1	R_2		R_k

The large sample distribution of the test statistic TS_{K-W} is chi-square with $k-1$ degrees of freedom. The p-values are obtained by the expression

$$p = P(\chi^2 \geq \text{Observed value of } TS_{K-W})$$

with small p-values leading to rejection of the null hypothesis of equal means (medians).

Computational Note:

Suppose the computations for an analysis of variance are applied to the ranks in Table 10.12 above. The computational form for SS(T) using R_j in place of T_j, after simplification, results in the expression

$$\left(\sum_{j=1}^{k} \frac{R_j^2}{n_j} - N\frac{(N+1)^2}{4} \right)$$

Similarly, the computation for SS(Total) using the ranks in place of the X_{ij}'s give the expression

$$\left(\sum_{all\ ij} R(X_{ij})^2 - N\frac{(N+1)^2}{4} \right)$$

Therefore, the Kruskal-Wallis test presents no new computations that have to be learned. Computations are used with which you are already familiar. In fact, if you have a computer package that performs statistical computations and analysis of variance computations specifically, that package can be easily used to obtain the Kruskall-Wallis test statistic above.

In summary, assume that you have replaced all of the data in a completely randomized design with their ranks from 1 to N. Then

$$TS_{K-W} = \frac{SS(T)}{\left[\dfrac{SS(Total)}{(N-1)}\right]}.$$

Exact tables can be obtained using the same logic as used for the Mann-Whitney test. However, you should anticipate that as the n_j's get large and k gets larger than 2, a table of probabilities would become very large and cumbersome. Therefore, the asymptotic properties are generally used for testing hypotheses and a chi-square table is consulted to obtain p-values.

Next, consider a few examples of the application of the Kruskall-Wallis test.

EXAMPLES

EXAMPLE 2: An experiment was conducted to compare the wearing qualities of three types of paint when subjected to the abrasive action of a slowly rotating cloth-surfaced wheel. Ten paint specimens were tested for each paint type and the number of hours until visible abrasion was apparent was recorded. Using the data in Table 10.13 below is there evidence to indicate a difference in the three paint types?

Table 10.13 Abrasion-resistant measures for three types of paint.

	Paint Type		
	1	2	3
	148	513	335
	76	264	643
	393	433	216
	520	94	536
	236	535	128
	134	327	723
	55	214	258
	166	135	380
	415	280	594
	153	304	465
Totals, T_i	2296	3099	4278

SOLUTION: First we take all 30 measurements in Table 10.13 and replace them by their ranks from 1 to 30. The table of ranks is shown below as Table 10.14. The form of the test statistic is:

$$TS_{K-W} = \frac{1}{S^2}\left(\sum_{j=1}^{k}\frac{R_j^2}{n_j} - N\frac{(N+1)^2}{4}\right)$$

where

$$\left(\sum_{j=1}^{k}\frac{R_j^2}{n_j} - N\frac{(N+1)^2}{4}\right) = \frac{110^2}{10} + \frac{153^2}{10} + \frac{202^2}{10} - 30\cdot\frac{(30+1)^2}{4} = 423.80$$

and

$$S^2 = \frac{1}{N-1}\left(\sum_{all\ ij}R(X_{ij})^2 - N\frac{(N+1)^2}{4}\right) = \frac{1}{29}\left(9455 - 30\cdot\frac{(31^2)}{4}\right) = 77.5$$

Table 10.14 Ranks of abrasion-resistant measures for three types of paint.

	Paint Type		
	1	2	3
	7	24	18
	2	14	29
	20	22	11
	25	3	27
	12	26	4
	5	17	30
	1	10	13
	9	6	19
	21	15	28
	8	16	23
Total of Ranks	110	153	202

Therefore,

$$TS_{K-W} = \frac{423.80}{77.5} = 5.47$$

This test statistic is approximately chi-squared with 2 degrees of freedom. Therefore, the approximate p-value is

$$p = P(\chi^2 > 5.47) = 0.0649$$

This p-value is borderline significant. Therefore, there is fairly strong evidence that there is a difference in the abrasion resistance, but it wouldn't be declared significant if $\alpha = 0.05$. (Compared to the analysis of these data using the analysis of variance in section 10.1, the p-value obtained there was $p = 0.0452$, which is slightly smaller than $p = 0.0649$.)

EXAMPLE 3: Repeat the solution of Example 2 using the computations of analysis of variance.

SOLUTION: The data of Table 10.14 were analyzed with an analysis of variance package. The analysis of variance table is produced as Table 10.15.

Table 10.15 Analysis of variance of ranks of abrasion resistant measures for three types of paint.

Analysis of Variance On Ranks of Paint Types					
Source	df	Sum of Squares	Mean Square	F-ratio	Prob
Type	2	**423.800**	211.900	3.14	0.0596
Error	27	1823.70	67.5444		
Total	**29**	**2247.50**			

Selecting the appropriate entries from the table (shown in bold-face type) we get the test statistic value:

$$TS_{K-W} = \frac{SS(T)}{\left(\frac{SS(Total)}{(N-1)}\right)} = \frac{423.800}{\left(\frac{2247.50}{29}\right)} = 5.47$$

which is identical with the results previously obtained.

EXERCISES

1. Four brands of tires were tested under controlled conditions for wear resistance. Four tires of each brand were randomly selected from the production line and tested. Use the following data to test whether there is a difference in average wear resistance among brands. Use $\alpha = 0.05$ for your level of significance.   Use the Kruskall-Wallis test on this data.

	Brand		
1	2	3	4
17	13	14	12
12	11	14	8
14	13	11	8
13	10	12	9

2. An experiment was conducted to compare the compressive strengths of two types of plastic. Eight specimens of plastic were tested for each type and the compressive strengths (in thousands of pounds per square inch) were recorded with the results listed in the accompanying table. Analyze these data using the Mann-Whitney test.

Plastic Type	Compressive Strength							
1	14.1	14.3	13.8	14.2	14.0	14.5	13.9	13.7
2	14.2	13.9	13.8	14.3	14.1	13.4	13.8	14.0

3. Analyze the data of exercise 2 using the Kruskall-Wallis test. Compare your results with the results of exercise 2.

4. We want to determine whether there is a significant difference between an Ace Company Product and the products of Ace's two chief competitors. We randomly select five identical units from each of the three companies and independently rate them on the basis of their performance. Do the data in the following table suggest that the products rate equally well? Use $\alpha = 0.05$.

Ace	Competitor I	Competitor 2
28	34	29
31	33	33
32	30	37
26	31	36
25	27	37

5. To reduce the time spent in transferring materials from one location to another, three methods have been devised. With no previous information available on the effectiveness of these three approaches, a study is performed. Each approach is tried several times, and the amount of time to completion (in hours) is recorded in the table. Analyze these data to determine if there is a difference in methods.

	Method	
A	B	C
8.2	7.9	7.1
7.1	8.1	7.4
7.8	8.3	6.9
8.9	8.5	6.8
8.8	7.6	
	8.5	

6. A tire manufacturer wishes to begin production operations in a mid-western city, and is considering four cities in which to locate the new plant. An important feature to be considered in the placement of the plant is the reliability of the workers in similar plants already in operation in these cities. The manufacturer takes a random sample of six plants in each of the four cities and investigates the number of days absent per 1000 work-days. The results are listed in the table below. Is the average number of days absent the same from city to city? Use $\alpha = 0.05$.

City 1	City 2	City 3	City 4
124	126	144	183
131	144	141	189
157	151	127	145
163	129	1111	161
132	127	104	157
105	98	131	136

7. Suppose from other considerations the tire manufacturer of exercise 6 narrowed the choices of cities in which to locate his new plant to city 2 and city 3. Use the data of exercise 6 to compare the absentee rate of the two cities. Make this comparison using the Mann-Whitney test.

8. Repeat exercise 7 using the Kruskall-Wallis test. Compare your results with the results of exercise 7.

10.4 LINEAR REGRESSION ANALYSIS

OBJECTIVE

In this section we present a rudimentary view of a very important topic in applied and theoretical statistics called linear regression. The concept of a regression equation has already been introduced in Chapter 3, but that notion is extended in a more general setting and with fewer assumptions than those made in Chapter 3. We review that presentation first and then look briefly at the logic behind the principle of least squares and its application to linear regression.

Definitions

Regression

DEFINITION 10.12: Assume that the random variables X and Y have a joint probability density function $f(x,y)$ which is known. Therefore, the conditional distribution of Y given X is given as $f_Y(y|x) = \dfrac{f(x,y)}{f_X(x)}$. The regression of Y on X is then defined to be $E(Y|X)$. That is,

$$\mu_{Y|X} = \int_{-\infty}^{\infty} y f_Y(y|x)\,dy$$

If $\mu_{Y|X} = \beta_0 + \beta_1 X$ then the regression of Y on X is said to be a simple linear regression. (See Chapter 3, section 3.7 to review the concept of a conditional probability density function.)

DEFINITION 10.13 (THE PRINCIPLE OF LEAST SQUARES): Suppose a data set of n X,Y pairs is observed and denoted by $(x_1,y_1), (x_2,y_2), \ldots, (x_n,y_n)$ and there is an assumed relationship between Y and X of the form

$$y_i = h(x_i;\omega) + \varepsilon_i$$

Let ω denote a vector of one or more unknown parameters that defines a model which describes the relationship between Y and X, and ε_i is the error in the fit of $h(x_i;\omega)$ to the observed value y_i such that $E(\varepsilon_i) = 0$ and $V(\varepsilon_i) = \sigma^2$. Then

$$E(y_i) = E[h(x_i;\omega) + \varepsilon_i] = h(x_i;\omega)$$

If we let $\hat{Y}_i$ denote the predicted value of $E(y_i)$ at the point x_i based on the model, using $\hat{\omega}$ as the predicted values of ω, then $\varepsilon_i = Y_i - \hat{Y}_i$. The values of the parameters ω that minimize the "sum of squared errors"

$$\sum_{i=1}^{n} \varepsilon_i^2 = \sum_{i=1}^{n} (Y_i - \hat{Y}_i)^2$$

are called the *least squares estimators*, and the equation

$$E(Y) = \hat{Y}_i = h(x_i;\hat{\omega})$$

is called the *least squares regression equation*. If inferential properties are desired, we also make the assumption that the error ε_i is *normally distributed* with a mean of 0 and variance σ^2 for all i.

DEFINITION 10.14 (SIMPLE LINEAR REGRESSION): If $h(x_i;\beta_o,\beta_1)$ is of the form $\beta_o + \beta_1 x_i$ so that the theoretical model is

$$y_i = \beta_o + \beta_1 x_i + \varepsilon_i$$

then the regression equation is called a *simple linear regression*.

THEOREM 10.1: The least squares estimators of β_1 and β_o are

$$\hat{\beta}_1 = \frac{\displaystyle\sum_{i=1}^{n} (x_i - \bar{x})(y_i - \bar{y})}{\displaystyle\sum_{i=1}^{n} (x_i - \bar{x})^2} = \frac{\displaystyle\sum_{i=1}^{n} x_i y_i - \frac{1}{n}\left(\sum_{i=1}^{n} x_i\right)\left(\sum_{i=1}^{n} y_i\right)}{\displaystyle\sum_{i=1}^{n} x_i^2 - \frac{1}{n}\left(\sum_{i=1}^{n} x_i\right)^2}$$

$$\hat{\beta}_0 = \bar{y} - \hat{\beta}_1 \bar{x}$$

And the least squares equation can be written as:

$$\hat{Y}_i = \hat{\beta}_o + \hat{\beta}_1 x_i$$

DEFINITION 10.15 (MULTIPLE LINEAR REGRESSION): Suppose that the assumed linear relationship between y and a set of x's, $x_1, x_2,\ldots, x_k$ is

$$y_i = \beta_o + \beta_1 x_{1i} + \beta_2 x_{2i} + \beta_3 x_{3i} + \varepsilon_i$$

Such a regression model is called a multiple linear regression model. It is assumed that $E(\varepsilon_i) = 0$ and $V(\varepsilon_i) = \sigma^2$ and that the x's are fixed or controlled.

DISCUSSION

Introduction

Suppose a local natural gas company is concerned about meeting peak demands during the winter heating season. Under the expectation that total energy consumption is going to be related to the average daily temperature, a study was done to collect data on total energy usage versus average daily temperature for a random sample of days throughout the winter months. A typical house was used for the study. The expectation is that if an association is present, this association might be quantified so that as daily weather predictions are made, estimates of total energy consumption could also be made, and peak load days could be planned for.

In the above situation, we have a set of bivariate random variables, X, the average daily temperature, and Y, the energy consumption. These random variables would be assumed to come from some joint bivariate pdf, $f(x,y)$. The objective of the problem defined is to find some equation relating Y to X in some fashion so that by knowing X, the average daily temperature, a prediction of the total energy used, Y, could be made. Such an equation is called the regression of Y on X. Whether such a prediction will be any good will depend upon how much association or correlation there is between Y and X.

A simple model that we might choose to predict Y from X is called a linear model which we might write as:

$$y_i = \beta_o + \beta_1 x_i + \varepsilon_i$$

where y_i is the energy used, x_i is the average daily temperature and ε_i is the "error" in the model which recognizes that we have an imperfect model and will not predict y exactly knowing x.

However, as we think about this problem a little more, we recognize there are other factors besides temperature that would affect the energy consumption. For instance, we may suspect that humidity affects energy use; also we would expect to use more energy on a windy day than a calm day as the wind dissipates the energy in a home more quickly. In addition, the day of the week may also be an important factor in light of people being home on the weekends and at work during the day on weekdays. Therefore, we may immediately think of a more complicated model such as

$$y_i = \beta_o + \beta_1 x_{1i} + \beta_2 x_{2i} + \beta_3 x_{3i} + \beta_4 x_{4i} + \varepsilon_i$$

where x_{1i} is average daily temperature, x_{2i} is average daily humidity, x_{3i} is average wind speed and x_{4i} is the day of the week. Notice that if in fact the second model is a better model than the first, we would expect the error term ε_i in the first model to be larger than the error term in the second because the first error term included humidity and wind speed which have been removed from the error term in the second model. We might continue this approach, adding more terms to the model, until we have a model whose error cannot be efficiently reduced any further.

The first model is what we call a simple regression model which we shall discuss next. The second model is an example of a multiple linear regression model which will also be discussed in this section.

There are two approaches that are used to find a regression equation. In simple linear regression, the first assumes the joint distribution function $f(x,y)$ is known, and the conditional distribution is created as given in Chapter 3, section 3.7. The second approach, applicable to either simple or multiple regression, makes no such assumption but uses a principle called least squares to find the equation. In the former, no data are needed to determine the regression equation; in the latter, it is entirely a data-based procedure. This discussion is limited to the least squares approach to developing a regression equation.

Simple Linear Regression

Suppose we have a set of bivariate data (x_i, y_i) and we assume that y_i is related to x_i by the equation $y_i = \beta_o + \beta_1 x_i + \varepsilon_i$ where we assume the value of x_i can be controlled or determined exactly by the experimenter in advance and ε_i is a random variable. As a result, y_i is also a random variable since it is a function of a random variable through the model. Suppose when the data is plotted on a 2-dimensional plane, it looks like Figure 10.2. The pattern of points strongly suggests (as it should) a straight-line relationship; as X increases there is a corresponding increase in Y. (However, it is also obvious that there is "error" associated with such an argument since a straight line will not pass through all the points in the plot.) The question is: "What straight line (what slope and intercept value) would provide a "best" fit to the plot of points?" (We need not be restricted to straight-line relationships, but for this brief introduction, will restrict ourselves to that case.)

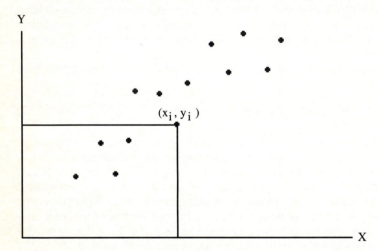

Figure 10.2 Two-dimensional scatterplot.

Suppose we draw a line through the points as indicated in Figure 10.3. This line can be represented in the usual intercept-slope form, $Y = \beta_o + \beta_1 X$, where β_o and β_1 are the intercept and slope of the line, respectively.

If the vertical distance, ε_i, from each point to the line is measured, then a rational argument is that the best line would be one which minimizes the average distance. However, some of the distances are positive and some are negative such that the average could be zero. Therefore, let us square the distances and then choose the line that will minimize the sum of the squared distances,

$$\sum_{i=1}^{n} \varepsilon_i^2.$$

The *least squares regression line* refers to the line that satisfies this condition.

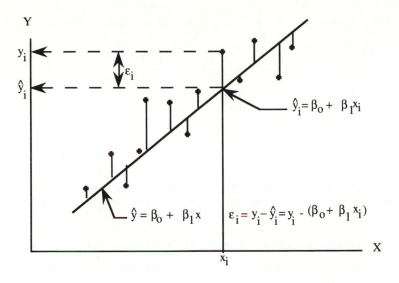

Figure 10.3 Least squares regression line plotted through the data.

This whole process could be carried out by using trial and error, with carefully drawn lines on graph paper, though it would be a tedious and time-consuming process. However, a mathematical solution to this problem follows by minimizing the expression

$$\sum_{i=1}^{n} \varepsilon_i^2 = \sum_{i=1}^{n} [y_i - (\beta_o + \beta_1 x_i)]^2$$

We minimize this expression by taking partial derivatives relative to β_o and β_1 respectively   and setting the resulting two equations equal to zero.   This results in two equations in two unknowns which are:

$$2\sum_{i=1}^{n} [y_i - (\beta_o + \beta_1 x_i)](-1) = 0 \quad \text{or} \quad \sum_{i=1}^{n} y_i = n\beta_o + \beta_1 \sum_{i=1}^{n} x_i$$

$$2\sum_{i=1}^{n} [y_i - (\beta_o + \beta_1 x_i)](-x_i) = 0 \quad \text{or} \quad \sum_{i=1}^{n} x_i y_i = \beta_o \sum_{i=1}^{n} x_i + \beta_1 \sum_{i=1}^{n} x_i^2$$

The solution of these two equations in terms of the two unknowns β_o and β_1 gives the least squares estimators of the slope and intercept as follows (the development of these equations is left as an exercise at the end of this section):

$$\hat{\beta}_1 = \frac{\displaystyle\sum_{i=1}^{n}(x_i-\bar{x})(y_i-\bar{y})}{\displaystyle\sum_{i=1}^{n}(x_i-\bar{x})^2} = \frac{\displaystyle\sum_{i=1}^{n}x_iy_i-\frac{1}{n}\left(\sum_{i=1}^{n}x_i\right)\left(\sum_{i=1}^{n}y_i\right)}{\displaystyle\sum_{i=1}^{n}x_i^2-\frac{1}{n}\left(\sum_{i=1}^{n}x_i\right)^2}$$

$$\hat{\beta}_0 = \bar{y} - \hat{\beta}_1\bar{x}$$

An examination of these results requires computations with which you are already familiar. The computations needed to produce the needed quantities can be summarized in a table such as Table 10.16 below.

Table 10.16 Table of data and summary computations to compute the least squares estimates.

Observation	x_i	y_i	x_i^2	y_i^2	x_iy_i
1	x_1	y_1	x_1^2	y_1^2	x_1y_1
2	x_2	y_2	x_2^2	y_2^2	x_2y_2
...	...	...	...	...	...
n	x_n	y_n	x_n^2	y_n^2	x_ny_n
Sum	$\displaystyle\sum_{i=1}^{n}x_i$	$\displaystyle\sum_{i=1}^{n}y_i$	$\displaystyle\sum_{i=1}^{n}x_i^2$	$\displaystyle\sum_{i=1}^{n}y_i^2$	$\displaystyle\sum_{i=1}^{n}x_iy_i$

If we assume that in the model $y_i = \beta_o + \beta_1 x_i + \varepsilon_i$ that $E(\varepsilon_i) = 0$ and $V(\varepsilon_i) = \sigma^2$, then we may want an estimator of σ^2. An intuitive estimator would be based on the following argument. First, remember the definition of the variance σ^2 as:

$$\sigma^2 = V(\varepsilon_i) = E[(\varepsilon_i - E(\varepsilon_i))^2] = E[(\varepsilon_i - 0)^2] = E[(\varepsilon_i)^2] = \text{"the mean of the squared errors"}$$

Thus, it would seem reasonable to use the mean of the squared errors in the sample to estimate the variance. This would produce the estimator

$$\hat{\sigma}^2 = \frac{\displaystyle\sum_{i=1}^{n}\varepsilon_i^2}{n} = \frac{\displaystyle\sum_{i=1}^{n}[y_i-\hat{y}_i]^2}{n} = \frac{\displaystyle\sum_{i=1}^{n}[y_i-(\hat{\beta}_o+\hat{\beta}_1 x_i)]^2}{n}$$

$$= \frac{\displaystyle\sum_{i=1}^{n}[y_i-\bar{y}_i]^2-\hat{\beta}_1\sum_{i=1}^{n}(x_i-\bar{x})(y_i-\bar{y})}{n}$$

However, the usual way to find an estimator of a parameter would be to apply the method of moments or the method of maximum likelihood. But a distributional assumption is necessary before either of these techniques can be applied. It can be shown that if the errors are assumed to be normally distributed with mean zero and variance σ^2, the maximum likelihood estimator of the variance produces the same estimator as given above. In addition, with the normality assumption, tests of hypotheses and confidence interval techniques can applied to the parameters and estimators from the equations. The specifics for these procedures will not be derived here but can be found in many books on linear regression methods. There is no new theory needed to develop those results.

[Note: Even though the above estimator of the variance agrees with the maximum likelihood estimator under appropriate assumptions of normality for the distribution of the errors, the estimator is biased. The statistical literature uses the divisor "$n - 2$" to make the estimator unbiased. Thus, the more commonly encountered expression for the estimator of the variance is

$$\hat{\sigma}^2 = \frac{\sum_{i=1}^{n} [y_i - \bar{y}_i]^2 - \hat{\beta}_1 \sum_{i=1}^{n} (x_i - \bar{x})(y_i - \bar{y})}{n - 2}$$

If we are willing to assume the errors are normally distributed it is possible to develop tests of hypotheses and derive expressions for confidence intervals. Rather than delve much more into those topics in this brief presentation let is simply note that if the slope β_1 is zero then we would conclude that y does not depend on x. The model would then be $y_i = \beta_0 + \varepsilon_i$ and a change in x would not predict a change in y. To test the hypothesis that the slope is zero we can form an analysis of variance table leading to an F-test as shown in Table 10.17 below.

Table 10.17 Layout of an ANOVA table for testing $H_o: \beta_1 = 0$ in simple linear regression.

Source	Sum of Squares	Degrees of Freedom	Mean Square (MS)	F-Ratio
Regression	$SS(R) =$ $\hat{\beta}_1 \sum_{i=1}^{n} (x_i - \bar{x})(y_i - \bar{y})$	1	$MS(R) = \dfrac{SS(R)}{1}$	$\dfrac{MS(R)}{MS(E)}$
Error	$SS(E) = SS(Total) - SS(R)$	$n - 2$	$MS(E) = \dfrac{SS(E)}{n-2}$	
Total	$SS(Total) = \sum_{i=1}^{n} [y_i - \bar{y}_i]^2$	$n - 1$		

The F-ratio in the above table is compared with values from the F distribution having 1 and $n - 2$ degrees of freedom. Also note that MS(E) is equal to the unbiased estimator of error referred to above.

Consider the following examples for illustration of the computations of simple linear regression.

EXAMPLES

EXAMPLE 1: Suppose that a developer of a new insulation material wants to know the amount of compression that would occur when a 2-inch thick specimen of the material is subjected to various amounts of pressure. Five pieces of the material were selected and tested under different pressures. Find the least squares regression equation to describe the relationship between compression y and pressure x. The data are shown below.

Specimen	Compression y	Pressure x
1	1	1
2	1.5	2
3	2	3
4	2	4
5	3	5

SOLUTION: A plot of these data is given first as Figure 10.4 below.   To fit the data we choose the model $y_i = \beta_o + \beta_1 x_i + \varepsilon_i$.

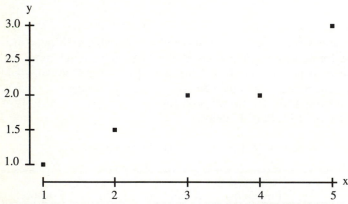

Figure 10.4 Scatterplot of compression (y) versus pressure (x).

The summary of computations is shown next with the least squares equations and solution:

Table 10.18 Data table and summary computations for Example 1.

	y	y^2	x	x^2	xy
	1	1	1	1	1
	1.5	2.25	2	4	3
	2	4	3	9	6
	2	4	4	16	8
	3	9	5	25	15
SUM	9.5	20.25	15	55	33

$$\sum_{i=1}^{n} y_i = n\beta_o + \beta_1 \sum_{i=1}^{n} x_i$$

$$\sum_{i=1}^{n} x_i y_i = \beta_o \sum_{i=1}^{n} x_i + \beta_1 \sum_{i=1}^{n} x_i^2$$

or

$$9.5 = 5\beta_0 + \beta_1 15$$

$$33 = 15\beta_0 + \beta_1 55$$

Solving these two equations fore β_0 and β_1 yields

$$\hat{\beta}_1 = 0.45 \qquad \text{and} \qquad \hat{\beta}_0 = 0.55$$

giving us the simple linear regression equation

$$\hat{y}_i = 0.55 + 0.45x_i$$

EXAMPLE 2: Using the regression equation of Example 1, estimate the "compression" if the pressure were set at 3.5.

SOLUTION: The value of 3.5 is substituted into the regression equation, $\hat{Y}_i = 0.55 + 0.45x_i$ to provide the estimate

$$\hat{Y}_i = 0.55 + 0.45(3.5) = 2.125$$

EXAMPLE 3: Solve the regression problem of Example 1 using the general solution equations previously given.

SOLUTION: Using the general equation and the summary computations from Table 10.16 we get

$$\hat{\beta}_1 = \frac{\sum\limits_{i=1}^{n} x_i y_i - \dfrac{\left(\sum\limits_{i=1}^{n} x_i\right)\left(\sum\limits_{i=1}^{n} y_i\right)}{n}}{\sum\limits_{i=1}^{n} x_i^2 - \dfrac{\left(\sum\limits_{i=1}^{n} x_i\right)^2}{n}} = \frac{33 - \dfrac{(15)(9.5)}{5}}{55 - \dfrac{(15)^2}{5}} = 0.45$$

$$\hat{\beta}_0 = \frac{9.5}{5} - 0.45\left(\frac{15}{5}\right) = 0.55$$

the same results as obtained in Example 1.

EXAMPLE 4: Provide an estimate of the variance σ^2 for the data of Example 1.

SOLUTION: The easiest computational formula to estimate σ^2 is

$$\hat{\sigma}^2 = \frac{\sum_{i=1}^{n} [y_i - \bar{y}_i]^2 - \hat{\beta}_1 \sum_{i=1}^{n} (x_i - \bar{x})(y_i - \bar{y})}{n} = \frac{\left(20.25 - \frac{(9.5)^2}{5}\right) - 0.45\left(33 - \frac{(15)(9.5)}{5}\right)}{5}$$

$$= \frac{2.2 - 0.45(4.5)}{5} = 0.0350$$

or if the form for the "unbiased" estimator is used we divide by $5 - 2$ which gives

$$\hat{\sigma}^2 = \frac{2.2 - 0.45(4.5)}{3} = 0.0583$$

Remember that

$$\sum_{i=1}^{n} [y_i - \bar{y}_i]^2 = \sum_{i=1}^{n} y_i^2 - \frac{1}{n}\left(\sum_{i=1}^{n} y_i\right)^2$$

and

$$\sum_{i=1}^{n} (x_i - \bar{x})(y_i - \bar{y}) = \sum_{i=1}^{n} x_i y_i - \frac{1}{n}\left(\sum_{i=1}^{n} x_i\right)\left(\sum_{i=1}^{n} y_i\right)$$

EXAMPLE 5: Use an analysis of variance table to test whether the slope β_1 is equal to zero.

SOLUTION: The computations needed to insert into the table are already available. The table is filled out as follows:

Source	Sum of Squares	Degrees of Freedom	Mean Square (MS)	F-Ratio
Regression	SS(R) = 0.45(4.5) = 2.025	1	2.025	34.7143
Error	SS(E) = 2.2 – 0.45(4.5) = 0.1750	5 – 2 = 3	0.0583	
Total	SS(Total) = 2.2	5 – 1 = 4		

The p-value for an F-ratio of 34.7143 with 1 and 3 degrees of freedom is 0.0098 which strongly suggests that the slope is not zero.

Multiple Linear Regression

Suppose we have a random variable Y that we believe to be a function of several X's, namely X_1, X_2, ..., X_k and is linear in its relationships. We may then write a linear model as

$$Y_i = \beta_o + \beta_1 x_{1i} + \beta_2 x_{2i} + ... \beta_k x_{ki} + \varepsilon_i$$

This is linear in the sense that the powers on the X's are 1's and the different terms in the model are added together. (This model is not as restrictive as it may seem. For instance suppose that $X_1 = X$, $X_2 = X^2$ and finally $X_k = X^k$. In such an event the model would be

$$Y_i = \beta_o + \beta_1 x_i + \beta_2 x_i^2 + \ldots + \beta_k x_i^k + \varepsilon_i$$

a polynomial in x. Though the powers of the X's are no longer 1, we still refer to this as a linear regression model and the techniques that follow will apply.)

Graphically, this model produces a surface in $k + 1$ dimensional space and our objective is to fit the "best" surface through the set of points $(y, x_{1i}, x_{2i}, \ldots, x_{ki})$ that are plotted in the $k + 1$ dimensional space that is defined. As before, the principle of least squares suggests we minimize the sum of the squared errors,

$$\sum_{i=1}^{n} \varepsilon_i^2 = \sum_{i=1}^{n} [y_i - (\beta_o + \beta_1 x_{1i} + \beta_2 x_{2i} + \ldots + \beta_k x_{ki})]^2 \, .$$

To do this we take $k + 1$ partial derivatives with respect to the $k + 1$ unknown parameters, set each resulting equation to 0 and solve. This gives the following $k + 1$ equations in $k + 1$ unknowns (known as the *normal equations*).

$$\sum_{i=1}^{n} y_i = n\beta_o + \beta_1 \sum_{i=1}^{n} x_{1i} + \beta_2 \sum_{i=1}^{n} x_{2i} + \ldots + \beta_k \sum_{i=1}^{n} x_{ki}$$

$$\sum_{i=1}^{n} x_{1i} y_i = \beta_o \sum_{i=1}^{n} x_{1i} + \beta_1 \sum_{i=1}^{n} x_{1i}^2 + \beta_2 \sum_{i=1}^{n} x_{1i} x_{2i} + \ldots + \beta_k \sum_{i=1}^{n} x_{1i} x_{ki}$$

$$\sum_{i=1}^{n} x_{2i} y_i = \beta_o \sum_{i=1}^{n} x_{2i} + \beta_1 \sum_{i=1}^{n} x_{2i} x_{1i} + \beta_2 \sum_{i=1}^{n} x_{2i}^2 + \ldots + \beta_k \sum_{i=1}^{n} x_{2i} x_{ki}$$

$$\ldots$$

$$\sum_{i=1}^{n} x_{ki} y_i = \beta_o \sum_{i=1}^{n} x_{ki} + \beta_1 \sum_{i=1}^{n} x_{ki} x_{1i} + \beta_2 \sum_{i=1}^{n} x_{ki} x_{2i} + \ldots + \beta_k \sum_{i=1}^{n} x_{ki}^2$$

To obtain a solution to these equations is not a simple task, and becomes especially difficult if k, the number of x's is very large. Developing procedures to solve these quations are best handled using matrix algebra methods. But rather than divert ourselves from our purpose here by presenting a discussion on matrix notation and methods, we simply turn to the technology of the day. We don't solve these equations using pen and paper anymore, even with our understanding of matrix algebra. Instead, there are hand calculators as well as computer software programs available on most types of machines that very simply and quickly give us the least squares estimators of β_o, β_1, β_2, $\ldots$, and β_k. In the examples that follow, we will rely on such computational aids rather than concern ourselves with the time-consuming steps to solve the equations by hand.

An estimator of the variance σ^2 follows the same argument as for simple regression and since the unbiased form for the estimator is preferred over the maximum likelihood form, for a variety of reasons, we have:

$$\hat{\sigma}^2 = \frac{\sum_{i=1}^{n} \varepsilon_i^2}{n-(k+1)} = \frac{\sum_{i=1}^{n}[y_i - \hat{y}_i]^2}{n-(k+1)} = \frac{\sum_{i=1}^{n}[y_i - (\beta_o + \beta_1 x_{1i} + \beta_2 x_{2i} + \ldots + \beta_k x_{ki})]^2}{n-(k+1)}$$

Note that the above "normal equations" have a general pattern to handle some of the specific cases that we will look at. In Table 10.19 below these equations are written again and partitioned to show the equations if Y depends on (a) one independent variable, X_1; (b), two independent variables X_1, and X_2, and so forth.

Table 10.19 The normal equations.

$$\text{k Independent Variables, } X_1, X_2, \ldots, X_k$$
$$\text{Two Independent Variables, } X_1 \text{ and } X_2$$
$$\text{One Independent Variable } X_1$$

$$\sum_{i=1}^{n} y_i = n\beta_o + \beta_1\sum_{i=1}^{n} x_{1i} + \beta_2\sum_{i=1}^{n} x_{2i} + \ldots + \beta_k\sum_{i=1}^{n} x_{ki}$$

$$\sum_{i=1}^{n} x_{1i}y_i = \beta_o\sum_{i=1}^{n} x_{1i} + \beta_1\sum_{i=1}^{n} x_{1i}^2 + \beta_2\sum_{i=1}^{n} x_{1i}x_{2i} + \ldots + \beta_k\sum_{i=1}^{n} x_{1i}x_{ki}$$

$$\sum_{i=1}^{n} x_{2i}y_i = \beta_o\sum_{i=1}^{n} x_{2i} + \beta_1\sum_{i=1}^{n} x_{2i}x_{1i} + \beta_2\sum_{i=1}^{n} x_{2i}^2 + \ldots + \beta_k\sum_{i=1}^{n} x_{2i}x_{ki}$$

$$\vdots$$

$$\sum_{i=1}^{n} x_{ki}y_i = \beta_o\sum_{i=1}^{n} x_{ki} + \beta_1\sum_{i=1}^{n} x_{ki}x_{1i} + \beta_2\sum_{i=1}^{n} x_{ki}x_{2i} + \ldots + \beta_k\sum_{i=1}^{n} x_{ki}^2$$

EXAMPLES

EXAMPLE 6: The following are data on the percent effectiveness of a pain relief and the amounts of two different pain relief medications (in milligrams) present in each capsule.

Medication A x_1	Medication B x_2	Percent Effective y
15	20	47
15	20	54
30	20	59
30	30	71
45	20	72
45	30	85

Assuming the regression is linear, estimate the regression coefficients, β_o, β_1 and β_2 and write the expression for the regression equation.

SOLUTION: Looking at the normal equations for two independent variables we need the following summary computations

x_1	x_2	y	x_1^2	x_2^2	x_1x_2	x_1y	x_2y
15	20	47	225	400	300	705	940
15	20	54	225	400	300	810	1080
30	20	59	900	400	600	1770	1180
30	30	71	900	900	900	2130	2130
45	20	72	2025	400	900	3240	1440
45	30	85	2025	900	1350	3825	2550
Totals 180	140	388	6300	3400	4350	12480	9320

The normal equations for this case (see Table 10.19 above) are:

$$
\begin{aligned}
388 &= 6\beta_0 + 180\beta_1 + 140\beta_2 \\
12480 &= 180\beta_0 + 6300\beta_1 + 4350\beta_2 \\
9320 &= 140\beta_0 + 4350\beta_1 + 3400\beta_2
\end{aligned}
$$

Solving these equations (actually I used a regression software package to get the solution) we get:

$$\beta_0 = 15.2308 \qquad \beta_1 = 0.738462 \qquad \beta_2 = 1.16923$$

which produces the regression equation

$$\hat{Y} = 15.2308 + 0.738462x_1 + 1.16923x_2$$

EXAMPLE 7:   Using the regression equation of Example 6 what would you estimate the percent effective pain relief to be if the dosages for medications A and B were both 20 milligrams? Which medication seems to contribute most to the pain relief?

SOLUTION: The estimate requested is obtained by inserting $x_1 = 20$ and $x_2 = 20$ in the above regression equation getting the result

$$\hat{Y} = 15.2308 + 0.738462(20) + 1.16923(20) = 53.3846\%$$

Since the coefficient on x_2 is larger than the coefficient on x_1, it would appear that medication B contributes more to the pain relief.

EXAMPLE 8: The following data pertain to the demand for a product (in thousands of units) and its price (in cents) charged in five different market areas. Fit a parabola to these data using the model

$$Y_i = \beta_0 + \beta_1 x_i + \beta_2 x_i^2$$

SOLUTION: The data and a scatterplot of the data are given in Figure 10.5. The normal equations are obtained by thinking of $x_1 = x$, $x_2 = x^2$ so that the table headings as shown in the solution to Example 6 would be x, x^2, y, x^2, x^4, x^3, xy, and finally x^2y.

Using a software package to fit this "polynomial" model to the data we get

$$\hat{Y} = 384.393 - (35.9975)x + (0.896422)x^2$$

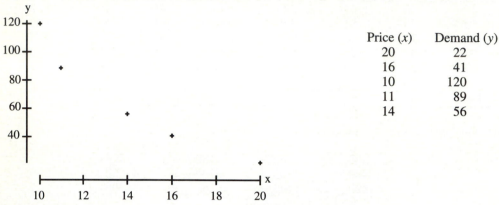

Price (x)	Demand (y)
20	22
16	41
10	120
11	89
14	56

Figure 10.5 A scatterplot of price versus demand. Data are shown at right.

EXERCISES

1. Show that the solution to the two equations

$$\sum_{i=1}^{n} y_i = \beta_0 n + \beta_1 \sum_{i=1}^{n} x_i$$

$$\sum_{i=1}^{n} x_i y_i = \beta_0 \sum_{i=1}^{n} x_i + \beta_1 \sum_{i=1}^{n} x_i^2$$

is

$$\hat{\beta}_1 = \frac{\sum_{i=1}^{n} (x_i - \bar{x})(y_i - \bar{y})}{\sum_{i=1}^{n} (x_i - \bar{x})^2} = \frac{\sum_{i=1}^{n} x_i y_i - \frac{1}{n}\left(\sum_{i=1}^{n} x_i\right)\left(\sum_{i=1}^{n} y_i\right)}{\sum_{i=1}^{n} x_i^2 - \frac{1}{n}\left(\sum_{i=1}^{n} x_i\right)^2}$$

and

$$\hat{\beta}_0 = \bar{y} - \hat{\beta}_1 \bar{x}$$

2. Suppose Y and X are jointly distributed as indicated below. Find the regression equation of Y on X and plot it on a graph.

$$f(x,y) = \begin{cases} x\,e^{-x(1+y)} & x > 0,\ y > 0 \\ 0 & \text{elsewhere} \end{cases}$$

3. The following data were obtained relative to average daily temperature (in degrees centigrade) vs. energy usage for fifteen randomly selected days for a typical residential home.

Temperature (oC)	0.0	8.0	7.5	13.5	14.0	18.5	4.5	−11.0
Energy usage	0.0	57.0	60.0	63.0	57.0	66.0	67.0	107.0

Temperature (oC)	−7.5	−8.5	1.5	0.5	2.0	−6.0	−4 .0
Energy usage	96.0	88.0	80.0	64.0	79.0	82.0	97.0

  a. Plot a scatterplot of these data. Is a linear association suggested?
  b. For a simple linear model, determine the estimated regression equation and plot it on your graph of part a.

4. For the regression equation of exercise 3, interpret the estimated regression coefficients. What would you recommend to the power company to improve the empirical model?

5. For the following set of data find the simple linear regression equation and compute an estimate of σ^2. Form a plot and show the regression line on the plot.

x	0.4	1.5	2.5	3.5	4.0
y	1.0	0.9	1.3	2.0	3.9

6. For the data of exercise 5 use the regression model

$$Y_i = \beta_o + \beta_1 x_i + \beta_2 x_i^2$$

  to fit a line through the data.   Find an estimate of σ^2 and see if it is smaller than the variance estimate of exercise 5. Draw the line for this equation on the plot of exercise 5.

7. Various doses (in mgs) of a poison were given to groups of 25 mice and the following results were observed.

Dose	4	6	8	10	12	14	16	18
No. of deaths	1	3	6	8	14	16	20	21

  Plot the graph of these data. Is a linear association suggested? For a simple linear model, determine the estimated regression equation and plot it on your graph.

8. For the regression equation and data of exercise 7, interpret the estimated regression coefficients. Estimate the number of deaths in a group of 25 mice who receive a 7-mg dose of this poison.

9. (Optional) The following small data set consist of the dependent variable Y, college GPA and two independent variables X_1, and X_2, high school math grade, and the math SAT score. Find the multiple regression equation of Y on X_1 and X_2. Solve using the normal equations or use a statistical software package if one is available.

GPA Y	HSM X_1	SAT-M X_2
5.32	10	670
5.14	9	630
3.84	9	610
5.34	10	570
4.08	9	670
4.57	9	417
5.06	8	540
3.60	4	460
5.47	10	720
5.18	9	670

10. For the multiple regression equation and data of exercise 9, find an estimate of the variance σ^2.

11. Find the simple regression equations of Y on X_1 and Y on X_2 for the data of exercise 9. Compute the variances for both cases and compare with the variance of exercise 10.

CHAPTER SUMMARY

This chapter is unlike most of the other chapters. Its purpose is to provide a collection of common applications of the statistical principles that can be justified using the theorems and principles of the text covered in Chapters 1 through 8. Thus, a light treatment of analysis of variance is given, some distribution-free techniques that are based on ranks or "order statistics" are presented, and some ideas of linear regression and the principle of *least squares* are included. All of these methods use concepts that can be justified by theory that has been presented throughout the course.

Because of the nature of this text, the coverage of the techniques in each of these broad areas is necessarily brief. But having read this chapter, you should have a better conceptual and theoretical idea of their foundations. If you were to pick a text devoted solely to analysis of variance, or distribution-free tests, or linear and multiple regression, you would at least have a theoretical perspective for their application.

To conclude this chapter we give you a list of possible references that you might pursue for a more complete coverage of the topics presented here as your interest and time permits.

References

ANALYSIS OF VARIANCE:

Anderson,V.L., and McLean, R. A., *Design of Experiments: A Realistic Approach*. New York: Marcel Dekker, Inc. 1974

Cochran, W. G., and Cox, G. M., *Experimental Design.* 2nd ed. New York: John Wiley & Sons, Inc., 1957

Hicks, C. R., *Fundamental Concepts in the Design of Experiments.* 2nd ed. New York: Holt, Rinehart and Winston, Inc., 1973

DISTRIBUTION-FREE/NONPARAMETRIC METHODS:

Conover, W. J., *Practical Nonparametric Statistics.* 2nd ed. New York: John Wiley & Sons, Inc., 1980

Daniel, W. W., *Applied Nonparametric Statistics.* 2nd ed. Boston, PWS-Kent Publishing Co., 1990

Gibbons, J. D., *Nonparametric Statistical Inference.* New York: McGraw-Hill Book Company, 1971

Lehmann, E. L., *Nonparametrics: Statistical Methods Based on Ranks.* San Francisco: Holden-Day, Inc., 1975

MULTIPLE REGRESSION:

Draper, N. R., and Smith, H., *Applied Regression Analysis.* 2nd ed., New York: John Wiley & Sons, Inc., 1981

Mendenhall, W., and Sincich, T., *A Second Course in Business Statistics: Regression Analysis.* San Francisco: Dellen Publishing Co. 1986

Wonnacott, T. H., and Wonnacott, R. J., *Regression: A Second Course in Statistics.* New York: John Wiley & Sons, Inc., 1981

APPENDIX: STATISTICAL TABLES

Table A Cumulative Values of the Binomial Distribution

Table entries represent:
$$F(x) = \sum_{i=0}^{x} \frac{n!}{i! \cdot (n-i)!} p^i (1-p)^{n-i}$$

Table B Cumulative Values of the Poisson Distribution

Table entries represent:
$$F(x) = \sum_{i=0}^{x} \frac{e^{-\lambda} \lambda^i}{i!}$$

Table C Cumulative Values of the Standard Normal Distribution

Table entries represent:
$$F(z) = \int_{-\infty}^{z} \frac{1}{\sqrt{2\pi}} e^{-\frac{[z]^2}{2}} dz.$$

Table D Quantile Values of the Chi-Square Distribution with ν Degress of Freedom

Table entries represent quantiles χ_q such that:
$$q = F(\chi_q) = \int_{0}^{\chi_q} \frac{1}{\Gamma(\nu/2)2^{\nu/2}} x^{\nu/2-1} e^{-\frac{x}{2}} dx$$

Table E Quantile Values of the t-Distribution with ν Degrees of Freedom

Table entries represent quantiles t_q such that:

$$q = F(t_q) = \int_{-\infty}^{t_q} \frac{\Gamma[\left(\frac{\nu+1}{2}\right)]}{\Gamma\left(\frac{\nu}{2}\right)\sqrt{\pi\nu}} \left(1 + \frac{t^2}{\nu}\right)^{-\frac{(\nu+1)}{2}} dt$$

Table F Quantile Values of the F-Distribution with ν_1 and ν_2 Degrees of Freedom

Table entries represent quantiles F_q such that:

$$q = F(F_q) = \int_{0}^{F_q} \frac{\Gamma[(\nu_1+\nu_2)/2](\nu_1/\nu_2)^{\nu_1/2}}{\Gamma(\frac{\nu_1}{2})\Gamma(\frac{\nu_2}{2})} \frac{F^{\nu_1/2 - 1}}{(1 + \nu_1 F/\nu_2)^{(\nu_1+\nu_2)/2}} dF$$

Table G Table of Random Numbers

Table A Cumulative Values of the Binomial Distribution

n	x	0.01	0.05	0.10	0.15	0.20	0.25	0.30	0.35	0.40	0.45	0.50
2	0	0.9801	0.9025	0.8100	0.7225	0.6400	0.5625	0.4900	0.4225	0.3600	0.3025	0.2500
	1	0.9999	0.9975	0.9900	0.9775	0.9600	0.9375	0.9100	0.8775	0.8400	0.7975	0.7500
	2	1.0000	1.0000	1.0000	1.0000	1.0000	1.0000	1.0000	1.0000	1.0000	1.0000	1.0000
3	0	0.9703	0.8574	0.7290	0.6141	0.5120	0.4219	0.3430	0.2746	0.2160	0.1664	0.1250
	1	0.9997	0.9928	0.9720	0.9392	0.8960	0.8438	0.7840	0.7182	0.6480	0.5748	0.5000
	2	1.0000	0.9999	0.9990	0.9966	0.9920	0.9844	0.9730	0.9571	0.9360	0.9089	0.8750
	3	1.0000	1.0000	1.0000	1.0000	1.0000	1.0000	1.0000	1.0000	1.0000	1.0000	1.0000
4	0	0.9606	0.8145	0.6561	0.5220	0.4096	0.3164	0.2401	0.1785	0.1296	0.0915	0.0625
	1	0.9994	0.9860	0.9477	0.8905	0.8192	0.7383	0.6517	0.5630	0.4752	0.3910	0.3125
	2	1.0000	0.9995	0.9963	0.9880	0.9728	0.9492	0.9163	0.8735	0.8208	0.7585	0.6875
	3	1.0000	1.0000	0.9999	0.9995	0.9984	0.9961	0.9919	0.9850	0.9744	0.9590	0.9375
	4	1.0000	1.0000	1.0000	1.0000	1.0000	1.0000	1.0000	1.0000	1.0000	1.0000	1.0000
5	0	0.9510	0.7738	0.5905	0.4437	0.3277	0.2373	0.1681	0.1160	0.0778	0.0503	0.0313
	1	0.9990	0.9774	0.9185	0.8352	0.7373	0.6328	0.5282	0.4284	0.3370	0.2562	0.1875
	2	1.0000	0.9988	0.9914	0.9734	0.9421	0.8965	0.8369	0.7648	0.6826	0.5931	0.5000
	3	1.0000	1.0000	0.9995	0.9978	0.9933	0.9844	0.9692	0.9460	0.9130	0.8688	0.8125
	4	1.0000	1.0000	1.0000	0.9999	0.9997	0.9990	0.9976	0.9947	0.9898	0.9815	0.9688
	5	1.0000	1.0000	1.0000	1.0000	1.0000	1.0000	1.0000	1.0000	1.0000	1.0000	1.0000
6	0	0.9415	0.7351	0.5314	0.3771	0.2621	0.1780	0.1176	0.0754	0.0467	0.0277	0.0156
	1	0.9985	0.9672	0.8857	0.7765	0.6554	0.5339	0.4202	0.3191	0.2333	0.1636	0.1094
	2	1.0000	0.9978	0.9841	0.9527	0.9011	0.8306	0.7443	0.6471	0.5443	0.4415	0.3438
	3	1.0000	0.9999	0.9987	0.9941	0.9830	0.9624	0.9295	0.8826	0.8208	0.7447	0.6563
	4	1.0000	1.0000	0.9999	0.9996	0.9984	0.9954	0.9891	0.9777	0.9590	0.9308	0.8906
	5	1.0000	1.0000	1.0000	1.0000	0.9999	0.9998	0.9993	0.9982	0.9959	0.9917	0.9844
	6	1.0000	1.0000	1.0000	1.0000	1.0000	1.0000	1.0000	1.0000	1.0000	1.0000	1.0000
7	0	0.9321	0.6983	0.4783	0.3206	0.2097	0.1335	0.0824	0.0490	0.0280	0.0152	0.0078
	1	0.9980	0.9556	0.8503	0.7166	0.5767	0.4449	0.3294	0.2338	0.1586	0.1024	0.0625
	2	1.0000	0.9962	0.9743	0.9262	0.8520	0.7564	0.6471	0.5323	0.4199	0.3164	0.2266
	3	1.0000	0.9998	0.9973	0.9879	0.9667	0.9294	0.8740	0.8002	0.7102	0.6083	0.5000
	4	1.0000	1.0000	0.9998	0.9988	0.9953	0.9871	0.9712	0.9444	0.9037	0.8471	0.7734
	5	1.0000	1.0000	1.0000	0.9999	0.9996	0.9987	0.9962	0.9910	0.9812	0.9643	0.9375
	6	1.0000	1.0000	1.0000	1.0000	1.0000	0.9999	0.9998	0.9994	0.9984	0.9963	0.9922
	7	1.0000	1.0000	1.0000	1.0000	1.0000	1.0000	1.0000	1.0000	1.0000	1.0000	1.0000
8	0	0.9227	0.6634	0.4305	0.2725	0.1678	0.1001	0.0576	0.0319	0.0168	0.0084	0.0039
	1	0.9973	0.9428	0.8131	0.6572	0.5033	0.3671	0.2553	0.1691	0.1064	0.0632	0.0352
	2	0.9999	0.9942	0.9619	0.8948	0.7969	0.6785	0.5518	0.4278	0.3154	0.2201	0.1445
	3	1.0000	0.9996	0.9950	0.9786	0.9437	0.8862	0.8059	0.7064	0.5941	0.4770	0.3633
	4	1.0000	1.0000	0.9996	0.9971	0.9896	0.9727	0.9420	0.8939	0.8263	0.7396	0.6367
	5	1.0000	1.0000	1.0000	0.9998	0.9988	0.9958	0.9887	0.9747	0.9502	0.9115	0.8555
	6	1.0000	1.0000	1.0000	1.0000	0.9999	0.9996	0.9987	0.9964	0.9915	0.9819	0.9648
	7	1.0000	1.0000	1.0000	1.0000	1.0000	1.0000	0.9999	0.9998	0.9993	0.9983	0.9961
	8	1.0000	1.0000	1.0000	1.0000	1.0000	1.0000	1.0000	1.0000	1.0000	1.0000	1.0000

Table A (continued)

n	x	0.01	0.05	0.10	0.15	0.20	*p* 0.25	0.30	0.35	0.40	0.45	0.50
9	0	0.9135	0.6302	0.3874	0.2316	0.1342	0.0751	0.0404	0.0207	0.0101	0.0046	0.0020
	1	0.9966	0.9288	0.7748	0.5995	0.4362	0.3003	0.1960	0.1211	0.0705	0.0385	0.0195
	2	0.9999	0.9916	0.9470	0.8591	0.7382	0.6007	0.4628	0.3373	0.2318	0.1495	0.0898
	3	1.0000	0.9994	0.9917	0.9661	0.9144	0.8343	0.7297	0.6089	0.4826	0.3614	0.2539
	4	1.0000	1.0000	0.9991	0.9944	0.9804	0.9511	0.9012	0.8283	0.7334	0.6214	0.5000
	5	1.0000	1.0000	0.9999	0.9994	0.9969	0.9900	0.9747	0.9464	0.9006	0.8342	0.7461
	6	1.0000	1.0000	1.0000	1.0000	0.9997	0.9987	0.9957	0.9888	0.9750	0.9502	0.9102
	7	1.0000	1.0000	1.0000	1.0000	1.0000	0.9999	0.9996	0.9986	0.9962	0.9909	0.9805
	8	1.0000	1.0000	1.0000	1.0000	1.0000	1.0000	1.0000	0.9999	0.9997	0.9992	0.9980
	9	1.0000	1.0000	1.0000	1.0000	1.0000	1.0000	1.0000	1.0000	1.0000	1.0000	1.0000
10	0	0.9044	0.5987	0.3487	0.1969	0.1074	0.0563	0.0282	0.0135	0.0060	0.0025	0.0010
	1	0.9957	0.9139	0.7361	0.5443	0.3758	0.2440	0.1493	0.0860	0.0464	0.0233	0.0107
	2	0.9999	0.9885	0.9298	0.8202	0.6778	0.5256	0.3828	0.2616	0.1673	0.0996	0.0547
	3	1.0000	0.9990	0.9872	0.9500	0.8791	0.7759	0.6496	0.5138	0.3823	0.2660	0.1719
	4	1.0000	0.9999	0.9984	0.9901	0.9672	0.9219	0.8497	0.7515	0.6331	0.5044	0.3770
	5	1.0000	1.0000	0.9999	0.9986	0.9936	0.9803	0.9527	0.9051	0.8338	0.7384	0.6230
	6	1.0000	1.0000	1.0000	0.9999	0.9991	0.9965	0.9894	0.9740	0.9452	0.8980	0.8281
	7	1.0000	1.0000	1.0000	1.0000	0.9999	0.9996	0.9984	0.9952	0.9877	0.9726	0.9453
	8	1.0000	1.0000	1.0000	1.0000	1.0000	1.0000	0.9999	0.9995	0.9983	0.9955	0.9893
	9	1.0000	1.0000	1.0000	1.0000	1.0000	1.0000	1.0000	1.0000	0.9999	0.9997	0.9990
	10	1.0000	1.0000	1.0000	1.0000	1.0000	1.0000	1.0000	1.0000	1.0000	1.0000	1.0000
11	0	0.8953	0.5688	0.3138	0.1673	0.0859	0.0422	0.0198	0.0088	0.0036	0.0014	0.0005
	1	0.9948	0.8981	0.6974	0.4922	0.3221	0.1971	0.1130	0.0606	0.0302	0.0139	0.0059
	2	0.9998	0.9848	0.9104	0.7788	0.6174	0.4552	0.3127	0.2001	0.1189	0.0652	0.0327
	3	1.0000	0.9984	0.9815	0.9306	0.8389	0.7133	0.5696	0.4256	0.2963	0.1911	0.1133
	4	1.0000	0.9999	0.9972	0.9841	0.9496	0.8854	0.7897	0.6683	0.5328	0.3971	0.2744
	5	1.0000	1.0000	0.9997	0.9973	0.9883	0.9657	0.9218	0.8513	0.7535	0.6331	0.5000
	6	1.0000	1.0000	1.0000	0.9997	0.9980	0.9924	0.9784	0.9499	0.9006	0.8262	0.7256
	7	1.0000	1.0000	1.0000	1.0000	0.9998	0.9988	0.9957	0.9878	0.9707	0.9390	0.8867
	8	1.0000	1.0000	1.0000	1.0000	1.0000	0.9999	0.9994	0.9980	0.9941	0.9852	0.9673
	9	1.0000	1.0000	1.0000	1.0000	1.0000	1.0000	1.0000	0.9998	0.9993	0.9978	0.9941
	10	1.0000	1.0000	1.0000	1.0000	1.0000	1.0000	1.0000	1.0000	1.0000	0.9998	0.9995
	11	1.0000	1.0000	1.0000	1.0000	1.0000	1.0000	1.0000	1.0000	1.0000	1.0000	1.0000
12	0	0.8864	0.5404	0.2824	0.1422	0.0687	0.0317	0.0138	0.0057	0.0022	0.0008	0.0002
	1	0.9938	0.8816	0.6590	0.4435	0.2749	0.1584	0.0850	0.0424	0.0196	0.0083	0.0032
	2	0.9998	0.9804	0.8891	0.7358	0.5583	0.3907	0.2528	0.1513	0.0834	0.0421	0.0193
	3	1.0000	0.9978	0.9744	0.9078	0.7946	0.6488	0.4925	0.3467	0.2253	0.1345	0.0730
	4	1.0000	0.9998	0.9957	0.9761	0.9274	0.8424	0.7237	0.5833	0.4382	0.3044	0.1938
	5	1.0000	1.0000	0.9995	0.9954	0.9806	0.9456	0.8822	0.7873	0.6652	0.5269	0.3872
	6	1.0000	1.0000	0.9999	0.9993	0.9961	0.9857	0.9614	0.9154	0.8418	0.7393	0.6128
	7	1.0000	1.0000	1.0000	0.9999	0.9994	0.9972	0.9905	0.9745	0.9427	0.8883	0.8062
	8	1.0000	1.0000	1.0000	1.0000	0.9999	0.9996	0.9983	0.9944	0.9847	0.9644	0.9270
	9	1.0000	1.0000	1.0000	1.0000	1.0000	1.0000	0.9998	0.9992	0.9972	0.9921	0.9807
	10	1.0000	1.0000	1.0000	1.0000	1.0000	1.0000	1.0000	0.9999	0.9997	0.9989	0.9968
	11	1.0000	1.0000	1.0000	1.0000	1.0000	1.0000	1.0000	1.0000	1.0000	0.9999	0.9998
	12	1.0000	1.0000	1.0000	1.0000	1.0000	1.0000	1.0000	1.0000	1.0000	1.0000	1.0000

Table A (continued)

n	x	0.01	0.05	0.10	0.15	0.20	0.25	0.30	0.35	0.40	0.45	0.50
13	0	0.8775	0.5133	0.2542	0.1209	0.0550	0.0238	0.0097	0.0037	0.0013	0.0004	0.0001
	1	0.9928	0.8646	0.6213	0.3983	0.2336	0.1267	0.0637	0.0296	0.0126	0.0049	0.0017
	2	0.9997	0.9755	0.8661	0.6920	0.5017	0.3326	0.2025	0.1132	0.0579	0.0269	0.0112
	3	1.0000	0.9969	0.9658	0.8820	0.7473	0.5843	0.4206	0.2783	0.1686	0.0929	0.0461
	4	1.0000	0.9997	0.9935	0.9658	0.9009	0.7940	0.6543	0.5005	0.3530	0.2279	0.1334
	5	1.0000	1.0000	0.9991	0.9925	0.9700	0.9198	0.8346	0.7159	0.5744	0.4268	0.2905
	6	1.0000	1.0000	0.9999	0.9987	0.9930	0.9757	0.9376	0.8705	0.7712	0.6437	0.5000
	7	1.0000	1.0000	1.0000	0.9998	0.9988	0.9944	0.9818	0.9538	0.9023	0.8212	0.7095
	8	1.0000	1.0000	1.0000	1.0000	0.9998	0.9990	0.9960	0.9874	0.9679	0.9302	0.8666
	9	1.0000	1.0000	1.0000	1.0000	1.0000	0.9999	0.9993	0.9975	0.9922	0.9797	0.9539
	10	1.0000	1.0000	1.0000	1.0000	1.0000	1.0000	0.9999	0.9997	0.9987	0.9959	0.9888
	11	1.0000	1.0000	1.0000	1.0000	1.0000	1.0000	1.0000	1.0000	0.9999	0.9995	0.9983
	12	1.0000	1.0000	1.0000	1.0000	1.0000	1.0000	1.0000	1.0000	1.0000	1.0000	0.9999
	13	1.0000	1.0000	1.0000	1.0000	1.0000	1.0000	1.0000	1.0000	1.0000	1.0000	1.0000
14	0	0.8687	0.4877	0.2288	0.1028	0.0440	0.0178	0.0068	0.0024	0.0008	0.0002	0.0001
	1	0.9916	0.8470	0.5846	0.3567	0.1979	0.1010	0.0475	0.0205	0.0081	0.0029	0.0009
	2	0.9997	0.9699	0.8416	0.6479	0.4481	0.2811	0.1608	0.0839	0.0398	0.017	0.0065
	3	1.0000	0.9958	0.9559	0.8535	0.6982	0.5213	0.3552	0.2205	0.1243	0.0632	0.0287
	4	1.0000	0.9996	0.9908	0.9533	0.8702	0.7415	0.5842	0.4227	0.2793	0.1672	0.0898
	5	1.0000	1.0000	0.9985	0.9885	0.9561	0.8883	0.7805	0.6405	0.4859	0.3373	0.2120
	6	1.0000	1.0000	0.9998	0.9978	0.9884	0.9617	0.9067	0.8164	0.6925	0.5461	0.3953
	7	1.0000	1.0000	1.0000	0.9997	0.9976	0.9897	0.9685	0.9247	0.8499	0.7414	0.6047
	8	1.0000	1.0000	1.0000	1.0000	0.9996	0.9978	0.9917	0.9757	0.9417	0.8811	0.7880
	9	1.0000	1.0000	1.0000	1.0000	1.0000	0.9997	0.9983	0.9940	0.9825	0.9574	0.9102
	10	1.0000	1.0000	1.0000	1.0000	1.0000	1.0000	0.9998	0.9989	0.9961	0.9886	0.9713
	11	1.0000	1.0000	1.0000	1.0000	1.0000	1.0000	1.0000	0.9999	0.9994	0.9978	0.9935
	12	1.0000	1.0000	1.0000	1.0000	1.0000	1.0000	1.0000	1.0000	0.9999	0.9997	0.9991
	13	1.0000	1.0000	1.0000	1.0000	1.0000	1.0000	1.0000	1.0000	1.0000	1.0000	0.9999
	14	1.0000	1.0000	1.0000	1.0000	1.0000	1.0000	1.0000	1.0000	1.0000	1.0000	1.0000
15	0	0.8601	0.4633	0.2059	0.0874	0.0352	0.0134	0.0047	0.0016	0.0005	0.0001	0.0000
	1	0.9904	0.8290	0.5490	0.3186	0.1671	0.0802	0.0353	0.0142	0.0052	0.0017	0.0005
	2	0.9996	0.9638	0.8159	0.6042	0.3980	0.2361	0.1268	0.0617	0.0271	0.0107	0.0037
	3	1.0000	0.9945	0.9444	0.8227	0.6482	0.4613	0.2969	0.1727	0.0905	0.0424	0.0176
	4	1.0000	0.9994	0.9873	0.9383	0.8358	0.6865	0.5155	0.3519	0.2173	0.1204	0.0592
	5	1.0000	0.9999	0.9977	0.9832	0.9389	0.8516	0.7216	0.5643	0.4032	0.2608	0.1509
	6	1.0000	1.0000	0.9997	0.9964	0.9819	0.9434	0.8689	0.7548	0.6098	0.4522	0.3036
	7	1.0000	1.0000	1.0000	0.9994	0.9958	0.9827	0.9500	0.8868	0.7869	0.6535	0.5000
	8	1.0000	1.0000	1.0000	0.9999	0.9992	0.9958	0.9848	0.9578	0.9050	0.8182	0.6964
	9	1.0000	1.0000	1.0000	1.0000	0.9999	0.9992	0.9963	0.9876	0.9662	0.9231	0.8491
	10	1.0000	1.0000	1.0000	1.0000	1.0000	0.9999	0.9993	0.9972	0.9907	0.9745	0.9408
	11	1.0000	1.0000	1.0000	1.0000	1.0000	1.0000	0.9999	0.9995	0.9981	0.9937	0.9824
	12	1.0000	1.0000	1.0000	1.0000	1.0000	1.0000	1.0000	0.9999	0.9997	0.9989	0.9963
	13	1.0000	1.0000	1.0000	1.0000	1.0000	1.0000	1.0000	1.0000	1.0000	0.9999	0.9995
	14	1.0000	1.0000	1.0000	1.0000	1.0000	1.0000	1.0000	1.0000	1.0000	1.0000	1.0000
	15	1.0000	1.0000	1.0000	1.0000	1.0000	1.0000	1.0000	1.0000	1.0000	1.0000	1.0000

Table A (continued)

n	x	0.01	0.05	0.10	0.15	0.20	0.25	0.30	0.35	0.40	0.45	0.50
16	0	0.8515	0.4401	0.1853	0.0743	0.0281	0.0100	0.0033	0.0010	0.0003	0.0001	0.0000
	1	0.9891	0.8108	0.5147	0.2839	0.1407	0.0635	0.0261	0.0098	0.0033	0.0010	0.0003
	2	0.9995	0.9571	0.7892	0.5614	0.3518	0.1971	0.0994	0.0451	0.0183	0.0066	0.0021
	3	1.0000	0.9930	0.9316	0.7899	0.5981	0.4050	0.2459	0.1339	0.0651	0.0281	0.0106
	4	1.0000	0.9991	0.9830	0.9209	0.7982	0.6302	0.4499	0.2892	0.1666	0.0853	0.0384
	5	1.0000	0.9999	0.9967	0.9765	0.9183	0.8103	0.6598	0.4900	0.3288	0.1976	0.1051
	6	1.0000	1.0000	0.9995	0.9944	0.9733	0.9204	0.8247	0.6881	0.5272	0.3660	0.2272
	7	1.0000	1.0000	0.9999	0.9989	0.9930	0.9729	0.9256	0.8406	0.7161	0.5629	0.4018
	8	1.0000	1.0000	1.0000	0.9998	0.9985	0.9925	0.9743	0.9329	0.8577	0.7441	0.5982
	9	1.0000	1.0000	1.0000	1.0000	0.9998	0.9984	0.9929	0.9771	0.9417	0.8759	0.7728
	10	1.0000	1.0000	1.0000	1.0000	1.0000	0.9997	0.9984	0.9938	0.9809	0.9514	0.8949
	11	1.0000	1.0000	1.0000	1.0000	1.0000	1.0000	0.9997	0.9987	0.9951	0.9851	0.9616
	12	1.0000	1.0000	1.0000	1.0000	1.0000	1.0000	1.0000	0.9998	0.9991	0.9965	0.9894
	13	1.0000	1.0000	1.0000	1.0000	1.0000	1.0000	1.0000	1.0000	0.9999	0.9994	0.9979
	14	1.0000	1.0000	1.0000	1.0000	1.0000	1.0000	1.0000	1.0000	1.0000	0.9999	0.9997
	15	1.0000	1.0000	1.0000	1.0000	1.0000	1.0000	1.0000	1.0000	1.0000	1.0000	1.0000
	16	1.0000	1.0000	1.0000	1.0000	1.0000	1.0000	1.0000	1.0000	1.0000	1.0000	1.0000
17	0	0.8429	0.4181	0.1668	0.0631	0.0225	0.0075	0.0023	0.0007	0.0002	0.0000	0.0000
	1	0.9877	0.7922	0.4818	0.2525	0.1182	0.0501	0.0193	0.0067	0.0021	0.0006	0.0001
	2	0.9994	0.9497	0.7618	0.5198	0.3096	0.1637	0.0774	0.0327	0.0123	0.0041	0.0012
	3	1.0000	0.9912	0.9174	0.7556	0.5489	0.3530	0.2019	0.1028	0.0464	0.0184	0.0064
	4	1.0000	0.9988	0.9779	0.9013	0.7582	0.5739	0.3887	0.2348	0.1260	0.0596	0.0245
	5	1.0000	0.9999	0.9953	0.9681	0.8943	0.7653	0.5968	0.4197	0.2639	0.1471	0.0717
	6	1.0000	1.0000	0.9992	0.9917	0.9623	0.8929	0.7752	0.6188	0.4478	0.2902	0.1662
	7	1.0000	1.0000	0.9999	0.9983	0.9891	0.9598	0.8954	0.7872	0.6405	0.4743	0.3145
	8	1.0000	1.0000	1.0000	0.9997	0.9974	0.9876	0.9597	0.9006	0.8011	0.6626	0.5000
	9	1.0000	1.0000	1.0000	1.0000	0.9995	0.9969	0.9873	0.9617	0.9081	0.8166	0.6855
	10	1.0000	1.0000	1.0000	1.0000	0.9999	0.9994	0.9968	0.9880	0.9652	0.9174	0.8338
	11	1.0000	1.0000	1.0000	1.0000	1.0000	0.9999	0.9993	0.9970	0.9894	0.9699	0.9283
	12	1.0000	1.0000	1.0000	1.0000	1.0000	1.0000	0.9999	0.9994	0.9975	0.9914	0.9755
	13	1.0000	1.0000	1.0000	1.0000	1.0000	1.0000	1.0000	0.9999	0.9995	0.9981	0.9936
	14	1.0000	1.0000	1.0000	1.0000	1.0000	1.0000	1.0000	1.0000	0.9999	0.9997	0.9988
	15	1.0000	1.0000	1.0000	1.0000	1.0000	1.0000	1.0000	1.0000	1.0000	1.0000	0.9999
	16	1.0000	1.0000	1.0000	1.0000	1.0000	1.0000	1.0000	1.0000	1.0000	1.0000	1.0000
	17	1.0000	1.0000	1.0000	1.0000	1.0000	1.0000	1.0000	1.0000	1.0000	1.0000	1.0000
18	0	0.8345	0.3972	0.1501	0.0536	0.0180	0.0056	0.0016	0.0004	0.0001	0.0000	0.0000
	1	0.9862	0.7735	0.4503	0.2241	0.0991	0.0395	0.0142	0.0046	0.0013	0.0003	0.0001
	2	0.9993	0.9419	0.7338	0.4797	0.2713	0.1353	0.0600	0.0236	0.0082	0.0025	0.0007
	3	1.0000	0.9891	0.9018	0.7202	0.5010	0.3057	0.1646	0.0783	0.0328	0.0120	0.0038
	4	1.0000	0.9985	0.9718	0.8794	0.7164	0.5187	0.3327	0.1886	0.0942	0.0411	0.0154
	5	1.0000	0.9998	0.9936	0.9581	0.8671	0.7175	0.5344	0.3550	0.2088	0.1077	0.0481
	6	1.0000	1.0000	0.9988	0.9882	0.9487	0.8610	0.7217	0.5491	0.3743	0.2258	0.1189
	7	1.0000	1.0000	0.9998	0.9973	0.9837	0.9431	0.8593	0.7283	0.5634	0.3915	0.2403
	8	1.0000	1.0000	1.0000	0.9995	0.9957	0.9807	0.9404	0.8609	0.7368	0.5778	0.4073
	9	1.0000	1.0000	1.0000	0.9999	0.9991	0.9946	0.9790	0.9403	0.8653	0.7473	0.5927
	10	1.0000	1.0000	1.0000	1.0000	0.9998	0.9988	0.9939	0.9788	0.9424	0.8720	0.7597
	11	1.0000	1.0000	1.0000	1.0000	1.0000	0.9998	0.9986	0.9938	0.9797	0.9463	0.8811
	12	1.0000	1.0000	1.0000	1.0000	1.0000	1.0000	0.9997	0.9986	0.9942	0.9817	0.9519

Table A (continued)

							p					
n	*x*	0.01	0.05	0.10	0.15	0.20	0.25	0.30	0.35	0.40	0.45	0.50
18	13	1.0000	1.0000	1.0000	1.0000	1.0000	1.0000	1.0000	0.9997	0.9987	0.9951	0.9846
	14	1.0000	1.0000	1.0000	1.0000	1.0000	1.0000	1.0000	1.0000	0.9998	0.9990	0.9962
	15	1.0000	1.0000	1.0000	1.0000	1.0000	1.0000	1.0000	1.0000	1.0000	0.9999	0.9993
	16	1.0000	1.0000	1.0000	1.0000	1.0000	1.0000	1.0000	1.0000	1.0000	1.0000	0.9999
	17	1.0000	1.0000	1.0000	1.0000	1.0000	1.0000	1.0000	1.0000	1.0000	1.0000	1.0000
	18	1.0000	1.0000	1.0000	1.0000	1.0000	1.0000	1.0000	1.0000	1.0000	1.0000	1.0000
19	0	0.8262	0.3774	0.1351	0.0456	0.0144	0.0042	0.0011	0.0003	0.0001	0.0000	0.0000
	1	0.9847	0.7547	0.4203	0.1985	0.0829	0.0310	0.0104	0.0031	0.0008	0.0002	0.0000
	2	0.9991	0.9335	0.7054	0.4413	0.2369	0.1113	0.0462	0.0170	0.0055	0.0015	0.0004
	3	1.0000	0.9868	0.8850	0.6841	0.4551	0.2631	0.1332	0.0591	0.0230	0.0077	0.0022
	4	1.0000	0.9980	0.9648	0.8556	0.6733	0.4654	0.2822	0.1500	0.0696	0.0280	0.0096
	5	1.0000	0.9998	0.9914	0.9463	0.8369	0.6678	0.4739	0.2968	0.1629	0.0777	0.0318
	6	1.0000	1.0000	0.9983	0.9837	0.9324	0.8251	0.6655	0.4812	0.3081	0.1727	0.0835
	7	1.0000	1.0000	0.9997	0.9959	0.9767	0.9225	0.8180	0.6656	0.4878	0.3169	0.1796
	8	1.0000	1.0000	1.0000	0.9992	0.9933	0.9713	0.9161	0.8145	0.6675	0.4940	0.3238
	9	1.0000	1.0000	1.0000	0.9999	0.9984	0.9911	0.9674	0.9125	0.8139	0.6710	0.5000
	10	1.0000	1.0000	1.0000	1.0000	0.9997	0.9977	0.9895	0.9653	0.9115	0.8159	0.6762
	11	1.0000	1.0000	1.0000	1.0000	1.0000	0.9995	0.9972	0.9886	0.9648	0.9129	0.8204
	12	1.0000	1.0000	1.0000	1.0000	1.0000	0.9999	0.9994	0.9969	0.9884	0.9658	0.9165
	13	1.0000	1.0000	1.0000	1.0000	1.0000	1.0000	0.9999	0.9993	0.9969	0.9891	0.9682
	14	1.0000	1.0000	1.0000	1.0000	1.0000	1.0000	1.0000	0.9999	0.9994	0.9972	0.9904
	15	1.0000	1.0000	1.0000	1.0000	1.0000	1.0000	1.0000	1.0000	0.9999	0.9995	0.9978
	16	1.0000	1.0000	1.0000	1.0000	1.0000	1.0000	1.0000	1.0000	1.0000	0.9999	0.9996
	17	1.0000	1.0000	1.0000	1.0000	1.0000	1.0000	1.0000	1.0000	1.0000	1.0000	1.0000
	18	1.0000	1.0000	1.0000	1.0000	1.0000	1.0000	1.0000	1.0000	1.0000	1.0000	1.0000
	19	1.0000	1.0000	1.0000	1.0000	1.0000	1.0000	1.0000	1.0000	1.0000	1.0000	1.0000
20	0	0.8179	0.3585	0.1216	0.0388	0.0115	0.0032	0.0008	0.0002	0.0000	0.0000	0.0000
	1	0.9831	0.7358	0.3917	0.1756	0.0692	0.0243	0.0076	0.0021	0.0005	0.0001	0.0000
	2	0.9990	0.9245	0.6769	0.4049	0.2061	0.0913	0.0355	0.0121	0.0036	0.0009	0.0002
	3	1.0000	0.9841	0.8670	0.6477	0.4114	0.2252	0.1071	0.0444	0.0160	0.0049	0.0013
	4	1.0000	0.9974	0.9568	0.8298	0.6296	0.4148	0.2375	0.1182	0.0510	0.0189	0.0059
	5	1.0000	0.9997	0.9887	0.9327	0.8042	0.6172	0.4164	0.2454	0.1256	0.0553	0.0207
	6	1.0000	1.0000	0.9976	0.9781	0.9133	0.7858	0.6080	0.4166	0.2500	0.1299	0.0577
	7	1.0000	1.0000	0.9996	0.9941	0.9679	0.8982	0.7723	0.6010	0.4159	0.2520	0.1316
	8	1.0000	1.0000	0.9999	0.9987	0.9900	0.9591	0.8867	0.7624	0.5956	0.4143	0.2517
	9	1.0000	1.0000	1.0000	0.9998	0.9974	0.9861	0.9520	0.8782	0.7553	0.5914	0.4119
	10	1.0000	1.0000	1.0000	1.0000	0.9994	0.9961	0.9829	0.9468	0.8725	0.7507	0.5881
	11	1.0000	1.0000	1.0000	1.0000	0.9999	0.9991	0.9949	0.9804	0.9435	0.8692	0.7483
	12	1.0000	1.0000	1.0000	1.0000	1.0000	0.9998	0.9987	0.9940	0.9790	0.9420	0.8684
	13	1.0000	1.0000	1.0000	1.0000	1.0000	1.0000	0.9997	0.9985	0.9935	0.9786	0.9423
	14	1.0000	1.0000	1.0000	1.0000	1.0000	1.0000	1.0000	0.9997	0.9984	0.9936	0.9793
	15	1.0000	1.0000	1.0000	1.0000	1.0000	1.0000	1.0000	0.9999	0.9997	0.9985	0.9941
	16	1.0000	1.0000	1.0000	1.0000	1.0000	1.0000	1.0000	1.0000	1.0000	0.9997	0.9987
	17	1.0000	1.0000	1.0000	1.0000	1.0000	1.0000	1.0000	1.0000	1.0000	1.0000	0.9998
	18	1.0000	1.0000	1.0000	1.0000	1.0000	1.0000	1.0000	1.0000	1.0000	1.0000	1.0000
	19	1.0000	1.0000	1.0000	1.0000	1.0000	1.0000	1.0000	1.0000	1.0000	1.0000	1.0000
	20	1.0000	1.0000	1.0000	1.0000	1.0000	1.0000	1.0000	1.0000	1.0000	1.0000	1.0000

Table A (continued)

							p					
n	*x*	0.01	0.05	0.10	0.15	0.20	0.25	0.30	0.35	0.40	0.45	0.50
25	0	0.7778	0.2774	0.0718	0.0172	0.0038	0.0008	0.0001	0.0000	0.0000	0.0000	0.0000
	1	0.9742	0.6424	0.2712	0.0931	0.0274	0.0070	0.0016	0.0003	0.0001	0.0000	0.0000
	2	0.998	0.8729	0.5371	0.2537	0.0982	0.0321	0.0090	0.0021	0.0004	0.0001	0.0000
	3	0.9999	0.9659	0.7636	0.4711	0.2340	0.0962	0.0332	0.0097	0.0024	0.0005	0.0001
	4	1.0000	0.9928	0.9020	0.6821	0.4207	0.2137	0.0905	0.0320	0.0095	0.0023	0.0005
	5	1.0000	0.9988	0.9666	0.8385	0.6167	0.3783	0.1935	0.0826	0.0294	0.0086	0.0020
	6	1.0000	0.9998	0.9905	0.9305	0.7800	0.5611	0.3407	0.1734	0.0736	0.0258	0.0073
	7	1.0000	1.0000	0.9977	0.9745	0.8909	0.7265	0.5118	0.3061	0.1536	0.0639	0.0216
	8	1.0000	1.0000	0.9995	0.9920	0.9532	0.8506	0.6769	0.4668	0.2735	0.1340	0.0539
	9	1.0000	1.0000	0.9999	0.9979	0.9827	0.9287	0.8106	0.6303	0.4246	0.2424	0.1148
	10	1.0000	1.0000	1.0000	0.9995	0.9944	0.9703	0.9022	0.7712	0.5858	0.3843	0.2122
	11	1.0000	1.0000	1.0000	0.9999	0.9985	0.9893	0.9558	0.8746	0.7323	0.5426	0.3450
	12	1.0000	1.0000	1.0000	1.0000	0.9996	0.9966	0.9825	0.9396	0.8462	0.6937	0.5000
	13	1.0000	1.0000	1.0000	1.0000	0.9999	0.9991	0.9940	0.9745	0.9222	0.8173	0.6550
	14	1.0000	1.0000	1.0000	1.0000	1.0000	0.9998	0.9982	0.9907	0.9656	0.9040	0.7878
	15	1.0000	1.0000	1.0000	1.0000	1.0000	1.0000	0.9995	0.9971	0.9868	0.9560	0.8852
	16	1.0000	1.0000	1.0000	1.0000	1.0000	1.0000	0.9999	0.9992	0.9957	0.9826	0.9461
	17	1.0000	1.0000	1.0000	1.0000	1.0000	1.0000	1.0000	0.9998	0.9988	0.9942	0.9784
	18	1.0000	1.0000	1.0000	1.0000	1.0000	1.0000	1.0000	1.0000	0.9997	0.9984	0.9927
	19	1.0000	1.0000	1.0000	1.0000	1.0000	1.0000	1.0000	1.0000	0.9999	0.9996	0.9980
	20	1.0000	1.0000	1.0000	1.0000	1.0000	1.0000	1.0000	1.0000	1.0000	0.9999	0.9995
	21	1.0000	1.0000	1.0000	1.0000	1.0000	1.0000	1.0000	1.0000	1.0000	1.0000	0.9999
	22	1.0000	1.0000	1.0000	1.0000	1.0000	1.0000	1.0000	1.0000	1.0000	1.0000	1.0000
	23	1.0000	1.0000	1.0000	1.0000	1.0000	1.0000	1.0000	1.0000	1.0000	1.0000	1.0000
	24	1.0000	1.0000	1.0000	1.0000	1.0000	1.0000	1.0000	1.0000	1.0000	1.0000	1.0000
	25	1.0000	1.0000	1.0000	1.0000	1.0000	1.0000	1.0000	1.0000	1.0000	1.0000	1.0000

Table B   Cumulative Values of the Poisson Distribution

λ

x	0.1	0.2	0.3	0.4	0.5	0.6	0.7	0.8	0.9	1
0	0.9048	0.8187	0.7408	0.6703	0.6065	0.5488	0.4966	0.4493	0.4066	0.3679
1	0.9953	0.9825	0.9631	0.9384	0.9098	0.8781	0.8442	0.8088	0.7725	0.7358
2	0.9998	0.9989	0.9964	0.9921	0.9856	0.9769	0.9659	0.9526	0.9371	0.9197
3	1.0000	0.9999	0.9997	0.9992	0.9982	0.9966	0.9942	0.9909	0.9865	0.9810
4	1.0000	1.0000	1.0000	0.9999	0.9998	0.9996	0.9992	0.9986	0.9977	0.9963
5	1.0000	1.0000	1.0000	1.0000	1.0000	1.0000	0.9999	0.9998	0.9997	0.9994
6	1.0000	1.0000	1.0000	1.0000	1.0000	1.0000	1.0000	1.0000	1.0000	0.9999
7	1.0000	1.0000	1.0000	1.0000	1.0000	1.0000	1.0000	1.0000	1.0000	1.0000

x	1.1	1.2	1.3	1.4	1.5	1.6	1.7	1.8	1.9	2
0	0.3329	0.3012	0.2725	0.2466	0.2231	0.2019	0.1827	0.1653	0.1496	0.1353
1	0.6990	0.6626	0.6268	0.5918	0.5578	0.5249	0.4932	0.4628	0.4337	0.4060
2	0.9004	0.8795	0.8571	0.8335	0.8088	0.7834	0.7572	0.7306	0.7037	0.6767
3	0.9743	0.9662	0.9569	0.9463	0.9344	0.9212	0.9068	0.8913	0.8747	0.8571
4	0.9946	0.9923	0.9893	0.9857	0.9814	0.9763	0.9704	0.9636	0.9559	0.9473
5	0.9990	0.9985	0.9978	0.9968	0.9955	0.9940	0.9920	0.9896	0.9868	0.9834
6	0.9999	0.9997	0.9996	0.9994	0.9991	0.9987	0.9981	0.9974	0.9966	0.9955
7	1.0000	1.0000	0.9999	0.9999	0.9998	0.9997	0.9996	0.9994	0.9992	0.9989
8	1.0000	1.0000	1.0000	1.0000	1.0000	1.0000	0.9999	0.9999	0.9998	0.9998
9	1.0000	1.0000	1.0000	1.0000	1.0000	1.0000	1.0000	1.0000	1.0000	1.0000

x	2.1	2.2	2.3	2.4	2.5	2.6	2.7	2.8	2.9	3
0	0.1225	0.1108	0.1003	0.0907	0.0821	0.0743	0.0672	0.0608	0.0550	0.0498
1	0.3796	0.3546	0.3309	0.3084	0.2873	0.2674	0.2487	0.2311	0.2146	0.1991
2	0.6496	0.6227	0.5960	0.5697	0.5438	0.5184	0.4936	0.4695	0.4460	0.4232
3	0.8386	0.8194	0.7993	0.7787	0.7576	0.7360	0.7141	0.6919	0.6696	0.6472
4	0.9379	0.9275	0.9162	0.9041	0.8912	0.8774	0.8629	0.8477	0.8318	0.8153
5	0.9796	0.9751	0.9700	0.9643	0.9580	0.9510	0.9433	0.9349	0.9258	0.9161
6	0.9941	0.9925	0.9906	0.9884	0.9858	0.9828	0.9794	0.9756	0.9713	0.9665
7	0.9985	0.9980	0.9974	0.9967	0.9958	0.9947	0.9934	0.9919	0.9901	0.9881
8	0.9997	0.9995	0.9994	0.9991	0.9989	0.9985	0.9981	0.9976	0.9969	0.9962
9	0.9999	0.9999	0.9999	0.9998	0.9997	0.9996	0.9995	0.9993	0.9991	0.9989
10	1.0000	1.0000	1.0000	1.0000	0.9999	0.9999	0.9999	0.9998	0.9998	0.9997
11	1.0000	1.0000	1.0000	1.0000	1.0000	1.0000	1.0000	1.0000	0.9999	0.9999
12	1.0000	1.0000	1.0000	1.0000	1.0000	1.0000	1.0000	1.0000	1.0000	1.0000

x	3.1	3.2	3.3	3.4	3.5	3.6	3.7	3.8	3.9	4
0	0.0450	0.0408	0.0369	0.0334	0.0302	0.0273	0.0247	0.0224	0.0202	0.0183
1	0.1847	0.1712	0.1586	0.1468	0.1359	0.1257	0.1162	0.1074	0.0992	0.0916
2	0.4012	0.3799	0.3594	0.3397	0.3208	0.3027	0.2854	0.2689	0.2531	0.2381
3	0.6248	0.6025	0.5803	0.5584	0.5366	0.5152	0.4942	0.4735	0.4532	0.4335
4	0.7982	0.7806	0.7626	0.7442	0.7254	0.7064	0.6872	0.6678	0.6484	0.6288
5	0.9057	0.8946	0.8829	0.8705	0.8576	0.8441	0.8301	0.8156	0.8006	0.7851
6	0.9612	0.9554	0.9490	0.9421	0.9347	0.9267	0.9182	0.9091	0.8995	0.8893
7	0.9858	0.9832	0.9802	0.9769	0.9733	0.9692	0.9648	0.9599	0.9546	0.9489
8	0.9953	0.9943	0.9931	0.9917	0.9901	0.9883	0.9863	0.9840	0.9815	0.9786
9	0.9986	0.9982	0.9978	0.9973	0.9967	0.9960	0.9952	0.9942	0.9931	0.9919
10	0.9996	0.9995	0.9994	0.9992	0.9990	0.9987	0.9984	0.9981	0.9977	0.9972
11	0.9999	0.9999	0.9998	0.9998	0.9997	0.9996	0.9995	0.9994	0.9993	0.9991

Table B (continued)

						λ				

x	3.1	3.2	3.3	3.4	3.5	3.6	3.7	3.8	3.9	4
12	1.0000	1.0000	1.0000	0.9999	0.9999	0.9999	0.9999	0.9998	0.9998	0.9997
13	1.0000	1.0000	1.0000	1.0000	1.0000	1.0000	1.0000	1.0000	0.9999	0.9999
14	1.0000	1.0000	1.0000	1.0000	1.0000	1.0000	1.0000	1.0000	1.0000	1.0000

x	4.1	4.2	4.3	4.4	4.5	4.6	4.7	4.8	4.9	5
0	0.0166	0.0150	0.0136	0.0123	0.0111	0.0101	0.0091	0.0082	0.0074	0.0067
1	0.0845	0.0780	0.0719	0.0663	0.0611	0.0563	0.0518	0.0477	0.0439	0.0404
2	0.2238	0.2102	0.1974	0.1851	0.1736	0.1626	0.1523	0.1425	0.1333	0.1247
3	0.4142	0.3954	0.3772	0.3594	0.3423	0.3257	0.3097	0.2942	0.2793	0.2650
4	0.6093	0.5898	0.5704	0.5512	0.5321	0.5132	0.4946	0.4763	0.4582	0.4405
5	0.7693	0.7531	0.7367	0.7199	0.7029	0.6858	0.6684	0.651	0.6335	0.6160
6	0.8786	0.8675	0.8558	0.8436	0.8311	0.8180	0.8046	0.7908	0.7767	0.7622
7	0.9427	0.9361	0.9290	0.9214	0.9134	0.9049	0.8960	0.8867	0.8769	0.8666
8	0.9755	0.9721	0.9683	0.9642	0.9597	0.9549	0.9497	0.9442	0.9382	0.9319
9	0.9905	0.9889	0.9871	0.9851	0.9829	0.9805	0.9778	0.9749	0.9717	0.9682
10	0.9966	0.9959	0.9952	0.9943	0.9933	0.9922	0.9910	0.9896	0.9880	0.9863
11	0.9989	0.9986	0.9983	0.9980	0.9976	0.9971	0.9966	0.9960	0.9953	0.9945
12	0.9997	0.9996	0.9995	0.9993	0.9992	0.9990	0.9988	0.9986	0.9983	0.9980
13	0.9999	0.9999	0.9998	0.9998	0.9997	0.9997	0.9996	0.9995	0.9994	0.9993
14	1.0000	1.0000	1.0000	0.9999	0.9999	0.9999	0.9999	0.9999	0.9998	0.9998
15	1.0000	1.0000	1.0000	1.0000	1.0000	1.0000	1.0000	1.0000	0.9999	0.9999
16	1.0000	1.0000	1.0000	1.0000	1.0000	1.0000	1.0000	1.0000	1.0000	1.0000

x	5.1	5.2	5.3	5.4	5.5	5.6	5.7	5.8	5.9	6
0	0.0061	0.0055	0.0050	0.0045	0.0041	0.0037	0.0033	0.0030	0.0027	0.0025
1	0.0372	0.0342	0.0314	0.0289	0.0266	0.0244	0.0224	0.0206	0.0189	0.0174
2	0.1165	0.1088	0.1016	0.0948	0.0884	0.0824	0.0768	0.0715	0.0666	0.0620
3	0.2513	0.2381	0.2254	0.2133	0.2017	0.1906	0.1800	0.1700	0.1604	0.1512
4	0.4231	0.4061	0.3895	0.3733	0.3575	0.3422	0.3272	0.3127	0.2987	0.2851
5	0.5984	0.5809	0.5635	0.5461	0.5289	0.5119	0.4950	0.4783	0.4619	0.4457
6	0.7474	0.7324	0.7171	0.7017	0.6860	0.6703	0.6544	0.6384	0.6224	0.6063
7	0.8560	0.8449	0.8335	0.8217	0.8095	0.7970	0.7841	0.7710	0.7576	0.7440
8	0.9252	0.9181	0.9106	0.9027	0.8944	0.8857	0.8766	0.8672	0.8574	0.8472
9	0.9644	0.9603	0.9559	0.9512	0.9462	0.9409	0.9352	0.9292	0.9228	0.9161
10	0.9844	0.9823	0.9800	0.9775	0.9747	0.9718	0.9686	0.9651	0.9614	0.9574
11	0.9937	0.9927	0.9916	0.9904	0.9890	0.9875	0.9859	0.9841	0.9821	0.9799
12	0.9976	0.9972	0.9967	0.9962	0.9955	0.9949	0.9941	0.9932	0.9922	0.9912
13	0.9992	0.9990	0.9988	0.9986	0.9983	0.9980	0.9977	0.9973	0.9969	0.9964
14	0.9997	0.9997	0.9996	0.9995	0.9994	0.9993	0.9991	0.9990	0.9988	0.9986
15	0.9999	0.9999	0.9999	0.9998	0.9998	0.9998	0.9997	0.9996	0.9996	0.9995
16	1.0000	1.0000	1.0000	0.9999	0.9999	0.9999	0.9999	0.9999	0.9999	0.9998
17	1.0000	1.0000	1.0000	1.0000	1.0000	1.0000	1.0000	1.0000	1.0000	0.9999
18	1.0000	1.0000	1.0000	1.0000	1.0000	1.0000	1.0000	1.0000	1.0000	1.0000

x	6.1	6.2	6.3	6.4	6.5	6.6	6.7	6.8	6.9	7
0	0.0022	0.0020	0.0018	0.0017	0.0015	0.0014	0.0012	0.0011	0.0010	0.0009
1	0.0159	0.0146	0.0134	0.0123	0.0113	0.0103	0.0095	0.0087	0.0080	0.0073
2	0.0577	0.0536	0.0498	0.0463	0.0430	0.0400	0.0371	0.0344	0.0320	0.0296
3	0.1425	0.1342	0.1264	0.1189	0.1118	0.1052	0.0988	0.0928	0.0871	0.0818

Table B (continued)

					λ					

x	6.1	6.2	6.3	6.4	6.5	6.6	6.7	6.8	6.9	7
4	0.2719	0.2592	0.2469	0.2351	0.2237	0.2127	0.2022	0.1920	0.1823	0.1730
5	0.4298	0.4141	0.3988	0.3837	0.3690	0.3547	0.3406	0.3270	0.3137	0.3007
6	0.5902	0.5742	0.5582	0.5423	0.5265	0.5108	0.4953	0.4799	0.4647	0.4497
7	0.7301	0.7160	0.7017	0.6873	0.6728	0.6581	0.6433	0.6285	0.6136	0.5987
8	0.8367	0.8259	0.8148	0.8033	0.7916	0.7796	0.7673	0.7548	0.7420	0.7291
9	0.9090	0.9016	0.8939	0.8858	0.8774	0.8686	0.8596	0.8502	0.8405	0.8305
10	0.9531	0.9486	0.9437	0.9386	0.9332	0.9274	0.9214	0.9151	0.9084	0.9015
11	0.9776	0.9750	0.9723	0.9693	0.9661	0.9627	0.9591	0.9552	0.9510	0.9467
12	0.9900	0.9887	0.9873	0.9857	0.9840	0.9821	0.9801	0.9779	0.9755	0.9730
13	0.9958	0.9952	0.9945	0.9937	0.9929	0.9920	0.9909	0.9898	0.9885	0.9872
14	0.9984	0.9981	0.9978	0.9974	0.9970	0.9966	0.9961	0.9956	0.9950	0.9943
15	0.9994	0.9993	0.9992	0.9990	0.9988	0.9986	0.9984	0.9982	0.9979	0.9976
16	0.9998	0.9997	0.9997	0.9996	0.9996	0.9995	0.9994	0.9993	0.9992	0.9990
17	0.9999	0.9999	0.9999	0.9999	0.9998	0.9998	0.9998	0.9997	0.9997	0.9996
18	1.0000	1.0000	1.0000	1.0000	0.9999	0.9999	0.9999	0.9999	0.9999	0.9999
19	1.0000	1.0000	1.0000	1.0000	1.0000	1.0000	1.0000	1.0000	1.0000	1.0000

x	7.1	7.2	7.3	7.4	7.5	7.6	7.7	7.8	7.9	8
0	0.0008	0.0007	0.0007	0.0006	0.0006	0.0005	0.0005	0.0004	0.0004	0.0003
1	0.0067	0.0061	0.0056	0.0051	0.0047	0.0043	0.0039	0.0036	0.0033	0.0030
2	0.0275	0.0255	0.0236	0.0219	0.0203	0.0188	0.0174	0.0161	0.0149	0.0138
3	0.0767	0.0719	0.0674	0.0632	0.0591	0.0554	0.0518	0.0485	0.0453	0.0424
4	0.1641	0.1555	0.1473	0.1395	0.1321	0.1249	0.1181	0.1117	0.1055	0.0996
5	0.2881	0.2759	0.2640	0.2526	0.2414	0.2307	0.2203	0.2103	0.2006	0.1912
6	0.4349	0.4204	0.4060	0.3920	0.3782	0.3646	0.3514	0.3384	0.3257	0.3134
7	0.5838	0.5689	0.5541	0.5393	0.5246	0.5100	0.4956	0.4812	0.4670	0.4530
8	0.7160	0.7027	0.6892	0.6757	0.6620	0.6482	0.6343	0.6204	0.6065	0.5925
9	0.8202	0.8096	0.7988	0.7877	0.7764	0.7649	0.7531	0.7411	0.7290	0.7166
10	0.8942	0.8867	0.8788	0.8707	0.8622	0.8535	0.8445	0.8352	0.8257	0.8159
11	0.9420	0.9371	0.9319	0.9265	0.9208	0.9148	0.9085	0.9020	0.8952	0.8881
12	0.9703	0.9673	0.9642	0.9609	0.9573	0.9536	0.9496	0.9454	0.9409	0.9362
13	0.9857	0.9841	0.9824	0.9805	0.9784	0.9762	0.9739	0.9714	0.9687	0.9658
14	0.9935	0.9927	0.9918	0.9908	0.9897	0.9886	0.9873	0.9859	0.9844	0.9827
15	0.9972	0.9969	0.9964	0.9959	0.9954	0.9948	0.9941	0.9934	0.9926	0.9918
16	0.9989	0.9987	0.9985	0.9983	0.9980	0.9978	0.9974	0.9971	0.9967	0.9963
17	0.9996	0.9995	0.9994	0.9993	0.9992	0.9991	0.9989	0.9988	0.9986	0.9984
18	0.9998	0.9998	0.9998	0.9997	0.9997	0.9996	0.9996	0.9995	0.9994	0.9993
19	0.9999	0.9999	0.9999	0.9999	0.9999	0.9999	0.9998	0.9998	0.9998	0.9997
20	1.0000	1.0000	1.0000	1.0000	1.0000	1.0000	0.9999	0.9999	0.9999	0.9999
21	1.0000	1.0000	1.0000	1.0000	1.0000	1.0000	1.0000	1.0000	1.0000	1.0000

x	8.1	8.2	8.3	8.4	8.5	8.6	8.7	8.8	8.9	9
0	0.0003	0.0003	0.0002	0.0002	0.0002	0.0002	0.0002	0.0002	0.0001	0.0001
1	0.0028	0.0025	0.0023	0.0021	0.0019	0.0018	0.0016	0.0015	0.0014	0.0012
2	0.0127	0.0118	0.0109	0.0100	0.0093	0.0086	0.0079	0.0073	0.0068	0.0062
3	0.0396	0.0370	0.0346	0.0323	0.0301	0.0281	0.0262	0.0244	0.0228	0.0212
4	0.0940	0.0887	0.0837	0.0789	0.0744	0.0701	0.0660	0.0621	0.0584	0.0550
5	0.1822	0.1736	0.1653	0.1573	0.1496	0.1422	0.1352	0.1284	0.1219	0.1157

Table B (continued)

						λ				

x	8.1	8.2	8.3	8.4	8.5	8.6	8.7	8.8	8.9	9
6	0.3013	0.2896	0.2781	0.2670	0.2562	0.2457	0.2355	0.2256	0.2160	0.2068
7	0.4391	0.4254	0.4119	0.3987	0.3856	0.3728	0.3602	0.3478	0.3357	0.3239
8	0.5786	0.5647	0.5507	0.5369	0.5231	0.5094	0.4958	0.4823	0.4689	0.4557
9	0.7041	0.6915	0.6788	0.6659	0.6530	0.6400	0.6269	0.6137	0.6006	0.5874
10	0.8058	0.7956	0.7850	0.7743	0.7634	0.7522	0.7409	0.7294	0.7178	0.7060
11	0.8807	0.8731	0.8652	0.8571	0.8487	0.8400	0.8311	0.8220	0.8126	0.8030
12	0.9313	0.9261	0.9207	0.9150	0.9091	0.9029	0.8965	0.8898	0.8829	0.8758
13	0.9628	0.9595	0.9561	0.9524	0.9486	0.9445	0.9403	0.9358	0.9311	0.9261
14	0.9810	0.9791	0.9771	0.9749	0.9726	0.9701	0.9675	0.9647	0.9617	0.9585
15	0.9908	0.9898	0.9887	0.9875	0.9862	0.9848	0.9832	0.9816	0.9798	0.9780
16	0.9958	0.9953	0.9947	0.9941	0.9934	0.9926	0.9918	0.9909	0.9899	0.9889
17	0.9982	0.9979	0.9977	0.9973	0.9970	0.9966	0.9962	0.9957	0.9952	0.9947
18	0.9992	0.9991	0.9990	0.9989	0.9987	0.9985	0.9983	0.9981	0.9978	0.9976
19	0.9997	0.9997	0.9996	0.9995	0.9995	0.9994	0.9993	0.9992	0.9991	0.9989
20	0.9999	0.9999	0.9998	0.9998	0.9998	0.9998	0.9997	0.9997	0.9996	0.9996
21	1.0000	1.0000	0.9999	0.9999	0.9999	0.9999	0.9999	0.9999	0.9998	0.9998
22	1.0000	1.0000	1.0000	1.0000	1.0000	1.0000	1.0000	1.0000	0.9999	0.9999
23	1.0000	1.0000	1.0000	1.0000	1.0000	1.0000	1.0000	1.0000	1.0000	1.0000

x	9.1	9.2	9.3	9.4	9.5	9.6	9.7	9.8	9.9	10
0	0.0001	0.0001	0.0001	0.0001	0.0001	0.0001	0.0001	0.0001	0.0001	0.0000
1	0.0011	0.0010	0.0009	0.0009	0.0008	0.0007	0.0007	0.0006	0.0005	0.0005
2	0.0058	0.0053	0.0049	0.0045	0.0042	0.0038	0.0035	0.0033	0.0030	0.0028
3	0.0198	0.0184	0.0172	0.0160	0.0149	0.0138	0.0129	0.0120	0.0111	0.0103
4	0.0517	0.0486	0.0456	0.0429	0.0403	0.0378	0.0355	0.0333	0.0312	0.0293
5	0.1098	0.1041	0.0986	0.0935	0.0885	0.0838	0.0793	0.0750	0.0710	0.0671
6	0.1978	0.1892	0.1808	0.1727	0.1649	0.1574	0.1502	0.1433	0.1366	0.1301
7	0.3123	0.3010	0.2900	0.2792	0.2687	0.2584	0.2485	0.2388	0.2294	0.2202
8	0.4426	0.4296	0.4168	0.4042	0.3918	0.3796	0.3676	0.3558	0.3442	0.3328
9	0.5742	0.5611	0.5479	0.5349	0.5218	0.5089	0.4960	0.4832	0.4705	0.4579
10	0.6941	0.6820	0.6699	0.6576	0.6453	0.6329	0.6205	0.6080	0.5955	0.5830
11	0.7932	0.7832	0.7730	0.7626	0.7520	0.7412	0.7303	0.7193	0.7081	0.6968
12	0.8684	0.8607	0.8529	0.8448	0.8364	0.8279	0.8191	0.8101	0.8009	0.7916
13	0.9210	0.9156	0.9100	0.9042	0.8981	0.8919	0.8853	0.8786	0.8716	0.8645
14	0.9552	0.9517	0.9480	0.9441	0.9400	0.9357	0.9312	0.9265	0.9216	0.9165
15	0.9760	0.9738	0.9715	0.9691	0.9665	0.9638	0.9609	0.9579	0.9546	0.9513
16	0.9878	0.9865	0.9852	0.9838	0.9823	0.9806	0.9789	0.9770	0.9751	0.9730
17	0.9941	0.9934	0.9927	0.9919	0.9911	0.9902	0.9892	0.9881	0.9870	0.9857
18	0.9973	0.9969	0.9966	0.9962	0.9957	0.9952	0.9947	0.9941	0.9935	0.9928
19	0.9988	0.9986	0.9985	0.9983	0.9980	0.9978	0.9975	0.9972	0.9969	0.9965
20	0.9995	0.9994	0.9993	0.9992	0.9991	0.9990	0.9989	0.9987	0.9986	0.9984
21	0.9998	0.9998	0.9997	0.9997	0.9996	0.9996	0.9995	0.9995	0.9994	0.9993
22	0.9999	0.9999	0.9999	0.9999	0.9999	0.9998	0.9998	0.9998	0.9997	0.9997
23	1.0000	1.0000	1.0000	1.0000	0.9999	0.9999	0.9999	0.9999	0.9999	0.9999
24	1.0000	1.0000	1.0000	1.0000	1.0000	1.0000	1.0000	1.0000	1.0000	1.0000

Table B (continued)

	λ									
x	11	12	13	14	15	16	17	18	19	20
0	0.0000	0.0000	0.0000	0.0000	0.0000	0.0000	0.0000	0.0000	0.0000	0.0000
1	0.0002	0.0001	0.0000	0.0000	0.0000	0.0000	0.0000	0.0000	0.0000	0.0000
2	0.0012	0.0005	0.0002	0.0001	0.0000	0.0000	0.0000	0.0000	0.0000	0.0000
3	0.0049	0.0023	0.0011	0.0005	0.0002	0.0001	0.0000	0.0000	0.0000	0.0000
4	0.0151	0.0076	0.0037	0.0018	0.0009	0.0004	0.0002	0.0001	0.0000	0.0000
5	0.0375	0.0203	0.0107	0.0055	0.0028	0.0014	0.0007	0.0003	0.0002	0.0001
6	0.0786	0.0458	0.0259	0.0142	0.0076	0.0040	0.0021	0.0010	0.0005	0.0003
7	0.1432	0.0895	0.0540	0.0316	0.0180	0.0100	0.0054	0.0029	0.0015	0.0008
8	0.2320	0.1550	0.0998	0.0621	0.0374	0.0220	0.0126	0.0071	0.0039	0.0021
9	0.3405	0.2424	0.1658	0.1094	0.0699	0.0433	0.0261	0.0154	0.0089	0.0050
10	0.4599	0.3472	0.2517	0.1757	0.1185	0.0774	0.0491	0.0304	0.0183	0.0108
11	0.5793	0.4616	0.3532	0.2600	0.1848	0.1270	0.0847	0.0549	0.0347	0.0214
12	0.6887	0.5760	0.4631	0.3585	0.2676	0.1931	0.1350	0.0917	0.0606	0.0390
13	0.7813	0.6815	0.5730	0.4644	0.3632	0.2745	0.2009	0.1426	0.0984	0.0661
14	0.8540	0.7720	0.6751	0.5704	0.4657	0.3675	0.2808	0.2081	0.1497	0.1049
15	0.9074	0.8444	0.7636	0.6694	0.5681	0.4667	0.3715	0.2867	0.2148	0.1565
16	0.9441	0.8987	0.8355	0.7559	0.6641	0.5660	0.4677	0.3751	0.2920	0.2211
17	0.9678	0.9370	0.8905	0.8272	0.7489	0.6593	0.5640	0.4686	0.3784	0.2970
18	0.9823	0.9626	0.9302	0.8826	0.8195	0.7423	0.6550	0.5622	0.4695	0.3814
19	0.9907	0.9787	0.9573	0.9235	0.8752	0.8122	0.7363	0.6509	0.5606	0.4703
20	0.9953	0.9884	0.9750	0.9521	0.9170	0.8682	0.8055	0.7307	0.6472	0.5591
21	0.9977	0.9939	0.9859	0.9712	0.9469	0.9108	0.8615	0.7991	0.7255	0.6437
22	0.9990	0.9970	0.9924	0.9833	0.9673	0.9418	0.9047	0.8551	0.7931	0.7206
23	0.9995	0.9985	0.9960	0.9907	0.9805	0.9633	0.9367	0.8989	0.8490	0.7875
24	0.9998	0.9993	0.9980	0.9950	0.9888	0.9777	0.9594	0.9317	0.8933	0.8432
25	0.9999	0.9997	0.9990	0.9974	0.9938	0.9869	0.9748	0.9554	0.9269	0.8878
26	1.0000	0.9999	0.9995	0.9987	0.9967	0.9925	0.9848	0.9718	0.9514	0.9221
27	1.0000	0.9999	0.9998	0.9994	0.9983	0.9959	0.9912	0.9827	0.9687	0.9475
28	1.0000	1.0000	0.9999	0.9997	0.9991	0.9978	0.9950	0.9897	0.9805	0.9657
29	1.0000	1.0000	1.0000	0.9999	0.9996	0.9989	0.9973	0.9941	0.9882	0.9782
30	1.0000	1.0000	1.0000	0.9999	0.9998	0.9994	0.9986	0.9967	0.9930	0.9865
31	1.0000	1.0000	1.0000	1.0000	0.9999	0.9997	0.9993	0.9982	0.9960	0.9919
32	1.0000	1.0000	1.0000	1.0000	1.0000	0.9999	0.9996	0.9990	0.9978	0.9953
33	1.0000	1.0000	1.0000	1.0000	1.0000	0.9999	0.9998	0.9995	0.9988	0.9973
34	1.0000	1.0000	1.0000	1.0000	1.0000	1.0000	0.9999	0.9998	0.9994	0.9985
35	1.0000	1.0000	1.0000	1.0000	1.0000	1.0000	1.0000	0.9999	0.9997	0.9992
36	1.0000	1.0000	1.0000	1.0000	1.0000	1.0000	1.0000	0.9999	0.9998	0.9996
37	1.0000	1.0000	1.0000	1.0000	1.0000	1.0000	1.0000	1.0000	0.9999	0.9998
38	1.0000	1.0000	1.0000	1.0000	1.0000	1.0000	1.0000	1.0000	1.0000	0.9999
39	1.0000	1.0000	1.0000	1.0000	1.0000	1.0000	1.0000	1.0000	1.0000	0.9999

Table C   **Cumulative Values of the Standard Normal Distribution**

Z	0.09	0.08	0.07	0.06	0.05	0.04	0.03	0.02	0.01	0.00
-3.0	0.0010	0.0010	0.0011	0.0011	0.0011	0.0012	0.0012	0.0013	0.0013	0.0013
-2.9	0.0014	0.0014	0.0015	0.0015	0.0016	0.0016	0.0017	0.0018	0.0018	0.0019
-2.8	0.0019	0.0020	0.0021	0.0021	0.0022	0.0023	0.0023	0.0024	0.0025	0.0026
-2.7	0.0026	0.0027	0.0028	0.0029	0.0030	0.0031	0.0032	0.0033	0.0034	0.0035
-2.6	0.0036	0.0037	0.0038	0.0039	0.0040	0.0041	0.0043	0.0044	0.0045	0.0047
-2.5	0.0048	0.0049	0.0051	0.0052	0.0054	0.0055	0.0057	0.0059	0.0060	0.0062
-2.4	0.0064	0.0066	0.0068	0.0069	0.0071	0.0073	0.0075	0.0078	0.0080	0.0082
-2.3	0.0084	0.0087	0.0089	0.0091	0.0094	0.0096	0.0099	0.0102	0.0104	0.0107
-2.2	0.0110	0.0113	0.0116	0.0119	0.0122	0.0125	0.0129	0.0132	0.0136	0.0139
-2.1	0.0143	0.0146	0.0150	0.0154	0.0158	0.0162	0.0166	0.0170	0.0174	0.0179
-2.0	0.0183	0.0188	0.0192	0.0197	0.0202	0.0207	0.0212	0.0217	0.0222	0.0228
-1.9	0.0233	0.0239	0.0244	0.0250	0.0256	0.0262	0.0268	0.0274	0.0281	0.0287
-1.8	0.0294	0.0301	0.0307	0.0314	0.0322	0.0329	0.0336	0.0344	0.0351	0.0359
-1.7	0.0367	0.0375	0.0384	0.0392	0.0401	0.0409	0.0418	0.0427	0.0436	0.0446
-1.6	0.0455	0.0465	0.0475	0.0485	0.0495	0.0505	0.0516	0.0526	0.0537	0.0548
-1.5	0.0559	0.0571	0.0582	0.0594	0.0606	0.0618	0.0630	0.0643	0.0655	0.0668
-1.4	0.0681	0.0694	0.0708	0.0721	0.0735	0.0749	0.0764	0.0778	0.0793	0.0808
-1.3	0.0823	0.0838	0.0853	0.0869	0.0885	0.0901	0.0918	0.0934	0.0951	0.0968
-1.2	0.0985	0.1003	0.1020	0.1038	0.1056	0.1075	0.1093	0.1112	0.1131	0.1151
-1.1	0.1170	0.1190	0.1210	0.1230	0.1251	0.1271	0.1292	0.1314	0.1335	0.1357
-1.0	0.1379	0.1401	0.1423	0.1446	0.1469	0.1492	0.1515	0.1539	0.1562	0.1587
-0.9	0.1611	0.1635	0.1660	0.1685	0.1711	0.1736	0.1762	0.1788	0.1814	0.1841
-0.8	0.1867	0.1894	0.1922	0.1949	0.1977	0.2005	0.2033	0.2061	0.2090	0.2119
-0.7	0.2148	0.2177	0.2206	0.2236	0.2266	0.2297	0.2327	0.2358	0.2389	0.2420
-0.6	0.2451	0.2483	0.2514	0.2546	0.2578	0.2611	0.2643	0.2676	0.2709	0.2743
-0.5	0.2776	0.2810	0.2843	0.2877	0.2912	0.2946	0.2981	0.3015	0.3050	0.3085
-0.4	0.3121	0.3156	0.3192	0.3228	0.3264	0.3300	0.3336	0.3372	0.3409	0.3446
-0.3	0.3483	0.3520	0.3557	0.3594	0.3632	0.3669	0.3707	0.3745	0.3783	0.3821
-0.2	0.3859	0.3897	0.3936	0.3974	0.4013	0.4052	0.4090	0.4129	0.4168	0.4207
-0.1	0.4247	0.4286	0.4325	0.4364	0.4404	0.4443	0.4483	0.4522	0.4562	0.4602
-0.0	0.4641	0.4681	0.4721	0.4761	0.4801	0.4840	0.4880	0.4920	0.4960	0.5000

Table C   (continued)

Z	0.00	0.01	0.02	0.03	0.04	0.05	0.06	0.07	0.08	0.09
0.0	0.5000	0.5040	0.5080	0.5120	0.5160	0.5199	0.5239	0.5279	0.5319	0.5359
0.1	0.5398	0.5438	0.5478	0.5517	0.5557	0.5596	0.5636	0.5675	0.5714	0.5753
0.2	0.5793	0.5832	0.5871	0.5910	0.5948	0.5987	0.6026	0.6064	0.6103	0.6141
0.3	0.6179	0.6217	0.6255	0.6293	0.6331	0.6368	0.6406	0.6443	0.6480	0.6517
0.4	0.6554	0.6591	0.6628	0.6664	0.6700	0.6736	0.6772	0.6808	0.6844	0.6879
0.5	0.6915	0.6950	0.6985	0.7019	0.7054	0.7088	0.7123	0.7157	0.7190	0.7224
0.6	0.7257	0.7291	0.7324	0.7357	0.7389	0.7422	0.7454	0.7486	0.7517	0.7549
0.7	0.7580	0.7611	0.7642	0.7673	0.7704	0.7734	0.7764	0.7794	0.7823	0.7852
0.8	0.7881	0.7910	0.7939	0.7967	0.7995	0.8023	0.8051	0.8078	0.8106	0.8133
0.9	0.8159	0.8186	0.8212	0.8238	0.8264	0.8289	0.8315	0.8340	0.8365	0.8389
1.0	0.8413	0.8438	0.8461	0.8485	0.8508	0.8531	0.8554	0.8577	0.8599	0.8621
1.1	0.8643	0.8665	0.8686	0.8708	0.8729	0.8749	0.8770	0.8790	0.8810	0.8830
1.2	0.8849	0.8869	0.8888	0.8907	0.8925	0.8944	0.8962	0.8980	0.8997	0.9015
1.3	0.9032	0.9049	0.9066	0.9082	0.9099	0.9115	0.9131	0.9147	0.9162	0.9177
1.4	0.9192	0.9207	0.9222	0.9236	0.9251	0.9265	0.9279	0.9292	0.9306	0.9319
1.5	0.9332	0.9345	0.9357	0.9370	0.9382	0.9394	0.9406	0.9418	0.9429	0.9441
1.6	0.9452	0.9463	0.9474	0.9484	0.9495	0.9505	0.9515	0.9525	0.9535	0.9545
1.7	0.9554	0.9564	0.9573	0.9582	0.9591	0.9599	0.9608	0.9616	0.9625	0.9633
1.8	0.9641	0.9649	0.9656	0.9664	0.9671	0.9678	0.9686	0.9693	0.9699	0.9706
1.9	0.9713	0.9719	0.9726	0.9732	0.9738	0.9744	0.9750	0.9756	0.976	0.9767
2.0	0.9772	0.9778	0.9783	0.9788	0.9793	0.9798	0.9803	0.9808	0.9812	0.9817
2.1	0.9821	0.9826	0.9830	0.9834	0.9838	0.9842	0.9846	0.9850	0.9854	0.9857
2.2	0.9861	0.9864	0.9868	0.9871	0.9875	0.9878	0.9881	0.9884	0.9887	0.9890
2.3	0.9893	0.9896	0.9898	0.9901	0.9904	0.9906	0.9909	0.9911	0.9913	0.9916
2.4	0.9918	0.9920	0.9922	0.9925	0.9927	0.9929	0.9931	0.9932	0.9934	0.9936
2.5	0.9938	0.9940	0.9941	0.9943	0.9945	0.9946	0.9948	0.9949	0.9951	0.9952
2.6	0.9953	0.9955	0.9956	0.9957	0.9959	0.9960	0.9961	0.9962	0.9963	0.9964
2.7	0.9965	0.9966	0.9967	0.9968	0.9969	0.9970	0.9971	0.9972	0.9973	0.9974
2.8	0.9974	0.9975	0.9976	0.9977	0.9977	0.9978	0.9979	0.9979	0.9980	0.9981
2.9	0.9981	0.9982	0.9982	0.9983	0.9984	0.9984	0.9985	0.9985	0.9986	0.9986
3.0	0.9987	0.9987	0.9987	0.9988	0.9988	0.9989	0.9989	0.9989	0.9990	0.9990
3.5	0.99977									
4.0	0.999968									
4.5	0.999997									
5.0	0.99999971									

Quantiles, Zq, of particular interest								
$Z_{0.500}$	$Z_{0.750}$	$Z_{0.800}$	$Z_{0.900}$	$Z_{0.950}$	$Z_{0.975}$	$Z_{0.990}$	$Z_{0.995}$	$Z_{0.999}$
0.000	0.674	0.842	1.282	1.645	1.960	2.326	2.576	3.090

Table D **Quantile Values of the Chi-Square Distribution with ν Degrees of Freedom**

ν	$\chi_{.001}$	$\chi_{.005}$	$\chi_{.010}$	$\chi_{.025}$	$\chi_{.050}$	$\chi_{.100}$	$\chi_{.200}$
1	0.000	0.000	0.000	0.001	0.004	0.016	0.064
2	0.002	0.010	0.020	0.051	0.103	0.211	0.446
3	0.024	0.072	0.115	0.216	0.352	0.584	1.005
4	0.091	0.207	0.297	0.484	0.711	1.064	1.649
5	0.210	0.412	0.554	0.831	1.145	1.610	2.343
6	0.381	0.676	0.872	1.237	1.635	2.204	3.070
7	0.598	0.989	1.239	1.690	2.167	2.833	3.822
8	0.857	1.344	1.646	2.180	2.733	3.490	4.594
9	1.152	1.735	2.088	2.700	3.325	4.168	5.380
10	1.479	2.156	2.558	3.247	3.940	4.865	6.179
11	1.834	2.603	3.053	3.816	4.575	5.578	6.989
12	2.214	3.074	3.571	4.404	5.226	6.304	7.807
13	2.617	3.565	4.107	5.009	5.892	7.042	8.634
14	3.041	4.075	4.660	5.629	6.571	7.790	9.467
15	3.483	4.601	5.229	6.262	7.261	8.547	10.307
16	3.942	5.142	5.812	6.908	7.962	9.312	11.152
17	4.416	5.697	6.408	7.564	8.672	10.085	12.002
18	4.905	6.265	7.015	8.231	9.390	10.865	12.857
19	5.407	6.844	7.633	8.907	10.117	11.651	13.716
20	5.921	7.434	8.260	9.591	10.851	12.443	14.578
21	6.447	8.034	8.897	10.283	11.591	13.240	15.445
22	6.983	8.643	9.542	10.982	12.338	14.041	16.314
23	7.529	9.260	10.196	11.689	13.091	14.848	17.186
24	8.085	9.886	10.856	12.401	13.848	15.659	18.062
25	8.649	10.520	11.524	13.120	14.611	16.473	18.940
26	9.222	11.160	12.198	13.844	15.379	17.292	19.820
27	9.803	11.808	12.879	14.573	16.151	18.114	20.703
28	10.391	12.461	13.565	15.308	16.928	18.939	21.588
29	10.986	13.121	14.256	16.047	17.708	19.768	22.475
30	11.588	13.787	14.953	16.791	18.493	20.599	23.364
40	17.916	20.707	22.164	24.433	26.509	29.051	32.345
50	24.674	27.991	29.704	32.354	34.761	37.687	41.451
60	31.736	35.531	37.481	40.478	43.185	46.457	50.642
70	39.034	43.272	45.438	48.754	51.736	55.327	59.900
80	46.517	51.168	53.536	57.149	60.388	64.276	69.209
90	54.153	59.192	61.750	65.642	69.122	73.289	78.560
100	61.915	67.323	70.060	74.217	77.925	82.356	87.947

Table D (continued)

ν	$\chi_{.500}$	$\chi_{.800}$	$\chi_{.900}$	$\chi_{.950}$	$\chi_{.975}$	$\chi_{.990}$	$\chi_{.995}$	$\chi_{.999}$
1	0.455	1.643	2.707	3.843	5.026	6.637	7.881	10.829
2	1.386	3.219	4.605	5.991	7.378	9.210	10.597	13.816
3	2.366	4.642	6.251	7.815	9.348	11.345	12.838	16.266
4	3.357	5.989	7.779	9.488	11.143	13.277	14.860	18.467
5	4.351	7.289	9.236	11.071	12.833	15.086	16.750	20.515
6	5.348	8.558	10.645	12.592	14.449	16.812	18.548	22.458
7	6.346	9.803	12.017	14.067	16.013	18.475	20.278	24.322
8	7.344	11.030	13.362	15.507	17.535	20.090	21.955	26.124
9	8.343	12.242	14.684	16.919	19.023	21.666	23.589	27.877
10	9.342	13.442	15.987	18.307	20.486	23.213	25.188	29.588
11	10.341	14.631	17.275	19.678	21.923	24.728	26.757	31.264
12	11.340	15.812	18.549	21.029	23.340	26.221	28.300	32.910
13	12.340	16.985	19.813	22.365	24.739	27.692	29.819	34.528
14	13.339	18.151	21.065	23.687	26.122	29.145	31.319	36.123
15	14.339	19.311	22.308	24.999	27.492	30.582	32.801	37.697
16	15.339	20.465	23.543	26.299	28.849	32.004	34.267	39.252
17	16.338	21.615	24.770	27.590	30.195	33.413	35.718	40.790
18	17.338	22.758	25.991	28.872	31.530	34.809	37.156	42.313
19	18.338	23.899	27.205	30.147	32.856	36.195	38.582	43.820
20	19.337	25.036	28.413	31.413	34.173	37.570	39.997	45.315
21	20.337	26.170	29.617	32.674	35.483	38.936	41.401	46.797
22	21.337	27.300	30.815	33.928	36.785	40.294	42.796	48.268
23	22.337	28.428	32.008	35.176	38.080	41.643	44.181	49.728
24	23.337	29.552	33.198	36.418	39.368	42.984	45.559	51.179
25	24.337	30.674	34.383	37.656	40.651	44.319	46.928	52.620
26	25.336	31.793	35.565	38.888	41.927	45.646	48.290	54.052
27	26.336	32.910	36.743	40.117	43.199	46.968	49.645	55.476
28	27.336	34.025	37.918	41.341	44.465	48.283	50.993	56.892
29	28.336	35.138	39.089	42.560	45.727	49.593	52.336	58.301
30	29.336	36.249	40.258	43.777	46.984	50.897	53.672	59.703
40	39.335	47.267	51.807	55.762	59.347	63.696	66.766	73.402
50	49.335	58.162	63.169	67.509	71.426	76.160	79.490	86.661
60	59.335	68.970	74.399	79.087	83.304	88.386	91.958	99.612
70	69.334	79.713	85.529	90.536	95.029	100.432	104.221	112.322
80	79.334	90.403	96.581	101.885	106.635	112.336	116.328	124.844
90	89.334	101.051	107.568	113.151	118.143	124.124	128.306	137.214
100	99.334	111.664	118.501	124.348	129.568	135.814	140.177	149.455

Table E   **Quantile Values of the t-Distribution with ν Degrees of Freedom**

ν	$t_{0.500}$	$t_{0.800}$	$t_{0.900}$	$t_{0.950}$	$t_{0.975}$	$t_{0.990}$	$t_{0.995}$	$t_{0.999}$
1	0	1.376	3.078	6.314	12.706	31.821	63.657	318.313
2	0	1.061	1.886	2.920	4.303	6.965	9.925	22.327
3	0	0.978	1.638	2.353	3.182	4.541	5.841	10.215
4	0	0.941	1.533	2.133	2.777	3.747	4.604	7.173
5	0	0.919	1.476	2.016	2.571	3.366	4.032	5.893
6	0	0.906	1.440	1.944	2.448	3.144	3.708	5.208
7	0	0.896	1.415	1.895	2.365	2.999	3.500	4.786
8	0	0.889	1.397	1.860	2.307	2.897	3.356	4.502
9	0	0.883	1.383	1.834	2.263	2.822	3.251	4.297
10	0	0.879	1.372	1.813	2.229	2.764	3.170	4.144
11	0	0.875	1.364	1.796	2.202	2.719	3.106	4.025
12	0	0.872	1.356	1.783	2.179	2.682	3.055	3.930
13	0	0.870	1.350	1.771	2.161	2.651	3.013	3.853
14	0	0.868	1.345	1.762	2.145	2.625	2.977	3.788
15	0	0.866	1.341	1.753	2.132	2.603	2.947	3.733
16	0	0.864	1.337	1.746	2.120	2.584	2.921	3.687
17	0	0.863	1.334	1.740	2.110	2.568	2.899	3.646
18	0	0.862	1.331	1.734	2.101	2.553	2.879	3.611
19	0	0.861	1.328	1.730	2.094	2.540	2.861	3.580
20	0	0.860	1.326	1.725	2.086	2.529	2.846	3.552
21	0	0.859	1.323	1.721	2.080	2.518	2.832	3.528
22	0	0.858	1.321	1.718	2.074	2.509	2.819	3.505
23	0	0.857	1.320	1.714	2.069	2.500	2.808	3.485
24	0	0.857	1.318	1.711	2.064	2.493	2.797	3.467
25	0	0.856	1.317	1.709	2.060	2.486	2.788	3.451
26	0	0.855	1.315	1.706	2.056	2.479	2.779	3.435
27	0	0.855	1.314	1.704	2.052	2.473	2.771	3.421
28	0	0.854	1.313	1.702	2.049	2.468	2.764	3.409
29	0	0.854	1.312	1.700	2.046	2.463	2.757	3.397
30	0	0.854	1.311	1.698	2.043	2.458	2.750	3.386
35	0	0.852	1.306	1.690	2.031	2.438	2.724	3.340
40	0	0.851	1.303	1.684	2.022	2.424	2.705	3.307
45	0	0.850	1.301	1.680	2.015	2.413	2.690	3.282
50	0	0.849	1.299	1.676	2.009	2.404	2.678	3.262
60	0	0.847	1.296	1.671	2.001	2.391	2.661	3.232
70	0	0.847	1.294	1.667	1.995	2.381	2.648	3.211
80	0	0.846	1.292	1.664	1.991	2.374	2.639	3.196
90	0	0.845	1.291	1.662	1.987	2.369	2.632	3.184
100	0	0.845	1.290	1.661	1.984	2.365	2.626	3.174
∞	0	0.841	1.282	1.645	1.960	2.327	2.576	3.091

Table F Quantile Values of the *F*-Distribution with ν_1 and ν_2 Degrees of Freedom

$P(F \le f)$	Den. d. f., ν_2	Numerator Degrees of Freedom, ν_1										
		1	**2**	**3**	**4**	**5**	**6**	**7**	**8**	**9**	**10**	**11**
0.950	**1**	161	199	216	225	230	234	236	237	239	239	240
0.975		648	797	855	885	903	916	924	930	935	938	941
0.990		3980	4843	5192	5377	5492	5570	5613	5653	5683	5706	5712
0.950	**2**	18.51	19.00	19.16	19.25	19.30	19.33	19.35	19.37	19.38	19.4	19.40
0.975		38.51	39.00	39.17	39.25	39.30	39.33	39.35	39.37	39.39	39.4	39.41
0.990		98.50	99.00	99.17	99.25	99.30	99.33	99.35	99.37	99.38	99.4	99.41
0.950	**3**	10.13	9.55	9.28	9.12	9.01	8.94	8.89	8.85	8.81	8.79	8.76
0.975		17.44	16.04	15.44	15.10	14.89	14.73	14.62	14.54	14.47	14.42	14.37
0.990		34.12	30.82	29.46	28.71	28.24	27.91	27.67	27.49	27.35	27.23	27.13
0.950	**4**	7.71	6.94	6.59	6.39	6.26	6.16	6.09	6.04	6.00	5.96	5.94
0.975		12.22	10.65	9.98	9.60	9.36	9.20	9.07	8.98	8.90	8.84	8.79
0.990		21.20	18.00	16.69	15.98	15.52	15.21	14.98	14.80	14.66	14.55	14.45
0.950	**5**	6.61	5.79	5.41	5.19	5.05	4.95	4.88	4.82	4.77	4.74	4.70
0.975		10.01	8.43	7.76	7.39	7.15	6.98	6.85	6.76	6.68	6.62	6.57
0.990		16.26	13.27	12.06	11.39	10.97	10.67	10.46	10.29	10.16	10.05	9.96
0.950	**6**	5.99	5.14	4.76	4.53	4.39	4.28	4.21	4.15	4.10	4.06	4.03
0.975		8.81	7.26	6.60	6.23	5.99	5.82	5.70	5.60	5.52	5.46	5.41
0.990		13.75	10.92	9.78	9.15	8.75	8.47	8.26	8.10	7.98	7.87	7.79
0.950	**7**	5.59	4.74	4.35	4.12	3.97	3.87	3.79	3.73	3.68	3.64	3.60
0.975		8.07	6.54	5.89	5.52	5.29	5.12	4.99	4.90	4.82	4.76	4.71
0.990		12.25	9.55	8.45	7.85	7.46	7.19	6.99	6.84	6.72	6.62	6.54
0.950	**8**	5.32	4.46	4.07	3.84	3.69	3.58	3.5	3.44	3.39	3.35	3.31
0.975		7.57	6.06	5.42	5.05	4.82	4.65	4.53	4.43	4.36	4.30	4.24
0.990		11.26	8.65	7.59	7.01	6.63	6.37	6.18	6.03	5.91	5.81	5.7
0.950	**9**	5.12	4.26	3.86	3.63	3.48	3.37	3.29	3.23	3.18	3.14	3.10
0.975		7.21	5.71	5.08	4.72	4.48	4.32	4.2	4.1	4.03	3.96	3.91
0.990		10.56	8.02	6.99	6.42	6.06	5.8	5.61	5.47	5.35	5.26	5.18
0.950	**10**	4.96	4.1	3.71	3.48	3.33	3.22	3.14	3.07	3.02	2.98	2.94
0.975		6.94	5.46	4.83	4.47	4.24	4.07	3.95	3.85	3.78	3.72	3.66
0.990		10.04	7.56	6.55	5.99	5.64	5.39	5.2	5.06	4.94	4.85	4.77
0.950	**11**	4.84	3.98	3.59	3.36	3.20	3.09	3.01	2.95	2.90	2.85	2.82
0.975		6.72	5.26	4.63	4.28	4.04	3.88	3.76	3.66	3.59	3.53	3.47
0.990		9.65	7.21	6.22	5.67	5.32	5.07	4.89	4.74	4.63	4.54	4.46

Table F (continued)

$P(F \leq f)$	Den. d.f. v_2	\multicolumn{11}{c}{Numerator Degrees of Freedom, v_1}
		1	2	3	4	5	6	7	8	9	10	11
0.950	12	4.75	3.89	3.49	3.26	3.11	3.00	2.91	2.85	2.80	2.75	2.72
0.975		6.55	5.10	4.47	4.12	3.89	3.73	3.61	3.51	3.44	3.37	3.32
0.990		9.33	6.93	5.95	5.41	5.06	4.82	4.64	4.50	4.39	4.30	4.22
0.950	15	4.54	3.68	3.29	3.06	2.90	2.79	2.71	2.64	2.59	2.54	2.51
0.975		6.200	4.77	4.15	3.80	3.58	3.41	3.29	3.20	3.12	3.06	3.01
0.990		8.68	6.36	5.42	4.89	4.56	4.32	4.14	4.00	3.89	3.80	3.73
0.950	20	4.35	3.49	3.10	2.87	2.71	2.60	2.51	2.45	2.39	2.35	2.31
0.975		5.87	4.46	3.86	3.51	3.29	3.13	3.01	2.91	2.84	2.77	2.72
0.990		8.10	5.85	4.94	4.43	4.10	3.87	3.70	3.56	3.46	3.37	3.29
0.950	24	4.26	3.40	3.01	2.78	2.62	2.51	2.42	2.36	2.30	2.25	2.22
0.975		5.72	4.32	3.72	3.38	3.15	2.99	2.87	2.78	2.70	2.64	2.59
0.990		7.82	5.61	4.72	4.22	3.90	3.67	3.50	3.36	3.26	3.17	3.09
0.950	30	4.17	3.32	2.92	2.69	2.53	2.42	2.33	2.27	2.21	2.16	2.13
0.975		5.57	4.18	3.59	3.25	3.03	2.87	2.75	2.65	2.57	2.51	2.46
0.990		7.56	5.39	4.51	4.02	3.70	3.47	3.30	3.17	3.07	2.98	2.91
0.950	40	4.08	3.23	2.84	2.61	2.45	2.34	2.25	2.18	2.12	2.08	2.04
0.975		5.42	4.05	3.46	3.13	2.90	2.74	2.62	2.53	2.45	2.39	2.33
0.990		7.31	5.18	4.31	3.83	3.51	3.29	3.12	2.99	2.89	2.80	2.73
0.950	50	4.03	3.18	2.79	2.56	2.40	2.29	2.20	2.13	2.07	2.03	1.99
0.975		5.34	3.97	3.39	3.05	2.83	2.67	2.55	2.46	2.38	2.32	2.26
0.990		7.17	5.06	4.20	3.72	3.41	3.19	3.02	2.89	2.78	2.70	2.63
0.950	80	3.96	3.11	2.72	2.49	2.33	2.21	2.13	2.06	2.00	1.95	1.91
0.975		5.22	3.86	3.28	2.95	2.73	2.57	2.45	2.35	2.28	2.21	2.16
0.990		6.96	4.88	4.04	3.56	3.26	3.04	2.87	2.74	2.64	2.55	2.48
0.950	120	3.92	3.07	2.68	2.45	2.29	2.18	2.09	2.02	1.96	1.91	1.87
0.975		5.15	3.80	3.23	2.89	2.67	2.52	2.39	2.30	2.22	2.16	2.10
0.990		6.85	4.79	3.95	3.48	3.17	2.96	2.79	2.66	2.56	2.47	2.40
0.950	∞	3.84	3.00	2.60	2.37	2.22	2.10	2.01	1.94	1.88	1.83	1.79
0.975		6.98	3.81	3.12	2.82	2.62	2.44	2.29	2.19	2.12	2.06	2.00
0.990		7.60	4.60	3.86	3.43	3.05	2.81	2.65	2.53	2.43	2.33	2.25

Table F (continued)

P(F≤f)	Den. d.f. v₂	12	15	20	24	30	40	50	80	120	∞
		Numerator Degrees of Freedom, v₁									
0.950	1	244	246	248	249	250	251	252	253	253	254
0.975		977	985	993	997	1001	1006	1008	1012	1014	1018
0.990		6106	6157	6209	6235	6261	6287	6303	6326	6339	6366
0.950	2	19.41	19.43	19.45	19.45	19.46	19.47	19.48	19.48	19.49	19.51
0.975		39.41	39.43	39.45	39.46	39.46	39.47	39.48	39.48	39.49	39.50
0.990		99.41	99.43	99.45	99.46	99.47	99.47	99.48	99.48	99.49	98.50
0.950	3	8.74	8.70	8.66	8.64	8.62	8.59	8.58	8.56	8.55	8.53
0.975		14.34	14.25	14.17	14.12	14.08	14.04	14.01	13.97	13.95	13.90
0.990		27.05	26.87	26.69	26.60	26.51	26.41	26.35	26.27	26.22	26.13
0.950	4	5.91	5.86	5.80	5.77	5.75	5.72	5.70	5.67	5.66	5.63
0.975		8.75	8.66	8.56	8.51	8.46	8.41	8.38	8.34	8.31	8.26
0.990		14.37	14.20	14.02	13.93	13.84	13.75	13.69	13.61	13.56	13.46
0.950	5	4.68	4.62	4.56	4.53	4.50	4.46	4.44	4.41	4.40	4.36
0.975		6.52	6.43	6.33	6.28	6.23	6.18	6.14	6.10	6.07	6.02
0.990		9.89	9.72	9.55	9.47	9.38	9.29	9.24	9.16	9.11	9.02
0.950	6	4.00	3.94	3.87	3.84	3.81	3.77	3.75	3.72	3.70	3.67
0.975		5.37	5.27	5.17	5.12	5.07	5.01	4.98	4.93	4.90	4.85
0.990		7.72	7.56	7.40	7.31	7.23	7.14	7.09	7.01	6.97	6.88
0.950	7	3.57	3.51	3.44	3.41	3.38	3.34	3.32	3.29	3.27	3.23
0.975		4.67	4.57	4.47	4.41	4.36	4.31	4.28	4.23	4.20	4.14
0.990		6.47	6.31	6.16	6.07	5.99	5.91	5.86	5.78	5.74	5.65
0.950	8	3.28	3.22	3.15	3.12	3.08	3.04	3.02	2.99	2.97	2.93
0.975		4.20	4.10	4.00	3.95	3.89	3.84	3.81	3.76	3.73	3.67
0.990		5.67	5.52	5.36	5.28	5.20	5.12	5.07	4.99	4.95	4.86
0.950	9	3.07	3.01	2.94	2.90	2.86	2.83	2.80	2.77	2.75	2.71
0.975		3.87	3.77	3.67	3.61	3.56	3.51	3.47	3.42	3.39	3.33
0.990		5.11	4.96	4.81	4.73	4.65	4.57	4.52	4.44	4.40	4.31
0.950	10	2.91	2.85	2.77	2.74	2.70	2.66	2.64	2.60	2.58	2.54
0.975		3.62	3.52	3.42	3.37	3.31	3.26	3.22	3.17	3.14	3.08
0.990		4.71	4.56	4.41	4.33	4.25	4.17	4.12	4.04	4.00	3.91
0.950	11	2.79	2.72	2.65	2.61	2.57	2.53	2.51	2.47	2.45	2.41
0.975		3.43	3.33	3.23	3.17	3.12	3.06	3.03	2.97	2.94	2.89
0.990		4.4	4.25	4.1	4.02	3.94	3.86	3.81	3.73	3.69	3.61

Table F (continued)

$P(F \leq f)$	Den. d.f. v_2	Numerator Degrees of Freedom, v_1									
		12	15	20	24	30	40	50	80	120	∞
0.950	12	2.69	2.62	2.54	2.51	2.47	2.43	2.40	2.36	2.34	2.30
0.975		3.28	3.18	3.07	3.02	2.96	2.91	2.87	2.82	2.79	2.72
0.990		4.16	4.01	3.86	3.78	3.70	3.62	3.57	3.49	3.45	3.36
0.950	15	2.48	2.40	2.33	2.29	2.25	2.20	2.18	2.14	2.11	2.07
0.975		2.96	2.86	2.76	2.70	2.64	2.59	2.55	2.49	2.46	2.40
0.990		3.67	3.52	3.37	3.29	3.21	3.13	3.08	3.00	2.96	2.87
0.950	20	2.28	2.20	2.12	2.08	2.04	1.99	1.97	1.92	1.90	1.84
0.975		2.68	2.57	2.46	2.41	2.35	2.29	2.25	2.19	2.16	2.09
0.990		3.23	3.09	2.94	2.86	2.78	2.69	2.64	2.56	2.52	2.42
0.950	24	2.18	2.11	2.03	1.98	1.94	1.89	1.86	1.82	1.79	1.73
0.975		2.54	2.44	2.33	2.27	2.21	2.15	2.11	2.05	2.01	1.94
0.990		3.03	2.89	2.74	2.66	2.58	2.49	2.44	2.36	2.31	2.21
0.950	30	2.09	2.01	1.93	1.89	1.84	1.79	1.76	1.71	1.68	1.62
0.975		2.41	2.31	2.20	2.14	2.07	2.01	1.97	1.90	1.87	1.79
0.990		2.84	2.70	2.55	2.47	2.39	2.30	2.25	2.16	2.11	2.01
0.950	40	2.00	1.92	1.84	1.79	1.74	1.69	1.66	1.61	1.58	1.51
0.975		2.29	2.18	2.07	2.01	1.94	1.88	1.83	1.76	1.72	1.64
0.990		2.66	2.52	2.37	2.29	2.20	2.11	2.06	1.97	1.92	1.80
0.950	50	1.95	1.87	1.78	1.74	1.69	1.63	1.60	1.54	1.51	1.45
0.975		2.22	2.11	1.99	1.93	1.87	1.80	1.75	1.68	1.64	1.56
0.990		2.56	2.42	2.27	2.18	2.10	2.01	1.95	1.86	1.80	1.70
0.950	80	1.88	1.79	1.70	1.65	1.60	1.54	1.51	1.45	1.41	1.33
0.975		2.11	2.00	1.88	1.82	1.75	1.68	1.63	1.55	1.51	1.41
0.990		2.42	2.27	2.12	2.03	1.94	1.85	1.79	1.69	1.63	1.51
0.950	120	1.83	1.75	1.66	1.61	1.55	1.50	1.46	1.39	1.35	1.25
0.975		2.05	1.94	1.82	1.76	1.69	1.61	1.56	1.48	1.43	1.31
0.990		2.34	2.19	2.03	1.95	1.86	1.76	1.70	1.60	1.53	1.38
0.950	∞	1.75	1.67	1.57	1.52	1.46	1.39	1.35	1.28	1.22	1.00
0.975		1.95	1.84	1.71	1.64	1.57	1.49	1.43	1.33	1.27	1.00
0.990		2.19	2.05	1.88	1.79	1.70	1.60	1.53	1.41	1.33	1.00

Table G Table of Random Numbers

51350	09452	02182	67144	81412	13280	20447	30855	50467	29163
49856	82515	34193	24493	98347	34272	70349	63696	48816	55978
16813	68029	17118	07956	24248	07171	58883	42206	18189	37713
98077	16558	15052	18630	25908	92928	50991	81600	10771	82304
77843	47811	27463	63378	32237	06992	22945	04658	62603	33653
68940	71800	33449	11205	42362	09005	04611	90544	09978	54957
39752	21336	81258	10869	21663	61159	28423	55140	74045	63810
68144	35410	60534	36730	85595	94687	96973	61813	04355	65108
12872	43394	30921	35276	47889	64732	34074	56374	39436	49724
17362	45561	55700	34137	94785	19775	64749	18871	20777	61314
41783	71577	78934	07577	92120	21331	51150	52467	30730	09592
19396	66504	94640	32537	55999	62007	74286	15945	22484	95195
12331	37036	98318	42747	76099	84177	68137	29267	34366	21103
80489	81947	65806	22546	37858	02418	94543	43990	99015	07391
79867	85397	75854	64695	86738	94981	65135	69319	63305	30437
03576	34061	90292	81407	04899	86929	95271	49032	37511	90091
75483	99321	19865	08460	64925	59720	57758	21326	38625	59077
09467	88921	59450	92467	07607	62507	57050	70229	35099	64220
40828	49877	33560	16916	99557	15675	03297	83602	61563	51611
47357	17307	67007	03170	86886	03133	05758	35417	71193	63743
75611	99865	88920	18093	30713	68285	80367	56290	52060	88747
42085	63949	29590	33142	52028	50242	14875	08272	59328	64303
97668	07906	58654	37042	66801	39339	47031	55289	31280	07290
72316	33961	41846	59832	27506	81655	34831	08730	03996	37607
96733	54959	96226	38221	58349	92319	15973	96635	01486	76093
64564	75807	93381	51390	16697	76142	07217	04046	27569	76667
68230	04749	15285	06805	02366	27015	15332	13168	94330	41525
70975	75710	62443	06143	24383	43092	94620	28358	02523	62942
20191	86845	08179	67383	02714	13200	11476	35073	88128	63715
90295	53465	34639	37313	50270	29342	33155	26852	80529	46635
35522	42833	82196	62294	72709	71140	67209	38008	67987	57821
91000	64605	91017	25877	53040	66161	05252	62694	75308	06788
85086	25738	65983	61770	59955	53243	98109	37324	17938	66822
09118	62199	24763	52297	29453	69991	69559	92097	25319	82182
99507	53674	49876	38520	54183	84380	98955	95725	24449	56208
12775	02544	18475	39617	28726	96437	19455	79415	95311	02412
98128	45653	58793	34175	30018	05630	59347	73382	93159	86282
44694	96257	60796	96790	74826	89764	91855	14647	93189	56652
69639	74161	49975	14130	21838	20578	17188	04678	32214	37943
71318	29930	88018	11343	40859	08513	95065	35129	06320	91462
06700	38900	89293	84565	75040	39526	42929	78866	91133	27008
91358	45891	50722	10443	33749	81140	51266	02455	61558	50749
16645	78884	24129	13915	69961	99986	75155	50142	19183	14447
87871	95411	73962	69190	38768	71940	82609	58922	19907	47444
10910	93640	67446	38798	34595	63958	81932	02966	28283	00836

INDEX

A priori probabilities, 97
Absolute value, 278
Acceptance region
    size of, 438
Addition rule, 106
Addition rule of probability, 85
Additive effects, 246
Additive rule
    of probability, 85
Alternative hypothesis, 424, 426, 512, 538
Analysis of variance, 534, 539
ANOVA (*See* analysis of variance)
Approximation, 289
    normal for binomial, 334
    Poisson for binomial, 212, 214
Area, 272
    center of, 142
Area as probability, 120, 123
Association
    linear, 52, 171
    measure of, 171
    pattern of, 172
Average deviation, 39
Balanced design, 536
Bar graphs, 12, 14, 59, 120
Bayes' theorem, 97, 106
Before/after experiment, 466
Bell-shaped distribution, 236, 235, 249
Bernoulli distribution, 194, 295, 510, 512
    mean of, 194-196
    method of moment estimators, 358
    moment-generating

function, 194
    parameter of, 195
    sum of random
    variables, 285
    variance of, 194-196
Bernoulli probability density function, 153
Bernoulli trial, 194, 197, 204, 209, 254, 429, 501
    sequence of, 197
Bias, 367, 369
Bimodal, 32
Binomial distribution, 197, 259, 264, 402, 500, 512, 548, 550
    generating random variates, 257
    graphs, 199
    mean of, 198
    moment-generating function, 198
    normal approximation to, 334-337
    parameters of, 198, 234
    tables of, 200
    variance of, 198
Births, 411
Bivariate data, 47-53, 155
Bivariate histogram, 47
Bivariate normal distribution, 248, 264
    graph of, 249
    parameters of, 248
    properties of conditional distributions, 249
    properties of marginal distributions, 248
Bivariate probability density function, 155
Block, 536, 544
Bound on error, 378

Boundary value, 433, 438, 439
Boxplot, 59-63, 542
Bureau of the Census, 353
Calculator formula, 40
Calculus, 142
Carl Friedrich Gauss, 264
Categorical data, 402, 497, 549
Census, 409
Center of Gravity (*See* mean), 39, 134
Central Limit Theorem, 235, 294, 298, 303, 304, 335, 343, 378, 399, 411, 458, 498, 510, 513, 549
    proof of, 311
Central moments, 142, 152, 153
Centroid, 170
Chance (*See* probability)
Change of variable technique, 278-282, 329
Chebyshev's inequality, 148-151, 186, 189
Chi-square distribution, 224, 226, 312, 329, 394, 396, 473, 474, 498
    definition of, 313
Chi-square goodness-of-fit test, 499
Class interval, 60
Cluster sample, 417
Cluster sampling, 414
Coefficient of skewness, 142
Combination, 109, 209
Complement, 77, 86, 106
    probability, 86
Completely randomized design, 534, 535, 536, 537, 549, 563

Composite hypothesis, 424, 426
Computer simulation, 254, 298, 383
Computer software, 579
Computers, 253
Conditional distribution, 114, 160-168, 175, 186
    discrete example, 165
    mean, 162, 164
    variance, 162, 164
Conditional probability, 74, 97, 106, 569
Confidence
    measure of, 5
Confidence coefficient, 379, 381, 382, 389
Confidence interval, 64, 378, 379, 383, 395, 396, 400, 402, 406, 411, 422, 493, 575
    Bernoulli parameter, 400
    lower limit, 378
    table of formulas, 407
    upper limit, 378
Confidence interval for difference
    Bernoulli parameters, 400
Confidence interval for ratio of variances, 396
Confidence interval for variance, 396
Confidence level, 354, 379, 401, 402
Consistency, 353
Consistent, 406
Consistent estimators, 372
Continuity correction
    normal approximation, 335
Continuous, 117
Continuous data, 13
Continuous random variable, 193
    expected value of, 134
    theoretical mean, 134
    theoretical variance, 134
Control chart, 347
Control group, 425

Correlation, 48, 52, 114, 170-174, 186, 467, 535
    definition, 48
    symbol for, 48
Correlation coefficient, 171
Counting
    fundamental rule of, 108
    number of combinations, 108
    number of permutations, 108
    rules of, 108-111
Covariance, 48, 51, 52, 114, 170-174, 186, 249, 537
    definition, 48
    symbol for, 48
Cramer-Rao inequality, 367, 370
Critical region, 481, 485
    best, 421
    most powerful, 479
Critical value, 433, 438, 439
Cumulative distribution, 23, 25, 200, 213, 255, 560
    graph, 122
Cumulative distribution function, 116, 117, 121, 186, 271, 272
    bivariate, 155-159
    continuous, 124
    continuous random variable, 118
    discrete random variable, 117
Cumulative distribution table, 255
Cumulative table, 254
Current Population Survey, 409
Data
    bivariate, 155
    ordering, 339
Deaths, 411
Deciles, 22
Decision rule, 424
Degrees of freedom, 226, 313, 321, 329, 390, 502, 503, 539, 540
Delta method, 289
Department of Agriculture,

409
Dependent variable, 535
Derivative, 272, 362
    first, 142, 278, 290
    rth, 152
    second, 152, 290
Deviation, 37-39, 52, 171
    mean, 141
Df (See degrees of freedom)
Dichotomous data, 497
Difference in proportions, 401
Discrete, 117
Discrete data, 13
Discrete random variable, 193, 214
    expected value of, 134
    theoretical mean, 134
    theoretical variance, 134
Discrete uniform distribution, 217-218
    graph of, 217
    mean of, 217
    moment-generating function, 217
    parameters of, 217
    variance of, 217
Dispersion, 36
Distribution, 148
    bell-shaped, 322
    chi-square, 312
    conditional, 160-168, 248
    cumualtive, 25
    empirical, 300, 303, 315
    Fisher's F, 328
    function of random variable, 269-286
    hypergeometric, 112
    marginal, 50, 160-168, 248
    of data, 115
    of sample mean, 296
    random sample from, 253
    sampling, 294
    student t, 321
    symmetric, 16, 221, 303, 322
Distribution function (See

cumulative distribution
function), 253
Distribution function
technique, 271-277, 278,
313, 321
Distribution-free methods,
535
Distributions
    Bernoulli ,194
    binomial, 197
    bivariate normal, 248
    continuous uniform, 221
    discrete uniform, 217
    gamma, 224
    geometric, 204
    hypergeometric, 205
    lognormal, 245
    negative binomial, 204
    negative exponential,
    230
    normal, 235
    Poisson, 212
    symmetric, 32
    table of properties, 265
Double integral, 157, 191
    region of integration,
    158
Efficiency, 353, 370, 406
Empirical data, 13
Empirical distribution, 303
Empirical moments, 27
Empirical rule, 242
Empty set, 76
Equal likelihood, 81, 108
Erlang probability model,
225
Error, 5, 76, 378, 540, 572
    margin of, 354
    type I, 433
    type II, 433
Error of estimation, 377, 378
Estimator
    bias of, 377
    efficiency of, 367
    interval, 354
    maximum likelihood,
    361
    method of moments,
    356
    minimum variance

unbiased, 367
    notation for, 356
    point, 354
    sufficient, 368
    unbiased, 367
    variance of, 377
Event, 79, 81, 212
    complement of, 86
    equal likelihood, 81
    independent, 89, 93
    intersection of, 89, 93
    mutually exclusive, 86
    rare, 149
    union of, 86
Expectation, 142
Expected payoff, 133
Expected value, 131, 134-
138, 152, 163, 171, 180,
367
        properties of, 132, 135
Experiment, 76, 79, 427
    repeatable, 82
Experimental design, 466,
535, 536
Experimental error, 536
Experimental group, 425,
426
Experimental unit, 9, 536
Explanatory variable, 535,
536
Exponential distribution,
224, 225, 230, 264, 275,
278, 303
    distribution function,
    232
    generating random
    variates, 257
    graph of, 232
    maximum likelihood
    estimator, 363
    mean of, 231
    moment-generating
    function, 231
    parameters of, 231
    quantiles, 232
    sum of random
    variables, 285
    table of, 232
    variance of, 231
F distribution, 328-333, 397,

474, 475, 539
    definition of, 328
    graph of, 330
    parameters of, 329
    table of, 331
F-distribution, 395
Factor, 536
Factor level, 536
Factorization theorem, 368
Failure,194
Failure rates, 213
Finite population correction,
411
Finite population sampling,
410
First central moment, 134
First empirical moment, 22
Floor, 158
Frequency, 14
    theory of probability, 74
Frequency, table 14, 15
Function, 117
    continuous, 118
    expected value of, 132
    gamma, 136
    inverse, 254, 278
    likelihood, 361
    mathematical, 116
    monotonic, 278
    non-decreasing, 118, 124
    step, 122
Function of random variables
    approximating mean and
    variance, 289-291
    expected value, 290
    variance of 290
Fundamental rule of
counting, 108
Game of chance, 133
Gamma distribution, 224-
229, 231, 264, 276, 285,
303
    graph of, 226
    mean of, 225
    moment-generating
    function, 225
    parameter estimation,
    356
    parameters of, 225
    table of ,227

variance of, 225
Gamma distribution
function, 228
Gamma function, 136, 225
Gaussian distribution, 235, 264
Generating random numbers, 221, 223, 232
Generating random variates, 252-261, 279
Geometric distribution, 204-207, 264
    mean of, 204, 206
    moment-generating
    function, 204
    variance of, 204, 206
Geometric random variable, 204
Goodness-of-fit test, 498, 499-508
    background
    assumptions, 499
Gosset, 320
Grade on the curve, 236
Grand total ,540
Graphical methods, 59-63
Histogram, 12, 50, 59, 61, 115, 259
    bivariate, 50
Hypergeometric, 112
Hypergeometric distribution, 205, 209, 264
    mean of, 209
    variance of, 209
Hypothesis, 422, 423
    alternative, 424
    composite, 424
    null, 424
    research, 423, 424
    simple, 424
    statistical, 423
IID (*See* independent and identically distributed), 297, 335
Independence, 89-93, 106, 175-178, 180, 186, 249
    assumption of, 176
Independence and mutually exclusive events, 93
Independence and random

sampling, 92
Independent and identically distributed, 176, 181, 197
Independent events, 89-93
Independent variable, 536
Inference, 354
    statistical, 5, 354
Integral, 164, 290
    definite, 136
    double, 157, 191
Integration, 157
    by parts, 136, 227
    repeated, 191
Intercept, 573
Interdecile range, 37
Interquartile range, 37, 38, 142, 144, 232, 243
    theoretical, 141
Intersection, 77
    probability of, 92
Interval estimate, 353
Interval estimator, 377
    Bernoulli parameter, 399
    difference of Bernoulli
    parameters, 399
    difference of means, 387-392
    for one variance, 394
    for ratio of variances, 394
Interval estimator difference of Bernoulli's
    assumptions for,401
Interval estimator for one variance
    assumptions, 394
Interval estimator of Bernoulli parameter
    assumptions, 400
Interval estimator of mean, 379-386
    assumptions of, 380, 381
Interval estimator of ratio of variances
    assumptions, 395
Interval measurement, 7
Inverse function, 254, 266, 279
Jacobian, 278, 279

JMP-IN, 64
Joint distribution, 170, 175, 248
Joint probability (*See*
Probability density
function:bivariate)
Joint probability density
function, 248
Kruskal-Wallis, 549
Kruskal-Wallis test, 534, 559, 563-567
    analysis of variance
    computations, 564
    test statistic,564
Large sample confidence
interval ,399
Least squares, 535, 569
Least squares estimators, 570, 579
Least squares regression line, 572
Level of confidence, 385
Level of measurement, 552, 557
Level of significance, 424, 425, 427, 433, 453, 467, 473, 474, 499, 510
Likelihood function, 361, 479, 484
Likelihood ratio test, 421, 478-489
Linear association, 48, 171
Linear combination, 143, 179, 183, 186, 287
    covariance of,180
    mean of, 181
    variance of, 181
Linear contrast, 183
Linear model, 571, 578
Linear regression, 569-578
Linear trend, 52
Location,61, 237
    measure of, 21, 30, 32
    measures of, 141-146
Location parameter, 232
Lognormal distribution, 245-246, 264
    mean of, 246
    parameters of, 245
    variance of, 246

Lower control limit, 347
Mann-Whitney test, 534, 549, 559, 560-562
    assumptions of, 560
    distribution of, 561
    test statistic, 560
Margin of error, 353, 377, 378, 379, 381, 382, 385, 389, 395, 401, 402, 409, 411
Marginal, 50
Marginal distribution, 114, 160-168, 175, 186
    discrete example, 162
    mean, 161, 163
    variance, 161, 163
Matrix algebra,579
Maximum, 61, 141, 271
Maximum likelihood, 380, 503
    method of, 353, 360
Maximum-likelihood estimate, 479, 484
Mean, 6, 30-32, 114, 147, 186, 236, 269
    definition of, 30
    expected value, 132
    graphically, 32
    symbols for, 30
    theoretical, 132
Mean deviation, 37, 39, 142, 144
    definition, 37
    symbols for, 37
    theoretical, 141
Mean square, 540
Mean square error, 367, 369, 377, 406, 574
Measure of location, 18, 134
    mean, 30-32
    median, 30-32
    mode, 30-32
Measure of variability, 21, 36-42, 134, 141-146
    range, 42
    (*See* variance)
Measurement, 5, 6, 11, 118
    continuous, 7-8, 119
    discrete, 7-8, 119
    interval, 7-9

level of, 7-9
    nominal, 7-9
    ordinal, 7-9, 338
    qualitative, 7-8
    quantitative, 7-8, 116
    standardized, 149
Measures of location ,21, 30-32, 114, 141-146
Median, 60, 61, 141, 142, 143, 232, 236, 339, 553
    definition of, 31
    graphically, 32
    symbols for, 31
    theoretical, 141
Method of maximum likelihood, 360-365, 378, 406, 574
Method of moments, 355-359, 378, 380, 406, 503, 574
Minimum, 61, 270, 271
Minimum variance, 353
Minimum variance unbiased estimator, 367, 371,406
Minitab, 60
Modal class, 32
Mode, 30-32, 62, 141, 144, 236, 246
    definition of, 31
    graphically, 32
    symbols for, 31
    theoretical, 141
Model
    mathematical, 114
    probability, 114
Modelling,
    probability, 192
Moment
    about the mean, 141-146
    about the origin, 141-146,355
    central, 141
    empirical, 356
    first empirical, 27
    product, 170
    sample, 355
    standardized, 142
    theoretical, 355
Moment-generating function, 152, 186, 236, 311

uniqueness, 153
Moment-generating function technique, 283-286, 299, 313
Moments, 186
    about the mean, 142
    central, 142
    empirical, 27
    method of, 353, 355
Monotonic function, 278, 279
Most powerful critical region, 479,481
Most powerful test, 551
MS (*See* mean square)
Multinomial distribution, 500
Multiple comparison test, 542
Multiple linear regression, 570, 578-582
    least squares estimators for, 579
Multiplication rule, 106
Multiplication rule of probability, 89-93
Multiplicative effects, 246
Multivariate statistics, 249
Mutually exclusive, 77, 85, 106, 195, 413
Mutually exclusive events, 81
MVU (*See* minimum variance unbiased estimators)
Natural log, 362
Negative binomial, 264
Negative binomial distribution ,204, 207, 264
    mean of, 205, 208
    moment-generating function, 205
    tables of, 208
    variance of, 205, 208
Negative exponential distribution, 225, 230
Neyman, 409
Neyman-Pearson Lemma, 421, 478-489
Nominal data, 402, 549
Nominal measurement, 7, 497

Non-integrable, 289
Non-random influences, 347
Nonparametric methods, 338
Nonparametric statistics, 498
Nonparametric test, 548, 557
Normal approximation to
binomial, 550
Normal distribution, 235-
243, 245, 259, 264, 294,
295, 299, 300, 314, 323,
347, 427, 501, 536, 537
    cumulative distribution,
    241
    generating random
    variates, 257
    graph of, 237
    interval estimator of
    mean, 379
    maximum likelihood
    estimators, 364
    mean of, 235
    method of moment
    estimators, 357
    moment-generating
    function, 235
    non-integrable, 241
    notation, 235
    parameters of, 235
    table of, 239
    variance of, 235
Normal equations, 579
Notation
    product, 284
Null hypothesis , 424, 426,
538
Null set, 78, 81
Numerical integration, 256
Numerical methods
    measures of variability,
    36
OC curve, 443-451
One sample test of a mean
    normal distribution,
    452-457
One-factor experiment, 536
One-stage cluster sampling,
409, 414
One-tailed test, 444, 512
Order statistics, 26, 338-341,
549, 560

Ordered pairs, 548
Ordinal data, 534, 549
Ordinal measurement, 7, 338
Outlier, 18, 61
P-value 428, 438, 444, 454,
459, 460, 467, 473, 474,
500, 511, 512, 552
Paired comparison, 467, 537
Paired comparison
experiment, 466
Paired sample test, 549
Parameter ,503
    location, 265
    scale, 232, 238, 239,
    240, 241, 250
    shape, 232, 238, 239,
    240, 241, 250, 265
Parameter space, 484
Partition, 97
Pdf (See probability density
function)
Pearson, Karl, 502
Percentile, 22, 24
Periodicities, 417
Permutation, 108
    with identical items, 108
Pie charts, 12
Pivotal quantity, 377, 380,
388, 389, 394, 395, 400,
401, 427
Point estimate, 379, 381,
382, 389, 401, 402
Point estimator, 354, 377,
380
Poisson distribution, 212-
214, 231, 259, 264, 334
    generating random
    variates, 258
    graph of, 213
    mean of, 212
    moment-generating, 212
    parameter of, 212
    sum of random
    variables, 284
    table of, 213
    variance of, 212
Poisson distribution
function, 228
Political polls, 2
Pooled estimate of variance,

389
Pooled variance, 391
Population, 4-6, 75, 115,
294
    offspring, 301
    parent, 300
Population mean, 133, 411
Population parameter, 22-24,
115, 411
Population total, 411
Postulates of probability, 90
Power, 557
Power curve, 443
Power function, 443-451,
481
Power series, 152
Pps sampling, 410
Predictor variable, 536
Principle of least squares,
572, 579
Probability
    a priori, 99
    conditional, 89-93, 97
    frequentist view, 82
    model, 114
    postulates of, 80-82
    rare, 149
    relative frequency
    approach, 81
    revised, 99
    theory of, 75
Probability density function,
115, 117, 120-125, 186, 253
    bivariate, 155-159
    continuous, 123-125
    continuous random
    variable, 118
    discrete, 120-123
    graphically, 120
    joint, 155
    moment-generating
    function of, 153
Probability model, 192
Probability postulates, 74
Probability sampling, 409
Product moment, 170, 171
Projection, 164
Proof by contradiction, 428
Quality control, 2, 346, 534
Quantile, 22, 24-28, 141,

186, 200, 243, 322, 329
   theoretical, 141
Quantitative measurement, 119
Quartile, 22, 24, 28, 60, 61, 232
   first, 24-28
   second, 24-28
   third, 24-28
Queues, 252
Random number generation, 252
Random number generator 221, 253, 410
Random order, 417
Random sample, 4-6, 82, 115, 176, 252, 296, 409
   definition of, 297
   of n, 270
Random starting point, 416
Random value, 254
Random variable, 114, 116, 117, 118-129, 156
   Bernoulli, 197
   continuous, 119, 193, 221, 231, 237, 264, 271, 278
   discrete, 114, 119, 155, 193, 194, 217, 264
   function of, 269
   integer-valued, 214
   moment-generating function of, 152
   sum of, 270, 283
   transformation of, 269, 270, 271, 279
   uniform, 282
Randomized block design, 534, 536, 543-546
Range, 37, 38, 339
   interquartile, 141
Rank, 553
Ranks,
   sum of, 553
Rare event, 149, 186, 424
Rare population, 211
Ratio measurement, 7
Region of integration, 158, 191
Regression of Y on X, 161,

186
Rejection region, 434, 438, 481, 512
   area of, 445
   size of, 438
Relationship
   one-to-one, 283
Relative frequency, 14, 82, 108
Relative frequency table, 115
Reliability, 96, 213
Repeated sampling, 294, 298, 382
Replication, 536, 544
Research hypothesis, 423, 424
Run, 347
Salk vaccine, 425
Sample, 4, 23, 75, 115, 294
   random, 5
   size of, 385
   without replacement, 205
Sample maximum, 339
Sample mean, 22, 271
   expected value, of 182
   variance of, 182
Sample median, 339
Sample minimum, 339
Sample moment, 356
Sample outcome, 76
Sample point, 76, 120
Sample size, 253, 385
Sample size calculation
   test of hypothesis, 450
Sample space, 76, 90, 120, 121
Sample statistic, 22, 23-24, 294, 296, 422, 427
   definition of, 297
Sample Survey of Unemployment, 409
Sample survey, 294
Sample variance, 271
Sampling
   without replacement, 92, 109
Sampling distribution, 294-336, 343, 370, 377, 427
   empirical, 294

   of the mean, 298-310
Sampling error, 377
Sampling fraction, 410, 413
Sampling frame, 411
Sampling inspection, 346
Sampling variability, 307
Sampling without replacement, 97, 209, 410
SAS Institute, 64
Scale parameter, 226, 232
Scatterplot 47, 48, 49, 52, 572
Scientific method, 421, 423-432
Score
   standardized, 149
Second empirical moment, 22
Set theory, 76
Shape, 238
Shape parameter, 226, 232
Sign test, 534, 548, 549, 550-551, 553
Signed rank, 549, 553
Signed rank test, 534
Significance level, 424
SIM (*See* Statistical-inferential model), 115
Simple alternative hypothesis, 443, 484
Simple hypothesis, 424, 426
Simple linear regression, 534, 569, 571, 572-578
   ANOVA for, 575
   least squares estimators, 573
   test of hypothesis of slope, 575
Simple null hypothesis, 443, 484
Simple random sample, 409, 410, 412, 413, 414, 417
Simulation, 252, 383
Size of Critical Region, 481
Skewed ,16, 193
   left, 16, 61, 94
   negative, 16, 198
   positive, 16, 198, 212, 233, 246
   right, 16, 194, 227

right or positive, 16
Skewness, 114, 141
    coefficient of, 142
    measure of (theoretical), 143
Slope, 573
    negative, 171
    positive, 171
Sociological study, 2
Software
    statistical, 259
Space
    3-dimensional, 156
Spread, 18, 36, 61, 148
Squared error, 132
    expected value of, 132
SS (*See* sum of squares)
Standard deviation, 6, 37, 39, 136, 148, 269, 385
    definition, 37
    symbol for, 37
Standard error, 417, 546
Standard error of estimator, 368
Standard normal distribution, 236, 237, 239, 323, 334, 388, 401, 453, 459, 512, 549
    graph of, 239
    mean of, 236
    variance of, 236
Standard normal random variable ,510
Standard score, 237, 324
Standardized score, 149
State of nature, 434
Statistic, 269
Statistical hypothesis, 423, 424, 427
Statistical independence, 175-178
Statistical inference, 4, 80, 354, 406, 498
Statistical quality control, 346-350
Statistical software, 60, 259
Statistical-inferential model, 4-6, 75, 193
Statistics, 2, 23
    order, 338

profession, 2
Stem-and-leaf, 59-63
Step function, 122
Strata, 413
Stratified random sampling, 409, 413, 414
Stratified sampling, 413
Success, 194
Sufficiency, 353
Sufficient, 406
Sufficient estimator, 373, 368
Sufficient statistic, 374
Sum, 284
Sum of squared errors, 569, 579
Sum of Squares, 539
Summarizing data
    graphical methods, 12
    graphically, 115
    numerical methods, 21
    numerically, 115
Survey sampling, 409
Symmetric, 193, 194, 553
Symmetric distribution, 32, 221, 236
Symmetry, 16, 61, 141, 167
Systematic random sampling, 409, 416-418
t distribution, 321-326, 390, 454, 460, 467, 536
    definition of, 321
    table of, 322
t-test (*See* test of one mean or test of two means)
Table D, 241
Taylor's series, 288, 289
Test of hypothesis, 421, 423-432, 512
Test of means for paired observations, 466-470
Test of one mean
    background
    assumptions, 453, 454
    normal distribution, known standard deviation, 453
    normal distribution,unknown standard deviation, 454

Test of one proportion, 510-516
    background assumptions, 510
Test of one variance
    background assumptions, 472
    normal distribution, 472
Test of two means
    background assumptions, 458, 459, 466
    normal distribution, known standard deviations, 458
    normal distribution, paired observations, 466
    normal distribution, unknown standard deviations, 459
Test of two proportions, 511, 513-516
Test of two variances
    background assumptions, 473
    normal distributions, 473
Test of variances
    normal distribution, 472-477
Test statistic(s), 424, 426, 427, 453, 454, 459, 460, 467, 473, 474, 499, 502, 510, 511, 512, 544, 548
    table of, 493
Theoretical moment, 116
Three-dimensional, 249
Ties in rank tests, 553
Time-to-failure, 225, 231, 271
Transformation, 246
Treatment, 536, 540
Treatment combination, 536
Trimodal, 32
Two independent sample t-test, 536
Two independent samples, 397, 402, 510
Two sample test of equal means

normal distribution ,458-464

Two-independent sample t-test, 543

Two-independent sample test, 549

Two-parameter family, 237

Two-tailed test, 444, 512

Type I error, 433-442, 505
  probability, 435

Type II error, 433-442, 505
  probability, 435

Unbalanced design, 536

Unbiased, 406, 353, 370

Unbiased estimator, 367, 538

Uncertainty, 3

Unexplained variability, 534

Uniform continuous distribution, 221-222
  distribution function, 222
  graph of, 222
  mean of, 221
  moment-generating function, 221
  parameters of, 221
  quantiles of, 222
  variance of, 221

Uniform distribution, 147, 221, 253, 254, 259, 274, 282
  generating random variates, 257
  maximum likelihood estimators, 363

Uniformly most powerful test, 479, 483

Union, 77, 85, 106
  probability of, 85

Unit of observation, 9

Univariate normal distribution, 249

Upper control limit, 347

Variability, 3, 18, 114
  measure of, 21, 134
  measures of (See measures of variability), 141-146
  unexplained ,534

Variable

random, 117
  continuous, 117
  discrete, 117

Variance,37, 39, 114, 134, 186
  as expected value, 132
  definition, 37
  properties, of 133, 135
  symbol for, 37
  theoretical, 132

Variation, 36
  process, 346

Vector, 270

Venn Diagram, 77

Volume, 159, 191

Volume as probability, 157

Waiting time distributions, 225

Weather prediction, 2

Weibull distribution, 279

Wilcoxon matched-pairs signed rank test, 548, 549

Wilcoxon signed-rank test, 553-557

Z-score, 119, 149, 384, 411, 510